Lecture Notes in Computer Science

Commenced Publication in 1973
Founding and Former Series Editors:
Gerhard Goos, Juris Hartmanis, and Jan van Leeuwen

More information about this series at http://www.springer.com/series/7409

Laurent Amsaleg · Gylfi Þór Guðmundsson
Cathal Gurrin · Björn Þór Jónsson
Shin'ichi Satoh (Eds.)

MultiMedia Modeling

23rd International Conference, MMM 2017
Reykjavik, Iceland, January 4–6, 2017
Proceedings, Part II

 Springer

Editors
Laurent Amsaleg
CNRS–IRISA
Rennes
France

Gylfi Þór Guðmundsson
Reykjavík University
Reykjavik
Iceland

Cathal Gurrin
Dublin City University
Dublin
Ireland

Björn Þór Jónsson
Reykjavik University
Reykjavik
Ireland

Shin'ichi Satoh
National Institute of Informatics
Tokyo
Japan

ISSN 0302-9743   ISSN 1611-3349 (electronic)
Lecture Notes in Computer Science
ISBN 978-3-319-51813-8   ISBN 978-3-319-51814-5 (eBook)
DOI 10.1007/978-3-319-51814-5

Library of Congress Control Number: 2016962021

LNCS Sublibrary: SL3 – Information Systems and Applications, incl. Internet/Web, and HCI

Printed on acid-free paper

This Springer imprint is published by Springer Nature
The registered company is Springer International Publishing AG
The registered company address is: Gewerbestrasse 11, 6330 Cham, Switzerland

Preface

These proceedings contain the papers presented at MMM 2017, the 23rd International Conference on MultiMedia Modeling, held at Reykjavik University during January 4–6, 2017. MMM is a leading international conference for researchers and industry practitioners for sharing new ideas, original research results, and practical development experiences from all MMM related areas, broadly falling into three categories: multimedia content analysis; multimedia signal processing and communications; and multimedia applications and services.

MMM conferences always include special sessions that focus on addressing new challenges for the multimedia community. The following four special sessions were held at MMM 2017:

- SS1: Social Media Retrieval and Recommendation
- SS2: Modeling Multimedia Behaviors
- SS3: Multimedia Computing for Intelligent Life
- SS4: Multimedia and Multimodal Interaction for Health and Basic Care Applications

MMM 2017 received a total 198 submissions across four categories; 149 full-paper submissions, 34 special session paper submissions, eight demonstration submissions, and seven submissions to the Video Browser Showdown (VBS 2017). Of all submissions, 68% were from Asia, 27% from Europe, 3% from North America, and 1% each from Oceania and Africa.

Of the 149 full papers submitted, 35 were selected for oral presentation and 33 for poster presentation, which equates to a 46% acceptance rate overall. Of the 34 special session papers submitted, 24 were selected for oral presentation and two for poster presentation, which equates to a 76% acceptance rate overall. In addition, five demonstrations were accepted from eight submissions, and all seven submissions to VBS 2017. The overall acceptance percentage across the conference was thus 54%, but 46% for full papers and 23% of full papers for oral presentation.

The submission and review process was coordinated using the ConfTool conference management software. All full-paper submissions were reviewed by at least three members of the Program Committee. All special session papers were reviewed by at least three reviewers from the Program Committee and special committees established for each special session. All demonstration papers were reviewed by at least three reviewers, and VBS papers by two reviewers. We owe a debt of gratitude to all these reviewers for providing their valuable time to MMM 2017.

We would like to thank our invited keynote speakers, Marcel Worring from the University of Amsterdam, The Netherlands, and Noriko Kando from the National Institute of Informatics, Japan, for their stimulating contributions.

We also wish to thank our organizational team: Demonstration Chairs Esra Acar and Frank Hopfgartner; Video Browser Showdown Chairs Klaus Schoeffmann, Werner Bailer, Cathal Gurrin and Jakub Lokoč; Sponsorship Chairs Yantao Zhang and Tao Mei;

Proceedings Chair Gylfi Þór Guðmundsson; and Local Organization Chair Marta Kristín Lárusdóttir.

We would like to thank Reykjavik University for hosting MMM 2017. Finally, special thanks go to our supporting team at Reykjavik University (Arnar Egilsson, Ýr Gunnlaugsdóttir, Þórunn Hilda Jónasdóttir, and Sigrún Heba Ómarsdóttir) and CP Reykjavík (Kristjana Magnúsdóttir, Elísabet Magnúsdóttir and Ingibjörg Hjálmfríðardóttir), as well as to student volunteers, for all their contributions and valuable support.

The accepted research contributions represent the state of the art in multimedia modeling research and cover a very diverse range of topics. A selection of the best papers will be invited to submit extended versions to a special issue of *Multimedia Tools and Applications*. We wish to thank all authors who spent their valuable time and effort to submit their work to MMM 2017. And, finally, we thank all those who made the (sometimes long) trip to Reykjavík to attend MMM 2017 and VBS 2017.

January 2017   Björn Þór Jónsson
                                                                          Cathal Gurrin
                                                                          Laurent Amsaleg
                                                                          Shin'ichi Satoh
                                                                          Gylfi Þór Guðmundsson

Organization

Organizing Committee

General Chairs

Björn Þór Jónsson	Reykjavik University, Iceland
Cathal Gurrin	Dublin City University, Ireland

Program Chairs

Laurent Amsaleg	CNRS–IRISA, France
Shin'ichi Satoh	NII, Japan

Demonstration Chairs

Frank Hopfgartner	University of Glasgow, UK
Esra Acar	Technische Universität Berlin, Germany

VBS 2017 Chairs

Klaus Schöffmann	Klagenfurt University, Austria
Werner Bailer	Joanneum Research, Austria
Cathal Gurrin	Dublin City University, Ireland
Jakub Lokoč	Charles University in Prague, Czech Republic

Sponsorship Chairs

Yantao Zhang	Snapchat
Tao Mei	Microsoft Research Asia

Proceedings Chair

Gylfi Þ. Guðmundsson	Reykjavik University, Iceland

Local Chair

Marta K. Lárusdóttir	Reykjavik University, Iceland

Local Support

Reykjavik University Event Services, CP Reykjavik

Steering Committee

Phoebe Chen (Chair)	La Trobe University, Australia
Tat-Seng Chua	National University of Singapore, Singapore

Kiyoharu Aizawa   University of Tokyo, Japan
Cathal Gurrin   Dublin City University, Ireland
Benoit Huet   EURECOM, France
R. Manmatha   University of Massachusetts, USA
Noel E. O'Connor   Dublin City University, Ireland
Klaus Schöffmann   Klagenfurt University, Austria
Yang Shiqiang   Tsinghua University, China
Cees G.M. Snoek   University of Amsterdam, The Netherlands
Meng Wang   Hefei University of Technology, China

Special Session Organizers

SS1: Social Media Retrieval and Recommendation

Liqiang Nie   National University of Singapore, Singapore
Yan Yan   University of Trento, Italy
Benoit Huet   EURECOM, France

SS2: Modeling Multimedia Behaviors

Peng Wang   Tsinghua University, China
Frank Hopfgartner   University of Glasgow, UK
Liang Bai   National University of Defense Technology, China

SS3: Multimedia Computing for Intelligent Life

Zhineng Chen   Chinese Academy of Sciences, China
Wei Zhang   Chinese Academy of Sciences, China
Ting Yao   Microsoft Research Asia, China
Kai-Lung Hua   National Taiwan University of Science
                         and Technology, Taiwan, R.O.C.
Wen-Huang Cheng   Academia Sinica, Taiwan, R.O.C.

SS4: Multimedia and Multimodal Interaction for Health and Basic Care Applications

Stefanos Vrochidis   ITI-CERTH, Greece
Leo Wanner   Pompeu Fabra University, Spain
Elisabeth André   University of Augsburg, Germany
Klaus Schöffmann   Klagenfurt University, Austria

Program Committee

Esra Acar   Technische Universität Berlin, Germany
Amin Ahmadi   DCU/Insight Centre for Data Analytics, Ireland
Le An   UNC Chapel Hill, USA
Ognjen Arandjelović   University of St. Andrews, UK
Anant Baijal   SAMSUNG Electronics, South Korea

Werner Bailer	Joanneum Research, Austria
Ilaria Bartolini	University of Bologna, Italy
Jenny Benois-Pineau	University of Bordeaux/LABRI, France
Milan Bjelica	Unoiversity of Belgrade, Serbia
Laszlo Böszörmenyi	Klagenfurt University, Austria
Benjamin Bustos	University of Chile, Chile
K. Selçuk Candan	Arizona State University, USA
Shiyu Chang	UIUC, USA
Savvas Chatzichristofis	Democritus University of Thrace, Greece
Edgar Chávez	CICESE, Mexico
Wen-Huang Cheng	Academia Sinica, Taiwan, R.O.C.
Gene Cheung	National Institute of Informatics, Japan
Wei-Ta Chu	National Chung Cheng University, Taiwan, R.O.C.
Vincent Claveau	IRISA-CNRS, France
Kathy M. Clawson	University of Sunderland, UK
Claudiu Cobarzan	Babes-Bolyai University, Romania
Michel Crucianu	Cnam, France
Peng Cui	Tsinghua University, China
Rossana Damiano	Università di Torino, Italy
Petros Daras	Centre for Research and Technology Hellas, Greece
Wesley De Neve	Ghent University, Belgium
François Destelle	DCU Insight, Ireland
Cem Direkoglu	Middle East Technical University, Northern Cyprus Campus, Turkey
Lingyu Duan	Peking University, China
Jianping Fan	UNC Charlotte, USA
Mylène Farias	University of Brasília, Brazil
Gerald Friedland	ICSI/UC Berkeley, USA
Weijie Fu	Hefei University of Technology, China
Lianli Gao	University of Electronic Science and Technology, China
Yue Gao	Tsinghua University, China
Guillaume Gravier	CNRS, IRISA, and Inria Rennes, France
Ziyu Guan	Northwest University of China, China
Gylfi Þór Guðmundsson	Reykjavik University, Iceland
Silvio Guimarães	PUC Minas, Brazil
Allan Hanbury	TU Wien, Austria
Shijie Hao	Hefei University of Technology, China
Alex Hauptmann	Carnegie Mellon University, USA
Andreas Henrich	University of Bamberg, Germany
Nicolas Hervé	Institut National de l'Audiovisuel, France
Richang Hong	Hefei University of Technology, China
Frank Hopfgartner	University of Glasgow, UK
Michael Houle	National Institute of Informatics, Japan
Jun-Wei Hsieh	National Taiwan Ocean University, Taiwan, R.O.C.
Zhenzhen Hu	Nanyang Technological University, Singapore

Kai-Lung Hua	National Taiwan University of Science and Technology, Taiwan, R.O.C.
Jen-Wei Huang	National Cheng Kung University, Taiwan, R.O.C.
Benoit Huet	EURECOM, France
Wolfgang Hürst	Utrecht University, The Netherlands
Ichiro Ide	Nagoya University, Japan
Adam Jatowt	Kyoto University, Japan
Rongrong Ji	Xiamen University, China
Peiguang Jing	Tianjin University, China
Håvard Johansen	University of Tromsø, Norway
Mohan Kankanhalli	National University of Singapore, Singapore
Jiro Katto	Waseda University, Japan
Yoshihiko Kawai	NHK, Japan
Yiannis Kompatsiaris	CERTH-ITI, Greece
Harald Kosch	University of Passau, Germany
Markus Koskela	University of Helsinki, Finland
Duy-Dinh Le	National Institute of Informatics, Japan
Michael Lew	Leiden University, The Netherlands
Haojie Li	Dalian University of Technology, China
Teng Li	Anhui University, China
Yingbo Li	Tripnester, France
Rainer Lienhart	University of Augsburg, Germany
Suzanne Little	Dublin City University, Ireland
Bo Liu	Rutgers, USA
Xueliang Liu	HFUT, China
Zhenguang Liu	National University of Singapore, Singapore
Guojun Lu	Federation University Australia, Australia
Changzhi Luo	Hefei University of Technology, China
Stéphane Marchand-Maillet	University of Geneva, Switzerland
Jean Martinet	University of Lille, France
José M. Martínez	Universidad Autónoma de Madrid, Spain
Kevin McGuinness	Dublin City University, Ireland
Robert Mertens	HSW University of Applied Sciences, Germany
Vasileios Mezaris	CERTH, Greece
Rui Min	Google, USA
Dalibor Mitrović	mediamid digital services GmbH, Austria
Henning Muller	HES-SO, Switzerland
Phivos Mylonas	Ionian University, Greece
Chong-Wah Ngo	City University of Hong Kong, SAR China
Liqiang Nie	Shandong University, China
Naoko Nitta	Osaka University, Japan
Noel O'Connor	Dublin City University, Ireland
Neil O'Hare	Yahoo, USA
Vincent Oria	New Jersey Institute of Technology, USA
Tse-Yu Pan	National Cheng Kung University, Taiwan, R.O.C.
Fernando Pereira	Instituto Superior Técnico, Portugal

Yannick Prié   University of Nantes, France
Jianjun Qian   Nanjing University of Science and Technology, China
Xueming Qian   Xi'an Jiaotong University, China
Georges Quénot   LIG-CNRS, France
Miloš Radovanović   University of Novi Sad, Serbia
Michael Riegler   Simula Research Lab, Norway
Stevan Rudinac   University of Amsterdam, The Netherlands
Mukesh Kumar Saini   Indian Institute of Technology Ropar, India
Jitao Sang   Chinese Academy of Sciences, China
Klaus Schöffmann   Klagenfurt University, Austria
Pascale Sébillot   IRISA/INSA Rennes, France
Jie Shao   University of Electronic Science and Technology,
                              China
Xiaobo Shen   Nanjing University of Science and Technology, China
Koichi Shinoda   Tokyo Institute of Technology, Japan
Mei-Ling Shyu   University of Miami, USA
Alan Smeaton   Dublin City University, Ireland
Lifeng Sun   Tsinghua University, China
Yongqing Sun   NTT Media Intelligence Laboratories, Japan
Sheng Tang   Institute of Computing Technology, Chinese Academy
                              of Sciences, China
Shuhei Tarashima   NTT, Japan
Wei-Guang Teng   National Cheng Kung University, Taiwan, R.O.C.
Georg Thallinger   Joanneum Research, Austria
Qi Tian   University of Texas at San Antonio, USA
Christian Timmerer   Alpen-Adria-Universität Klagenfurt, Austria
Dian Tjondronegoro   Queensland University of Technology, Australia
Shingo Uchihashi   Fuji Xerox Co., Ltd., Japan
Feng Wang   East China Normal University, China
Jinqiao Wang   Chinese Academy of Sciences, China
Shikui Wei   Beijing Jiaotong University, China
Lai Kuan Wong   Multimedia University, Malaysia
Marcel Worring   University of Amsterdam, The Netherlands
Hong Wu   University of Electronic Science and Technology
                              of China, China
Xiao Wu   Southwest Jiaotong University, China
Changsheng Xu   Chinese Academy of Sciences, China
Toshihiko Yamasaki   The University of Tokyo, Japan
Bo Yan   Fudan University, China
Keiji Yanai   University of Electro-Communications, Tokyo, Japan
Kuiyuan Yang   Microsoft Research, China
You Yang   HUST, China
Jun Yu   Hangzhou Dianzi University, China
Maia Zaharieva   Vienna University of Technology, Austria
Matthias Zeppelzauer   University of Applied Sciences St. Poelten, Austria
Zheng-Jun Zha   University of Science and Technology of China, China

Cha Zhang	Microsoft Research, USA
Hanwang Zhang	National University of Singapore, Singapore
Tianzhu Zhang	CASIA, Bangladesh
Cairong Zhao	Tongji University, China
Ye Zhao	Hefei University of Technology, China
Lijuan Zhou	Dublin City University, Ireland
Shiai Zhu	University of Ottawa, Canada
Xiaofeng Zhu	Guangxi Normal University, China
Arthur Zimek	University of Southern Denmark, Denmark
Roger Zimmermann	National University of Singapore, Singapore

Demonstration, Special Session and VBS Reviewers

Shanshan Ai	Beijing Jiaotong University, China
Alberto Messina	RAI CRIT, Italy
François Brémond	Inria, France
Houssem Chatbri	Dublin City University, Ireland
Jingyuan Chen	National University of Singapore, Singapore
Yi Chen	Helsinki Institute for Information Technology, Finland
Yiqiang Chen	Chinese Academy of Sciences, China
Zhineng Chen	Chinese Academy of Sciences, China
Zhiyong Cheng	National University of Singapore, Singapore
Mariana Damova	Mozaika, Romania
Stamatia Dasiopoulou	Pompeu Fabra University, Spain
Monika Dominguez	Pompeu Fabra University, Spain
Ling Du	Tianjin Polytechnic University, China
Jana Eggink	BBC R&D, UK
Bailan Feng	Chinese Academy of Sciences, China
Fuli Feng	Chinese Academy of Sciences, China
Min-Chun Hu	National Cheng Kung University, Taiwan, R.O.C.
Lei Huang	Ocean University of China, China
Marco A. Hudelist	Klagenfurt University, Austria
Bogdan Ionescu	Politehnica University of Bucharest, Romania
Eleni Kamateri	CERTH, Greece
Hyowon Lee	Singapore University of Technology and Design, Singapore
Andreas Leibetseder	Alpen-Adria-Universität Klagenfurt, Austria
Na Li	Dublin City University, Ireland
Xirong Li	Renmin University of China, China
Wu Liu	Beijing University of Posts and Telecommunications, China
Jakub Lokoč	Charles University in Prague, Czech Republic
Mathias Lux	Klagenfurt University, Austria
Georgios Meditskos	CERTH, Greece
Wolfgang Minker	Ulm University, Germany
Bernd Münzer	Klagenfurt University, Austria

Adrian Muscat   University of Malta, Malta
Yingwei Pan   University of Science and Technology of China, China
Zhengyuan Pang   Tsinghua University, China
Stefan Petscharnig   Alpen-Adria Universität Klagenfurt, Austria
Manfred Jürgen Primus   Alpen-Adria-Universität Klagenfurt, Austria
Zhaofan Qiu   University of Science and Technology of China, China
Amon Rapp   University of Toronto, Canada
Fuming Sun   Liaoning University of Technology, China
Xiang Wang   National University of Singapore, Singapore
Hongtao Xie   Chinese Academy of Sciences, China
Yuxiang Xie   National University of Defense Technology, China
Shiqiang Yang   Tsinghua University, China
Yang Yang   University of Electronic Science and Technology
                           of China, China
Changqing Zhang   Tianjin University, China

External Reviewers

Duc Tien Dang Nguyen   Dublin City University, Ireland
Yusuke Matsui   National Institute of Informatics, Japan
Sang Phan   National Institute of Informatics, Japan
Jiang Zhou   Dublin City University, Ireland

Contents – Part II

Demonstrations

Video Browser Showdown

Contents – Part I

SS4: Multimedia and Multimodal Interaction for Health and Basic Care Applications

Full Papers Accepted for Poster Presentation

A Comparative Study for Known Item Visual Search Using Position Color Feature Signatures

Jakub Lokoč$^{(\boxtimes)}$, David Kuboň, and Adam Blažek

SIRET Research Group, Department of Software Engineering,
Faculty of Mathematics and Physics, Charles University in Prague,
Prague, Czech Republic
{lokoc,blazek}@ksi.mff.cuni.cz, KubonDavid@seznam.cz

Abstract. According to the results of the Video Browser Showdown competition, position-color feature signatures proved to be an effective model for visual known-item search tasks in BBC video collections. In this paper, we investigate details of the retrieval model based on feature signatures, given a state-of-the-art known item search tool – Signature-based Video Browser. We also evaluate a preliminary comparative study for three variants of the utilizes distance measures. In the discussion, we analyze logs and provide clues for understanding the performance of our model.

Keywords: Similarity search · Feature extraction · Known item search · Color sketch

1 Introduction

Nowadays, video data are present almost everywhere which challenges state-of-the-art video management systems and their query interfaces. Whereas traditional approaches rely on text-based query formulation or query by example paradigm [6], novel video retrieval scenarios require more sophisticated query interfaces [16]. An example of such scenario is known-item search (or mental query retrieval), where users cannot perfectly materialize their search intents and try to iteratively interact with the system to find a desired scene.

Systems for solving known-item search tasks rely on intuitive, interactive and multi-modal query interfaces. In the known-item search process, the user is in the center of the retrieval process and controls the intermediate actions navigating him towards the results. This feature complicates development and evaluation of known-item search systems. Therefore, competitions like the Video Browser Showdown [15] or evaluation campaigns (e.g., TRECVID [17]) are organized, where participating teams compete in predefined known-item search tasks. Two popular tasks are visual and textual known item search. In the visual known item search task, users see and memorize a short video clip (recording is not allowed), while in the textual known item search task, users receive a short text describing the desired scene.

© Springer International Publishing AG 2017
L. Amsaleg et al. (Eds.): MMM 2017, Part II, LNCS 10133, pp. 3–14, 2017.
DOI: 10.1007/978-3-319-51814-5_1

During the last five years of the Video Browser Showdown, several promising approaches have been revealed by winning teams. Generally, the winning tools [1,3,4,7,12] pointed on several features to be highly competitive in known-item search tasks:

Effective query initialization – instead of starting the search from the scratch, a preliminary initialization is necessary. For example, position-color sketches have proved to be highly effective for visual known-item search [3,4,11].

Concept based filtering – restricting the collection to scenes containing recognized concepts significantly helps with the retrieval. However, the technique relies on the effectiveness of concept detectors and their ability to recognize a search concept. In this direction, novel deep learning based approaches achieve promising results [2,9,18].

Visualization of the results – after query initialization and concept based filtering, the results require a suitable form of visualization enabling fast detection of desired scenes. A popular approaches are color sorted image maps [1] and/or results accompanied with temporal context from the corresponding video.

Effective browsing – the correct results are often not present on the first page and so scrolling between pages becomes necessary. Furthermore, scrolling can be performed also within the temporal context of a detected results, using similarity search techniques given a candidate key frame, or other video interaction scenarios [16].

Available work [5] investigates overall performance of most of the available video search approaches in detailed manner. In this paper we focus on one of such approaches based on position-color sketches introduced by the Signature-based Video Browser [13]. We summarize the tool in the next section and highlight the features investigated in this paper. Then we introduce three different distance measures that can be utilized in our model. To select the most appropriate one, we carried out a user study which is then analyzed and discussed. Alongside the distance measure selection, we provide a number of observations and clues on user behaviour leading to successful search.

2 Signature-Based Video Browser

The Signature-based Video Browser tool (SBVB) [4], has been successfully introduced at the Video Browser Showdown in 2014 (first place out of six teams). The first version relied solely on position color feature signatures [10,14] extracted from uniformly sampled key frames. This representation enables simple sketch based query interface, where users draw colored circles and place/modify them on the sketching canvas (depicted in Fig. 1 right). Assuming that the users are able to recognize and memorize rare/unique color stimuli, the color based ranking of the key frames provides a powerful tool to localize desired scenes. In order to improve the ranking, two time ordered color sketches can be provided.

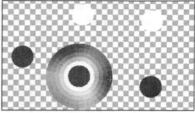

Fig. 1. Sketch-based Video Browser user interface (left) and a detail of sketching canvas (right). (Color figure online)

The details of the ranking model are presented in the following section. The results are visualized as a list of matched key frames, accompanied with a temporal context from the corresponding video (see Fig. 1 left). Hence, similar key frames from different parts of the video can be distinguished.

Since 2014, the tool has been significantly extended by several new features [3,11]. The tool supports compact visualizations of the results and interactive navigation summaries. Colored circles can be picked from or aligned to promising key frames. Since the video collection often consists of a set of clips, some clips can be excluded from the search or users can fix to just one particular video. All these features have been included in our user study as traditional browsing approaches. There are also other extensions that were not included in the user study as the study focuses solely on the color-based retrieval. The reason is that the position color feature signatures proved to be highly competitive in visual search also in the following years (the first place in 2015 and the third place in 2016). Therefore, in the user study users were not allowed to provide multi-modal sketches comprising also edges, perform similarity search using DeCAF features [8], enter keywords or query by example object. Note also that the results of color-sketch queries were presented just as a sorted list of matched keyframes (marked by red rectangle) accompanied by its temporal context occupying the whole line (depicted in Fig. 1 left). No other visualization technique was used in the study.

3 Feature Signatures Video Retrieval Model

In this section, details of the retrieval model for a set of video files are presented.

To index video files, roughly one key frame for every second is selected, resulting in the set of key frames $\mathbb{F} = \{F_1.F_2 \ldots F_N\}$. For all the key frames $F_i \in \mathbb{F}$, feature signatures $FS_i = \{r_{ij}\}$ are extracted where r_{ij} denotes the j-th centroid of the i-th feature signature. The centroid is defined as a tuple comprising x, y coordinates of the circle, L, a, b color coordinates from the CIE Lab color space and r denoting radius of the circle. The feature signatures are extracted using an adaptive k-means clustering. As mentioned earlier, users are enabled to define several sketch centroids that are matched to the extracted feature signatures. Since users may memorize only the most distinct color regions from the searched scene, only few query centroids are expected to be specified; hence, the model uses local instead of global matching between two feature signatures.

For a user defined query $FS_u = \{r_{uv}\}_{v=1}^m$ and a centroid distance measure δ, we start with calculating distances to all the key frames for each of the sketch centroids separately. For the sketch centroid r_{uv} the distance to the i-th key frame is defined as:

$$dist_{uvi} = \min_{\forall r_{ij} \in FS_i} \delta(r_{ij}, r_{uv}) \tag{1}$$

I.e., $dist_{uvi}$ is the distance to the closest of the key frame centroids. For a sketch centroid r_{uv} we denote the set of distances to all the key frames as D_{uv}. Formally,

$$D_{uv} = \{dist_{uvi} \mid i = 1 \ldots N\} \tag{2}$$

Now, the distances for the sketch centroid r_{uv} are scaled to $[0,1]$ interval according to the minimal and maximal value in D_{uv}.

$$rank_{uvi} = \frac{(dist_{uvi} - \min D_{uv})}{(\max D_{uv} - \min D_{uv})} \tag{3}$$

Thanks to the scaling, all the centroid rankings are comparable and we might obtain the overall ranking by simply averaging them. In particular, the rank of the key frame i is

$$rank_{ui} = \operatorname*{avg}_{\forall r_{uv} \in FS_u} rank_{uvi} \tag{4}$$

To complete our ranking model, we need to define the centroid distance measure δ. Given a user-defined centroid q and a database centroid o, $\delta(o, q)$ shall define their distance or dissimilarity. As both spatial and color spaces are suitable for L_p metrics, the first choice for δ is the regular Euclidean distance:

$$\textbf{A} \quad \delta_a(o, q) = L_2(o, q) = \sqrt{\sum_{d \in \{x,y,L,a,b\}} (d_o - d_q)^2}$$

Nonetheless, δ_a actually ignores one part of the extracted feature signatures information – the centroid size/radius. As we fixed the sketch circles sizes[1], including r in δ_a would only favor centroids of a particular size. For these reasons, we introduce two additional distance measures:

$$\textbf{B} \quad \delta_b(o, q) = \sqrt{\sum_{d \in \{L,a,b\}} (d_o - d_q)^2 + \max\left(0, \sqrt{\sum_{d \in \{x,y\}} (d_o - d_q)^2} - r_o\right)^2}$$

$$\textbf{C} \quad \delta_c(o, q) = \sqrt{\sum_{d \in \{L,a,b\}} (d_o - d_q)^2 + \max\left(0, \sum_{d \in \{x,y\}} (d_o - d_q)^2 - r_o^2\right)}$$

---

[1] Specifying sketch circle sizes was rather confusing for users. In practice, users were placing multiple circles of the same color to capture large color areas.

The idea is to take into account the database centroid radius r_o. The fundamental observation is that larger database centroids are effectively closer to the query centroid q. Hence, we subtract the radius r_o from the spatial part of the Euclidean distance. To elucidate the idea, consider the proposed distance functions without the color coordinates:

$$\delta_a'(o,q) = \sqrt{\sum_{d\in\{x,y\}} (d_o - d_q)^2}$$

$$\delta_b'(o,q) = \max\left(0, \sqrt{\sum_{d\in\{x,y\}} (d_o - d_q)^2} - r_o\right)$$

$$\delta_c'(o,q) = \max\left(0, \sqrt{\sum_{d\in\{x,y\}} (d_o - d_q)^2 - r_o^2}\right)$$

Given only position coordinates, the distance δ_b' measures just the distance from the query centroid to the border of the colored circle representing the database centroid. Hence, the distance enables users to match more likely centroids with larger radius.

4 User Study

We carried out a user study with total of 33 participants[2] that searched for previously presented video segments using SBVB tool. Our aim was to identify the most suitable distance measure as well as to determine whether there are certain search patterns and strategies providing higher chances of finding the searched scene.

4.1 Conditions and Collected Data

The user study focuses on the visual known item search, where users see a nine-second video clip, memorize it and then try to find the clip in a video database using the SBVB tool with a limited functionality. The employed database contains almost 30-hour subset of diverse video content randomly selected from the Video Browser Showdown 2016 dataset, including, but not limited to, TV shows, sports, indoor and outdoor activities, etc. Example key frames from the searched video segments are displayed in Fig. 2.

As up to date SBVB is rather complex tool, we selected only a subset of features (listed bellow) to be available for users in the study.

- Color based retrieval using position-color sketches and feature signatures.
- Temporal color-based queries using two color sketches.
- Fitting and updating the currently used centroids to a specific key frame.

---

[2] 17 out of 33 participant received a university education in computer science.

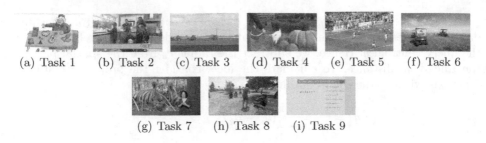

(a) Task 1   (b) Task 2   (c) Task 3   (d) Task 4   (e) Task 5   (f) Task 6

(g) Task 7   (h) Task 8   (i) Task 9

Fig. 2. Example key frames from the searched video segments which were selected randomly from the dataset. Note that both natural and artificial contents are covered.

- Excluding a video from the search.
- Fixing the search to a specific video.
- Horizontal (e.g. temporal order of key frames within one video) and vertical (e.g. between different pages of results) scrolling.

After brief introduction, users were asked to perform 9 visual KIS tasks as fast as possible using the tool with a limit of 2 min per task. The study was carried out on a notebook with Intel Core i7-4710HQ CPU, 8 GB RAM and FullHD 15,4" screen. It has been set up in the same way for all the participants. The participants performed the tasks in a room without direct sunlight and always during the day.

For each task one of the distance measures (A, B, C) was selected randomly so that each participant searched three times with each of the three distance measures. The experiments were observed by an administrator and each user action was stored in a log file. The system monitored the ranking of the searched scene and its adjacent key frames depending on the participant's actions. In particular, we recorded any query modification (adding, moving, removing of sketch centroids etc.), results displaying, browsing and even events induced by mouse movements.

Some of the tasks were composed of parts of longer scenes; hence, it could happen that the correct scene is found yet the exact key frames to be searched are missed by a second or two. Therefore, we decided to keep track not only of the exact key frames from the task, but also the adjacent key frames up to a tolerance of 10 s.

In the study, we were interested whether the search was successful or not, if any of the searched key frames were displayed and how the participants interacted with the tool. These properties are – **Successfully submitted** denoting the number of successful searches (user submitted a correct key frame), **First page appearance** representing how many participants received (not necessarily spotted) the searched scene on the first page and **Number of searches** showing the number of searches where participants used a given feature (e.g. the second sketch).

4.2   Color Sketching Observations

The main way to interact with the tool was to sketch the searched scene. On average, participants added 11 centroids, deleted 4.1 of them, moved 3.3 and changed the color of 5.4 centroids. Restricting the count to successful tasks only, 8.7 centroids were added, 2 deleted, 1.7 moved and 3.8 had their color changed, which is about the decrease that can be expected in a successful search. The analysis has also shown that there is no optimal number of centroids that would lead to a success, as that depends on the complexity of the searched scene and the similarity to other scenes in the dataset. According to our empirical observations, experienced users with a prior knowledge about the dataset can select fewer colored centroids and find the scene faster. In all tasks, some users tried both color-sketch canvases (Fig. 3). Overall, two canvases were used only in one fourth of all searches.

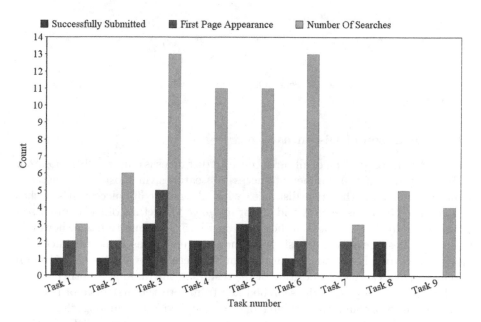

Fig. 3. The effects of using the second sketch. Only searches where the second sketch was used are considered. (Color figure online)

Fitting and updating the currently used centroids to a specific key frame (Fig. 4) was used mainly for task 7 and 9, where the initial color sketching effectiveness was not high for all three compared distances (see Fig. 6).

Fixing the search to a specific video was mostly counterproductive for the search. In many cases, users fixed the search to a wrong video and then were unable to retrieve the desired scene.

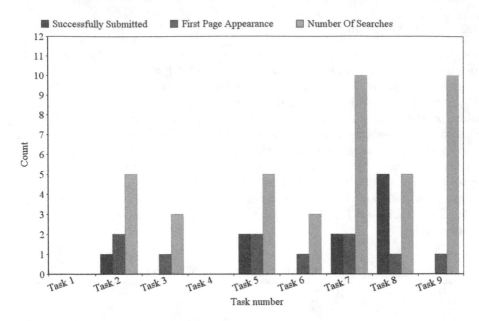

Fig. 4. The effects of fixing a scene. Only searches where the scene was fixed are considered.

4.3   Comparison of Distances

In this section, comparison of all three centroid distances is empirically evaluated.

In the terms of the number of successful searches, in most tasks distance B outperforms the other two distances with altogether 36 successful searches compared to 25 for distance A and 29 for distance C. Details of the comparison with regards to single scenes can be found in Fig. 5. It is questionable, however, if the overall success is the best performance measure as it is influenced by additional factors; hence, we have investigated whether the number of successful searches correlates with the position of the results in the ranking. In other words, whether distance B pushes the searched key frames to the first page of results.

Taking into account the ranking of the searched scene among the search results instead of the successful submit, the results do not show distance B to be the most effective variant. We measured the rankings for the matched key frame with tolerance of 0 to 10 s. Thorough analysis of the results has shown that tolerance of 3 s is the highest length of tolerance that doesn't compromise the results. In this case, the chance that a ranking of a completely different scene that followed a cut that happened right after the end of the searched is taken into account is still reasonably low; and the likelihood of the user's sketch matching a part of the searched scene is fairly high.

In the terms of the total number of the first page appearances, all distances show different behavior for different tasks, without a clear winner technique (see Fig. 6). This surprising empirical observation demonstrates that the appearance on the first page does not necessarily guarantee a successful search

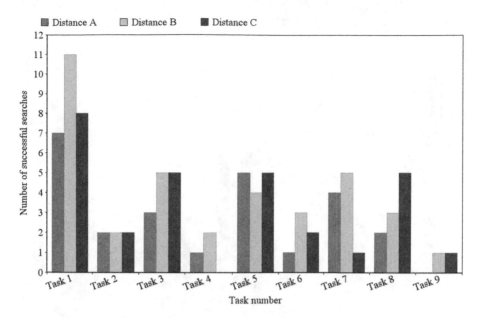

Fig. 5. The success rate per scene comparison.

(e.g., Tasks 1, 2 and 3 for distances A and B). According to the experience of the tool administrator controlling the users and tasks, often the users had the correct results in front of their eyes but did not recognize them in the list of results. First page appearance of a searched frame without consequent successful submit occurred 18 times for distance A, 15 times for distance B and 16 times for distance C.

4.4 Discussion

It is not easy to explain why one distance has higher success rate without dominating the other two distances in the first page ranking. In the following we present just several observations from the administrator of the user study and several observations from the logs.

The number of logged operations after the best rank was achieved showed that the distances differ in the browsing phase. Using distance B leads to a lower number of logged operations (i.e., browsing efforts) comparing to distances A and C. This could explain the higher number of successful searches in our time-limited test settings. The reason why users browse less with distance B is the subject of our future investigations.

There are several general issues when users try to solve the known-item search task. One group of users focuses too much on drawing a sketch and miss very promising results that were shown during the process (observed by the administrator). Another group looks only for a certain part of the video and when another part appears in the results, they miss it. Some users scroll

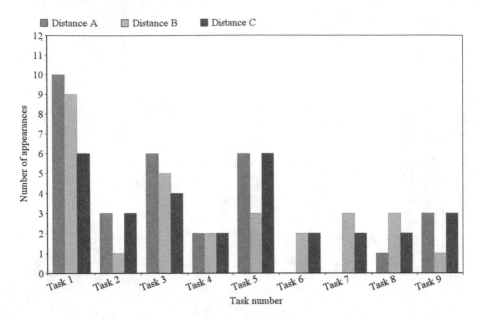

Fig. 6. The number of first page appearances of the searched scene with 3 s tolerance. (Color figure online)

randomly through a promising video. It was also observed that some users were able to locate the searched key frame through noticing visually dissimilar, but semantically relevant scenes appearing close to the result. Thus early in the task restricted the search to that particular video. Scrolling through three or more pages usually doesn't help very much, because people tend to miss the searched scene when overwhelmed by the rest of the content. The other search techniques, such as the second sketch and fixing a scene are indeed helpful, but only when used properly. Hence more difficult for novice users.

5   Conclusions

In this paper, we have focused on a technique for visual known-item search based on position-color feature signatures. The video retrieval model was presented and a preliminary comparative user study evaluated. The results of the study show that the ranking of the results is not necessarily the most important criterion for visual known-item search tasks. Based on this preliminary observations, we would like to organize a more thorough user study with a higher number of participants and perform a more sophisticated statistical analysis of the results.

Acknowledgments. This research was supported by grant SVV-2016-260331, Charles University project P46 and GAUK project no. 1134316.

References

1. Barthel, K.U., Hezel, N., Mackowiak, R.: Navigating a graph of scenes for exploring large video collections. In: Tian, Q., Sebe, N., Qi, G.-J., Huet, B., Hong, R., Liu, X. (eds.) MMM 2016. LNCS, vol. 9517, pp. 418–423. Springer, Heidelberg (2016). doi:10.1007/978-3-319-27674-8_43
2. Bengio, Y.: Learning deep architectures for AI. Found. Trends Mach. Learn. **2**(1), 1–127 (2009)
3. Blažek, A., Lokoč, J., Matzner, F., Skopal, T.: Enhanced signature-based video browser. In: He, X., Luo, S., Tao, D., Xu, C., Yang, J., Hasan, M.A. (eds.) MMM 2015. LNCS, vol. 8936, pp. 243–248. Springer, Heidelberg (2015). doi:10.1007/978-3-319-14442-9_22
4. Blažek, A., Lokoč, J., Skopal, T.: Video retrieval with feature signature sketches. In: Traina, A.J.M., Traina, C., Cordeiro, R.L.F. (eds.) SISAP 2014. LNCS, vol. 8821, pp. 25–36. Springer, Heidelberg (2014). doi:10.1007/978-3-319-11988-5_3
5. Cobârzan, C., Schoeffmann, K., Bailer, W., Hürst, W., Blažek, A., Lokoč, J., Vrochidis, S., Barthel, K.U., Rossetto, L.: Interactive video search tools: a detailed analysis of the video browser showdown 2015. Multimedia Tools Appl., 1–33 (2016)
6. Datta, R., Joshi, D., Li, J., Wang, J.Z.: Image retrieval: Ideas, influences, and trends of the new age. ACM Comput. Surv. **40**(2), 5:1–5:60 (2008)
7. Fabro, M., Böszörmenyi, L.: AAU video browser: non-sequential hierarchical video browsing without content analysis. In: Schoeffmann, K., Merialdo, B., Hauptmann, A.G., Ngo, C.-W., Andreopoulos, Y., Breiteneder, C. (eds.) MMM 2012. LNCS, vol. 7131, pp. 639–641. Springer, Heidelberg (2012). doi:10.1007/978-3-642-27355-1_63
8. Donahue, J., Jia, Y., Vinyals, O., Hoffman, J., Zhang, N., Tzeng, E., Darrell, T.: DeCAF: a deep convolutional activation feature for generic visual recognition. CoRR abs/1310.1531 (2013)
9. Krizhevsky, A., Sutskever, I., Hinton, G.E.: ImageNet classification with deep convolutional neural networks. In: Pereira, F., Burges, C., Bottou, L., Weinberger, K. (eds.) Advances in Neural Information Processing Systems 25: 26th Annual Conference on Neural Information Processing Systems 2012, Lake Tahoe, Nevada, US, 3–6 December 2012, pp. 1097–1105. Curran Associates, Inc. (2012)
10. Kruliš, M., Lokoč, J., Skopal, T.: Efficient extraction of clustering-based feature signatures using GPU architectures. Multimedia Tools Appl. **75**(13), 8071–8103 (2016)
11. Kuboň, D., Blažek, A., Lokoč, J., Skopal, T.: Multi-sketch semantic video browser. In: Tian, Q., Sebe, N., Qi, G.-J., Huet, B., Hong, R., Liu, X. (eds.) MMM 2016. LNCS, vol. 9517, pp. 406–411. Springer, Heidelberg (2016). doi:10.1007/978-3-319-27674-8_41
12. Le, D.-D., Lam, V., Ngo, T.D., Tran, V.Q., Nguyen, V.H., Duong, D.A., Satoh, S.: NII-UIT-VBS: a video browsing tool for known item search. In: Li, S., Saddik, A., Wang, M., Mei, T., Sebe, N., Yan, S., Hong, R., Gurrin, C. (eds.) MMM 2013. LNCS, vol. 7733, pp. 547–549. Springer, Heidelberg (2013). doi:10.1007/978-3-642-35728-2_65
13. Lokoč, J., Blažek, A., Skopal, T.: Signature-based video browser. In: Gurrin, C., Hopfgartner, F., Hurst, W., Johansen, H., Lee, H., O'Connor, N. (eds.) MMM 2014. LNCS, vol. 8326, pp. 415–418. Springer, Heidelberg (2014). doi:10.1007/978-3-319-04117-9_49
14. Rubner, Y., Tomasi, C., Guibas, L.J.: The earth mover's distance as a metric for image retrieval. Int. J. Comput. Vis. **40**(2), 99–121 (2000)

15. Schoeffmann, K.: A user-centric media retrieval competition: the video browser showdown 2012–2014. IEEE MultiMedia **21**(4), 8–13 (2014)
16. Schoeffmann, K., Hudelist, M.A., Huber, J.: Video interaction tools: a survey of recent work. ACM Comput. Surv. **48**(1), 14 (2015)
17. Smeaton, A.F., Over, P., Kraaij, W.: Evaluation campaigns and TRECVid. In: Proceedings of the 8th ACM International Workshop on Multimedia Information Retrieval, MIR 2006, pp. 321–330. ACM Press, New York (2006)
18. Szegedy, C., Liu, W., Jia, Y., Sermanet, P., Reed, S., Anguelov, D., Erhan, D., Vanhoucke, V., Rabinovich, A.: Going deeper with convolutions. In: 2015 IEEE Conference on Computer Vision and Pattern Recognition (CVPR), pp. 1–9, June 2015

A Novel Affective Visualization System for Videos Based on Acoustic and Visual Features

Jianwei Niu$^{1(\boxtimes)}$, Yiming Su^1, Shasha Mo^1, and Zeyu Zhu^2

^{1} State Key Laboratory of Virtual Reality Technology and Systems,
School of Computer Science and Engineering, Beihang University,
Beijing 100191, China
{niujianwei,sy1406126}@buaa.edu.cn, mo06211225@163.com
^{2} School of Electronics and Information,
Xi'an Jiaotong University, Xi'an 710049, China
asxxzzy@qq.com

Abstract. With the fast development of social media in recent years, affective video content analysis has become a hot research topic and the relevant techniques are adopted by quite a few popular applications. In this paper, we firstly propose a novel set of audiovisual movie features to improve the accuracy of affective video content analysis, including seven audio features, eight visual features and two movie grammar features. Then, we propose an iterative method with low time complexity to select a set of more significant features for analyzing a specific emotion. And then, we adopt the BP (Back Propagation) network and circumplex model to map the low-level audiovisual features onto high-level emotions. To validate our approach, a novel video player with affective visualization is designed and implemented, which makes emotion visible and accessible to audience. Finally, we built a video dataset including 2000 video clips with manual affective annotations, and conducted extensive experiments to evaluate our proposed features, algorithms and models. The experimental results reveals that our approach outperforms state-of-the-art methods.

Keywords: Affective analysis · Novel features · Feature selection · Emotion visualization

1 Introduction

In the last decades, thousands of videos are produced every day. Since the number of videos is enormous, various applications based on affective video content analysis have become more and more popular in recent years, such as video segmentation [1], video recommendation [2,3], highlights extraction [4], and video retrieval [5]. In affective video content analysis, audiovisual features extracted from video play an important role. To some extent, significant features can bridge the gap between a video stream and the emotions the video stream may elicit.

© Springer International Publishing AG 2017
L. Amsaleg et al. (Eds.): MMM 2017, Part II, LNCS 10133, pp. 15–27, 2017.
DOI: 10.1007/978-3-319-51814-5_2

In [5], Zhang et al. extracted music related features like tempo, beat strength, and rhythm regularity to analyze music videos. Several low-level features like saturation, lighting, pitch, and zero crossing rate were also utilized in their work. In [6], Hanjalic et al. extracted four features (motion, shot change rate, sound energy and audio pitch) to compute "arousal" and "valence". Besides the features mentioned in the above two papers, abundant features have been introduced in previous work. Most of these features can be classified into several categories: (1) Features extracted directly from raw data, like sound energy, saturation, lighting, etc. [2,7,8]. (2) Classical features frequently used in signal processing like MFCC (Mel Frequency Cepstrum Coefficient), ZCR (Zero Crossing Rate), LPCC (Linear Prediction Cepstrum Coefficient), spectral rolloff, etc. [9–13]. (3) Features used for special analysis, for instance, rhythm strength and tempo for music analysis [5], and music emotion and face features for highlight extraction [4]. (4) Shot features, such as shot types, shot change rate, and motion [6,14]. In this paper, a novel feature is proposed to describe the "harmony" in audio signal. This feature measures the amount of harmonious elements in a piece of audio. Besides, four "color emotions" in the color psychology field are also utilized [15]. These four color emotion features contain more affective information since they are the result of psychology research, and therefore, the four features are supposed to be more useful than other features. Together with another 12 common features introduced in previous work, 17 features make up the feature set used in our work.

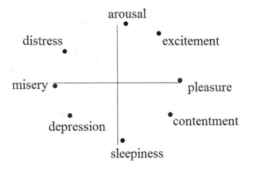

Fig. 1. Russell's circumplex model

In order to achieve affective visualization, we adopt a categorical affective model since dimensional affective models are usually obscure to people. As shown in Fig. 1, the categorical affective model (Russell's circumplex model) contains eight emotions (i.e., arousal, excitement, pleasure, contentment, sleepiness, depression, misery, and distress) [16]. The intensity of these eight intuitive emotions will be computed using the extracted features and BP (back propagation) network. In order to obtain significant features for computing certain emotion in movies, a novel feature selection algorithm with quadratic time complexity is proposed. Based on the affective content analysis of movies, a video player with affective visualization is implemented, which shows the emotions of videos while playing videos. Extensive experiments are conducted to prove the effect of the novel features we used and the validity of the proposed feature selection method. The major contributions of this work include:

1. We propose a novel feature set including an audio feature depicting the "harmony" of the sound in videos and four color emotion features from the color psychology field. The features are proved to be helpful for affective video content analysis.
2. We propose a quadratic-time method to select significant features for computing a certain emotion in a movie.
3. We design and implement a player that can display video emotions while playing videos. Extensive experiments are conducted to validate our methods and player.

The rest of this paper is organized as follows. Section 2 introduces the overall methodology of this work. Section 3 explains the specific methods of extracting the novel features proposed in this work and also discusses the proposed feature selection method. Section 4 presents the data collection in this work and the experiment results. Finally, conclusions are drawn in Sect. 5.

2 Overall Methodology

In this paper, we aim to improve the accuracy of affective video content analysis. Figure 2 describes the framework of our work.

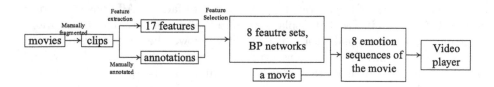

Fig. 2. The framework of our work

Data collection and annotation: We segment movies into short clips, and each movie clips is annotated with the intensity indices of eight emotions shown in Fig. 1. The movie clips are collected as the training or testing sets for experiments.

Feature extraction: Seventeen features are extracted from each movie clip and used for predicting emotions.

Feature selection: Apparently, it is not appropriate to use all the seventeen features to predict an emotion, and we proposed a novel method to select significant features for predicting each emotion.

Emotion prediction: BP networks are adopted to map features to emotions. The BP network used in this work contains one hidden layer. The training function

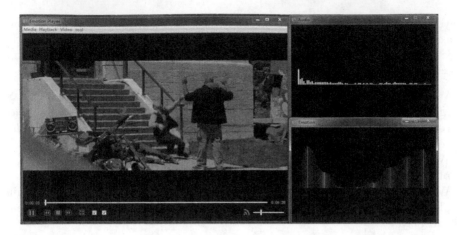

Fig. 3. User Interface of AffePlayer

is "trainlm", and the transfer function used in the hidden and output layers is "tansig". Eight BP networks are trained for predicting eight emotions respectively.

Affective visualization: Based on the affective content analysis, a video player, we term it *AffePlayer*[1], is developed to exhibit the emotions in the playing video.

Emotion annotations of movie clips and the corresponding features are used to train eight BP networks, for the aim of computing the eight emotions of a movie. Since a movie is long, the emotions of the movie may vary greatly over time. We extract the emotions of a movie every 10 s (using windows of ten seconds and 9 s overlap between neighboring windows) and thus obtain eight emotions sequences of the movie. The UI (User Interface) of AffePlayer is shown in Fig. 3. As shown in Fig. 3, it displays video in the left window, and the emotions (bottom) and audio signals (top) in the right window. The 8 light bars represent arousal, excitement, pleasure, contentment, sleepiness, depression, misery, and distress from left to right, respectively. The dim bars between two light bars represent the transition between two emotions, which can be interpreted as the mixture of the two emotions. The height of these bars represents the intensity of emotions. When the player is playing a video, the height of these bars will change with the intensity fluctuations of emotions.

3   Feature Extraction and Selection

Totally 17 features are extracted, including seven acoustic features, eight visual features, and two movie grammar features. A novel method is proposed to select significant features from these 17 features for analysing emotions in movies.

---

[1] www.ldmc.buaa.edu.cn/AffePlayer/AffePlayer.html.

3.1   Features

One of the novel features is proposed in this work to measure the "harmony" in audio signal. Figure 4 shows the spectrogram of noise sound and a piece of piano music, i.e., the changes of frequency over time. The lighter is the color in the spectrogram, the stronger is the amplitude of the corresponding frequency. Harmony sound show continuity in spectrogram while noisy sound show only disorder. As is shown in Fig. 4, there are continuous peak in the spectrogram of piano music but none in the spectrogram of noise sound. Based on this observation, we propose a method to compute "harmony" in audio signal by measuring the peak continuity in the spectrogram.

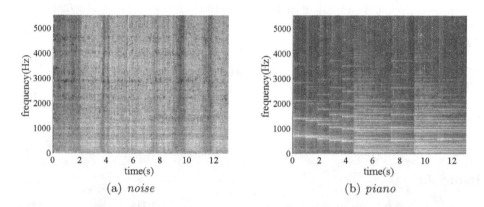

(a) *noise*   (b) *piano*

Fig. 4. Spectrogram of noise and a piece of piano music

Firstly, the Fourier transform of the audio segment is obtained every 1024 samples, and the spectrogram of the audio segment can be obtained. Assume the spectrogram of an audio segment is denoted by F_{ij}, where i is the index of frequency and j is the index of time. Then the local maximum of each j is found, and a matrix M is used to record the local maximum. In matrix M, the element M_{ij} equals 1 if F_{ij} is bigger than F_{i-2j}, F_{i-1j}, F_{i+1j}, and F_{i+2j}; otherwise M_{ij} equals 0. Matrix M is used to compute the harmony of the audio. "Harmony sequence" is defined as (M_{ia}, M_{ib}), where $a \leqslant b$, $M_{ia-1} = 0$, $M_{ib+1} = 0$, and $M_{ik} = 1$ $(a \leqslant k \leqslant b)$. It is assumed that there are n "harmony sequences" in total: $(M_{ia}, M_{ib})_s$ $(s = 1, 2, \cdots, n)$, then the harmony of the audio segment can be denoted as

$$H = \frac{1}{Size(M)} \sum_{1 \leqslant s \leqslant n} ||(M_{ia}, M_{ib})_s|| \tag{1}$$

where $||(M_{ia}, M_{ib})_s|| = (b - a)^2$, and $Size(M)$ denotes the size of matrix M which equals the product of the height and the width of M.

Color is important in affective video content analysis since existing research results suggest that color directly affects human's emotion. The research results

from color psychology reveal that different color induce different feelings of people, and based on the result, four novel color features (Warm-Cool, Transparent-Turbid, Vivid-Sombre, and Hard-Soft) are utilized in this work. Nakamura et al. investigated twelve pairs of color emotions and quantify these color emotions. As the results of their experiments, empirical formulae were obtained, and we select four typical color emotions as the features used in our work. The four color emotions are calculated in the CIELAB color space and the specific calculation methods are:
Warm-Cool:

$$WC = 3.5[cos(h - 50) + 1]B - 80 \qquad (2)$$

where B represents the brightness calculated by $B = 50C^*(1 - \Delta h_{290}/360)/D$, and D is the color depth calculated by $D = (100 - L^*) + (0.1 + \Delta h_{290}/360)(1 - \Delta h_{290}/360)C^*$.
Transparent-Turbid:

$$TT = [\{5(L^* - 40)\}^2 + [5.8\{1 + 0.35cos(\Delta h_{220})\}(1 - \Delta h_{290}/360)C^*]^2]^{1/2}$$
$$- 180 \qquad (3)$$

Vivid-Sombre:

$$VS = [\{2.5(L^* - 50)\}^2 + \{5.4(1 - \Delta h_{290}/360)\}^2]^{1/2} - 130 \qquad (4)$$

Hard-Soft:

$$HS = [(3.2L^*)^2 + \{2.4(1 - \Delta h_{290}/360)C^*\}^2]^{1/2} - 180 \qquad (5)$$

In the above equations, L^* is the CIELAB lightness, C^* is the CIELAB chroma, h is the CIELAB hue-angle, and Δh_{290} is the CIELAB hue-angle difference from $h = 290$ [15].

Besides the novel features mentioned before, other twelve common features are also extracted. Six acoustic features including: (1) Sound energy (the average amplitude of audio signals), (2) Sound centroid (the average frequency of the fourier transform of audio signals), (3) Spectral contrast of an audio segment (extracted in three sub-bands resembling that in Lu's work) [17], (4) Silence ratio (the proportion of the time duration of silent sound to the entire time length). Four visual features including: (1) Darkness ratio and brightness ratio extracted in the HSV (Hue-Saturation-Value) color space according to the value of V (Value) dimension, (2) Saturation, which has strong effects on emotions [18], is the average saturation of all frames. (3) Color energy is extracted following that in [19]. Two movie grammar features are extracted in this paper: motion and shot change rate.

3.2   Feature Selection

It is a simple way to feed all seventeen features into a BP network and get the emotions of a video, but is not rational. Since some features are not relevant to a certain emotion, that is, some features may be "detrimental" for calculating a certain emotion. Thus, it is necessary to find a way to select significant features and exclude redundant features.

If we are going to obtain the optimal combination of features for analysing emotions, the computational complexity will be $O(2^n)$ (where n is the number of features extracted from videos), which is unpractical. Therefore, we propose a method to select significant features with quadratic time complexity. Though it is hard to get the optimal feature combination, we can obtain a near-optimal solution with low time complexity. For emotion e_1, assume that S_1 is the set of selected features to predict e_1 and S_0 is the set of features that are not selected. Initially, S_0 contains all the 17 extracted features, while S_1 contains none. Then one feature f in S_0 is chosen and added into set S_1, where f meets the condition:

$$mse(f, D) \leqslant mse(f', D)|_{f' \in S_0 \& f' \neq f} \tag{6}$$

D denotes the training set, f' denotes the feature in S_0. f' is different from the chosen feature f. $mse(f, D)$ means the Mean Square Error (MSE) while using feature f to predict emotion e_1 on data set D. Equation 6 means that feature f performs better than any other feature in S_0 on predicting e_1. The Mean Square Error is recorded as m. The chosen feature f is moved from S_0 to S_1. Next, features in S_1 (now there is only one selected feature) are used to predict emotion e_1, and 80% of the movie clips, which are worse predicted than the other 20% movie clips, make up a data set named D'. Then, a feature f in S_0 is chosen and added into set S_1 if f meet the condition:

$$mse(f, D') \leqslant mse(f', D')|_{f' \in S_0 \& f' \neq f} \tag{7}$$

Emotion e_1 in set D is predicted using features in S_1 and the Mean Square Error m' is obtained. If m' is greater than m, then the selecting process is terminated and remove the last selected feature from S_1; or else assign m' to m and continually choosing a feature and move it from S_0 to S_1 until there are no features in S_0. At last, the features in set S_1 are the features selected to predict emotion e_1. There are eight emotions used in this paper, and therefore, eight feature sets will be obtained to predict the eight emotions of videos respectively.

4   Experimental Results

Experiments are conducted on both our own collected database and the LIRIS-ACCEDE database [20]. In the first subsection, we introduce the process of building our own database. In the second subsection, experiments conducted on the LIRIS-ACCEDE database are presented. The significant features for valence and arousal are obtained and the results of our method are compared to that in [20]. In the third subsection, experiments are conducted on our own database. The most relevant single feature of each emotion is found from the 17 extracted features, and the feature combinations for predicting eight emotions are obtained using the method we proposed in Sect. 3.

4.1   Data Collection and Annotation

Our database consists of 2000 video clips manually segmented from 112 movies which can be download from our website[2]. The movies are various in genres, including action, comedy, thriller, romance, etc. Languages are English, Chinese, Spanish, Japanese, and Korean. Each movie clip lasts about 10 s and all the movie clips are annotated with the intensity of eight emotions from the Russell's circumplex model [16]. The intensity of eight emotions are divided into 5 grades: 1, 2, 3, 4, and 5 correspond to very weak, weak, normal, strong, and very strong, respectively. The annotators are fifty students from our university, including 13 females and 37 males. A web site is built for students to complete their annotating work. After the students watch a movie clip, they are requested to estimate the intensity of eight emotions and annotate the movie clip with eight values denoting the intensity of eight emotions respectively. Thus, for each emotion of a movie clip, there are 50 values annotated by 50 different persons. A video clip is excluded from our dataset if the Mean Square Error of any emotion annotations is bigger than 1. Fifty values are averaged and rounded to an integer, and the average value is taken as the final annotation of the emotion.

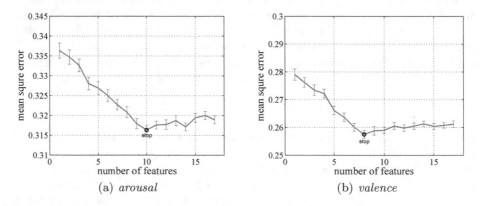

(a) *arousal*   (b) *valence*

Fig. 5. Mean Square Error of feature selection for arousal and valence (95% confidence interval)

Table 1. Features used to calculate arousal and valence in LIRIS-ACCEDE database

Emotion	Features
Arousal	**Warm-cool**, **vivid-sombre**, **harmony**, centroid, low contrast, mid contrast, high contrast, color energy, dark color ratio, silence ratio
Valence	**Warm-cool**, **transparent-turbid**, **vivid-sombre**, centroid, **hard-soft**, saturation, color energy, dark color ratio

---

[2] www.ldmc.buaa.edu.cn/AffePlayer/database.rar.

4.2   Experiments on LIRIS-ACCEDE Database

The results of experiment on the LIRIS-ACCEDE database are presented in this subsection. Figure 5 shows the variation of Mean Square Error while selecting features for predicting arousal and valence. At first, only one feature is selected for predicting emotions with relatively big error. With the increase of the number of features, the Mean Square Error become smaller. According to the algorithm introduced in Sect. 3, the process end at the black circle marked in Fig. 5. That is, ten features are selected for arousal and eight features for valence as shown in Table 1. As can be seen from Fig. 5, the prediction error will increase slightly if all 17 features are used. Although there may exist better feature combinations for predicting arousal or valence, the process of the feature selection shown in Fig. 5 proves that the results in Table 1 are at least near-optimal solutions. The proposed feature "harmony" and two color emotions are in the feature combination for arousal and four color emotions are included in the feature combination for valence, which demonstrate the effectiveness of the novel features we introduced in Sect. 3. Experiments are conducted based on protocol A and protocol B introduced in [20]. The results of experiments are shown in Fig. 6. It can be seen from Fig. 6 that the method proposed in this paper performs better on the prediction of valence and the prediction of valence is as good as that in [20].

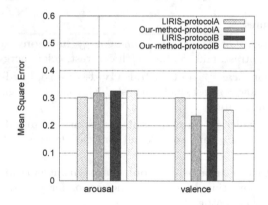

Fig. 6. Results of experiments on LIRIS-ACCEDE database.

4.3   Experiments on Our Own Database

In this subsection, the database collected in this work is used in several experiments. Seventy percent of the 2000 movie clips are used as the training set and the rest thirty percent are used as the testing set. We apply the feature selection method on the training set, and the variation of Mean Square Errors in the process of feature selection are shown in Fig. 7. Eight feature combinations are selected for predicting eight emotions using the method proposed in Sect. 3. The black circles marked in the figure represent the terminations of feature selection, and the selected features are shown in Table 2. As can be seen from Table 2, the novel features we proposed and introduced from color psychology play important roles in emotion predictions.

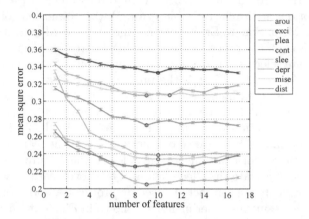

Fig. 7. MSE of feature selection for 8 emotions (95% confidence interval)

Table 2. Features used to estimate the 8 emotions in our work

Emotion	Features
Arousal	Motion, shot change rate, sound energy, **harmony**, sound centroid, low contrast, mid contrast, high contrast, color energy, silence ratio
Excitement	Motion, **transparent-turbid**, **vivid-sombre**, **hard-soft**, silence ratio, low contrast, mid contrast, high contrast, brightness ratio, darkness ratio
Pleasure	Shot change rate, **warm-cool**, **transparent-turbid**, **vivid-sombre**, color energy, **hard-soft**, sound centroid, darkness ratio, brightness ratio
Contentment	Shot change rate, **warm-cool**, **transparent-turbid**, **vivid-sombre**, **hard-soft**, sound centroid, silence ratio, darkness ratio, brightness ratio, low contrast
Sleepiness	Motion, shot change rate, sound energy, **harmony**, high contrast, sound centroid, low contrast, mid contrast, silence ratio
Depression	**Warm-cool**, **transparent-turbid**, **vivid-sombre**, **hard-soft**, brightness ratio, low contrast, mid contrast, color energy, darkness ratio
Misery	Shot change rate, **warm-cool**, **transparent-turbid**, color energy, **vivid-sombre**, **hard-soft**, sound centroid, saturation, darkness ratio, brightness ratio, silence ratio
Distress	Motion, shot change rate, **vivid-sombre**, darkness ratio, low contrast, mid contrast, color energy, silence ratio

The first feature chosen for a certain emotion can be regard as the most relevant feature of the emotion. Color emotion vivid-sombre is most relevant to pleasure, contentment, and misery. The feature most relevant to sleepiness is silence ratio, which is the same as common sense. Silence ration is also most relevant to distress and motion, mid-band spectral contrast and bright ratio are most relevant to arousal, excitement and depression respectively. It can be seen from Fig. 7 that excitement is best predicted with single feature, which means mid-band spectral is really suitable for predicting excitement.

The feature combinations for predicting eight emotions are shown in Table 2. The feature combinations for predicting each emotion are different from each other. As shown in Fig. 7, the Mean Square Errors of the computation are between 0.205 and 0.333. It can be seen from Fig. 7 that the computation results vary with emotions, and the computations of arousal, excitement, sleepiness, and misery are more precise than that of the other four emotions, which means the features extracted are more suitable for predicting arousal, excitement, sleepiness and distress. Among these emotions, sleepiness is the best estimated emotions with MSE of 0.205 and contentment is the worst estimated with MSE of 0.333. The average MSE of computing eight emotions is 0.2654. Therefore, we could say that compared to the result in [20], the features and the framework in this work perform better.

Another experiment is conducted to find the impact of feature types on analyzing different video emotions. The 17 features contain three types of feature, that is, acoustic features, visual features, and movie grammar features. Every feature is used to compute the eight emotions respectively. The MSEs are averaged within each type of features and the results are shown in Fig. 8. Visual features perform best in computing pleasure, contentment, and misery, which means features about color and lighting have greater impact on emotions pleasure and unpleasure. Acoustic features and movie grammar features are more suitable to compute the intensity of arousal and sleepiness.

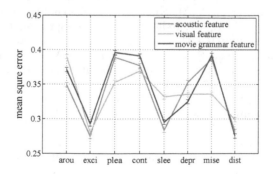

Fig. 8. MSE of computing emotions using different type of features (95% confidence interval)

5   Conclusion

In this work, a novel feature set is used in affective video content analysis, including an acoustic feature depicting the "harmony" in audio signal and four color emotions. The "harmony" feature performs well in predicting emotion arousal and sleepiness. And the four color emotion features which contain the psychology information are useful in predicting almost all the emotions. A feature selection method is proposed to obtain significant features, and experimental results proves the effectiveness of the feature selection method.

Acknowledgement. This work was supported by the National Natural Science Foundation of China (Grant Nos. 61572060, 61190125, 61472024) and CERNET Innovation Project 2015 (Grant No. NGII20151004).

References

1. Arifin, S., Cheung, P.: Affective level video segmentation by utilizing the pleature-arousal-dominance information. IEEE Trans. Multimed. **10**(7), 1325–1341 (2008)
2. Niu, J., Zhao, X., Zhu, L., Li, H.: Affivir: an affect-based Internet video recommendation system. Neurocomputing **120**, 422–433 (2013)
3. Arapakis, I., Moshfeghi, Y., Joho, H., Ren, R., Hannah, D., Jose, J.M.: Enriching user profiling with affective features for the improvement of a multimodal recommender system. In: Proceedings of the ACM International Conference on Image and Video Retrieval, p. 29. ACM (2009)
4. Lin, K.S., Lee, A., Yang, Y.H., Lee, C.T., Chen, H.: Automatic highlight extraction for drama video using music emotion and human face features. Neurocomputing **119**, 111–117 (2013)
5. Zhang, S., Huang, Q., Jiang, S., Gao, W., Tian, Q.: Affective visualization and retrieval for music video. IEEE Trans. Multimed. **12**(6), 510–522 (2010)
6. Hanjalic, A., Xu, L.Q.: Affective video content representation and modeling. IEEE Trans. Multimed. **7**(1), 143–154 (2005)
7. Chan, C.H., Jones, G.J.F.: An affect-based video retrieval system with open vocabulary querying. In: Detyniecki, M., Knees, P., Nürnberger, A., Schedl, M., Stober, S. (eds.) AMR 2010. LNCS, vol. 6817, pp. 103–117. Springer, Heidelberg (2012). doi:10.1007/978-3-642-27169-4_8
8. Zhang, S., Tian, Q., Jiang, S., Gao, W.: Affective MTV analysis based on arousal and valence features. In: 2008 IEEE International Conference on Multimedia and Expo, pp. 1369–1372. IEEE (2008)
9. Canini, L., Benini, S., Leonardi, R.: Affective recommendation of movies based on selected connotative features. IEEE Trans. Circ. Syst. Video Technol. **23**(4), 636–647 (2013)
10. Xu, M., Wang, J., He, X., et al.: A three-level framework for affective content analysis and its case studies. Multimed. Tools Appl. **70**(4), 757–779 (2014)
11. Cui, Y., Jin, J.S., Zhang, S., Tian, Q.: Music video affective understanding using feature importance analysis. In: Proceedings of the ACM International Conference on Image and Video Retrieval, pp. 213–219. ACM (2010)
12. Yazdani, A., Kappeler, K., Ebrahimi, T.: Affective content analysis of music video clips. In: Proceedings of the 1st International ACM Workshop on Music Information Retrieval with User-Centered and Multimodal Strategies, pp. 7–12. ACM (2011)

13. Acar, E., Hopfgartner, F., Albayrak, S.: Understanding affective content of music videos through learned representations. In: Gurrin, C., Hopfgartner, F., Hurst, W., Johansen, H., Lee, H., O'Connor, N. (eds.) MMM 2014. LNCS, vol. 8325, pp. 303–314. Springer, Heidelberg (2014). doi:10.1007/978-3-319-04114-8_26

14. Canini, L., Benini, S., Leonardi, R.: Affective analysis on patterns of shot types in movies. In: International Symposium on Image and Signal Processing and Analysis, pp. 253–258 (2011)

15. Xin, J.H., Cheng, K.M., Chong, T.F.: Quantifying colour emotion-what has been achieved. Res. J. Text. Apparel **2**(1), 46–54 (1998)

16. Russell, J.A.: A circumplex model of affect. J. Pers. Soc. Psychol. **39**(6), 1161–1178 (1980)

17. Lu, L., Liu, D., Zhang, H.: Automatic mood detection and tracking of music audio signals. IEEE Trans. Audio Speech Lang. Process. **14**(1), 5–18 (2006)

18. Valdez, P., Mehrabian, A.: Effects of color on emotions. J. Exp. Psychol. Gen. **123**(4), 394–409 (1994)

19. Wang, H.L., Cheong, L.-F.: Affective understanding in film. IEEE Trans. Circ. Syst. Video Technol. **16**(6), 689–704 (2006)

20. Baveye, Y., Dellandrea, E., Chamaret, C., Chen, L.: LIRIS-ACCEDE: a video database for affective content analysis. IEEE Trans. Affect. Comput. **6**(1), 43–55 (2015)

A Novel Two-Step Integer-pixel Motion Estimation Algorithm for HEVC Encoding on a GPU

Keji Chen[1], Jun Sun[1,2], Zongming Guo[1,2(✉)], and Dachuan Zhao[3]

[1] Institute of Computer Science and Technology of Peking University, Beijing, China
{chenkeji,jsun,guozongming}@pku.edu.cn
[2] Cooperative Medianet Innovation Center, Shanghai, China
[3] Advanced Micro Devices Co., Ltd., Beijing, China
Dachuan.Zhao@amd.com

Abstract. Integer-pixel Motion Estimation (IME) is one of the fundamental and time-consuming modules in encoding. In this paper, a novel two-step IME algorithm is proposed for High Efficiency Video Coding (HEVC) on a Graphic Processing Unit (GPU). First, the whole search region is roughly investigated with a predefined search pattern, which is analyzed in detail to effectively reduce the complexity. Then, the search result is further refined in the zones only around the best candidates of the first step. By dividing IME into two steps, the proposed algorithm combines the advantage of one-step algorithms in synchronization and the advantage of multiple-step algorithms in complexity. According to the experimental results, the proposed algorithm achieves up to 3.64 times speedup compared with previous representative algorithms, and the search accuracy is maintained at the same time. Since IME algorithm is independent from other modules, it is a good choice for different GPU-based encoding applications.

Keywords: High Efficiency Video Coding (HEVC) · Graphics Processing Unit (GPU) · Integer-pixel Motion Estimation (IME) · Two-step algorithm

1 Introduction

High Efficiency Video Coding (HEVC) [1] is the latest generation video coding standard. With enhanced coding tools, HEVC achieves about 50% bit-rate reductions at similar Mean Opinion Score (MOS) compared with H.264/AVC [2]. Among the enhanced coding tools, HEVC introduces a quadtree-like coding structure. The frames are first partitioned into Coding Tree Units (CTUs) with a maximum size of 64×64, then the CTUs are further partitioned into Coding Units (CUs) and Prediction Units (PUs). For each CU, there are up to 8 different kinds of PU partitions.

© Springer International Publishing AG 2017
L. Amsaleg et al. (Eds.): MMM 2017, Part II, LNCS 10133, pp. 28–36, 2017.
DOI: 10.1007/978-3-319-51814-5_3

Within the encoding process, Motion Estimation (ME) is applied to find proper Motion Vectors (MVs) for different PUs. ME investigates the MV Candidates (MVCs) within a specific region determined by the search range and the start MV, and tries to find the MVC with the least cost. Generally, ME is partitioned into Integer-pixel ME (IME) and Fractional-pixel ME (FME). IME investigates the integer-pixel MVCs among the whole search region and find out the best integer-pixel MVC for FME. Since IME is the topic of this paper, the MVCs in the following part means the integer-pixel MVCs only.

The cost of MVCs determines which MVC is better. In typical IME methods, Rate Distortion Optimization (RDO) [3] framework is applied with the following formula:

$$cost = D + \lambda * R, \tag{1}$$

where D can be measured with Sum of Absolute Difference (SAD), R can be derived from the distance between the MVC and the start MV, and λ is the Lagrange multiplier.

It is complex to retrieve the best MVC for different PUs in a quad-tree. One possible way to increase the speed is to use Graphic Processing Units (GPUs). GPUs are essential for modern personal computers. With many-core architecture, numerous tasks can be gathered as a kernel and be processed simultaneously on a GPU. Synchronizations of GPU are quite expensive. Besides, sufficient independent tasks should be provided to make full use of the cores. Compared with Central Processing Units (CPUs), GPUs are more energy saving and powerful when processing numerous independent tasks.

GPU-based IME algorithms are extensively applied in previous works. Since synchronizations are quite expensive, CPU-based multiple-step IME algorithms are not adopted. Early GPU-based IME methods [4,5] are usually based on Full Search algorithm, which introduces only one synchronization and achieves the best accuracy. However, since Full Search is complex, these methods are not satisfactory. S. Radicke et al. propose an algorithm based on Frayed Diamond Search Pattern (FDSP) [6]. Only the MVCs of a predefined pattern are searched in this work. And Recursive Sum of Absolute Differences (RSAD), which accumulates the SADs of basic blocks for the large blocks, is introduced to avoid redundant SAD calculations. Although the complexity is significantly reduced with FDSP, the synchronization cost is overrated for high motion videos. C. Jiang et al. propose an algorithm called DZfast [7]. By judging the possible zone of the best MVC, DZfast reduces the complexity for static area with lower search ranges. However, DZfast is designed for parallelizing within a macroblock with the multiview extension of H.264/AVC. The feature of independent views is utilized, which makes it unsuitable to be applied to HEVC directly. Since IME algorithm is fundamental in GPU-based encoding, it should be carefully studied and further optimized.

The main contribution of this paper is to propose a novel two-step IME algorithm for GPU-based encoding, which combines the advantage of one-step algorithms in synchronization and the advantage of multiple-step algorithms in complexity. According to experimental results over the widely used encoder x265

[8], the proposed algorithm achieves up to 9.75 times and 3.64 times speedup compared with Full Search and FDSP respectively, while the search accuracy is still maintained.

The rest of this paper is organized as follows. Section 2 introduces the proposed two-step IME algorithm. Section 3 presents the analysis of different IME algorithms. The detailed performance evaluation of the proposed algorithm is given in Sect. 4. And finally, the conclusions are made in Sect. 5.

2   The Proposed Two-Step IME Algorithm

The greedy algorithm is applied in the traditional CPU-based IME algorithms. The IME process is first divided into multiple steps, and only the MVCs around the start MV are investigated in each step. After a step, the cost of the MVCs in this step is summarized, and the best MVC among them is found as the start MV of the next step. Thus, regions far from the best MVC of each step will be neglected. To apply similar algorithms on a GPU, synchronizations will be introduced after each step. In summary, the greedy algorithm reduces the complexity. But the many synchronizations become a heavy burden for GPU-based encoding.

To avoid the synchronizations, previous GPU-based methods [4–6] apply one-step IME algorithms. The costs of the MVCs are only summarized once when all the MVCs are investigated. And the best MVC is found directly. Since there's only one step, the MVC distribution of this step should be quite dense, so that the search region can be well covered and good search accuracy can be achieved. As a result, although there's only one synchronization in one-step algorithms, the MVC count is large, and the complexity is significant.

By combining the advantage of multiple-step and one-step algorithms, we propose a two-step IME algorithm for GPU-based encoding. In the proposed algorithm, the IME process is divided into two steps. Only the MVCs around the best MVC of the first step are investigated in the second step. As a result, although an extra synchronization is introduced, the extra cost is limited and the complexity reduction is significant. Thus, the overall speed can be increased.

The two steps of the proposed IME algorithm are based on corresponding Search Patterns (SPs). The SP of the first step changes with the search range, while the SP of the second step is a fixed 8×8 pattern. The SPs of the two steps are carefully designed, so that every MVC within the search region has chance to be selected as the best MVC. Within a specific region, the denser the MVC of the first step, the higher the possibility it can be investigated in the second step. Considering that the MVCs closer to the start MV are more likely to be the best MVC, the SP of the first step is designed denser around the start MV. To reduce the complexity, the SP of the first step does not need to be too dense since the second step is always applied. Figure 1 shows the SP of the first step when search range is 16, and Fig. 2 shows the SP of the second step. In these two figures, the black point represents the start MV and other points represent the MVCs investigated in each step.

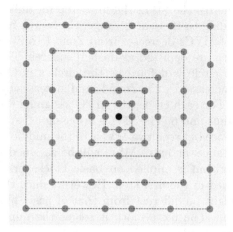

Fig. 1. The SP of the first step in the proposed two-step IME algorithm when search range is 16.

Fig. 2. The SP of the second step in the proposed two-step IME algorithm.

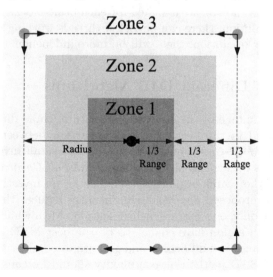

Fig. 3. The algorithm to derive the SP of the first step for different search range in the proposed two-step IME algorithm.

The algorithm to derive the SP of the first step for different search range is shown in Fig. 3. According to the search range, the search region is divided into 3 zones. The density of MVCs decreases from Zone 1 to Zone 3. The MVCs lie on several circles, which are depicted as dashed squares. The distance between the start MV and the circle is defined as the circle's radius. In the proposed algorithm, the radius increases by 2 in Zone 1, and increases by 4 and 6 separately in Zone 2 and Zone 3. For each circle, the MVC assigning distance is set as the minor between the radius and 6. The MVCs are assigned starting from the four angles. When the distance between two MVCs in the middle of an edge is greater than the assigning distance, an extra MVC will be assigned between them.

The proposed algorithm is applied on basic blocks first. For large blocks, RSAD method is applied in the first step. However, since the second step start MV of large blocks might be different from that of basic blocks, separate SAD calculations are applied. The basic block is set as the minimum PU, which is 8×8 in the experiment of this paper.

To implement the proposed algorithm on a GPU, the device memory consumption should be considered. Before applying IME algorithms, the frames and their referenced frames should be stored on device memory. The generated IME results should also be stored. Besides, if RSAD is applied, the SADs of the basic blocks should be stored to derive the SADs of large blocks. For the proposed two-step algorithm, RSAD is only applied in the first step. Thus, only the SADs of the first step should be stored. Meanwhile, the best MVCs of the first step are utilized as the start MV of the second step, and they should also be stored.

To make full use of the GPU platform, the IME of a whole frame is applied together. For the proposed algorithm, IME is divided into 3 kernels: (1) First step IME for the basic blocks, (2) First step IME for the large blocks, (3) Second step of every block size. 8 threads are assigned for each basic block. Thus, for a frame with 416×240 resolution, there can be 12480 threads working independently in a kernel. For higher resolution, there will be more independent threads.

3 Analysis of Different IME Algorithms

To present the advantage of the proposed algorithm, three different IME algorithms are analyzed in this section. The comparison between them is shown in Table 1, where *MVC Count* means the count of overall investigated MVCs, *Sync. Count* means the synchronization count, and *Full Covered* means whether all the MVCs in the search region have chance to be investigated. It can be observed that the proposed algorithm significantly reduces the MVC count of basic blocks with only one extra synchronization. Meanwhile, only a part of MVCs in the search region have chance to be the best MVC in FDSP, and all the MVCs have chance in Full Search and the proposed algorithm.

The effect of RSAD method on complexity should be considered. For large blocks, RSAD is applied in Full Search and FDSP, but it can only be applied in the first step of the proposed algorithm. However, there are always 64 MVCs in the second step, thus the complexity increment is constant for a specific PU

Table 1. Analysis of different IME algorithms

Algorithm	Search Range	MVC Count	Sync Count	Full Covered
Full Search	16	1089	1	Yes
	32	4225	1	Yes
	48	9409	1	Yes
	64	16641	1	Yes
FDSP	16	241	1	No
	32	865	1	No
	48	2001	1	No
	64	3521	1	No
Proposed	16	125	2	Yes
	32	233	2	Yes
	48	433	2	Yes
	64	709	2	Yes

partitioning strategy. It will be shown in Sect. 4 that for high motion videos, this increment is not significant compared with the complexity reduction from the MVC count.

The device memory consumption is an advantage of the proposed algorithm. Since RSAD is applied for all MVCs with Full Search and FDSP, the SADs of basic blocks should be all stored. Meanwhile, only the SADs of the first step with the proposed algorithm should be stored. When search range is 64, Full Search will store 16641 SADs for each basic unit. Suppose each SAD is 2 bytes, for a video with 2560×1600 resolution, about 2 G bytes will be utilized. For FDSP, which stores 3521 SADs for each basic unit, about 450 M bytes are necessary. For the proposed algorithm, only the 645 SADs of the first step need to be stored for each basic unit. Then, only about 83 M bytes are utilized. Extra device memory is necessary to store the best MVCs of the first step with the proposed algorithm. But it is only about 500 K bytes, and can be ignored.

4   Experimental Results

Detailed experimental results are given in this section. According to our experiments, λ in (1) does not affect the result much, and it is set to 10 (corresponding Quantization Parameter is 32) in IME process. The experimental platform is described in Table 2. And the results are collected under HEVC common test conditions [9] by applying the default parameter preset of x265.

The speed comparison between different IME algorithms using different sequences is shown in Fig. 4. The speed is measured in Frames Per Second (FPS), and is derived from the IME time of each algorithm on the GPU. It can be observed that the FPS reduction speed of the proposed algorithm is much

Table 2. Experimental platform

GPU	AMD Radeon R9-290X
Maximum power	250 w
Sreaming processors	2816
Core freq.	1 GHz
Device memory	4 GB
Device memory freq.	5 GHz
Computing framework	OpenCL 1.0

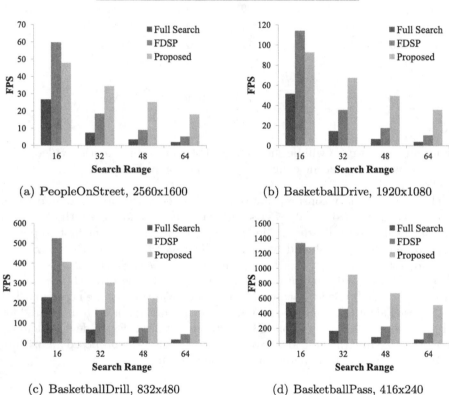

(a) PeopleOnStreet, 2560x1600

(b) BasketballDrive, 1920x1080

(c) BasketballDrill, 832x480

(d) BasketballPass, 416x240

Fig. 4. The speed comparison between different IME algorithms using different sequence with search range of 16, 32, 48 and 64.

slower than that of Full Search and FDSP with the increment of the search range, which is an obvious advantage when search range is large.

The detailed results are illustrated in Table 3, where SR represents the search range, and ΔS represents the speedup compared with Full Search algorithm. Each result is summarized by averaging the results of all 8-bit HEVC standard sequences in that class listed in [9]. By replacing the original IME algorithm

Table 3. Experimental results of different IME algorithms

SR	Class	Algorithm	FPS	BD-Rate	ΔS	SR	Class	Algorithm	FPS	BD-Rate	ΔS
16	A	Full Search	26.8	0.00%	1.00	32	A	Full Search	7.4	0.00%	1.00
		FDSP	59.6	1.24%	2.22			FDSP	18.4	1.63%	2.48
		Proposed	47.5	0.75%	1.77			Proposed	34.6	1.07%	4.65
	B	Full Search	51.5	0.00%	1.00		B	Full Search	14.4	0.00%	1.00
		FDSP	114.0	0.90%	2.22			FDSP	35.6	0.91%	2.46
		Proposed	92.4	0.54%	1.79			Proposed	67.5	0.64%	4.68
	C	Full Search	229.7	0.00%	1.00		C	Full Search	67.7	0.00%	1.00
		FDSP	522.6	1.38%	2.28			FDSP	165.3	1.49%	2.44
		Proposed	408.3	1.01%	1.78			Proposed	298.9	1.21%	4.41
	D	Full Search	583.9	0.00%	1.00		D	Full Search	164.7	0.00%	1.00
		FDSP	1369.7	1.02%	2.54			FDSP	461.1	1.20%	2.80
		Proposed	1259.6	0.74%	2.34			Proposed	904.9	0.85%	5.49
	E	Full Search	114.8	0.00%	1.00		E	Full Search	32.7	0.00%	1.00
		FDSP	255.3	0.48%	2.22			FDSP	81.4	0.51%	2.48
		Proposed	202.8	0.24%	1.77			Proposed	151.0	0.38%	4.61
48	A	Full Search	3.5	0.00%	1.00	64	A	Full Search	2.0	0.00%	1.00
		FDSP	9.0	1.84%	2.59			FDSP	5.2	2.09%	2.59
		Proposed	25.3	1.23%	7.28			Proposed	18.1	1.53%	9.01
	B	Full Search	6.7	0.00%	1.00		B	Full Search	3.8	0.00%	1.00
		FDSP	17.4	0.97%	2.61			FDSP	10.4	1.03%	2.70
		Proposed	49.6	0.75%	7.42			Proposed	35.6	0.80%	9.28
	C	Full Search	32.3	0.00%	1.00		C	Full Search	18.8	0.00%	1.00
		FDSP	74.5	1.67%	2.31			FDSP	44.9	1.76%	2.39
		Proposed	222.5	1.33%	6.88			Proposed	163.6	1.42%	8.70
	D	Full Search	84.6	0.00%	1.00		D	Full Search	52.3	0.00%	1.00
		FDSP	222.1	1.39%	2.63			FDSP	138.5	1.41%	2.65
		Proposed	662.6	1.07%	7.84			Proposed	503.8	1.09%	9.63
	E	Full Search	15.2	0.00%	1.00		E	Full Search	8.2	0.00%	1.00
		FDSP	39.2	0.64%	2.57			FDSP	23.3	0.64%	2.83
		Proposed	112.1	0.49%	7.36			Proposed	80.1	0.49%	9.75

of x265 and keeping other modules the same, PSNR and bitrate of different algorithms are derived and summarized as BD-Rate. BD-Rate is also derived by comparing with Full Search. The effect that does not use RSAD in the second step can be observed from the results. When the search range is 16, the complexity increment caused by RSAD is significant and the proposed algorithm is slower than FDSP. However, when the search range is larger, the proposed algorithm outperforms the other two algorithms. Up to 9.75 times speedup of the proposed algorithm is observed compared with Full Search, and up to 3.64 times speedup is observed compared with FDSP. Meanwhile, the results show that the BD-Rate of the proposed algorithm is better than FDSP and is close to Full Search. Considering that Full Search always investigates every MVC and precisely finds out the best one, the results indicate that the search accuracy with the proposed algorithm is well maintained.

5   Conclusion

In this paper, a two-step IME algorithm is proposed for HEVC encoding on a GPU. By dividing IME into two steps with the proposed algorithm, the MVC count as well as complexity is significant reduced. According to the experimental results, up to 9.75 times and 3.64 times speedup is achieved compared with Full Search and FDSP respectively, and the search accuracy is well maintained at the same time. Since IME algorithm is fundamental and independent from other modules in GPU-based encoding, the proposed two-step IME algorithm is a good choice for different GPU-based encoding applications.

Acknowledgment. This work was supported by National Natural Science Foundation of China under contract No. 61671025 and National Key Technology R&D Program of China under Grant 2015AA011605.

References

1. Sullivan, G.J., Ohm, J.-R., Han, W.-J., Wiegand, T.: Overview of the High Efficiency Video Coding (HEVC) standard. IEEE Trans. Circuits Syst. Video Technol. **22**(12), 1648–1667 (2012)
2. Wiegand, T., Sullivan, G.J., Bjntegaard, G., Luthra, A.: Overview of the H.264/AVC video coding standard. IEEE Trans. Circ. Syst. Video Technol. **13**(7), 560–576 (2003)
3. Sullivan, G.J., Wiegand, T.: Rate-distortion optimization for video compression. IEEE Signal Process. Mag. **15**(6), 74–90 (1998)
4. Chen, W.-N., Hang, H.-M.: H.264/AVC motion estimation implmentation on Compute Unified Device Architecture (CUDA). IEEE International Conference on Multimedia and Expo, pp. 697–700, June 2008
5. Rodríguez-Sánchez, R., Martínez, J.L., Fernández-Escribano, G., Claver, J.M., Sánchez, J.L.: Reducing complexity in H.264/AVC motion estimation by using a GPU. IEEE International Workshop on Multimedia Signal Processing, pp. 1–6, October 2011
6. Radicke, S., Hahn, J.-U., Wang, Q., Grecos, C.: Bi-predictive motion estimation for HEVC on a Graphics Processing Unit (GPU). IEEE Trans. Consumer Electron. **60**(4), 728–736 (2014)
7. Jiang, C., Nooshabadi, S.: A scalable massively parallel motion and disparity estimation scheme for multiview video coding. IEEE Trans. Circ. Syst. Video Technol. **26**, 346–359 (2016)
8. x265 Developers: x265 HEVC Encoder/H.265 Video Codec (2015). http://www.x265.org/
9. Bossen, F.: Common test conditions and software reference configurations. Document JCTVC-L1100, January 2013

A Scalable Video Conferencing System Using Cached Facial Expressions

Fang-Yu Shih, Ching-Ling Fan, Pin-Chun Wang, and Cheng-Hsin Hsu$^{(\boxtimes)}$

Department of Computer Science, National Tsing Hua University, Hsin Chu, Taiwan
chsu@cs.nthu.edu.tw

Abstract. We propose a scalable video conferencing system that streams High-Definition videos (when bandwidth is sufficient) and ultra-low-bitrate (<0.25 kbps) cached facial expressions (when the bandwidth is scarce). Our solution consists of optimized approaches to: (i) choose representative facial expressions from training video frames and (ii) match an incoming Webcam frame against the pre-transmitted facial expressions. To the best of our knowledge, such approach has never been studied in the literature. We evaluate the implemented video conferencing system using Webcam videos captured from 9 subjects. Compared to the state-of-the-art scalable codec, our solution: (i) reduces the bitrate by about 130 times when the bandwidth is scarce, (ii) achieves the same coding efficiency when the bandwidth is sufficient, (iii) allows exercising the tradeoff between initialization overhead and coding efficiency, (iv) performs better when the resolution is higher, and (v) runs reasonably fast before extensive code optimization.

Keywords: Compression · Cache · Codec · Facial landmarks · Facial models

1 Introduction

Video conferences using commodity computers and over the Internet has become increasingly popular in many sectors, including enterprises, healthcare, governments, defense, and educations. Reports indicate that the market of video conferencing systems is going to double between 2013 and 2020, reaching 6.4 billion USD annually [16]. Providing good video conferencing experience, however, is quite challenging, because video conferencing systems are resource demanding and delay sensitive. Commercial video conferencing systems recommend at least 700 kbps for 720p video calls [12], which is non-trivial for the best-effort Internet and wireless networks. In fact, *video conferencing systems are vulnerable to insufficient and fluctuating network bandwidth*, which must be addressed for good video conferencing experience. One way to cope with this problem is to reduce the encoded video bitrate, in order to reduce the demands on network bandwidth. Doing so, unfortunately, leads to lower video quality and degraded user experience, which may drive users away. Hence, a better solution is required to support video conferences under heterogeneous and dynamic network conditions.

© Springer International Publishing AG 2017
L. Amsaleg et al. (Eds.): MMM 2017, Part II, LNCS 10133, pp. 37–49, 2017.
DOI: 10.1007/978-3-319-51814-5_4

We argue that the existing commercial video conferencing systems fail to leverage a unique feature of conferencing videos: *each video usually contains a talking head[1] with stationary (and often simple) background.* Since the talking head and background are quasi-static, we believe some video frames may be *cached* for future reuse, so as to reduce the bitrate. The way we choose the cached video frames is based on the *facial expression* of the talking head. We refer to the cached video frames as *Cached Facial Expressions* (CFE), whose index can be encoded in a byte or two. The video frames chosen as CFEs are transmitted from the sender to the receiver before each video conference starts. That is, the CFEs exist at both the sender and the receiver, which can be used for encoding and decoding video frames. Therefore, the sender only needs to transmit the CFE indexes when the bandwidth is scarce, so the participants can still see *approximated* facial expressions. When the network conditions are good, the sender encodes and streams the Webcam video at high quality on top of the basic CFEs.

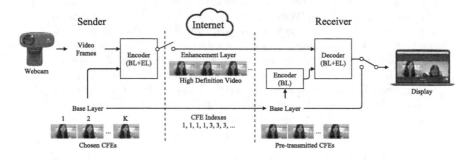

Fig. 1. Overview on the proposed scalable video conferencing system.

Figure 1 illustrates the proposed scalable video conferencing system that supports High-Definition (HD) videos and CFEs. We refer to the CFE indexes as the *base layer* (BL) and the HD videos as the *enhancement layer* (EL), following the terminologies used in scalable video coding [4]. The base layer contains a sequence of CFE indexes pointing to the cached, pre-transmitted video frames, while the enhancement layer encodes the actual video frames captured by the Webcam. We notice that during video conferences, although the encoder at the sender side compresses both the base layer (chosen CFEs) and the enhancement layer (HD videos), the enhancement layer dominates the traffic amount over the Internet. This is because the number of chosen CFEs is typically a small integer, so the CFE indexes can be largely compressed. This leads to the unique feature of our system: *the base layer stream is at extremely low bitrate, while the enhancement layer can optionally be transmitted for enhancing user experience.*

---

[1] For the sake of our discussion, we assume that each video call is between two people. Our system can be readily extended to video conferences with multiple participants at each site.

Realizing the proposed scalable video conferencing system is no easy task, because CFEs need to be carefully chosen to approximate the actual facial expressions during video conferences. Doing so not only improves the user experience (when only base layer is transmitted) but also increases the coding efficiency of the enhancement layer (redundancy across base and enhancement layers is exploited). Upon the CFEs are chosen, we need to extract the facial expression from each Webcam image during video conferences. Furthermore, the process needs to be efficient enough, for real-time executions on commodity computers. In this paper, we rigorously solve the problem of choosing CFEs, extracting facial expression, and managing selected CFEs. Our experiments demonstrate the merits of our proposed scheme and optimization techniques, e.g., compared to the state-of-the-art scalable codec, our proposed solution: (i) achieves higher video quality for most subjects, while consumes less than $1/130$ of the bandwidth, (ii) achieves the same coding efficiency when the bandwidth is sufficient, (iii) leads to more bitrate reduction and video quality increase as the resolution increases.

2   Related Work

Ultra-Low-Bitrate Video Conferencing. Wang et al. [18] design an ultra-low-bitrate video conferencing system by analyzing the structure of video frames and aggressively skipping redundant information. They adopt warping techniques for interpolating unsent frames, and achieves a bitrate at about 25 kbps. Zeng et al. [20] utilize facial models to reduce the transmission bitrate and encode the region of eyes and mouth by Differential Pulse-Code Modulation (DPCM) for better quality during video conferences. Koufakis et al. [10] propose that every human face is constructed by linear combination of three base images, and they utilize Principle Component Analysis (PCA) on regions such as eyes and mouth to identify these base images. Allen et al. [1] describe a model-based video coding system, using statistical shapes and models of faces to construct facial models. They utilize Active Appearance Models (AAM) to extract facial feature points from images for low-bitrate video transmission. The references above give human faces higher priority, which is similar to our proposed system. We use the concept of landmarks, which emphasizes importance of the areas around eyes and mouth. However, instead of sending facial landmarks directly [18], we send the index of the CFE that matches the best in terms of landmarks. This action lowers the bitrate further: <1 kbps is observed. Different from our work, Qi et al. [13] do not make an effort on analyzing facial expressions. Instead, they adaptively set the frame rate at the sender side, and interpolate the missing frames at the receiver side.

CFE Selection Mechanisms. For choosing CFEs, one straightforward approach is using facial expression extraction to classify frames. Fasel and Luettin [6] point out that facial expression extraction can be divided into three steps including face acquisition, facial feature extraction, and facial expression classification. Ari et al. [2] and Suk and Prabhakaran [15] propose a system architecture that first locates facial landmarks and uses those as features to train a facial expression classifier. However, using discrete facial expression classes is less general, and may hurt the coding efficiency. For example, the expression *joy* may include smile, laugh, grin, smirk, chuckle, which may not be distinguished by the facial expression classifiers [2,6,15]. Wang and Cohen [17] take open eyes as an important condition in their selection method, which is not that rigorous and comprehensive. In contrast to above studies, we take a more general approach by directly using facial landmarks. That is, we take a more *data-driven* approach.

3   Architecture

3.1   Overview

Figure 2 presents the architecture of our proposed solution, which has two phases. The *training phase* solves the problem of selecting CFEs using either a historical or a warm-up Webcam video at the sender *before* actual video conferences. The selected CFEs are stored at the sender side and sent to the receiver side. We next enter the *conferencing phase*, in which we stream CFE indexes as the BL and coded video as the optional EL. The receiver either renders the CFEs if only BL is received or decode the EL if both BL and EL are received. Details on individual components are given in the following.

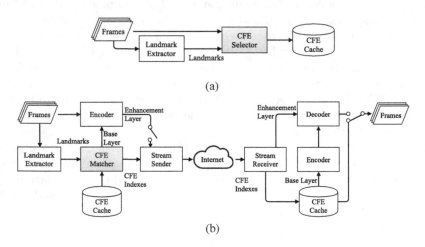

Fig. 2. The architecture of our system: (a) The training phase and (b) The conferencing phase. The highlighted CFE selector and CFE matcher are two key components.

Landmark extractor is responsible to extract the coordinates of facial features, including eyes, nose, and mouth from a talking-head Webcam frame. The extracted coordinates are referred to as landmarks, which represent the participant's facial expressions [6]. In this work, we adopt a low complexity facial model, called Constrained Local Model (CLM) [5], which captures the coordinates of 68 landmarks.

CFE selector analyzes the landmarks and frames of the historical or the warm-up video, and chooses K representative CFEs. We note that K is a system parameter. Larger K values allow us to send CFEs that better approximate the actual Webcam frames, while larger K values also lead to higher transmission and storage overhead. We will study such tradeoff in Sect. 5.

CFE cache stores the selected video frames and landmark coordinates. It also assigns a unique index to each video frame and corresponding landmark coordinates. The CFE cache is built at the sender side, and sent to the receiver side before the conferencing phase.

CFE matcher is responsible to find the CFE that is the best approximation of each Webcam frame. To achieve this, it compares the landmark coordinates of the current Webcam frame against those of the CFEs (in the CFE cache). The resulting CFE frames are sent to the encoder as the BL, and the CFE indexes are sent to the stream sender.

Encoder is a scalable video coder, such as SHVC [4], which takes the CFEs as the approximated BL and uses the HD Webcam frames as the EL. At the sender side, the encoder generates a BL and an EL, but only passes the EL to the stream sender. The BL (coded video stream) is *not* transmitted in our design; rather, it is regenerated at the receiver using the same encoder and CFEs. More precisely, at the receiver side, the received EL and regenerated BL are sent to the **decoder**, which renders the HD Webcam videos.

Stream sender/receiver are responsible for segmentation, packetization, and transmission of the coded video stream (EL) and CFE indexes (BL) using protocols like lightweight RTP [14] or reliable HTTP [9].

Last, we note that the CFE selector and CFE matcher are two unique components of our proposed system. We carefully design them in Sect. 4 to optimize our system.

4   Design Decisions

The design goal is to maximize the coding efficiency of the reconstructed CFEs (BL) and HD Webcam videos (EL). We achieve this in two steps. First, we assume the K CFEs are given, and for each Webcam frame, we identify the CFE that best approximates the Webcam frame. Second, we choose K CFEs from a set of historical or warm-up Webcam frames to best represent all possible facial expressions.

4.1   CFE Matcher

We use $x_{v,f,l}$ and $y_{v,f,l}$ to represent the coordinates of landmark l of video v, frame f. We define the landmark displacement $d_{i,j}^{v}$ of frames i,j in video v as:

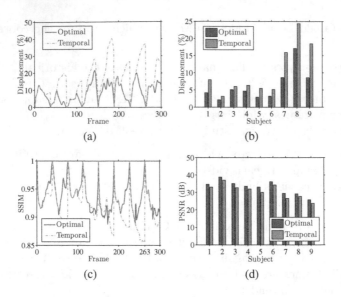

Fig. 3. Our optimal CFE matcher leads to: (a), (b) Smaller displacement, and (c), (d) Higher BL video quality.

$d_{i,j}^v = \frac{\sqrt{\sum_{l=1}^{L}[(x_{v,i,l}-x_{v,j,l})^2+(y_{v,i,l}-y_{v,j,l})^2]}}{L}$, where L is the number of landmarks. To achieve the design goal of the CFE matcher, it is rational to use the landmark displacement $d_{i,j}^v$ to guide the matching process. In particular, for frame f, we find the optimal CFE $e^* = \arg\min_{e\in\mathbf{E}} d_{f,e}^v$, where $\mathbf{E}$ is the given set of CFEs.

We conduct an experiment using 10-sec conferencing videos of 9 subjects (details are given in Sect. 5) with $K = 8$. In particular, we chose 8 CFEs from equally-spaced video frames among the 300 total frames. We compare the performance of our proposed CFE matcher (denoted as Optimal in the figures) and a temporal-based approach, in which the most recent (equally-spaced) CFE is matched. More specifically, for each 300-frame video, we replace the video frames that were not chosen as CFEs using Optimal and temporal-based approaches, to get two approximated BL videos. We then compute and report the landmark displacement and video quality in PSNR (Peak Signal-to-Noise Ratio) and SSIM (Structural Similarity Index) [19]. We consider 50 dB as the perfect PSNR value. Figure 3 shows that our proposed CFE matcher outperforms the temporal-based approach in all aspects. Figures 3(a) (from subject 7) and (b) (overall) reveal that our CFE matcher significantly reduces the landmark displacement: cutting the landmark displacement as high as half, e.g., frame 180 of Fig. 3(a) and subject 9 of Fig. 3(b). Such improvement then leads to higher video quality, as shown in Fig. 3(c) (from subject 7) and (d) (overall). The SSIM gap could be as high as 0.1351 (frame 263 in Fig. 3(c)), and the PSNR gap could be as high as 2.8777 dB (subject 7 in Fig. 3(d)). In summary, Fig. 3 demonstrates the merits of our CFE matcher.

4.2 CFE Selector

Our design goal is to maximize the diversity among the selected CFEs. The rationale is: the more distinguishable the selected CFEs, the higher chance for our CFE matcher to find a better approximation. To maximize the diversity, we transform all landmark coordinates of a video frame into an one-dimensional space as the features of that frame. Then, we use these features to perform K-means clustering [7] to divide F frames in the video into K groups. The outputs are the center points of each cluster, and we take the nearest frame of each cluster as our CFEs. In total, we get K CFEs.

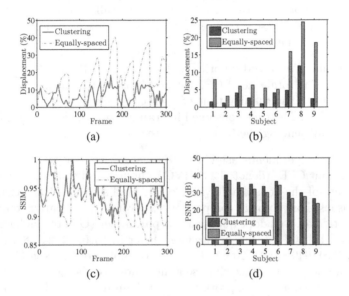

Fig. 4. Our clustering-based CFE selector leads to: (a), (b) smaller displacement, and (c), (d) higher BL video quality.

We also conduct an experiment to compare our proposed CFE selector (denoted as Clustering in the figures), and equally-spaced video frames with $K = 8$. Figure 4 presents some sample results. This figure reveals that our proposed CFE selector outperforms the equally-spaced approaches in all aspects. Figures 4(a) (from subject 7) and (b) (overall) show our CFE selector cuts the landmark displacement by as high as a half. Figures 4(c) and (d) depict the video quality improvements: up to 0.1114 in SSIM and 3.4061 dB are observed. In summary, Fig. 4 shows the merits of our CFE selector.

5 Experiments

5.1 Setup

We have implemented our video conferencing system using CLM [3] and Matlab libraries. Our prototype system is constructed by five major components: (i) landmark extractor, which extracts the facial landmarks from input frames, (ii) CFE selector, for choosing CFEs from the frames of training videos[2], (iii) CFE matcher, which matches the frames against the CFE cache for the most suitable CFE, (iv) encoder, which encodes the base layer (constructed from CFEs) and the enhancement layer (streamed when the bandwidth is sufficient) together, and (v) decoder, which reconstructs the videos for display. Using the five components, we build our system as illustrated in Fig. 2. Some components appear in both sender and receiver sides. Our prototype system consists of 346 lines of C/C++ code, 339 lines of Matlab code, and 65 lines of Shell script. We record videos of 9 subjects at 1280×720 and 30 frame-per-second using off-the-shelf Webcams and computers. We ask the subjects to talk naturally as if they are in video conferences. We select 10-second video from each subject. We measure the following performance metrics.

- **Video quality.** We consider both PSNR and SSIM [19].
- **Cache size.** The total size of CFEs stored in the CFE cache.
- **Running time.** The running time of each component.
- **Training delay.** The time consumed in training phase, including training for CFE cache and sending the CFEs to the receiver.

We run the experiments using the 9 videos with our solution and pure SHVC [8] without CFEs (denoted as SHVC in figures and tables). The SHVC [4] codec is a state-of-the-art scalable video codec, and serves as the baseline of all our comparisons. For thorough comparisons, we vary several parameters for each video. In particular, we adopt quantization parameter (QP) of the base layer in $\{16, \mathbf{24}, 32, 40, 48\}$, QP of the enhancement layer in $\{3, 6, \mathbf{12}, 24, 48\}$, resolution in $\{\mathbf{1280 \times 720}, 640 \times 360, 320 \times 180\}$, and the number of CFEs in $\{4, \mathbf{8}, 16, 32\}$. The **bold** numbers in the preceding sentence indicate the default settings. We run the experiments on a Linux workstation with an Intel i7 CPU at 3.6 GHz and 8 GB RAM. For running time and training delay measurements, we repeat each experiment 10 times, and report the average, minimum, and maximum values.

5.2   Results

Our Solution Delivers Good Video Quality even when the Bitrate is <1 kbps. We first report the video quality achieved by our solution and the SHVC. Figure 5 plots the sample video quality under different bitrates. Notice that the x axis is in the logarithmic scale. This figure shows that our solution achieves more than 30 dB and 0.9 in PSNR and SSIM at a very low video bitrate of 0.24 kbps.[3] In contrast, the lowest bitrate that can be achieved by the SHVC codec is 29 kbps, about 130 times higher than our solution. Such improvement can be attributed to the fact that we only transmit the CFE indexes to the receiver.

---

[2] The selected CFEs are saved and transmitted as lossless PNG files.
[3] The bitrate mentioned throughout the article only consider the bitrate of video.

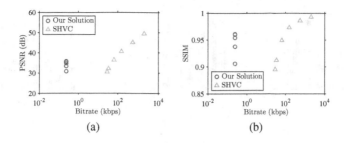

Fig. 5. Our solution achieves good video quality in: (a) PSNR and (b) SSIM at a very low bitrate (0.24 kbps), which is much lower than the lowest bitrate of SHVC (29 kbps). Sample results from subject 3 are shown.

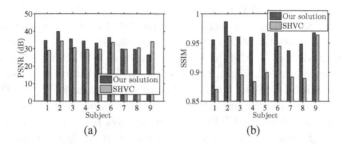

Fig. 6. Our solution outperforms the SHVC with most subjects in terms of: (a) PSNR and (b) SSIM.

Figure 6 compares the video quality of individual subjects under the default parameters. Illustrated in Fig. 6(a), our solution outperforms SHVC with most of the subjects except subjects 8 and 9 in PSNR. A deeper investigation indicates that this is due to the higher temporal displacements, which is the average landmark movement of adjacent frames, of subjects 8 and 9. Specifically, these two subjects have temporal displacements of 35.1% and 13.1% respectively, while the average value of the other videos is 8.91%. Figure 6(b) shows that our solution outperforms the SHVC in SSIM across all subjects.

In summary, the bitrate of our solution is lower than the SHVC by more than 130 times, yet our solution outperforms the SHVC in most cases (16 out of 18). Among all subjects, our solution achieves 33.38 dB in PSNR and 0.96 in SSIM, at about 0.25 kbps on average.

Our Solution Achieves Comparable Coding Efficiency as the SHVC does when the Bandwidth is Sufficient. Figure 7 plots the bitrate, PSNR, and SSIM values of the enhancement layer from individual subjects under the default QP settings. This figure reveals that our solution consumes slightly lower bitrate while achieving comparable PSNR and SSIM values. In particular, Fig. 7(a) shows that our solution saves bitrate by 250 kbps at least, and 595 kbps on average across all subjects, while Figs. 7(b) and (c) demonstrate that our solution still achieves almost the same video quality as the SHVC across all subjects.

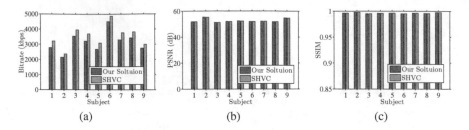

Fig. 7. Our solution saves (a) bitrate, and achieves almost the same video quality in (b) PSNR and (c) SSIM, compared to the SHVC with the enhancement layer.

Implications of the CFE Cache Size. Figure 8 plots the video quality under different cache sizes, where the cache size is the total PNG file size of the CFEs. Figure 8(a) shows that the video quality increases as the cache size increases, which can be attributed to the increased number of CFEs to match against in the conferencing phase. This figure shows the tradeoff between the storage and transmission overhead and the video quality of the base layer. Figure 8(b) shows that the cache size has little impact on the video quality of the enhancement layer. This demonstrates that CFEs in the base layer imposes virtually no negative impacts on the coding efficiency of the enhancement layer. Figure 9 reports

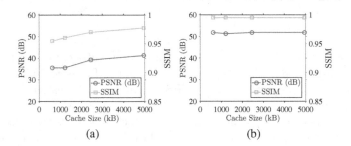

Fig. 8. The video quality under different cache sizes with (a) The base layer and (b) The enhancement layer. Sample results form subject 3 are shown.

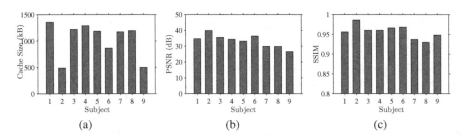

Fig. 9. Videos with the default parameters: (a) The cache size, (b) PSNR, and (c) SSIM.

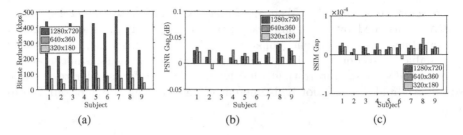

Fig. 10. The improvements of our solution over the SHVC in: (a) bitrate, (b) PSNR, and (c) SSIM.

the CFE cache size under the default settings. This figure shows that the average cache size is 824 kB across all subjects, which is manageable for modern computers and smartphones.

Implications of Resolutions. Figure 10 plots the improvements of our solution over the SHVC codec at different resolutions with the enhancement layer. Figure 10(a) shows that our solution always reduces the bitrate under all considered resolutions. Figure 10(b) and (c) show that our solution outperforms the SHVC in both PSNR and SSIM when the resolutions are high. There are 3 (out of 54) cases, where the SHVC performs better. However, we believe that is not a concern because these boundary cases all happen with the relatively low resolution of 320×180, which is less common on modern computers. Figure 10 reveals that our solution leads to more bitrate reduction and video quality improvement when the resolution is higher. This is a good property because the resolution of video conferences will increase as the hardware technologies advance.

Running Time and Training Delay. We report the average per-component running time in Table 1. For each 10-second video, the landmark extractor and CFE selector run in 11.40 s and 322.19 ms on average, which are quite manageable. For sending the chosen CFEs to the receiver, it takes 9.76–27.14 s at the bandwidth of 50 kbps. The CFE transfer time dominates the training delay, yet it is much shorter than the lengths of typical video conferences. For the

Table 1. The running time of each component

Component		Minimum	Average	Maximum
Landmark extractor		11.26 s	11.40 s	12.06 s
CFE selector		320.01 ms	322.19 ms	324.89 ms
CFE transfer	Bandwidth = 50 kbps	9.76 s	20.59 s	27.14 s
	Bandwidth = 100 kbps	4.38 s	10.30 s	13.57 s
	Bandwidth = 200 kbps	2.19 s	5.15 s	6.79 s
CFE matcher		0.019 ms	0.020 ms	0.020 ms

conferencing phase, for each Webcam frame, in addition to the $11.4\,\text{s}/300 = 38.00\,\text{ms}$ consumed by the landmark extractor, the CFE matcher also takes $0.02\,\text{ms}$. These two components consume negligible running time, compared to the scalable video codec [8], which is implemented in software and is not intended for real-time systems. Integrating a hardware SHVC codec, similar to the one used in Android TV [11], for real-time encoding/decoding is one of our future work.

6   Conclusion

We designed, implemented, and evaluated an ultra-low-bitrate scalable video conferencing system, which selects and transmits CFEs from a sender to a receiver in the training phase. In the conferencing phase, our system extracts landmarks, matches CFEs at the sender side, and transmits CFE indexes to the receiver side for constructing the base layer. The enhancement layer is transmitted along with the CFE indexes only if we have sufficient bandwidth. At the receiver side, the base and (optional) enhancement layers are decoded for display. We conducted extensive experiments using 9 subjects with commodity Webcams and computers. The experimental results show that, compared to the state-of-the-art SHVC, our proposed system: (i) lowers the bitrate by about 130 times, yet maintaining almost the same video quality at an ultra-low-bitrate of <1 kbps, (ii) achieves the same coding efficiency when the bandwidth is sufficient, (iii) provides a control knob to balance the CFE overhead and video quality, (iv) incurs more performance improvements when the resolution is higher, and (v) runs reasonably fast before extensive code optimization.

This study can be extended in several ways. First, we will conduct an user study to further investigate our design decisions, such as parameter selection, as well as the results compared with classic SHVC and audio-only video conferencing. Second, we will improve our CFE selection method by dynamically training and update CFE at the receiver when the bandwidth is sufficient. Last, our system could be more robust by supporting multi-participant, face obstruction, and no face situations. We plan to introduce a matching mechanism to handle multiple participants video conferencing.

References

1. Allen, N., Naidoo, B., McDonald, S.: Model-based compression for low-bitrate comms: a statistical approach to facial video encoding. In: Proceedings of Southern Africa Telecommunication Networks and Applications Conference (SATNAC), September 2006
2. Ari, I., Uyar, A., Akarun, L.: Facial feature tracking and expression recognition for sign language. In: Proceedings of International Symposium on Computer and Information Sciences (ISCIS), pp. 1–6, October 2008
3. Baltruvsaitis, T., Robinson, P., Morency, L.: 3D constrained local model for rigid and non-rigid facial tracking. In: Proceedings of IEEE Conference on Computer Vision and Pattern Recognition (CVPR), June 2012

4. Boyce, J., Ye, Y., Chen, J., Ramasubramonian, A.: Overview of SHVC: scalable extensions of the high efficiency video coding standard. IEEE Trans. Circuits Syst. Video Technol. **26**(1), 20–34 (2016)
5. Cristinacce, D., Cootes, T.: Feature detection and tracking with constrained local models. In: Proceedings of the British Machine Vision Conference (BMVC), September 2006
6. Fasel, B., Luettin, J.: Automatic facial expression analysis: a survey. Pattern Recognit. **36**(1), 259–275 (2003)
7. Hartigan, J.A., Wong, M.A.: Algorithm as 136: a k-means clustering algorithm. J. R. Stat. Soc. Ser. C (Appl. Stat.) **28**(1), 100–108 (1979)
8. HEVC Scalability Extension (SHVC) official site. https://hevc.hhi.fraunhofer.de/shvc
9. Hypertext transfer protocol (1999). http://www.rfc-base.org/rfc-2616.html
10. Koufakis, I., Buxton, B.: Very low bit rate face video compression using linear combination of 2D face views and principal components analysis. Image Vis. Comput. **17**(14), 1031–1051 (1999)
11. NVIDIA SHIELD: The best Android TV box (2016). https://shield.nvidia.com/android-tv
12. Plan network requirements for Skype for business (2015). https://technet.microsoft.com/en-us/library/gg425841.aspx, September 2015
13. Qi, X., Yang, Q., Nguyen, D., Zhou, G., Peng, G.: LBVC: towards low-bandwidth video chat on smartphones. In: Proceedings of ACM Multimedia System Conference (MMSys), March 2015
14. RTP: A transport protocol for real-time applications (1996). https://www.ietf.org/rfc/rfc1889.txt
15. Suk, M., Prabhakaran, B.: Real-time facial expression recognition on smartphones. In: Proceedings of the IEEE Applications of Computer Vision (WACV), January 2015
16. Video conferencing market to expand at 9.3% CAGR to 2020 thanks to increasing usage in healthcare and defense, July 2015. http://www.transparencymarketresearch.com/pressrelease/video-conferencing-market.htm
17. Wang, J., Cohen, M.: Very low frame-rate video streaming for face-to-face teleconference. In: Data Compression Conference, pp. 309–318. IEEE (2005)
18. Wang, P., Fan, C., Huang, C., Chen, K., Hsu, C.: Towards ultra-low-bitrate video conferencing using facial landmarks. In: Proceedings of ACM Multimedia Conference (MM), October 2016
19. Wang, Z., Lu, L., Bovik, A.: Video quality assessment based on structural distortion measurement. Signal Process. Image Commun. **19**(2), 121–132 (2004)
20. Zeng, W., Yang, M., Cui, Z.: Ultra-low bit rate facial coding hybrid model based on saliency detection. J. Image Graph. **3**(1), 25–29 (2015)

A Unified Framework for Monocular Video-Based Facial Motion Tracking and Expression Recognition

Jun Yu[✉]

Department of Automation, University of Science and Technology of China,
Hefei 230026, China
harryjun@ustc.edu.cn

Abstract. This paper proposes a unified facial motion tracking and expression recognition framework for monocular video. For retrieving facial motion, an online weight adaptive statistical appearance method is embedded into the particle filtering strategy by using a deformable facial mesh model served as an intermediate to bring input images into correspondence by means of registration and deformation. For recognizing facial expression, facial animation and facial expression are estimated sequentially for fast and efficient applications, in which facial expression is recognized by static anatomical facial expression knowledge. In addition, facial animation and facial expression are simultaneously estimated for robust and precise applications, in which facial expression is recognized by fusing static and dynamic facial expression knowledge. Experiments demonstrate the high tracking robustness and accuracy as well as the high facial expression recognition score of the proposed framework.

Keywords: Facial motion tracking · Facial expression recognition

1 Introduction

Facial motion and expression enable users to communicate with computers using natural skills. Constructing robust systems for facial motion tracking and expression recognition is an active research topic.

Generally, 2D approaches [1, 2] or 3D approaches [3, 4] can be conducted for this task. Compared with 2D methods, 3D methods are more qualified for the view-independent and illumination insensitive tracking and recognition situations [5]. For 3D methods, a 3D facial mesh model or a depth camera is often used [4–8]. Because the high cost-effectiveness of single video cameras, they are used as inputs here by a 3D facial mesh model, which is served as the priori knowledge and constraints.

For facial motion tracking [9], appearance-based techniques [12, 13] are more robust compared with feature-based ones [10, 11], and often implemented statistically to increase the robustness. Offline statistical appearance-based models [3, 14], such as 3D shape regression [15, 16], use a face image dataset taken under different conditions to learn the parameters of appearance model, while online statistical appearance models (OSAM) [17–19] are more flexible and efficient than the offline ones by updating the

© Springer International Publishing AG 2017
L. Amsaleg et al. (Eds.): MMM 2017, Part II, LNCS 10133, pp. 50–62, 2017.
DOI: 10.1007/978-3-319-51814-5_5

learned dataset progressively. In addition, an adequate motion filtering strategy should be adopted to obtain the true value. Particle filtering [19] has been widely used for the global optimization ability by using the Monte Carlo technique.

For facial expression recognition [20–22], static techniques use spatial ones or spatio-temporal features related to a single frame [23] to classify expressions by several statistical analysis tools [24–30], such as neural network, while dynamic techniques use the temporal variations of facial deformation to classify expressions by several statistical analysis tools [31–33], such as dynamic Bayesian networks.

In this paper, a framework (Fig. 1) is proposed for pose robust and illumination insensitive facial motion tracking and expression recognition on each video frame base on the work in [34].

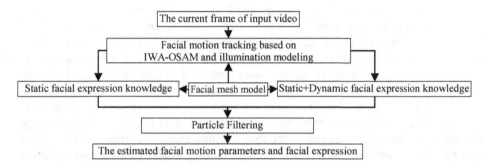

Fig. 1. Framework.

Firstly, facial animation and facial expression are obtained sequentially in particle filtering. To alleviate the illumination variation difficulty during facial motion tracking, OSAM is improved to illumination weight adaptive online statistical appearance model (IWA-OSAM), in which 13 basis point light positions are constructed to model the lighting condition of each video frame. Then facial expressions are recognized by the static facial expression knowledge learned from the anatomical definitions in [35].

Secondly, because both the temporal dynamics and static information are important for recognizing expressions [36], they are combined and fused here by tracking facial motion and recognizing facial expression simultaneously in particle filtering. Compared with the sequential approach discussed above, particles are not only generated by the resampling, but also predicted by the dynamic knowledge; thus resulting into a more accurate recognition result.

2   Facial Motion Tracking

2.1   OSM-Based Facial Motion Tracking

The model (Fig. 2(a)): *CANDIDE3* [37] is served as the priori knowledge and constraints for facial motion tracking, and defines the facial motion parameters as:

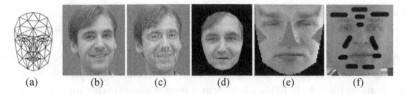

(a)   (b)   (c)   (d)   (e)   (f)

Fig. 2. (a) *CANDIDE3* model. (b) A frame of input video. (c) The projection of CANDIDE3 under *b*. (d) The GNFI. (e) The improved GNFI. (f) Selected facial areas.

$$\boldsymbol{b} = [\theta_x, \theta_y, \theta_z, t_x, t_y, t_z, \boldsymbol{\beta}^T, \boldsymbol{\alpha}^T]^T = [\boldsymbol{h}^T, \boldsymbol{\beta}^T, \boldsymbol{\alpha}^T]^T \qquad (1)$$

where $\boldsymbol{h} = [\theta_x, \theta_y, \theta_z, t_x, t_y, t_z]^T$ is global head motion parameters. $\boldsymbol{\beta}, \boldsymbol{\alpha}$ are shape and animation parameter. 10 shape parameter and 7 animation parameter are used.

A face texture is represented as a geometrically normalized facial image (GNFI) [37]. Figure 2(b)–(d) illustrate the process of obtaining a GNFI with an input image. Different facial areas may have different levels of influence on the tracking performance. Because the part above the eyebrows hardly take effect on the facial motion, and is often contaminated by the hair, it is removed from the GNFI. In addition, we found that the top part of the nose and the temples seldom undergo local motions. However, the appearance of these two facial areas is often influenced by head pose change and illumination variation. Therefore, the image regions corresponding to these two facial areas are removed from GNFI. The resulting image, called improved GNFI (Fig. 2(e)), is then used for measurements extraction.

Because the pixel color values are easily influenced by environment, and thus not robust for tracking, a more robust measurement is extracted from the improved GNFI as follows according to the discussion in [34]: for the improved GNFIs of the first and current frames, we obtain the illumination ratio images, and compute Gabor wavelet coefficients on the selected facial areas (Fig. 2(f)) where high frequency appearance changes more likely.

Moreover, illumination variation is one of the most important factors which reduce significantly the performance of face recognition system. It has been proved that the variations between images of the same face due to illumination are almost always larger than image variations due to change in face identity. So eliminating the effects due to illumination variations relates directly to the performance and practicality of face recognition. To alleviate this problem, a low-dimensional illumination space representation (LDISR) of human faces for arbitrary lighting conditions [38] is proposed for recognition. The key idea underlying the representation is that any lighting condition can be represented by 9 basis point light sources. The lighting subspace is constructed not using the eigenvectors from the training images with various lighting conditions directly but the light sources corresponding to the eigenvectors. The 9 basis light positions are shown in Table 1.

However, it is only used for the situation in which the face has only the 2D in-plane rotation, while the face with out-of-plane rotation is a common situation for the face in the real scene. Therefore, we extend the LDISR from 2D to 3D, in which the in-plane rotation and out-of-plane rotation are both considered. The training process is similar to that in the method discussed in [38], and the obtained 13 basis light positions are shown in Table 2.

Table 1. Positions of the 9 basis light sources.

Light	Pitch θ (degree)	Roll φ (degree)
1	0.0	0.0
2	17.5	−47.5
3	25.7	44.4
4	36.0	−108.0
5	44.0	88.0
6	68.6	−3.0
7	33.3	85.0
8	−35.0	−95.0
9	−70.0	22.5

Table 2. Positions of the 13 basis light sources.

Light	Pitch θ (degree)	Roll φ (degree)	Yaw ω (degree)
1	0.0	0.0	0.0
2	17.5	−47.5	−59.5
3	25.7	44.4	−60.7
4	36.0	−108.0	−55.7
5	44.0	88.0	−60.9
6	68.6	−3.0	55.6
7	−33.3	85.0	60.3
8	−35.0	−95.0	65.0
9	−70.0	22.5	60.1
10	−89.2	0.7	−91.3
11	−90.2	0.4	−90.4
12	90.1	−0.3	−90.6
13	91.1	−0.6	91.6

Because different human faces have similar 3D shapes, the LDISR of different faces is also similar. In addition, by using the normalization with GNFI (Fig. 2), it can be assumed that different persons have the same LDISR.

Suppose the 13 basis images obtained under 13 basis lights are $L_i, i = 1, \cdots, 13$, the LDISR of human face can be denoted as $A = [L_1, L_2, \cdots, L_{13}]$. Given an image of human face I_x under an arbitrary lighting condition, it can be expressed as:

$$I_x = A \cdot \lambda \tag{2}$$

where $\lambda = [\lambda_1, \lambda_2, \cdots, \lambda_{13}]^T, 1 \leq \lambda_i \leq 1$ is the lighting parameters of image I_x, and can be calculated by minimizing the energy function $E(\lambda)$ as:

$$E(\lambda) = \|A \cdot \lambda - I_x\|^2 \tag{3}$$

Then the lighting parameters can be obtained as:

$$\lambda = \left(A^T A\right)^{-1} A^T \cdot I_x \qquad (4)$$

In practice, the image pixel will be less influenced by lighting if the light positions are distributed more evenly. In this case, the values of $\lambda_1, \lambda_2, \cdots, \lambda_{13}$ should be close for the 13 basis light positions discussed above. With this fact, an index can be defined to evaluate the lighting influence on the pixel values of each triangular patch of the improved GNFI. To achieve this goal, these triangular patches are first split to 13 areas corresponding to 13 basis point light positions, and then the lighting influence weight of the $kth, k = 1, \cdots, 13$ area is given for the tth video frame as follows:

$$w_t^k(j) = abs\left(\lambda_k - \frac{\lambda_1 + \cdots + \lambda_{13}}{13}\right) \bigg/ \sum_{k=1}^{13} abs\left(\lambda_k - \frac{\lambda_1 + \cdots + \lambda_{13}}{13}\right) \qquad (5)$$

where j is the index of the pixel in the kth area of the triangular patches in Fig. 1(e).

Based on the lighting influence weight discussed above, OSAM is extended to illumination weight adaptive online statistical appearance model (IWA-OSAM). The details are as follows.

$m(b_t)$ with size d, abbreviated as m_t, is the concatenation of pixel color value at time t, and it is modeled as a Gaussian Mixture stochastic variable with 3 components, s, w, l, as Jepson et al. [17] does. $\left\{\mu_{i,t}; i = s, w, l\right\}$ is the mean vector. $\left\{\sigma_{i,t}; i = s, w, l\right\}$ is the vector composed of the square roots of the diagonal elements of the covariance matrix. $\left\{k_{i,t}; i = s, w, l\right\}$ is the mixed probability vector. The observation likelihood is $p(m_t/b_t)$, which is represented by the sum of the Gaussian distributions of 3 components, s, w, l, weighted by $\left\{k_{i,t}; i = s, w, l\right\}$.

The IWA-OSAM represents the stochastic process of all observations until time t-1: $m_{1:t-1}$. In order to enable IWA-OSAM to track target, $\left\{k_{i,t}; i = s, w, l\right\}$ and $\mu_{s,t}, \sigma_{s,t}$ are updated when b_t, m_t are got [18]. The following equations are valid for $j = 1, 2, \cdots, d$. $c = 0.2$ is forgetting factor.

$$k_{i,t}(j) = \left(\left(1 - c\right) + cN\left(w_{t-1}^k(j)m_{t-1}(j); \mu_{i,t-1}(j), \sigma_{i,t-1}^2(j)\right)\right)k_{i,t-1}(j)$$
$$\mu_{s,t}(j) = (1 - c)\mu_{s,t-1}(j)/k_{s,t}(j) + cw_{t-1}^k(j)m_{t-1}(j)k_{s,t-1}(j)/k_{s,t}(j)$$
$$\sigma_{s,t}^2(j) = (1 - c)\sigma_{s,t-1}^2(j)\big/k_{s,t}(j) + c\left(w_{t-1}^k(j)m_{t-1}(j)\right)k_{s,t-1}(j)/k_{s,t}(j) - \mu_{s,t-1}^2(j)$$
$$\qquad (6)$$

Moreover, the methods discussed in [19] are used to reduce the influences of occlusion and outlier here.

Once the solution b_t is solved, the corresponding pixels in the resulting synthesis texture will be used to update IWA-OSAM. While IWA-OSAM are not updated for outlier or occlusion pixels, thus the outlier and occlusion cannot deteriorate IWA-OSAM.

3   Facial Expression Recognition

Static knowledge and dynamic knowledge are extracted to cope with the complex variability of facial expression.

3.1   Static Facial Expression Knowledge

The retrieved α of one frame from the input video can be seen as a description of facial muscles activations of the person in that frame according to the definitions of action units in [36]. Therefore, the relationship between α and facial expression modes is established, namely 7 typical vectors $\left\{ \alpha_{su}, \alpha_{di}, \alpha_{fe}, \alpha_{ha}, \alpha_{sa}, \alpha_{an}, \alpha_{ne} \right\}$ are chosen as the representatives of 7 universal facial expressions: surprise, disgust, fear, happy, sad, angry and neutral. They are set as the static knowledge for facial expression recognition.

When α of one frame of input video is retrieved, the Euclidian distances between it and each of $\left\{ \alpha_{su}, \alpha_{di}, \alpha_{fe}, \alpha_{ha}, \alpha_{sa}, \alpha_{an}, \alpha_{ne} \right\}$ are computed, and the facial expression corresponding to the minimum distance is set as the recognition result.

3.2   Dynamic Facial Expression Knowledge

For each expression γ, a three layer Radial Basis Function (RBF) network is trained for describing the temporal evolution of facial animations α_t as:

$$\alpha_t = \boldsymbol{W} \cdot \boldsymbol{\Phi}(\alpha_{t-1}) + \boldsymbol{B} \tag{7}$$

The middle layer contains 400 nodes. $\boldsymbol{W}(7 \times 400)$, $\boldsymbol{B}(7 \times 1)$ are the weight matrix and bias vector of the output layer. The ith node of middle layer is given by RBF:

$$\boldsymbol{\Phi}_i(A) = exp(-(\|\alpha - \boldsymbol{IW}_i\| \times B_{mi})^2) \tag{8}$$

where $\boldsymbol{IW}_i$ is the ith row component of the weight matrix of the middle layer $\boldsymbol{IW}(400 \times 7)$, and represents the mean value of the ith RBF. $\|\alpha - \boldsymbol{IW}_i\|$ represents the distance between α and $\boldsymbol{IW}_i$. B_{mi} is the ith component of bias vector of the middle layer $\boldsymbol{B}_m(400 \times 1)$, and its reciprocal represents the variance of the ith RBF. The dynamic of facial animations associated with the neutral expression is simplified as $\alpha_t = \alpha_{t-1}$.

We define a transition matrix $\boldsymbol{T}$ whose entries $T_{\gamma',\gamma}$ describe the probability of transition between two expressions γ' and γ. The transition probabilities are learned from a database [39]. Then the RBF network is trained on the 60% of the Extended Cohn-Kanade (CK+) database 25 and the database [39]. The corresponding facial animations α_t are tracked by IWA-OSAM. Therefore, the RBF network is set as the dynamic knowledge for facial expression recognition.

3.3   Framework

Given a facial video, we would like to estimate b_t and the facial expression for each frame at time t, by particle filtering, given all the observations up to time t. Therefore, we create a mixed state $\left(b_t^T, \gamma_t\right)^T$, where $\gamma_t \in \{1, \cdots, 7\}$ is a discrete state, representing one of 7 universal expressions. For the estimation of $\left(b_t^T, \gamma_t\right)^T$, two schemes are proposed. The first scheme (Fig. 3) is to infer $\left(b_t^T, \gamma_t\right)^T$ sequentially, where facial expression is recognized by static facial expression knowledge. The second scheme (Fig. 4) is to infer $\left(b_t^T, \gamma_t\right)^T$ simultaneously, where facial expression is recognized by fusing the static and dynamic facial expression knowledge.

(a) **Initialization**: $t = 0$;

Based on priori distribution $p(b_0)$ and initial particle number N_0, a sampling set $S_0 = \left\{b_0^{(j)}, \gamma_0^{(j)}, \pi_0^{(j)}\right\}_{j=1}^{N_0}$ is initialized;

(b) In this procedure:

Resampling: start from b_{t-1}, new particle set $S_t = \left\{b_t^{(j)}, \gamma_t^{(j)}, \pi_t^{(j)}\right\}_{j=1}^{N_t}$ is obtained by particle number N_t and resampling;

(c) Synthesize improved GNFI from $b_t^{(j)} = \left[h_t^{(j)T}, \alpha_t^{(j)T}\right]^T$ to obtain $m\left(b_t^{(j)}\right)$;

(d) **Update**: particle weight is updated by $\pi_t^{(j)} = p\left(m\left(b_t^{(j)}\right) \mid b_t^{(j)}\right)$;

(e) α_t is obtained from $\sum_{j=1}^{N_t} \pi_t^{(j)} \alpha_t^{(j)} / \sum_{j=1}^{N_t} \pi_t^{(j)}$, and facial expression is classified based on static knowledge;

(f) IWA-OSAM are updated;

(g) $t = t+1$, iterate from *(b)* to *(g)* for the next frame.

Fig. 3.   Inferring the facial motion and facial expression sequentially.

(a) **Initialization**: $t = 0$;

Based on priori distribution $p(b_0)$ and initial particle number N_0, a sampling set $S_0 = \left\{b_0^{(j)}, \gamma_0^{(j)}, \pi_0^{(j)}\right\}_{j=1}^{N_0}$ is initialized;

(b) In this procedure:

(1) **Resampling**: start from b_{t-1}, new particle set $S_t = \left\{b_t^{(j)}, \gamma_t^{(j)}, \pi_t^{(j)}\right\}_{j=1}^{N_t}$ is obtained by particle number N_t and resampling;

(2) **Prediction**: Draw N_t particles according to dynamic facial expression model:

(i) Draw a expression label $\gamma_t^j = \gamma \in \{1, 2, \cdots, 7\}$ with probability $T_{\gamma', \gamma}$, where $\gamma' = \gamma_{t-1}^{(j)}$;

(ii) Compute $\alpha_t^{(j)} = W^{\gamma} \cdot \Phi\left(\alpha_{t-1}^{(j)}\right) + B^{\gamma}$, $\gamma = \gamma_t^{(j)}$;

(c) Synthesize improved GNFI from $b_t^{(j)} = \left[h_t^{(j)T}, \alpha_t^{(j)T}\right]^T$ to obtain $m\left(b_t^{(j)}\right)$;

(d) **Update**: particle weight is updated by $\pi_t^{(j)} = p\left(m\left(b_t^{(j)}\right) \mid b_t^{(j)}\right)$;

(e) In this procedure:

(1) α_t is obtained from $\sum_{j=1}^{N_t} \pi_t^{(j)} \alpha_t^{(j)} / \sum_{j=1}^{N_t} \pi_t^{(j)}$, and the probability of each facial expression is obtained based on static knowledge;

(2) Get the probability of each expression $P(\gamma^*) = \sum_{m=1}^{J} \begin{cases} \pi_t^{(m)} & \text{if } \gamma_t^{(m)} = \gamma^* \\ 0 & \text{otherwise} \end{cases}$, $\gamma^* \in \{1, 2, \cdots, N_r\}$ according to dynamic knowledge;

(3) The multiplication of the results of *(1)* and *(2)* is set as the expression probability, and the largest is the expression recognition result;

(f) IWA-OSAM is updated;

(g) $t = t+1$, iterate from *(b)* to *(g)* for the next frame.

Fig. 4.   Inferring the facial motion and facial expression simultaneously.

4 Evaluation

A workstation with Intel i7-6700K 4.0G, 8G memory and NVIDIA GTX960 is used.

4.1 Testing Dataset and Evaluation Methods for Facial Motion Tracking

A facial image sequence [5] and the IMM face database [40] with the ground truth landmarks available are used. They support point based comparison for errors, and a texture based test could also be performed on it. Besides, as the pose coverage and illumination condition variations of above databases are not large enough, the 13 videos, including Carphone and Forman image sequences, in the MPEG-4 testing database and 78 captured videos from 48 subjects with the resolution 352 × 288 are also used. The ground truth landmarks of them are obtained by manual adjustment.

Root Mean Square (RMS) landmark error measures the Root Mean Square Error (RMSE) between the ground truth landmark points and the fitted shape points after tracking, and is defined as:

$$\sum_{i=1}^{N} sqrt\left(C_{fit}^i - C_{grd}^i\right)\Big/N \tag{9}$$

where C_{est}^i, C_{grd}^i are the x or y coordinate of the ith fitted shape point and the ith ground truth landmark point.

4.2 Testing Dataset and Evaluation Methods for Facial Expression Recognition

The Extended Cohn-Kanade (CK+) database [25] and the database [39], not including the training part, are used. The database also presents the baseline results using AAM and a linear support vector machine classifier.

For evaluation, the facial expression recognition score is used, and a confusion matrix between different facial expression is used.

4.3 Facial Motion Tracking for Monocular Videos

Figure 5 shows the facial motion tracking results of several publically and captured videos. By computing the evaluation criteria, we can say that accurate tracking is

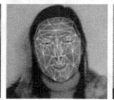

Fig. 5. Facial motion tracking results.

obtained even in the presence of perturbing factors including significant head pose and illumination as well as facial expression variations.

Based on the testing dataset, the comparison between different tracking algorithm is conducted. It should be stated that the work in [3] an offline method, and its performance is highly dependent on the training data. However, Active Shape Model (ASM)/ Active Appearance Model (AAM) based trackers, such as that in [3], are the mainstream, and after learning the model in a dataset, the tracker can track any face without further training. Therefore, we compare this work with that in [3]. The images of the training dataset, which approximately correspond to every 4th image in the sequences of the testing dataset, are used to construct the AAM in [3], and the number of bases in the constructed AAM is chosen so as to keep 95% of the variations. The performances is evaluated on the testing data except for the training images.

By computing the evaluation criteria, Table 3 shows the superiority of our algorithm. This is because the IWA-OSAM in our proposed approach can learn the variation of facial motion effectively, and our proposed approach has the ability to alleviate lighting influence. Moreover, this is also because the improved GNFI is less influenced by the head pose change and illumination variation.

Table 3. Performance evaluation of different facial motion tracking algorithms.

	RMS landmark error
Our facial motion tracking algorithm	2.54
Our facial motion tracking algorithm not using illumination modeling	4.23
The method in [3]	4.33
The method in [34]	3.65
The method in [37]	5.29

4.4   Facial Expression Recognition

Figure 6 shows the facial expression recognition results on the testing database. As can be seen from it, facial expressions can be recognized effectively by our proposed algorithm in the presence of perturbing factors including significant head pose and illumination.

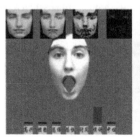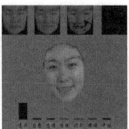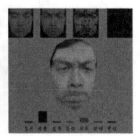

Fig. 6. Facial expression recognition results.

Based on the CK+ database, we compare our proposed algorithm to the methods in [20–22, 31, 34, 37] (Table 4). The recognition score of our facial expression recognition algorithm is higher than those of other algorithms, and also higher than those of our facial expression recognition algorithm not using illumination modeling.

Table 4. The accuracy comparison between several algorithms.

	Recognition score
Our facial expression recognition algorithm	87.4%
Our facial expression recognition algorithm not using illumination modeling	79.8%
The algorithm in [31]	82.2%
The algorithm in [31]	78.4%
The algorithm in [37]	80.1%
The algorithm in [21]	70.4%
The algorithm in [20]	72.7%
The algorithm in [22]	79.2%

According to the benchmarking protocol in the CK+ database, the leave-one-subject-out cross-validation configuration is used, and a confusion matrix is used to document the results. Table 5 shows the high recognition scores by our proposed algorithm. This is because that both static and dynamic knowledge are used, illumination is modeled and removed from improved GNFI to increase the accuracy and robustness of facial motion tracking.

Table 5. The confusion matrix of facial expression recognition by our algorithm.

	Angry	Disgust	Fear	Happy	Sad	Surprise
Angry	**89.7**	4.1	2.9	0.0	2.1	1.2
Disgust	0.9	**99.1**	0.0	0.0	0.0	0.0
Fear	3.1	0.0	**86.1**	4.2	0.0	6.6
Happy	0.0	0.0	0.0	**100.0**	0.0	0.0
Sad	6.1	2.8	2.7	0.0	**87.2**	1.2
Surprise	0.0	0.0	0.0	0.0	0.0	**100.0**

5   Conclusion

We propose a unified facial motion tracking and expression recognition framework for monocular video. For retrieving facial motion, an online weight adaptive statistical appearance method is embedded into the particle filtering strategy by using a deformable facial mesh model served as an intermediate to bring input images into correspondence by means of registration and deformation. For recognizing facial expression, facial animation and facial expression are estimated sequentially for fast and efficient applications, in which facial expression is recognized by static anatomical facial expression knowledge. In addition, facial animation and facial expression are simultaneously estimated for robust and precise applications, in which facial expression is recognized by fusing static and dynamic facial expression knowledge. Experiments demonstrate the high tracking robustness and accuracy as well as the high facial expression recognition score of the proposed framework.

In future, the recursive neural network will be used to learn the dynamic expression knowledge.

Acknowledgement. This work is supported by the National Natural Science Foundation of China (No. 61572450, No. 61303150), the Open Project Program of the State KeyLab of CAD&CG, Zhejiang University (No. A1501), the Fundamental Research Funds for the Central Universities (WK2350000002), the Open Funding Project of State Key Laboratory of Virtual Reality Technology and Systems, Beihang University (No. BUAA-VR-16KF-12), the Open Funding Project of State Key Laboratory of Novel Software Technology, Nanjing University (No. KFKT2016B08).

References

1. Black, M.J., et al.: Recognizing facial expressions in image sequences using local parameterized models of image motion. IJCV **25**(1), 23–28 (1997)
2. Gokturk, S., et al.: A data-driven model for monocular face tracking. In: ICCV, pp. 701–708 (2001)
3. Sung, J., Kanade, T., Kim, D.: Pose robust face tracking by combining active appearance models and cylinder head models. IJCV **80**(2), 260–274 (2008)
4. Dornaika, F., Davoine, F.: On appearance based face and facial action tracking. TCSVT **16**(9), 1107–1124 (2006)
5. Zeng, Z.H., et al.: A survey of affect recognition methods: audio, visual, and spontaneous expressions. TPAMI **31**(1), 31–58 (2009)
6. Wen, Z., Huang, T.S.: Capturing subtle facial motions in 3D face tracking. In: ICCV, pp. 1343–1350 (2003)
7. Sandbach, G., et al.: Static and dynamic 3D facial expression recognition: a comprehensive survey. IVS **30**(10), 683–697 (2012)
8. Fang, T., Zhao, X., et al.: 3D facial expression recognition: a perspective on promises and challenges. In: ICAFGR, pp. 603–610 (2011)
9. Marks, T.K., et al.: Tracking motion, deformation and texture using conditionally Gaussian processes. TPAMI **32**(2), 348–363 (2010)
10. Zhang, W., Wang, Q., Tang, X.: Real time feature based 3-D deformable face tracking. In: Forsyth, D., Torr, P., Zisserman, A. (eds.) ECCV 2008. LNCS, vol. 5303, pp. 720–732. Springer, Heidelberg (2008). doi:10.1007/978-3-540-88688-4_53
11. Liao, W.-K., Fidaleo, D., Medioni, G.: Integrating multiple visual cues for robust real-time 3D face tracking. In: Zhou, S.Kevin, Zhao, W., Tang, X., Gong, S. (eds.) AMFG 2007. LNCS, vol. 4778, pp. 109–123. Springer, Heidelberg (2007). doi:10.1007/978-3-540-75690-3_9
12. Cascia, M.L., et al.: Fast, reliable head tracking under varying illumination: an approach based on registration of texture mapped 3D models. TPAMI **22**(4), 322–336 (2000)

13. Fidaleo, D., Medioni, G., Fua, P., Lepetit, V.: An investigation of model bias in 3D face tracking. In: Zhao, W., Gong, S., Tang, X. (eds.) AMFG 2005. LNCS, vol. 3723, pp. 125–139. Springer, Heidelberg (2005). doi:10.1007/11564386_11
14. Liao, W.K., et al.: 3D face tracking and expression inference from a 2D sequence using manifold learning. In: CVPR, pp. 3597–3604 (2008)
15. Cao, C., Lin, Y., Lin, W.S., Zhou, K.: 3D shape regression for real-time facial animation. TOG 32(4), 149–158 (2013)
16. Cao, C., et al.: Displaced dynamic expression regression for real-time facial tracking and animation. In: SIGGRAPH, pp. 796–812 (2014)
17. Jepson, A.D., Fleet, D.J., et al.: Robust online appearance models for visual tracking. TPAMI 25(10), 1296–1311 (2003)
18. Lui, Y.M., et al.: Adaptive appearance model and condensation algorithm for robust face tracking. TSMC Part A 40(3), 437–448 (2010)
19. Yu, J., Wang, Z.F.: A video, text and speech-driven realistic 3-D virtual head for human-machine interface. IEEE Trans. Cybern. 45(5), 977–988 (2015)
20. Wang, Y., et al.: Realtime facial expression recognition with Adaboost. In: ICPR, pp. 30–34 (2004)
21. Bartlett, M., Littlewort, G., Lainscsek, C.: Machine learning methods for fully automatic recognition of facial expressions and facial actions. In: ICSMC, pp. 145–152 (2004)
22. Zhang, Y., Ji, Q.: Active and dynamic information fusion for facial expression understanding from image sequences. TPAMI 27(5), 699–714 (2005)
23. Tian, Y.L., et al.: Facial expression analysis. In: Li, S.Z., Jain, A.K. (eds.) Handbook of Face Recognition. Springer, New York (2005)
24. Chang, Y., et al.: Probabilistic expression analysis on manifolds. In: CVPR, pp. 520–527 (2004)
25. Lucey, P., et al.: The extended Cohn-Kande dataset (CK+): a complete facial expression dataset for action unit and emotion-specified expression. In: CVPR, pp. 217–224 (2010)
26. Tian, Y., Kanade, T., Cohn, J.F.: Recognizing action units for facial expression analysis. TPAMI 23, 97–115 (2001)
27. Hamester, D., et al.: Face expression recognition with a 2-channel convolutional neural network. In: IJCNN, pp. 12–17 (2015)
28. Wang, H., Ahuja, N.: Facial expression decomposition. In: ICCV, pp. 958–963 (2003)
29. Lee, C., Elgammal, A.: Facial expression analysis using nonlinear decomposable generative models. In: IWAMFG, pp. 958–963 (2005)
30. Zhu, Z., Ji, Q.: Robust realtime face pose and facial expression recovery. In: CVPR, pp. 1–8 (2006)
31. Cohen, L., Sebe, N., et al.: Facial expression recognition from video sequences: temporal and static modeling. CVIU 91(1–2), 160–187 (2003)
32. North, B., Blake, A., et al.: Learning and classification of complex dynamics. TPAMI 22(9), 1016–1034 (2000)
33. Zhou, S., Krueger, V., Chellappa, R.: Probabilistic recognition of human faces from video. CVIU 91(1–2), 214–245 (2003)
34. A Video-Based Facial Motion Tracking and Expression Recognition System. Multimed. Tools and Appl. (2016). doi:10.1007/s11042-016-3883-3
35. Ekman, P., Friesen, W., et al.: Facial Action Coding System: Research Nexus. Network Research Information, Salt Lake City (2002)
36. Schmidt, K., Cohn, J.: Dynamics of facial expression: normative characteristics and individual differences. In: ICME, pp. 728–731 (2001)
37. Dornaika, F., Davoine, F.: Simultaneous facial action tracking and expression recognition in the presence of head motion. IJCV 76(3), 257–281 (2008)

38. Hu, Y.K., Wang, Z.F.: A low-dimensional illumination space representation of human faces for arbitrary lighting conditions. Acta Automatica Sinica **33**(1), 9–14 (2007)
39. http://www.chineseldc.org/emotion.html
40. Nordstrøm, M.M., et al.: The IMM face database - an annotated dataset of 240 face images. Technical report, Technical University of Denmark (2004)

A Virtual Reality Framework for Multimodal Imagery for Vessels in Polar Regions

Scott Sorensen[1](✉), Abhishek Kolagunda[1], Andrew R. Mahoney[2],
Daniel P. Zitterbart[3], and Chandra Kambhamettu[1]

[1] University of Delaware, Newark, DE, USA
Sorensen@udel.edu
[2] University of Alaska Fairbanks, Fairbanks, AK, USA
[3] Alfred Wegener Institute for Polar and Marine Research, Bremerhaven, Germany

Abstract. Maintaining total awareness when maneuvering an ice-breaking vessel is key to its safe operation. Camera systems are commonly used to augment the capabilities of those piloting the vessel, but rarely are these camera systems used beyond simple video feeds. To aid in visualization for decision making and operation, we present a scheme for combining multiple modalities of imagery into a cohesive Virtual Reality application which provides the user with an immersive, real scale, view of conditions around a research vessel operating in polar waters. The system incorporates imagery from a 360° Long-wave Infrared camera as well as an optical band stereo camera system. The Virtual Reality application allows the operator multiple natural ways of interacting with and observing the data, as well as provides a framework for further inputs and derived observations.

1 Introduction and Related Works

As traffic and development in polar regions increases, systems to operate safely in this environment are becoming increasingly important to the vessels and the environment. Icebreakers like the RV Polarstern, are designed to operate in dense ice cover, and have a variety of camera systems for observing conditions. During the ARK-XXVII/3 research cruise two of the platforms in use were the FIRST-Navy IR system and the Polar Sea Ice Topography REconstruction System (PSITRES). These systems vary drastically in capabilities, field of view, and operational goals. These platforms have been used independently for a variety of tasks, but here we present a method for unifying these systems by leveraging their geometry and the scene. We then combine these modalities of imaging into a unified framework for visualization in a Virtual Reality application that provides a novel means of interacting with image data from multiple sources.

The two systems capture different aspects of the environment around the ship, from the coverage and type of ice, to the presence of melt ponds and algae, to animals. In long-wave infrared (LWIR), often referred to as thermal images, warm blooded animals like polar bears are readily visible. In the optical band algae and sediment can be seen. The differing field's of view and the different

© Springer International Publishing AG 2017
L. Amsaleg et al. (Eds.): MMM 2017, Part II, LNCS 10133, pp. 63–75, 2017.
DOI: 10.1007/978-3-319-51814-5_6

modalities can provide complimentary information for the operators of the ship, which should in turn help the ship maneuver in a safe and ecological manner.

The Polar Sea Ice Topography REconstruction System, or PSITRES, is a 3D camera system consisting of three cameras which obliquely observe a patch of ice and water greater than $1000\,\mathrm{m}^2$ adjacent to the vessel. PSITRES captures optical band images using machine vision cameras. PSITRES was mounted to the port side flying deck of the research icebreaker RV Polarstern in 2012 during the ARK-XXVII/3 cruise. As shown in Fig. 1, PSITRES consists of a stereo pair on a 2 m baseline which capture synchronized images, and an asynchronous center camera with a wider Field Of View (FOV) and higher resolution. The PSITRES system has been used for a variety of reconstruction techniques [14,15]. During the ARK-XXVII/3 cruise PSITRES collected stereo images at a rate of 1/3 Frames Per Second (FPS), and the center camera operated at 1 FPS.

The FIRST-Navy IR system is an omnidirectional gimbal stabilized long-wave Infrared (LWIR) camera system originally developed by Rheinmetall Defence for military use, but has been deployed aboard the RV Polarstern for marine mammal observation [27]. The camera system consists of a LWIR line scanner which is rotated on an axis atop a gimbal which stabilizes and isolates it from most of the ship's motion. It is mounted on the crow's nest of the ship and encompasses a circular area around the ship to the horizon. The sensor is a cooled LWIR sensor sensitive to electromagnetic radiation at 8 to 12 μm. The camera takes images at a resolution of 7200×576 as shown in Fig. 2, at 5 FPS.

Fig. 1. The FIRST-Navy IR camera system and PSITRES

Fig. 2. A sample image from the FIRST-Navy IR system. Note the edges of the image correspond to the rear of the ship and show a portion of the crow's nest

The two camera systems vary greatly in their imaging capabilities. The camera systems have different fields of view as illustrated in Fig. 3. The cameras have virtually no overlap in viewing area, they operate in radically different portions of the spectrum, and they operated at different frame rates. We aim to incorporate both of these camera systems into a unified Virtual Reality system, and to do this the cameras must be spatially and temporally aligned.

Aligning these two modalities is a difficult task even with overlap. Cross modality matching is inherently difficult because texture and edges in one modality may have no counterpart in other modalities. Our problem is further complicated by the lack of overlap between the two scenes, a dynamic environment, and different resolutions spatially and temporally. Clock drift further complicates temporal alignment as do dropped frames and non uniform frame rates. To overcome these limitations we leverage the geometry and configuration of the two systems, as well as the scene layout to align the 3D coordinate systems of each sensor. Temporal alignment is done by calculating a time offset and matching frames based on histogram binning. This allows for a common coordinate system and the reprojection.

The common scale and alignment of our reprojection allows us to create real time Virtual Reality (VR) visualizations of the conditions around the ship. The technique allows for video streams from each camera to be overlaid on corresponding real scale geometry. A Head Mounted Display (HMD) provides a real sense of scale, intuitive control and unprecedented interaction with data from the two camera systems. This system could help provide people aboard vessels like this to make informed decisions, and could be used for education and outreach purposes, allowing people to virtually explore the Arctic.

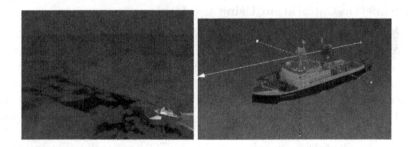

Fig. 3. (A) An illustration of the differing Fields of View of both camera systems. Both modalities of image are shown reprojected here in a rendering. (B) The Scaled axis aligned model used for alignment

There are many works related to using imagery of different modalities for a wide number of tasks. In this work we use both optical band stereo and omni-directional long-wave infrared, with applications to virtual reality. Here we will focus on a few related works with applications of cross modality, 3D calibration and alignment, and VR. Image based matching between modalities is a difficult task, and a number of works have attempted to solve this problem. [19] used trajectory-to-trajectory matching. [1] approximate the shape of the targets and align video sequences via affine transformation. Many techniques have attempted to match thermal and visible images of faces [2,7,17]. Unlike these works our sensors are spatially distant from one another, and have unshared fields of view. A number of works have developed techniques for calibrating cameras of different modalities, and many of these works use custom calibration objects. [6,20] have cut a calibration pattern to create thermal gradient. Other works have used grids of wire [23], and light bulbs [4,22], or external heating [16] and cooling [8]. Unlike these works, our calibration and 3D alignment approach uses scene alignment, and was performed asynchronously.

Virtual Reality systems have been developed for telepresence [3], therapy [12], surgical training [5], and teleoperation of vehicles [9]. Head Mounted Displays have been incorporated into an augmented reality system aboard the F-35 aircraft [10]. The system we have developed has potential applications in the operation of vessels in ice covered waters. It combines video feeds in different modalities from different areas and allows the user to view them in an accurate way.

2   Methods

To build a unified VR application that integrates both camera systems, we find geometric reprojections of the images, and align these models, and align the image sequences. In this section we discuss our methods for calibration, reprojection, spatial and temporal alignment to facilitate a useful VR application.

2.1   Altimetric Calibration Using the Horizon

While PSITRES was calibrated using Zhang's method [25], the FIRST-Navy system does not conform to the pinhole model of projection. Furthermore the configuration and scale of its viewing area means typical calibration techniques are ill suited. We advocate altimetric based calibration using the horizon, because for a given altitude the angle to the horizon can be determined. Unlike other horizon based calibration techniques [13,26] this does not use the pinhole camera model, and does not incorporate vanishing lines. The camera system has a 360° FOV in the horizontal image axis, and an 18° FOV vertically. This means we can model the projection spherically, with images projected from a segment of a sphere centered at the sensor. This segment is defined for every azimuth angle between a minimum and maximum angle of elevation. These angles, ϕ_{min} and ϕ_{max} are however unknown. To solve for these angles we assume a uniform pixel pitch in both the x and y dimensions. The size of our image is 7200×576 so

$$Pitch_x = \frac{360°}{7200} \qquad Pitch_y = \frac{18°}{576} \qquad (1)$$

where $Pitch_x$ and $Pitch_y$ are the angle pixel pitch in the horizontal and vertical direction respectively. By associating the horizon with its projection we can fix

ϕ_{min} and ϕ_{max}. We computed Sobel edge image [18] for 4413 images and summed the result to find consistent scene edges. While it is faint in a given image, the horizon stays consistent, because the sensor itself is gimbal stabilized. We found a common edge at $y_h = 40$ pixels. The angle of declination to the horizon is

$$d \approx 3.57\sqrt{h} \tag{2}$$

where d is the distance to the horizon in km, h is the height in meters [24]. The angle of declination to the horizon is therefore

$$\phi_{dec} = -1 * \arctan{(h/d)} \tag{3}$$

Using this angle of declination we can solve for ϕ_{min} and ϕ_{max} allowing for scene to pixel correspondences by

$$\phi_{min} = \phi_{dec} - ((576 - y_h) \cdot Pitch_y \phi_{max} = Pitch_y \cdot y_h + \phi_{dec}) \tag{4}$$

2.2 IR Reprojection

Visualizing the 360° images on a flat screen is akin to an extreme fish eye effect, and scene motion is unintuitive. To facilitate realtime playback in a manner that preserves geometry, we have generated a 3D mesh on which to apply imagery from the IR system. We present techniques for generating this mesh including the vertex locations, normals, and texture coordinates. We have developed a planar reprojection, and corresponding 3D mesh that follows naturally from the camera configuration. To reproject the images we define a UV texture parameterization with the mesh, allowing us to directly apply the images as texture.

Intuitively the planar reprojection is the projection of the images onto a plane at sea level. This is especially useful because it relates the images to real scale, and the projected images closely match the real world scene for objects near sea level (which is most of the scene). With the sensor at the origin, we can model sea level as a plane at $z = -h$ with surface normal $[0, 0, 1]$. To generate the planar mesh we uniformly sample $-\pi/2 \le \theta \le 3\pi/2$ and $\phi_{min} \le \phi \le \phi_{dec}$ using polar coordinates with $R = 1$. This allows us to tune the polygon count of the mesh by adjusting the sampling of θ and ϕ, allowing us to adjust the quality and computational load. This gives us a set of unit vectors from the sensor to the sea level plane for points on the image ranging to the horizon. Vertices in our model are the intersection of these vectors with the sea level plane computed by

$$i = t \cdot R \tag{5}$$

where

$$t = ((C_p \cdot N)/(R \cdot N) \tag{6}$$

and C_p is the plane center $[0, 0, -z]$, N is the plane normal $[0, 0, 1]$, R is the ray direction. Note that this is a specific case of ray plane intersection with ray origins at the global origin. Normalized texture coordinates are computed $((\theta + \pi/2)/2\pi, (\phi - \phi_{min}/(\phi_{max} - \phi_{min}))$. We define faces by explicitly indexing vertices and creating a series of upper an lower triangle faces connecting each vertex to its neighbors. This mesh provides a geometric canvases onto which the images can be overlaid using texture mapping based on the UV parameters.

2.3   PSITRES Reprojection

The PSITRES camera system uses calibrated stereo, capable of producing high polygon count 3D meshes. While directly integrating these meshes may seem intuitive at first, we advocate a planar reprojection of PSITRES imagery in this work for a number of reasons. Reconstructing each mesh is very time consuming, and takes roughly 2–5 min with the low texture stereo techniques outlined in [15]. This is time prohibitive for a real time application. Furthermore, using a low polygon reprojection means we can maintain a higher framerate in our VR application, which is critical for reducing simulation sickness and maximizing user comfort [21]. However we do not ignore 3D information, instead we advocate an offline approach to ensure geometrically accurate reprojection. To do this we use stereo information to identify the water and ice surface, model the surface, and reproject the images on a generated mesh. PSITRES looks obliquely at a patch of ice and water, so the scene is mostly planar with ice lying withing a few centimeters of the water surface. We took 100 stereo pairs captured by the PSITRES system sampled over the course of 6 weeks, reconstructed each pair, and fit a plane to the resulting point clouds using Principal Component Analysis.

With this plane we can generate a mesh based on our projection using a ray tracing similar to our technique for generating the planar mesh in Sect. 2.2. Instead of using a spherical projection we now use camera calibration parameters directly. For x_i uniformly sampled from 1 to our image width and y_i sample from 1 to our image height. We generate rays V_i by

$$V_i = C_0 + t \cdot \frac{\beta}{norm(\beta)} \tag{7}$$

where C_0 is coordinates of the camera center, and

$$\beta = R' \cdot A^{-1} \cdot [x_i, y_i, 1] \tag{8}$$

where A is the camera matrix, and R is the camera rotation matrix from stereo calibration. These rays are then intersected with the 3D plane by substituting solving P from the plane equation

$$P \cdot n + d = 0 \tag{9}$$

surface, with V_i. Where P is the plane origin, n the plane normal and d is a constant. These intersection points become the vertex coordinates of our 3D mesh, with vertex normal n defined by our scene plane, and $[x_i, y_i]$ serving as UV coordinates for texture mapping.

2.4   Temporal Alignment

While the goal of this work is a real time system with potential to operate on board a vessel while underway, development and testing have happened afterwards with prerecorded data. We have worked towards building a system that could operate with video or image streams over the ship's network. In testing we work with data that is sampled at different resolution spatially and temporally. While both camera systems record a timestamp, clock drift means that timestamps alone cannot be used. Temporally aligning the two modalities is further complicated by the fact that the area inside the near range of the infrared camera (the area in which PSITRES's field of view is contained) is very dynamic. This is the region where the ship breaks and moves ice. The scene goes through tremendous change between the time it exits the field of view of the IR camera system and enters into the field of view of PSITRES. Dropped frames and varying framerate further complicates aligning sequences. We decompose the problem into two steps, that of finding a time offset, and of frame matching.

Manually finding the initial offset is done once per month by using the wider FOV center camera on PSITRES and compare image sequences to a 40° wedge from the IR camera system. We manually compared patches of open water and different ice types in both modalities to facilitate minute scale alignment. Fine grain or second scale alignment was done by manually tracking ice features visible in both modalities. This was done by observing distinctive melt ponds and ridges. The result was an approximately 7.5 min offset for aligning sequences.

Matching sequences of frames is accomplished automatically by computing a single scalar for each adjusted timestamp (similar to Unix epoch time). Frame level matching can then done by computing a histogram with the bin values of the scalar times of the lower framerate optical camera. The counts of the bins correspond to the number of times these frames must be repeated for each corresponding frame of IR imagery. In this way dropped frames are simply repeated 0 times and ignored in the other modality. The result are synchronized sequences with drops occurring in both modalities.

2.5   Spatial Alignment

So far we have discussed each camera system and its 3D reprojection independently, but in order to facilitate a cohesive VR application we must spatially align the coordinate systems of both sensors. This means we must find $[R_{inter}|T_{inter}]$, or the rotation R_{inter} and translation T_{inter} from the FIRST-Navy system to PSITRES. Thus 3D points in the PSITRES coordinate system can be mapped to their correct relative position by

$$P_{IR} = (R_{inter} \cdot P_{stereo}) + T \tag{10}$$

where P_{IR} and P_{stereo} are 3D points in the IR system and PSITRES respectively. To solve for R_{inter} and T_{inter} we again leverage the configurations of the two systems.

We can directly observe T_{inter} by finding the left camera of PSITRES in the IR coordinate system. To do this we use a 3D model of the Polarstern. This model is a Structure From Motion (SFM) reconstruction that was made by flying around the ship in a helicopter and capturing images with a DSLR camera. We used Autodesk Memento to produce the original model, which was subsequently aligned to the axes and centered at the FIRST-Navy system. The model was scaled using technical drawings of the ship. We manually identified the left camera housing of PSITRES in this model, which is the origin for PSITRES, and therefore T_{inter}. Computing R_{inter} is more complicated. While the mounting system for PSITRES is rigid, with a set angle of declination, and a set vergence angle between the two cameras, in practice the cameras are mounted by hand, meaning there can be small rotations in multiple axes.

The stereo model is a mean plane generated from reconstruction, which represents sea level, the xy plane in our coordinate system. We use the known angles as a starting point and solve for a best rotation matrix to align the mean stereo plane to the sea level plane. To do this we create a rotation matrix R_{base} using the known angle of declination and vergence from the PSITRES mounts by

$$R_{base} = EulerToRot(\phi_c, 0, \theta_c) \tag{11}$$

where ϕ_c is the declination angle of camera mount, and θ_c is the angle relative to forward from the vergence angle of the camera mount, and the function $EulertoRot$ converts from Euler angles to a rotation matrix. This matrix R_{base} only roughly aligns the two planes however, and furthermore the value of θ_c has no affect on the plane alignment as this rotates points within the z axis, parallel to sea level. To precisely align the two planes we find

$$\arg \max_{\phi_d, \psi_d} ((EulerToRot(\phi_d, \psi_d, \theta_c) \cdot n_{stereo}) \cdot [0, 0, 1]) \tag{12}$$

This maximizes the cosine similarity with the sea level plane normal. To find ϕ_d and ψ_d we search a narrow wedge of $25°$ in both directions around ϕ_c and 0 for ϕ_d and ψ_d respectively. Thus our final rotation matrix becomes

$$R_{inter} = EulerToRot(\phi_d, \psi_d, \theta_c) \tag{13}$$

The resulting spatial alignment is shown in Fig. 3.

3   Experimental Verification

To verify the spatial alignment we use two metrics which are independent of the methods used for alignment, namely the position, and projected motion vectors. Both of these are independent of the normal and translation.

The transformation $[R_{inter}|T_{inter}]$ should align the mean plane mesh and sea level plane, but the Translation T_{inter} was computed using the observed position of the camera in the 3D ship model. To evaluate the transformation, we can compare the average distance between the two planes over the extent of the stereo model. We computed point plane distance for each vertex and arrived at mean distance of 3.4 m or 13.33% of the total distance to the plane. This number is somewhat high, however there are a number of compounding factors that contribute to it. The 3D model of the ship is somewhat noisy, and the PSITRES camera system in this model is a very small. Furthermore, the sea level plane is not a constant. The draft changes with ballast, fuel weight and a number of other factors. The stereo plane represents an average reconstruction of scenes with ice, not just water and this affects the estimated position of the plane relative to the sensor. The displacement of these two planes is not apparent from the perspective of a user of the VR system.

We further verify the alignment by looking at motion. Projected motion from the images in both modalities should be the same if the alignment is accurate. To compare projected motion we can compare motion vectors from tracking in both modalites. To do this we use sequences of images in both modalities with relatively uniform ship motion. In each modality we track points between consecutive images using SIFT matching [11]. These vectors can then be projected in the same way that we created the meshes in Sects. 2.2 and 2.3. We mask off portions the IR which are occluded by the ship, and place a threshold $\tau = 15$ pixels on the maximum distance between matches from frame to frame.

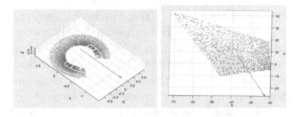

Fig. 4. A sampling of projected motion and mean motion vector for the IR and stereo cameras

We can readily compare the orientation of these projected motion vector by averaging vectors over the entire sequence and use cosine similarity to compare the vectors. A sample of these motion vectors as well as the mean is shown in Fig. 4. The result is a 91.81% similarity. These averaged motions vector could also be used to further optimize the rotation matrix in Sect. 2.5, as this error metric is more directly affected the θ_c term, which was not used in optimization.

4   Virtual Reality Application

We have used these alignment and reprojection techniques to develop a VR application using the Unreal Engine 4. The application allows a user to move around in a real scale 3D space and observe the reprojected imagery in a novel and intuitive ways. The infrared images have been colormapped to aid in visualization and enhance the aesthetics. The real benefit of the application is the ability to move around in 3 dimensions seamlessly. We are not limited to the view from the sensors, and can actually fly around in any direction and generate new views. If an interesting object in the scene is found, for example a polar bear, the user can fly out and watch it. The user can fly straight up and view a mixed modality bird's eye view of the area around the ship as seen in Fig. 5.

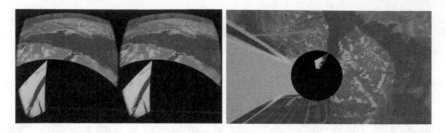

Fig. 5. (A) An example view through the HMD looking at both the planar thermal and stereo models. (B) A top down perspective from the VR app

The Unreal Engine has a powerful set of tools to allow us to build an application that works seamlessly with a variety of display and interface hardware. The Unreal Engine handles lighting, rendering, movement, and peripheral support, greatly accelerating development time. Using the unreal engine means that our application readily ports to different head mounted displays and natively supports a variety of common control schemes. We have tested our application with a standard HD monitor, and the Oculus Rift DK2, The Oculus Rift CV1, and the HTC Vive using either a keyboard and mouse, or a gamepad.

The Unreal Engine supports video textures in the form of both pre-recorded video files and video streams from across a local network or the internet. In a real world deployment this system would use networked video streams from each sensor across the ship network, which means little modification would be needed to the offline version of our application. We have experimented with video streams and give results below.

This application testifies to the above alignment and reprojection techniques. In Fig. 5 you can see a lead (a linear area of open water in an expanse of ice) which extends from the stereo reprojection into the infrared reprojection. We have developed the application to be easily extendable, and using the Unreal Engine means it is readily adaptable to different computing environments.

4.1   Evaluation

To evaluate the application we have compared several objective metrics related to performance and we have conducted a user evaluation with a small board of 5 experts in the fields of sea ice and biological science, ship's crew, thermal imaging experts and navigators. Quantitative evaluation was designed to evaluate the real time operation and feasibility of the application in a simulated environment.

Our application runs on a variety of displays ranging from traditional 2D monitors to a range of consumer HMD's. For evaluation we have used a portable workstation computer with an Intel Core i7 6700K, and Nvidia GTX 980. We have tested our application with the Oculus Rift DK2, The Oculus Rift CV1, and the HTC Vive. The application works at maximum resolution and operates at the maximum supported framerate of each device. Furthermore we have run the application using video feeds across a wired gigabit network, and it used between 1.5% and 5% of the available bandwidth on a single connection, with our application playing the video stream at 5 times the speed of recording.

To evaluate the efficacy of the system from a user standpoint we had each member of our board of experts look at static 2D images as well as their VR projections. Users were given 45 s to identify animals in the scene and quantify ice coverage and maximum floe length. The users were told they were not looking at the same images, however they were presented with mirrored reprojections. Every user preferred the VR application, and on average spotted more animals. Distance estimates varied greatly in both setups, indicating an area for possible improvement.

5 Conclusion

In this work we have presented a framework for multimodal virtual reality visualization using images captured from 360° thermal camera and a optical band stereo system. These camera systems were deployed simultaneously aboard a research vessel in the Arctic Ocean. We have developed a system that unifies these camera systems into a single VR application which allows users a novel means of viewing imagery. Such a system could be used by those onboard a vessel in ice covered waters to operate safely and ecologically. The application allows for video from both systems to be reprojected in real time in an accurate way.

To do this we have spatially aligned the two by leveraging the geometry of sensors, and the environment around the ship. We have created 3D models that allow for video streams to be applied directly to the model in real time. We have combined these aligned reprojections in Virtual reality application using the Unreal Engine 4, which enables users freedom to move around in 3D and view the reprojected images from different perspectives. This application was built to be readily deployable in a real environment on a ship. The VR application runs smoothly, and works with a variety of common control schemes and VR HMDs. This application allows for users to view conditions around the ship in every direction, and combines visible and thermal imaging.

In the future we will integrate more information as well as high level information extracted from the images themselves. Both camera systems have been the subject of computer vision research, and integrating detected marine mammals, ice concentration, and other derived values is a logical next step. Overlaying information such as position and heading information would improve the system for users in a real world scenario. This system could additionally be used for education, training, and outreach and distributing it would allow people from distant locations to virtually explore polar regions with those onboard.

Acknowledgments. We would like to thank the crew and scientific party of the Polarstern ARK-XXVII/3 research cruise. This work is funded by NSF CDI Type I grant 1124664.

References

1. Chakravorty, T., Bilodeau, G., Granger, E.: Automatic image registration in infrared-visible videos using polygon vertices. CoRR, abs/1403.4232 (2014)
2. Chen, C., Ross, A.: Matching thermal to visible face images using hidden factor analysis in a cascaded subspace learning framework. Pattern Recogn. Lett. **72**, 25–32 (2016). Special Issue on ICPR 2014 Awarded Papers
3. Edwards, J.: Telepresence: virtual reality in the real world [special reports]. IEEE Signal Process. Mag. **28**(6), 9–142 (2011)
4. Ellmauthaler, A., da Silva, E.A.B., Pagliari, C.L., Gois, J.N., Neves, S.R.: A novel iterative calibration approach for thermal infrared cameras. In: 2013 20th IEEE International Conference on Image Processing (ICIP), pp. 2182–2186, September 2013
5. Gurusamy, K.S., Aggarwal, R., Palanivelu, L., Davidson, B.R.: Virtual reality training for surgical trainees in laparoscopic surgery. Cochrane Database Syst. Rev. (1) (2009). Article Number CD006575, doi:10.1002/14651858.CD006575.pub2
6. Hilsenstein, V.: Surface reconstruction of water waves using thermographic stereo imaging (2005)
7. Hu, S., Choi, J., Chan, A.L., Schwartz, W.R.: Thermal-to-visible face recognition using partial least squares. J. Opt. Soc. Am. A **32**(3), 431–442 (2015)
8. Jung, H.H., Lyou, J.: Matching of thermal and color images with application to power distribution line fault detection. In: 2015 15th International Conference on Control, Automation and Systems (ICCAS), pp. 1389–1392, October 2015
9. Kadavasal, M.S., Oliver, J.H.: Sensor enhanced virtual reality teleoperation in dynamic environment. In: Virtual Reality Conference, VR 2007, pp. 297–298. IEEE, March 2007
10. Kipper, G., Rampolla, J.: Augmented Reality: An Emerging Technologies Guide to AR, 1st edn. Syngress Publishing, Rockland (2012)
11. Lowe, D.: Distinctive image features from scale-invariant keypoints. Int. J. Comput. Vis. **60**, 91–110 (2003)
12. Morina, N., Ijntema, H., Meyerbröker, K., Emmelkamp, P.M.: Can virtual reality exposure therapy gains be generalized to real-life? a meta-analysis of studies applying behavioral assessments. Behav. Res. Ther. **74**, 18–24 (2015)
13. Orghidan, R., Salvi, J., Gordan, M., Orza, B.: Camera calibration using two or three vanishing points. In: Proceedings of the Federated Conference on Computer Science and Information Systems (FedCSIS), Wrocław, Poland, pp. 9–12 (2012)
14. Rohith, M., Sorensen, S., Rhein, S., Kambhamettu, C.: Shape from stereo and shading by gradient constrained interpolation. In: 2013 20th IEEE International Conference on Image Processing (ICIP), pp. 2232–2236, September 2013
15. Rohith, M.V., Somanath, G., Kambhamettu, C., Geiger, C.A.: Stereo analysis of low textured regions with application towards sea-ice reconstruction. In: Arabnia, H.R., Schaefer, G. (eds.) IPCV, pp. 23–29. CSREA Press (2009)
16. Saponaro, P., Sorensen, S., Rhein, S., Kambhamettu, C.: Improving calibration of thermal stereo cameras using heated calibration board. In: 2015 IEEE International Conference on Image Processing (ICIP), pp. 4718–4722, September 2015
17. Sarfraz, M.S., Stiefelhagen, R.: Deep perceptual mapping for thermal to visible face recognition. CoRR, abs/1507.02879 (2015)
18. Sobel, I., Feldman, G.: A 3×3 Isotropic Gradient Operator for Image Processing (1968). Never published but presented at a talk at the Stanford Artificial Intelligence Project

19. Torabi, A., Massé, G., Bilodeau, G.-A.: An iterative integrated framework for thermal-visible image registration, sensor fusion, and people tracking for video surveillance applications. Comput. Vis. Image Underst. **116**(2), 210–221 (2012)
20. Vidas, S., Lakemond, R., Denman, S., Fookes, C., Sridharan, S., Wark, T.: A mask-based approach for the geometric calibration of thermal-infrared cameras. IEEE Trans. Instrum. Meas. **61**(6), 1625–1635 (2012)
21. O. VR. Oculus best practices (2016, online)
22. Yang, R., Yang, W., Chen, Y., Wu, X.: Geometric calibration of IR camera using trinocular vision. J. Lightwave Technol. **29**(24), 3797–3803 (2011)
23. Yiu-Ming, H., Ng, R.: Acquisition of 3D surface temperature distribution of a car body. In: 2005 IEEE International Conference on Information Acquisition, 5 pages, June 2005
24. Young, A.T.: Distance to the horizon (2012)
25. Zhang, Z.: A flexible new technique for camera calibration. IEEE Trans. Pattern Anal. Mach. Intell. **22**(11), 1330–1334 (2000)
26. Zheng, Y., Peng, S.: A practical roadside camera calibration method based on least squares optimization. IEEE Trans. Intell. Transp. Syst. **15**(2), 831–843 (2014)
27. Zitterbart, D.P., Kindermann, L., Burkhardt, E., Boebel, O.: Automatic round-the-clock detection of whales for mitigation from underwater noise impacts. PLoS ONE **8**(8), e71217 (2013)

Adaptive and Optimal Combination of Local Features for Image Retrieval

Neelanjan Bhowmik[1,2]($\boxtimes$), Valérie Gouet-Brunet[1], Lijun Wei[1], and Gabriel Bloch[2]

[1] University Paris-Est, LASTIG MATIS, IGN, ENSG, 94160 Saint-Mande, France
neelanjan.bhowmik@ign.fr
[2] Nicéphore Cité, 34 Quai Saint-Cosme, 71100 Chalon-sur-Saône, France

Abstract. With the large number of local feature detectors and descriptors in the literature of Content-Based Image Retrieval (CBIR), in this work we propose a solution to predict the optimal combination of features, for improving image retrieval performances, based on the spatial complementarity of interest point detectors. We review several complementarity criteria of detectors and employ them in a regression based prediction model, designed to select the suitable detectors combination for a dataset. The proposal can improve retrieval performance even more by selecting optimal combination for each image (and not only globally for the dataset), as well as being profitable in the optimal fitting of some parameters. The proposal is appraised on three state-of-the-art datasets to validate its effectiveness and stability. The experimental results highlight the importance of spatial complementarity of the features to improve retrieval, and prove the advantage of using this model to optimally adapt detectors combination and some parameters.

Keywords: CBIR · Interest points · Feature combination · Spatial complementarity · Regression model

1 Introduction

We are interested in content-based image similarity search, with the aim of better organizing and mining in the voluminous, complex image datasets. This work focuses on local image descriptors, where the extraction of feature points plays an imperative part in the process. The substantial number of available local feature descriptors in the present literature of Computer Vision and content-based image retrieval (CBIR), with respective advantages and drawbacks, makes it arduous to determine the most relevant descriptors for a given task and a given dataset. Thus it requires a framework making possible to evaluate the effectiveness of given descriptors on a specific dataset. It is also possible to combine descriptors to improve the content representation, such as in [5,17], or to learn the best combination of given descriptors from a representative dataset [6]. All the different descriptors involved may not have the same relevance, and in addition, their distinctiveness may be different from one image to another. We think

© Springer International Publishing AG 2017
L. Amsaleg et al. (Eds.): MMM 2017, Part II, LNCS 10133, pp. 76–88, 2017.
DOI: 10.1007/978-3-319-51814-5_7

that it is important to appraise the complementarity between such local features, and in this work we focus on the complementarity of the detected points in the image, by exploiting statistical criteria of spatial analysis, in order to give the possibility to combine several descriptors, optimally for each image of the dataset and not only globally for the dataset. We propose a regression model with multiple complementarity criteria of the feature detectors (which measure different properties of the detectors), such as distribution [7], contribution [9] and cluster-based measure [15]. Here, these criteria are evaluated for the combination of couples of detectors, but can easily be generalized to sets of detectors. Mean average precision (mAP) is incorporated to train the model and then anticipates the proper detector combination for a new dataset and new query images. Additionally, we demonstrate that this proposal allows to optimally fit some other parameters, here the best k during the k-nearest neighbor search.

The rest of the paper is organized as follows: Sect. 2 revisits the related work existing on the combination of descriptors, Sect. 3 explains the proposed methodology, Sect. 4 is dedicated to the experiments performed to evaluate our proposal, followed by conclusions in Sect. 5.

2 Related Work

Several approaches of descriptors combination are available in the literature of CBIR. They are usually categorized as *early* and *late* fusion approaches [2,23], based on the combination step position in the entire process according to the retrieval/learning step. The most common approach of early fusion is to combine multiple features into a single representation before exploiting it for retrieval/learning [23]. For instance, different shape properties are combined for image retrieval in [21], where genetic programing is used to find the suitable combination function for image descriptors, globally for the dataset. In the weighted early fusion approach proposed by [26], the weight values for different features such as color and texture are varied over the range of (0,1) to find the best appropriate weight values and then these features are combined depending on the assigned weight.

In the late fusion category, multiple features are learned or retrieved first separately, and then the responses or decisions are merged at the later stage [23]. In general for image retrieval, late fusion strategies are carried out in two primary ways such as, consolidate the rank responses and combining the different similarity scores for a query. The final output is obtained by cumulative ranked responses of the feature descriptors. In this context, a retrieval framework, based on genetic programing and relevance feedback, was proposed [8], where multiple sets of retrieved images are consolidated and then the rank list of the most relevant images to the query is returned. Other approaches, such as in short-term and long-term based learning [25], positive and negative feedbacks of the users are considered to construct semantic space and the final output. In the late fusion based image classification proposed by [27], each descriptor classifier output is combined by a weighted voting strategy where the voting weight is decided by the accuracy of the individual classifier.

Sometimes, fusion takes place during the retrieval/learning step, and is often called *intermediate* fusion. For example, in work [5], features fusion is performed during the retrieval step with multiple inverted index. Classification based on multiple kernels involves a learning step based on individual classifier and on the combination of weighted classifier [19].

In some of the previous approaches, the fusion step may not only rely on the fusion strategy but also on the selection of the features. For example, a hybrid method [18] is proposed for simultaneous feature adaptation and feature selection, for a given dataset; here the parameter optimization during feature extraction and feature selection are carried out on a subset of dataset images by employing mixed gravitational search algorithm. In the work of [28], a rank based graph fusion technique is proposed by combining deep learning features, global and local features and the best feature combination is selected globally for a dataset based on the retrieval performance. [24] proposed a method for local selection of image features for similarity search and similarity graph construction, by computing local laplacian score and feature sparsification and considering the importance of the local neighborhood of each image point with respect to the image. Note that in all these approaches, the optimal combination of features is carried out globally for a whole dataset and not locally for each image.

3   Prediction of the Complementarity Between Local Detectors

This section is dedicated to the presentation of our proposal. In Sect. 3.1, we revisit the criteria used to evaluate the complementarity of several point detectors, and Sect. 3.2 describes how they are integrated in the whole prediction framework with a regression model.

3.1   Evaluation Criteria of Complementarity Between Keypoints

Our hypothesis is that better the detections are spread in the image, better the content is described, first because the detections would have more chance to describe the many areas of the image, and second because distant detections should statistically increase the variety of the associated descriptions, making the whole content description more distinctive. Therefore, we exploit several detectors of various natures and measure the spatial complementarity of two detectors, which will be exploited in our prediction model. The presentation is restricted to pairs of detectors, but can easily be generalized to the complementarity of sets of detectors. Let us consider the sets of keypoints extracted from an image by two detectors, $D_a = \{d_a^1(x_a^1, y_a^1), \ldots, d_a^n(x_a^n, y_a^n)\}$, $|D_a| = n$ and $D_b = \{d_b^1(x_b^1, y_b^1), \ldots, d_b^m(x_b^m, y_b^m)\}$, $|D_b| = m$.

Analysis of the Spatial Coverage. One of the key measurement criteria is coverage [7], describing how well the sets of points are distributed over an image.

It is expected to gain a larger distribution if the points from two detectors occupy different locations in the image, which can traduce a better complementarity between the detections, producing a better representation of the content. First, a keypoint, e.g. $d_a^i(x_a^i, y_a^i)$, is considered as a reference point and Euclidean distances (ED_j^i) are calculated with other $(n + m - 1)$ points of $D_a \cup D_b$. If two points detected by the two detectors are the same, there is no effect on the overall distribution. In order to neutralize the effect of the extreme outliers on the overall spatial distribution of $D_a \cup D_b$, the coverage measure is based on harmonic mean. The mean of the distances is computed as:

$$EDMean_{nm}^i = \frac{n + m - 1}{\sum_{j=1, j \neq i}^{n+m-1}(1/ED_j^i)} \tag{1}$$

This step is reiterated for each keypoint of D_a and D_b considering each keypoint as a reference. The distribution complementarity score (DCS) is computed as:

$$DCS = \frac{n + m}{\sum_{i=1}^{n+m}(1/EDMean_{nm}^i)} \tag{2}$$

Higher distribution scores, which are normalized between 0 to 1, indicate the better distribution of the points in the image.

Contribution Measure. The contribution criterion [9] is a measure of the amount of dissimilar points detected by two detectors. It is possible that two detectors extract a certain number of same keypoints (p) for an image. The same detected points reduce the contribution measure of D_b over D_a and vice versa. The contribution of D_b over D_a is computed as:

$$Contribution_{D_b|D_a} = \frac{n - p}{n} \tag{3}$$

The overall complementarity between D_a and D_b is measured by:

$$CnCS = min(Contribution_{D_b|D_a}, Contribution_{D_a|D_b}) \tag{4}$$

If the detected points between the detectors are different, the score is 1; increasing number of common points reduces this score.

Cluster-Based Measurement of Complementarity. Based on spatial clustering, this measure [15] determines how the different detectors extract similar local structures in a cluster. The clusters are generated in the image space from extracted points of D_a and D_b, using a clustering algorithm (e.g. k-means). Each cluster $(c_j, j = 1 \ldots k)$ may contain points from D_a and/or D_b. Points from D_a and D_b in cluster c_j, i.e. resp. F_{jD_a} and F_{jD_b}, contribute to the total number of points (F_j) present in c_j. The frequency of the points from D_a and D_b in c_j is computed as:

$$p_{jD_a} = \frac{|F_{D_a}|}{|F_j|} \quad \& \quad p_{jD_b} = \frac{|F_{D_b}|}{|F_j|} \tag{5}$$

The whole complementarity score can be computed as:

$$ClCS = 1 - 2.\frac{1}{k} \sum_{j=1}^{k} min(p_{jD_a}, p_{jD_b}) \qquad (6)$$

When p_{jD_a} or p_{jD_b} is close to 1 and the other is close to 0, the score is close to 1; it implies a better complementarity of the detectors.

3.2   Image Retrieval Based on Regression Model and Complementarity Measures

To perform image retrieval, we propose to learn a regression model based on the three complementarity criteria revisited in Sect. 3.1, on the number of detected interest keypoints per image (Kp) and on mean Average Precision (mAP) as retrieval output, which is a summarized measure of quality across all the queries by averaging average precision. The objective is to predict the best detector combinations for an image dataset, and also to fit some parameters. Other configurations of criteria were tested, and we obtained the best results with this configuration.

To illustrate, in the Fig. 1 we consider two combinations of two detectors (hesaff-mser and har-colsym where hesaff-mser being generally more efficient on the considered dataset Paris_DB), and for each complementarity score and each query image, we plot the difference of scores between the two combinations vs. the corresponding difference in the mAP. We observe that globally, complementarity scores values increase with mAP values (most of the points are in the area related to positive axes). Therefore, We assume that the relationship between the complementarity criteria and the mAP is general for all image datasets, and we employ a classical linear regression model.

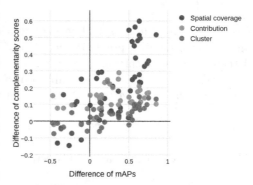

Fig. 1. Relationship between complementarity scores differences and mAP differences, for each query image and two combinations of two detectors.

Training of the Regression Model. The training step is decomposed into three steps involving complementary criteria and mAP:

1. Several detectors, *e.g.* $D_1, \ldots, D_x$, are used to extract keypoints from images, leading to x sets of keypoints. Here, we consider the C_x^2 couples of detectors $(D_i, D_j)_{i \neq j}$ and compute for them the three complementarity scores, for each image of the dataset. We also keep the number of keypoints per image.

2. One mAP is then computed for the images dataset described with a couple of detectors $(D_i, D_j)_{i \neq j}$, using a classical approach of query-by-example retrieval able to use several descriptors jointly, such as in [5]. We obtain C_x^2 mAPs.
3. Finally, the relationship between the different combination of complementarity scores and the retrieval output (mAP) is learned by a linear regression model. Regression coefficients, such as adjusted $R2$, are calculated to analyze the model fitness and to determine the best fitted model for the prediction for the given model inputs and the output.

Prediction of the Best Detector Combination. The prediction steps of the best detector pair for a new dataset are:

1. The detectors $D_1, \ldots, D_x$ extract keypoints on each image of the new dataset. The three complementarity scores of the detector pairs are computed, similarly to step 1 of Sect. 3.2.
2. For each detector combination $(D_i, D_j)_{i \neq j}$, we predict the mAP, called mAP', using previously trained regression model. The complementarity scores of each detector pair are the inputs for the regression model. The outputs mAP' are predicted using the model parameters and the inputs.
3. The detector pair with the highest mAP' is selected as the suitable detector pair for image retrieval on the new dataset.

These training and prediction steps are presented by considering pairs of detectors, but can be generalized to any sets of detectors, based on the generalization of the complementarity criteria. The approach of prediction presented above predicts the best detector combination *globally* for a given dataset. It can be directly employed to predict the best combination *for each query image*, which can be different from an image to another; in the experiment Sect. 4, we will see that the quality of image retrieval can be improved even more by considering such an image-by-image prediction. We will also see that the regression model can be employed to predict some other parameters, such as the k during k-NN retrieval.

4   Experiments and Evaluation

This section presents and discusses the experiments conducted to evaluate our contributions.

4.1   Framework of Evaluation

The experiments are conducted on three image datasets, illustrated in Fig. 2:

1. Paris_DB: this dataset is a public benchmark[1] consisting of 6412 images collected from Flickr by searching for Paris landmarks.

---

[1] http://www.robots.ox.ac.uk/~vgg/data/parisbuildings/.

2. Oxford_DB: this public benchmark[2] consists of 5062 images collected from Flickr by searching for particular 11 Oxford landmarks.
3. Holiday_DB: this dataset is a public benchmark[3] consisting of 1491 images includes a large variety of scene types.

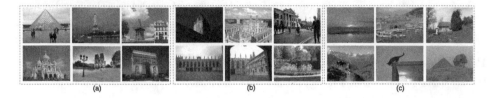

Fig. 2. Samples from the three benchmarks: (a) Paris_DB, (b) Oxford_DB and (c) Holiday_DB.

We have selected 7 detectors from characteristically diverse categories such as blob, corner, symmetry, etc.: Hessian affine (hesaff) [16], color symmetry (colsym) [10], MSER (mser) [14], Harris (har) [22], Star (star) [1], binary robust invariant scalable keypoints (brisk) [11] and oriented and rotated BRIEF (orb) [20]. Extracted keypoints are described by three complementarity local descriptors, *i.e.* SIFT [13], SURF [3] and SC [4], and used jointly in an image retrieval system designed for bag-of-word descriptors combination ('FII') [5]. Performances are presented with mean Average Precision (mAP). Codebook size and value of k during nearest neighbor (k-NN) search are two important parameters of the 'FII' system. Optimal codebook size used for Paris_DB and Oxford_DB is 1500000 words. For the Holiday_DB, 30% of the total description points of each detector combination is selected as codebook size. Parameter k is varied in between 2 to 10 for optimal combination of nearest neighbors and detectors.

4.2   Global Prediction of the Detectors Combination Performance

In this section, we present the prediction results of detector combinations using the linear regression model. It is trained with Paris_DB as described in Sect. 3.2. Model inputs, the complementarity scores, *i.e.* distribution, contribution, cluster and number of keypoints (Kp), of detectors pairs are computed for Paris_DB. The mAP is calculated using the 'FII' approach on Paris_DB. The best fitted model 'Kp-Distribution-Contribution-Cluster - mAP' is selected, based on the highest adjusted $R2$ value, for further prediction experiments on test datasets, *i.e.* Oxford_DB and Holiday_DB. For predictions on the test datasets, procedure of Sect. 3.2 is applied by computing complementarity scores of the detector pairs. The predictions of the detector pairs using previously trained 'Kp-Distribution-Contribution-Cluster - mAP' model are presented in the Table 1, with associated

---

[2] http://www.robots.ox.ac.uk/~vgg/data/oxbuildings/.
[3] https://lear.inrialpes.fr/~jegou/data.php.

mAP'. Due to the space limitation, we only present the selected representative detector pairs prediction results. We observe that detector pairs, 'hesaff-har' and 'hesaff-mser' are associated with the best predicted mAP. Thus, we consider them as the best combinations for image retrieval on these datasets.

Table 1. Different detector combinations and predicted mAP (mAP') using the regression model.

Dataset	Detector pair	mAP'	Detector pair	mAP'	Detector pair	mAP'
Paris_DB	hesaff-colsym	0.512	hesaff-mser	0.548	hesaff-har	0.481
	hesaff-star	0.547	hesaff-orb	0.501	hesaff-brisk	0.520
	colsym-mser	0.429	colsym-har	0.384	colsym-star	0.457
	colsym-orb	0.410	colsym-brisk	0.405	mser-har	0.481
	mser-star	0.526	mser-orb	0.510	mser-brisk	0.507
Oxford_DB	hesaff-colsym	0.501	hesaff-mser	0.537	hesaff-har	0.615
	hesaff-star	0.524	hesaff-orb	0.503	hesaff-brisk	0.523
	colsym-mser	0.482	colsym-har	0.554	colsym-star	0.459
	colsym-orb	0.360	colsym-brisk	0.504	mser-har	0.579
	mser-star	0.492	mser-orb	0.465	mser-brisk	0.527
Holiday_DB	hesaff-colsym	0.442	hesaff-mser	0.461	hesaff-har	0.427
	hesaff-star	0.450	hesaff-orb	0.392	hesaff-brisk	0.441
	colsym-mser	0.402	colsym-har	0.415	colsym-star	0.354
	colsym-orb	0.338	colsym-brisk	0.413	mser-har	0.400
	mser-star	0.395	mser-orb	0.376	mser-brisk	0.415

4.3   Effective Performances for Image Retrieval

In this section, the effective image retrieval results ($mAP^\natural$), for Paris_DB and the test datasets Oxford_DB and Holiday_DB, are presented. Due to space limitation, in Table 2 we only present results associated with the two best predicted pairs and one worst predicted pair using 'FII' retrieval. For Oxford_DB, the best effective result should be obtained with 'hesaff-har' pair (see Table 1). Indeed, the highest effective mAP ($mAP^\natural$) is achieved with this combination (see Table 2). Also, the $mAP^\natural$ of 0.269 is achieved with 'colsym-orb' which is the worst performing pair. For Holiday_DB, although the $mAP^\natural$ are not in the same range as the predicted mAP, the sorted sequence of mAP' reflects the one of effective retrieval results. This first set of experiments confirms us that the complementarity scores, employed with the linear regression model, are able to correctly estimate the performance of a detectors pair for image retrieval, then to enable the use of the best detector pair for a dataset.

We also compare our results (see Table 2, related to 'LF' rows) with one of the state-of-the-art late fusion ('LF') image retrieval technique [17]. We selected the two best performing detector pairs for 'LF' retrieval for each dataset. The comparison results demonstrate that our proposed detector combination selection approach and then 'FII' image retrieval outperforms the 'LF' retrieval. In Table 3, the retrieval results with selected single detector are presented in order to compare with detector pair results of Table 2. These results demonstrate the relevance of the use of several detectors in the representation of the content. In addition to 'FII' image retrieval system, the additional computation of our proposed framework includes computation of different features, complementarity between the features and regression model for prediction.

Table 2. Effective mAP ($mAP^\natural$) of detector pair using 'FII' and 'LF' [17] technique.

	Paris_DB			Oxford_DB			Holiday_DB		
	Detector pair	k-NN	$mAP^\natural$	Detector pair	k-NN	$mAP^\natural$	Detector pair	k-NN	$mAP^\natural$
FII	hesaff-mser	2	<u>0.589</u>	hesaff-har	2	<u>0.549</u>	hesaff-mser	2	<u>0.683</u>
	hesaff-star	2	0.570	mser-har	2	0.456	hesaff-star	2	0.666
	har-colsym	2	0.371	colsym-orb	2	0.269	colsym-orb	2	0.499
LF	hesaff-mser	2	0.541	hesaff-har	2	0.450	hesaff-mser	2	0.630
	hesaff-star	2	0.535	mser-har	2	0.334	hesaff-star	2	0.599

Table 3. $mAP^\natural$ of single detector using 'FII' for the different datasets.

Paris_DB			Oxford_DB			Holiday_DB		
Detector	k-NN	$mAP^\natural$	Detector	k-NN	$mAP^\natural$	Detector	k-NN	$mAP^\natural$
hesaff	2	0.546	hesaff	2	0.498	hesaff	2	0.646
mser	2	0.523	har	2	0.421	mser	2	0.505

4.4   Effect of k-NN Parameter on Retrieval and Its Prediction

In this section, we present retrieval results in Table 4 by varying k during k-NN retrieval of the closest neighbors ($k = 2, 5, 10$) and observe the consequence on $mAP^\natural$. The best $mAP^\natural$ is obtained with $k = 2$ for all datasets. The accuracy difference is 1.8% between $k = 2$ and $k = 5$ for 'hesaff-mser' in Paris_DB, while it is 1.6% between $k = 2$ and $k = 10$ for 'hesaff-har' with Oxford_DB. During the search for the nearest neighbors of the query point, higher values of k might include dissimilar neighbors in the k-NN lists. By using our model, it is possible to adapt the best value of k *for each query image* instead of finding it globally for the dataset. The procedure of Sect. 3.2 is applied for a detector combination, by varying k ($k = 2, 5, 10$) and the prediction mAP obtained allows to adapt k to each query. In Table 4, underlined mAP correspond to $mAP^\natural$ obtained by adapting k to each query. The accuracy is increased by 0.2%, 1.8% and 0.8% for Paris_DB, Oxford_DB and Holiday_DB resp. compared to the previous best with

$k = 2$. Figure 3 shows the distribution of the k selected adaptively across the queries for all datasets. The majority of the best results are associated with $k = 2$ followed by $k = 5$ and $k = 10$. For example with Oxford_DB, approximately 47% of the queries are executed with $k = 2$, 44% with $k = 5$, and only 9% with $k = 10$.

Table 4. $mAP^{\natural}$ for all datasets by varying k-NN and adapting it with prediction model.

Dataset	Detector pair	$mAP^{\natural}$		
		$k = 2$	$k = 5$	$k = 10$
Paris_DB	hesaff-mser	0.589	0.571	0.531
		0.591 (adaptive k 2,5 & 10)		
Oxford_DB	hesaff-har	0.549	0.547	0.533
		0.567 (adaptive k 2,5 & 10)		
Holiday_DB	hesaff-mser	0.683	0.677	0.670
		0.691 (adaptive k 2,5 & 10)		

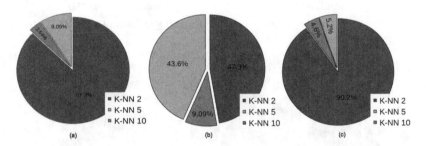

Fig. 3. Distribution of predicted k values across the queries: (a) 'hesaff-mser' for Paris_DB. (b) 'hesaff-har' for Oxford_DB (c) 'hesaff-mser' for Holiday_DB.

4.5 Image-by-Image Prediction of the Best Detector Combination

In this section, we refine the results obtained in Sects. 4.2, 4.3 and 4.4 by adapting the selection of the best detector combination to each image, by applying the prediction strategy of Sect. 3.2 to each query image. Six different combinations of $mAP^{\natural}$ obtained with two best detector pairs and three k-NN value ($k = 2, 5, 10$) are consolidated. In Table 5, we observe that $mAP^{\natural}$ is increased by 0.8%, 2.5% and 21.1% compared to previous best with $k = 2$ for Paris_DB, Oxford_DB and Holiday_DB. The achieved retrieval accuracy for Holiday_DB is 0.894, which is one of the best in the state-of-the-art to our knowledge, compared to Ref. [12] which is also based on bag of words. As depicted in Fig. 4, the majority of the selections are done with $k = 2$. For Paris_DB, 90.9% are selected for $k = 2$

of both pairs of detectors, while 5.49% are with $k = 5$. Most of the selections (85%) are done with 'hesaff-har' for Oxford_DB, while 15% are from 'mser-har'. Even if the statistical analysis (Figs. 3 and 4) has highlighted the dominance of some particular detectors pairs and values of k, we observe that using other ones adaptively allows to refine the results notably.

Table 5. $mAP^\natural$ obtained for all the datasets, by selecting optimal detector pairs and optimal value k for each query image.

Dataset	Detector pair	$mAP^\natural$		
		$k = 2$	$k = 5$	$k = 10$
Paris_DB	hesaff-mser	0.589	0.571	0.531
	hesaff-star	0.570	0.544	0.512
	Adaptive detector combination	0.597 (Adaptive k 2,5 & 10)		
Oxford_DB	hesaff-har	0.549	0.547	0.533
	mser-har	0.456	0.430	0.420
	Adaptive detector combination	0.574 (Adaptive k 2,5 & 10)		
Holiday_DB	hesaff-mser	0.683	0.677	0.670
	hesaff-star	0.666	0.661	0.650
	Adaptive detector combination	0.894 (Adaptive k 2,5 & 10)		

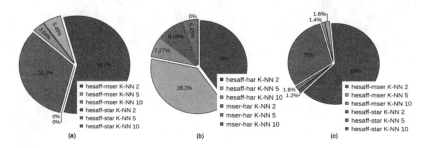

Fig. 4. Distribution of predicted values of k and detectors pairs across the queries: (a) 'hesaff-mser' & 'hesaff-star' for Paris_DB. (b) 'hesaff-har' & 'mser-har' for Oxford_DB (c) 'hesaff-mser' & 'hesaff-star' for Holiday_DB.

5   Conclusions

The main contribution of our approach is the possibility to select adaptively the best detector combination for each query in query-by-example image retrieval. The proposal rests on the use of spatial complementarity criteria between local features and on a linear regression model that models the relationship between complementarity and optimal performances during retrieval. Even if the statistical analysis highlights the dominance of some detectors pairs and values of k, we observe that using other ones adaptively allows to refine the results favorably. The conducted experiments clearly highlight the impact of the spatial complementarity of the selected features on the image retrieval performance: the higher complementarity scores imply a more distinctive representation of the content. The proposed framework can effectively reduce the overall experimental time by narrowing down the choice of detectors, and the adaptive selection of some parameters, such as k during the nearest neighbor retrieval, improves even more the retrieval accuracy. It is easily possible to extend this framework to the evaluation of the complementarity between multiple detectors.

Acknowledgments. The authors are grateful to Nicéphore Cité, Institut national de l'information géographique et forestière (IGN) and French project POEME ANR-12-CORD-0031 for the financial support.

References

1. Agrawal, M., Konolige, K., Blas, M.: Censure: center surround extremas for real-time feature detection and matching. In: Forsyth, D., Torr, P., Zisserman, A. (eds.) ECCV 2008. LNCS, vol. 5305, pp. 102–115. Springer, Heidelberg (2008)
2. Atrey, P.K., Hossain, M.A., Saddik, A.E., Kankanhalli, M.S.: Multimodal fusion for multimedia analysis: a survey. Multimed. Syst. **16**, 345–379 (2010)
3. Bay, H., Ess, A., Tuytelaars, T., Gool, L.V.: Speeded-up robust features (surf). Comput. Vis. Image Underst. **11**(3), 346–359 (2008)
4. Belongie, S., Malik, J., Puzicha, J.: Shape matching and object recognition using shape contexts. Pattern Anal. Mach. Intell. **24**(4), 509–522 (2002)
5. Bhowmik, N., Gonzalez, V.R., Gouet-Brunet, V., Pedrini, H., Bloch, G.: Efficient fusion of multidimensional descriptors for image retrieval. In: International Conference on Image Processing, pp. 5766–5770, October 2014
6. Deselaers, T., Keysers, D., Ney, H.: Features for image retrieval: an experimental comparison. Inf. Retr. **11**(2), 77–107 (2008)
7. Ehsan, S., Clark, A.F., McDonald-Maier, K.D.: Rapid online analysis of local feature detectors and their complementarity. Sensors **13**(8), 10876 (2013)
8. Ferreira, C.D., Santos, J.A., da Silva Torres, R., Goncalves, M.A., Rezende, R.C., Fan, W.: Relevance feedback based on genetic programming for image retrieval. Pattern Recogn. Lett. **32**(1), 27–37 (2011). Image Processing, Computer Vision and Pattern Recognition in Latin America
9. Gales, G., Crouzil, A., Chambon, S.: Complementarity of feature point detectors. In: Richard, P., Braz, J. (eds.) VISAPP(1), pp. 334–339. INSTICC Press, Setubal (2010)
10. Heidemann, G.: Focus-of-attention from local color symmetries. Pattern Anal. Mach. Intell. **26**(7), 817–830 (2004)
11. Leutenegger, S., Chli, M., Siegwart, R.: BRISK: binary robust invariant scalable keypoints. In: International Conference on Computer Vision, pp. 2548–2555, November 2011
12. Li, X., Larson, M., Hanjalic, A.: Pairwise geometric matching for large-scale object retrieval. In: Computer Vision and Pattern Recognition, pp. 5153–5161, June 2015

13. Lowe, D.: Distinctive image features from scale-invariant keypoints. Int. J. Comput. Vis. **60**(2), 91–110 (2004)
14. Matas, J., Chum, O., Urban, M., Pajdla, T.: Robust wide baseline stereo from maximally stable extremal regions. In: Proceedings of the British Machine Vision Conference, pp. 36.1–36.10 (2002)
15. Mikolajczyk, K., Leibe, B., Schiele, B.: Local features for object class recognition. In: International Conference on Computer Vision, vol. 2, pp. 1792–1799, October 2005
16. Mikolajczyk, K., Schmid, C.: Scale and affine invariant interest point detectors. Int. J. Comput. Vis. **60**(1), 63–86 (2004)
17. Neshov, N.: Comparison on late fusion methods of low level features for content based image retrieval. In: Mladenov, V., Koprinkova-Hristova, P., Palm, G., Villa, A.E., Appollini, B., Kasabov, N. (eds.) ICANN 2013. LNCS, vol. 8131, pp. 619–627. Springer, Heidelberg (2013)
18. Rashedi, E., Nezamabadi-pour, H., Saryazdi, S.: A simultaneous feature adaptation and feature selection method for content-based image retrieval systems. Knowl.-Based Syst. **39**, 85–94 (2013)
19. Risojevic, V., Babic, Z.: Fusion of global and local descriptors for remote sensing image classification. IEEE Geosci. Remote Sens. Lett. **10**(4), 836–840 (2013)
20. Rublee, E., Rabaud, V., Konolige, K., Bradski, G.: ORB: an efficient alternative to sift or surf. In: International Conference on Computer Vision, pp. 2564–2571, November 2011
21. da Silva Torres, R., Falcao, A.X., Goncalves, M.A., Papa, J.P., Zhang, B., Fan, W., Fox, E.A.: A genetic programming framework for content-based image retrieval. Pattern Recogn. **42**(2), 283–292 (2009). Learning Semantics from Multimedia Content
22. Schmid, C., Mohr, R.: Local grayvalue invariants for image retrieval. IEEE Trans. Pattern Anal. Mach. Intell. **19**(5), 530–534 (1997)
23. Snoek, C.G.M., Worring, M., Smeulders, A.W.M.: Early versus late fusion in semantic video analysis. In: Proceedings of the 13th Annual ACM International Conference on Multimedia. pp. 399–402. ACM, New York (2005)
24. Sun, J.: Local selection of features for image search and annotation. In: Proceedings of the 22nd ACM International Conference on Multimedia. pp. 655–658. ACM, New York (2014)
25. Wacht, M., Shan, J., Qi, X.: A short-term and long-term learning approach for content-based image retrieval. In: 2006 Proceedings of IEEE International Conference on Acoustics, Speech and Signal Processing, ICASSP 2006, vol. 2, p. II, May 2006
26. Yue, J., Li, Z., Liu, L., Fu, Z.: Content-based image retrieval using color and texture fused features. Math. Comput. Model. **54**(3–4), 1121–1127 (2011). Mathematical and Computer Modeling in Agriculture
27. Zhang, W., Qin, Z., Wan, T.: Image scene categorization using multi-bag-of-features. In: Proceedings of International Conference on Machine Learning and Cybernetics, vol. 4, pp. 1804–1808 (2011)
28. Zhou, Y., Zeng, D., Zhang, S., Tian, Q.: Augmented feature fusion for image retrieval system. In: Proceedings of the 5th ACM on International Conference on Multimedia Retrieval, pp. 447–450. ACM, New York (2015)

An Evaluation of Video Browsing on Tablets with the ThumbBrowser

Marco A. Hudelist and Klaus Schoeffmann[✉]

Klagenfurt University, Universitätsstrasse 65-67, 9020 Klagenfurt, Austria
{marco,ks}@itec.aau.at

Abstract. We present an extension and evaluation of a novel interaction concept for video browsing on tablets. It can be argued that the best user experience for watching video on tablets can be achieved when the device is held in landscape orientation. Most mobile video players ignore this fact and make the interaction unnecessarily hard when the tablet is held with both hands. Naturally, in this hand posture only the thumbs are available for interaction. Our *ThumbBrowser*-interface takes this into account and combines it in its latest iteration with content analysis information as well as two different interaction methods. The interface was already introduced in a basic form in earlier work. In this paper we report on extensions that we applied and show first evaluation results in comparison to standard video players. We are able to show that our video browser is superior in terms of search accuracy and user satisfaction.

Keywords: Video browsing · Mobile computing · Human computer interaction

1 Introduction

Using tablets in landscape orientation when watching videos can be quite cumbersome. User interfaces of mobile video players usually ignore the typical position of the hands at the sides of the device. Although mobile touchscreen devices offer manifold interaction options, video players still look and work very similar to their desktop counterparts, as can be seen in Fig. 5. Holding a device in landscape orientation for longer periods of time is usually more convenient, in particular for large tablets, e.g., 12-inch devices. As a result, interaction with the players' interface becomes unnecessarily hard. Using playback controls that are positioned in the middle of the screen like seeker bars, play/pause/fast forward/fast rewind buttons or similar, can cause strains of the hands and the fingers. This problem is ignored even though there is an increasing amount of video content that is watched with mobile video players. Furthermore, embedded camera systems and mobile content delivery have greatly improved and more video content – for professional and entertainment purposes – is consumed than ever before.

© Springer International Publishing AG 2017
L. Amsaleg et al. (Eds.): MMM 2017, Part II, LNCS 10133, pp. 89–100, 2017.
DOI: 10.1007/978-3-319-51814-5_8

To solve this problem an ergonomic video player interface should be designed that is tailored for the special needs in mobile multimedia interaction. As all eight fingers are occupied when holding the device in landscape orientation, the thumbs become especially important. In contrast to the fingers they are free to move at the left and right sides of the device. Therefore, UI controls need to be positioned at these areas on the screen, so that all functions can be easily reached with the thumbs. The idea of such a UI layout is in fact far from new. On-screen keyboards offer a special mode for such use cases already for a long time. The keyboard is split up into two pieces and placed at the left and right side of the screen. In Fig. 2 examples of default iOS and Windows keyboards can be seen. Nevertheless, this idea is completely ignored by user interfaces of current mobile video players.

Fig. 1. Interface of the ThumbBrowser.

In [9] we already proposed to use the thumbs for mobile video interaction. However, at that time we were only able to present a first, basic version of our ThumbBrowser. As we took part in the demo paper track we also were not able to provide extensive evaluation results and therefore had to leave the readers unclear about a definite benefit. Therefore, in this work we build on this basis and present extensions to our interface as well as results of a comparative user evaluation with a state-of-the-art mobile video player. The evaluation configuration is inspired by the work of Schoeffmann and Burgstaller [16], as they performed a similar experiment with a video player interface tailored for smartphone screens in portrait orientation. They utilize the idea of a scrubbing wheel, as used on Apple iPods to control music, for navigating in videos. Moreover, we are comparing our results to theirs in the analysis of the evaluation results.

Fig. 2. Examples of split keyboards on iOS and windows.

2   Related Work

Hürst and Darzentas [11] propose browsing videos on a tablet with a hierarchical storyboard interface. A videos' content is represented by a grid of thumbnails representing segments that can be tapped to be transferred to another, lower-level grid, representing the content of the specific segment. Hürst et al. [12] also show mobile browsing for timeline-based video browsing with focus on PDA-like devices that are operated with a stylus. For example, users are able to control seek-speed by varying the vertical position of the stylus. This approach is similar to the interaction paradigm later adopted by the default iOS video player. Furthermore, Hürst et al. [13] present a concept for video navigation on smartphones tailored for one-handed, i.e. thumb interaction.

Hudelist et al. [8] utilize the metaphor of a 3D filmstrip for browsing videos on tablet device, similar to the motivation of [1]. The content is represented by a floating filmstrip as it is used in analog film projectors and cameras. Each image on the strip represents a video segment that can be directly played in the strip visualization. Ganhör [5] shows ProPane, an interface for fast and very precise mobile browsing on smartphone-like devices. It enables users to control playback and seeking in a very precise manner, e.g. for video editing scenarios. Huber et al. [7] present Wipe'n'Watch, an interface for browsing interrelated video collections, similarly to the approach of De Rooij et al. [3] for desktop computers. Karrer et al. [15] propose an interface for mobile devices that utilizes direct manipulation of objects in a scene for navigation instead of traditional seeker bars as was shown in earlier work by Dragicevic et al. [4] for desktop PCs.

Moreover, Schoeffmann et al. [17] show a mobile video player that uses wipe gestures for controlling the seeking speed. Hudelist et al. [10] also show a video player for navigation in single videos on tablet devices that utilizes sub-shots and different levels of detail for browsing via keyframes. For this, the interface uses three synchronized filmstrips with which users can easily navigate in the content. A purely human-computational approach is shown by Hürst et al. [14] where users browse through videos by inspecting a large array of uniformly sampled keyframes. Zhang et al. [19] present a mobile interface for collaborative browsing of two users in a single video with the abilities to share sketches. The position in the video is controlled via simple touch gestures. Similarly, Cobarzan et al. [2] proposes a system for collaborative browsing with multiple mobile clients using tablets and a single server that manages communication and query requests of the clients. Finally, a general overview of the field of novel video browsing interfaces is given by Schoeffmann et al. [18].

3   Interface

The extended interface that is presented in this paper is based on our earlier work on the ThumbBrowser (see Hudelist et al. [9]). We extended our original concept with functionality regarding content analysis and user interaction, before we performed a comparative user study to prove its usefulness to users.

The interface tries to avoid interfering with users' watching experience as much as possible. Therefore, in normal playback mode all of the UI controls are hidden. When users want to interact with the player, e.g., to change playback position, they simply have to put one of their thumbs on either side on the screen. Depended on which side is touched different controls become visible, as can be seen in Fig. 1.

On the right hand side a vertical seeker control is activated. It is inspired by the classical layout of a traditional seeker bar but the timeline is positioned vertically. As a result, users can easily navigate to every position inside a video with their right thumb as every part of the timeline is reachable. Furthermore, a preview window appears in relation to users vertical thumb position on the screen. This feature is similar to the magnifying glass functionality shown by latest iterations of video players used by YouTube and similar websites. Therefore, to jump to any position in the video users have to place their right thumb on the screen, drag it to the wanted position on the timeline and then lift it up again. This activates the navigation process and the playback position is changed accordingly. It is possible to avoid this navigational jump by dragging the thumb all way to the right out of the screen, instead of lifting it.

We further extended the original interface concept by automatically analyzing the current video in the background and determining the dominant color in five second steps. This is done by creating a simple color histogram based on the HSB color space. Color values are assigned to one of eight bins. The bins cover value ranges of different size and were defined by results of a preliminary test (Fig. 3).

On the left side of the screen a radial menu can be activated. It offers options to play/pause the video, perform fast forwarding and fast rewinding.

Additionally, we added another timeline visualization mode to the interface, called the *filmstrip mode*.

The *filmstrip mode* provides a scroll-able list of keyframes at the right hand side of the screen, as can be seen in Fig. 4. The keyframes are uniformly sampled from the video in five second steps.

When users tap on one of the keyframes the video player adjusts accordingly. This feature is designed to help users refine their search in case of very long

Fig. 3. Visualization of dominant color directly in the vertical timeline. (Color figure online)

Fig. 4. Vertical seeker control after activation of the filmstrip mode.

videos. For example, they start a rather crude search with the seeker control and after some time notice a promising section of the video. As the seeker control is too sensitive to examine it in detail they are able to continue their browsing process by switching to the filmstrip mode.

4   Evaluation

In order to make the results of this study comparable we designed the evaluation to be very similar to the one used by Schoeffmann and Burgstaller [16]. We even used the same data set (we want to thank the authors for providing the whole data set with ground truth).

Participants of the user study had to search and mark all occurrences of pre-defined objects. Four videos had to be inspected and annotated. The first video was a documentary about gravity and planets in outer space with a duration of 35 min. In this video users had to find all scenes where images of the planet Earth were visible. The second video was an extended report about worldwide multi-cultural societies with a duration of 30 min. In this case users had to find scenes where glasses were shown. The third video was a documentary about cultivating fruits and vegetables with a duration of 25 min where study participants had to find all occurrences of bananas. Finally, video four was a report about *Gamification* with 40 min of length where all scenes with smartphones had to be marked by the participants.

Moreover, participants were told that they could spend as much time as they deemed appropriate to complete a task, but it was recommended not to spend more than up to seven minutes in each case. This was done in order to avoid putting too much stress on users by trying to find really every instance and thus spending unrealistic amounts of time.

To compare the performance of the ThumbBrowser to the performance of a default player we used the media player control of the iOS API, which offers a play/pause button and a seeker bar. Furthermore, two buttons were always available in both interfaces: one button to mark a scene and one button to finish the current task. As testing device we used an iPad Air (first generation) with iOS 9.3.

Fig. 5. Screenshot of the standard video player.

The order of interfaces and videos were alternated between each participant with the exception that two consecutive videos were always tested with the same interface, e.g., video one and two were processed with interface A, followed by video three and four processed with interface B. Furthermore, before a user could start the study, we asked them to provide us with their age, gender and smartphone/tablet experience level, e.g., beginner, advanced or professional. At the beginning of each task the system described the required objects textually. After completion of the first two tasks with the first interface a questionnaire was displayed where users had to give ratings about the interface according to the NASA Task-Load-Index (TLX) [6] and users had to rate them on a Likert-scale. The following questions were asked: (i) how mentally demanding the interaction was, (ii) how physically demanding the interaction was, (iii) how much they had the feeling that the interface supported them in solving the task, (iv) how much fun it was to use the interface, (v) how frustrating it was to use the interface, and (vi) how easy to understand and easy to use the interface was.

4.1 Experimental Results and Statistical Analysis

In total 26 participants took part in the study of which exactly the half were female. Average age of the participants was at 25.5 years. Moreover, eight indicated that they were smartphone/tablet beginners, 13 told us that they were advanced users and five selected that they were very experienced users.

A paired-samples t-test was used to determine whether there was a statistically significant mean difference between the search performance, e.g. how many scenes were found with the ThumbBrowser compared to the standard player. One outlier was detected that was more than 1.5 box-lengths from the edge of the box in a boxplot. Inspection of its value did not reveal it to be extreme and it was kept in the analysis. The difference scores for the search performance of the ThumbBrowser and standard player were normally distributed, as assessed by Shapiro-Wilk's test (p = 0.70). Data are mean ± standard deviation, unless otherwise stated. The percentage of found scenes was higher for the Thumb-Browser (66.19% ± 21.28%) than for the standard player (48.43% ± 18.68%). The test revealed a statistically significant difference between the two interfaces (t(51) = 4.53, p < 0.0005, d = 0.63). Please see Fig. 6 for a visualization of the performance differences.

This result is encouraging as it shows that the ThumbBrowser can in fact improve users performance significantly. Next, we analyzed the differences on how much time participants spent for each task.

A paired-samples t-test was used to determine if there was a statistically significant mean difference between the search times of the ThumbBrowser compared to the standard player. Three outliers were detected that were more than 1.5 box-lengths from the edge of the box in a boxplot. Inspection of their values did reveal that one of them was extreme and therefore was excluded from the analysis. The difference scores for the ThumbBrowsers' search times and the standard players' search times were normally distributed, as assessed by Shapiro-Wilk's test (p = 0.16). Data are mean ± standard deviation, unless otherwise

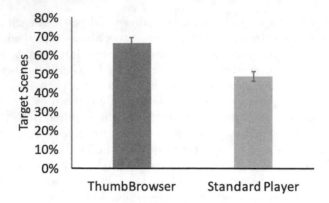

Fig. 6. Amount of retrieved target scenes (error bars: ± s.e. of the mean).

stated. The search times were higher for the ThumbBrowser (417.1s ± 217.4s) than for the standard player (315.7s ± 229.9s). The test revealed a statistically significant difference between the two interfaces ($t(50) = 2.529$, $p < 0.05$, $d = 0.35$). In Fig. 7 the difference in mean search times are visualized.

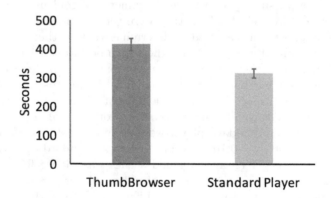

Fig. 7. Task solve time (error bars: 95% confidence interval).

This result is interestingly similar to the results of Schoeffmann and Burgstaller [16] and could indicate that users were more comfortable with the ThumbBrowser than with the standard player. Therefore, they invested more time in the tasks and this could also contribute to the overall better search performance.

To determine if there are statistical significant differences between the answers given in the questionnaires a Wilcoxon signed-rank test was performed. In case of mental demand the ThumbBrowser was significantly less demanding than the standard player ($Z = 2.22$, $p < 0.05$). Moreover, it was physically less demanding to use (Z = 2.373, p < 0.05), users had the feeling that it significantly supported them in their tasks more than the standard player (Z = −4.373, p < 0.005), it was significantly more fun to use (Z = −4.286, p < 0.005) and less frustrating (Z = 4,106, p < 0.005). In terms of usability both interface were equal (Z = 0.776, p = 0.438). In Fig. 8 the differences between the interfaces regarding the questionnaires are visualized.

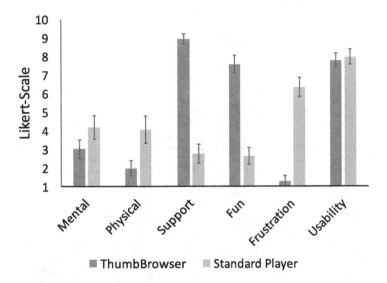

Fig. 8. Workload ratings according to NASA Task-Load-Index [6], with Likert-scale 1–10, for both interfaces (error bars: ± s.e. of the mean). Lower is better for *Mental*, *Physical* and *Frustration*.

When we visually examined the navigation behaviors of the participants it also became clear that the search strategies between the two interfaces were slightly different (see Fig. 9). In case of the ThumbBrowser users were much more likely to perform their search unidirectional, e.g., starting from the beginning and searching to the end of the video. This was not the case for the standard player, as users were more likely to restart their search again and again from the beginning when they reached the end of the video. Therefore, we can agree with the the findings of Schoeffmann and Burgstaller [16] who discovered the same behavior. Moreover, when we further compare their results to ours, we see that although the search performance as well as the average search duration were slightly higher in our study, the general trend is very similar. People spend significantly less time with the standard player and also find significantly less target scenes.

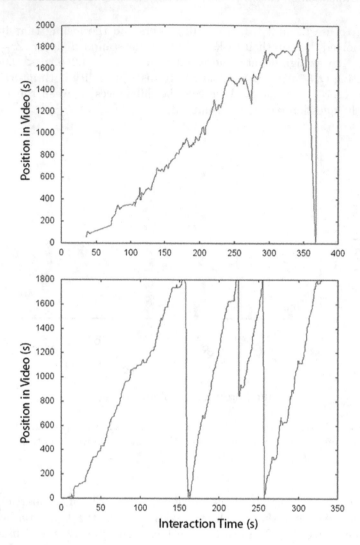

Fig. 9. Sample of the navigation behavior for one user in the documentary video. Users have to find scenes where planet earth is visible when using the ThumbBrowser (top) and the standard player (bottom).

4.2   User Feedback

Some additional comments that were given after completing the user study were that the ThumbBrowser *offered a much smoother seeking interaction* and that they would prefer *a different way to open and close the filmstrip UI.* Furthermore, they told us that *additional fast forwarding and fast rewinding speeds* would have also helped, as in its current iteration the ThumbBrowser only supports a fixed seeking speed of two times the normal playback speed. One participant who was editing a lot of videos professionally told us that she would love to have the interface for her work.

5   Conclusions

In this paper we have presented and evaluated an extended version of the Thumb-Browser - a video browsing and video search tool for tablets that is optimized for landscape operation.

As playback controls are often hard too reach when holding the device with both hands at the sides, using interface controls explicitly designed for the thumbs provides a much better user experience. The interface provides a vertical seeker control similar to a timeline on the right hand side and a radial menu with additional playback functionality on the left hand side of the screen. Moreover, the extended version of the interface uses an easy to understand visualization of dominant colors across the video for faster navigation to scenes with salient color patterns and provides means to switch between coarse and detailed browsing modes.

We tested the interface in a user study with 26 participants where users had to mark scenes containing predefined objects. The results of our study show that the ThumbBrowser could outperform a traditional standard player by letting users find significantly more target scenes. Also, it is significantly less demanding in terms of mental and physical needs, it supports users better in solving the tasks, is is more fun, less frustrating and as easy-to-use as a standard player.

References

1. Beecks, C., Skopal, T., Schöffmann, K., Seidl, T.: Towards large-scale multimedia exploration. In: Proceedings of the 5th International Workshop on Ranking in Databases (DBRank 2011), pp. 31–33 (2011)
2. Cobârzan, C., Hudelist, M.A., Fabro, M.: Content-based video browsing with collaborating mobile clients. In: Gurrin, C., Hopfgartner, F., Hurst, W., Johansen, H., Lee, H., O'Connor, N. (eds.) MMM 2014. LNCS, vol. 8326, pp. 402–406. Springer, Heidelberg (2014). doi:10.1007/978-3-319-04117-9_46
3. de Rooij, O., Snoek, C.G.M., Worring, M.: Mediamill: semantic video search using the rotorbrowser. In: Proceedings of the 6th ACM International Conference on Image and Video Retrieval, CIVR 2007, p. 649. ACM, New York (2007)
4. Dragicevic, P., Ramos, G., Bibliowitcz, J., Nowrouzezahrai, D., Balakrishnan, R., Singh, K.: Video browsing by direct manipulation. In: Proceedings of the SIGCHI Conference on Human Factors in Computing Systems, CHI 2008, pp. 237–246. ACM, New York (2008)
5. Ganhör, R.: Propane: last and precise video browsing on mobile phones. In: Proceedings of the 11th International Conference on Mobile and Ubiquitous Multimedia, MUM 2012, pp. 20:1–20:8. ACM, New York (2012)
6. Hart, S.G., Staveland, L.E.: Development of NASA-TLX (task load index): results of empirical and theoretical research. In: Hancock, P.A., Meshkati, N. (eds.) Human Mental Workload, Volume 52 of Advances in Psychology, pp. 139–183, North-Holland (1988)

7. Huber, J., Steimle, J., Lissermann, R., Olberding, S., Mühlhäuser, M., Wipe'n'watch: spatial interaction techniques for interrelated video collections on mobile devices. In: Proceedings of the 24th BCS Interaction Specialist Group Conference, BCS 2010, pp. 423–427. British Computer Society, Swinton (2010)
8. Hudelist, M.A., Schoeffmann, K., Boeszoermenyi, L.: Mobile video browsing with a 3D filmstrip. In: Proceedings of the 3rd ACM Conference on International Conference on Multimedia Retrieval, ICMR 2013, pp. 299–300. ACM, New York (2013)
9. Hudelist, M.A., Schoeffmann, K., Boeszoermenyi, L.: Mobile video browsing with the thumbbrowser. In: Proceedings of the 21st ACM International Conference on Multimedia, MM 2013, pp. 405–406. ACM, New York (2013)
10. Hudelist, M.A., Schoeffmann, K., Xu, Q.: Improving interactive known-item search in video with the keyframe navigation tree. In: He, X., Luo, S., Tao, D., Xu, C., Yang, J., Hasan, M.A. (eds.) MMM 2015. LNCS, vol. 8935, pp. 306–317. Springer, Heidelberg (2015). doi:10.1007/978-3-319-14445-0_27
11. Hürst, W., Darzentas, D., History: a hierarchical storyboard interface design for video browsing on mobile devices. In: Proceedings of the 11th International Conference on Mobile and Ubiquitous Multimedia, MUM 2012, pp. 17:1–17:4. ACM, New York (2012)
12. Hürst, W., Götz, G., Welte, M.: A new interface for video browsing on PDAS. In: Proceedings of the 9th International Conference on Human Computer Interaction with Mobile Devices and Services, MobileHCI 2007, pp. 367–369. ACM, New York (2007)
13. Hürst, W., Merkle, P.: One-handed mobile video browsing. In: Proceedings of the 1st International Conference on Designing Interactive User Experiences for TV and Video, UXTV 2008, pp. 169–178. ACM, New York (2008)
14. Hürst, W., Werken, R., Hoet, M.: A storyboard-based interface for mobile video browsing. In: He, X., Luo, S., Tao, D., Xu, C., Yang, J., Hasan, M.A. (eds.) MMM 2015. LNCS, vol. 8936, pp. 261–265. Springer, Heidelberg (2015). doi:10.1007/978-3-319-14442-9_25
15. Karrer, T., Wittenhagen, M., Borchers, J., Pocketdragon: a direct manipulation video navigation interface for mobile devices. In: Proceedings of the 11th International Conference on Human-Computer Interaction with Mobile Devices and Services, MobileHCI 2009, pp. 47:1–47:3. ACM, New York (2009)
16. Schoeffmann, K., Burgstaller, L.: Scrubbing wheel: an interaction concept to improve video content navigation on devices with touchscreens. In: IEEE Interenational Symposium on Multimedia (ISM), pp. 351–356, December 2015
17. Schoeffmann, K., Chromik, K., Boeszoermenyi, L.: Video navigation on tablets with multi-touch gestures. In: IEEE International Conference on Multimedia and Expo Workshops (ICMEW), pp. 1–6, July 2014
18. Schoeffmann, K., Hudelist, M.A., Huber, J.: Video interaction tools: a survey of recent work. ACM Comput. Surv. 48(1), 14:1–14:34 (2015)
19. Zhang, J.-K., Ma, C.-X., Liu, Y.-J., Fu, Q.-F., Fu, X.-L.: Collaborative interaction for videos on mobile devices based on sketch gestures. J. Comput. Sci. Technol. 28(5), 810–817 (2013)

Binaural Sound Source Distance Reproduction Based on Distance Variation Function and Artificial Reverberation

Jiawang Xu[1,2], Xiaochen Wang[2,3(✉)], Maosheng Zhang[2], Cheng Yang[2,4], and Ge Gao[2]

[1] State Key Laboratory of Software Engineering, Wuhan University, Wuhan, China
carrinton@163.com
[2] National Engineering Research Center for Multimedia Software, Computer School of Wuhan University, Wuhan, China
clowang@163.com, eterou@163.com, yangcheng41506@126.com
[3] Hubei Provincial Key Laboratory of Multimedia and Network Communication Engineering, Wuhan University, Wuhan, China
[4] School of Physics and Electronic Science, Guizhou Normal University, Guiyang, China

Abstract. In this paper, a method combining the distance variation function (DVF) and image source method (ISM) is presented to generate binaural 3D audio with accurate feeling of distance. The DVF is introduced to indicate the change in intensity and inter-aural difference when the distance between listener and source changes. Then an artificial reverberation simulated by ISM is added. The reverberation introduces the energy ratio of direct-to-reverberant, which provides an absolute cue to distance perception. The distance perception test results indicate improvement for distance perception when sound sources located within 50 cm. In addition, the variance of perceptual distance was much smaller than that using DVF only. The reduction of variance is a proof that the method proposed in this paper can generate 3D audio with more accurate and steadier feeling of distance.

Keywords: Distance perception · Distance variation function · Image source method

1 Introduction

With the rapid development of the virtual reality technique, the binaural 3D audio is now a heated topic due to its convenience and feeling of immersion. To generate a binaural 3D audio, Jens Blauert from Uni Bochum proposed the Head-related transfer function (HRTF) to simulate the propagation of sound from the sound source to the listener's ear drum in free filed [1]. Many organizations conducted measurements to get the HRTF: Algazi et al. released the CIPIC database measured in a semi-anechoic room in 2001, which included head-related

© Springer International Publishing AG 2017
L. Amsaleg et al. (Eds.): MMM 2017, Part II, LNCS 10133, pp. 101–111, 2017.
DOI: 10.1007/978-3-319-51814-5_9

impulse responses (HRIRs) for 45 subjects at 25 different azimuths and 50 different elevations [2]. While the HRIRs in the CIPIC database were measured under the condition that sound sources were fixed at the distance of 1 m from the receiver a semi-anechoic roomthe CIPIC database is not able to generate binaural 3D audio with different distances.

In 2009, the PKU-IOA database was published by Tianshu Qu et al. They conducted the measurements on a kemar mannequin to get the HRTFs of 6344 positions at different distances. The PKU-IOA database is the first database that contained HRTFs of different distances [3]. However, to generate binaural 3D audio with more accurate feeling of distance, the gap between measurements points at different distance should be small enough. To solve this problem, an increased number of HRIR measurements points are needed to cover the variation in the HRIRs at different distances which prolongs the measurement process and can be uncomfortable for human subjects. Due to these technical difficulties, it has so far been difficult to obtain HRTFs measured on human subjects to accurately synthesize near-field sound sources in virtual auditory displays [4].

To avoid the difficulties, Alan Kan proposed a method based on a distance variation function (DVF) to synthesize near-field HRTF from far-field HRTF [4]. To distinguish the region where the HRTFs show substantial variation with distance from the region where HRTFs are relatively constant, the term near-field and far-field will be used in our paper to refer to the two regions respectively. And a distance of 1 m is considered as the boundary between the two regions [5].

The DVF is an analytic function describing the change in the transfer function when the location of a sound source changes from far-field to near-field. It is calculated from a rigid sphere model and approximates the change in the frequency-dependent inter-aural level difference (ILD) cues as a function of the change in sound source distance. The distance perception improves for sounds at simulated distances of up to 60 cm using the DVF while compared to simple intensity adjustment. Although there is an improvement in distance perception in the DVF-generated stimulus condition compared to the intensity adjustment, there is still a high degree of overestimation of the distance and variation for different listeners.

While according to [6], reverberation is another significant factor which is not taken into account in the DVF method. Therefore, a method combined DVF with artificial reverberation is proposed in this paper. Firstly, the DVF is used to generate HRTFs at different distances, the original audio is convoluted with synthesized HRTFs using DVF at different distances. After the convolution, an artificial reverberation is calculated by image-source method (ISM). The ISM can help to simulate the room impulse response (RIR) that simulates the sound propagation corresponds to the reverberant settings and to the positions of listeners and sources. Finally, the reverberation is added to the convoluted audio.

Distance perception experiments were conducted to compare the perceptual fidelity of binaural 3D audio generated using the DVF-ISM method. The same experiment was also conducted with binaural 3D audio synthesized by simply using the DVF method. The results of the experiments are presented and compared in the section of distance perception experiment.

2   Distance Reproduction

The DVF-ISM method can be briefly described in the Fig. 1:

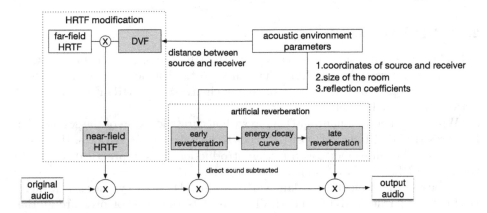

Fig. 1. The process of DVF-ISM is shown in this figure. The HRTF modification part uses DVF to synthesize near-field HRTF from far-field HRTF. The artificial reverberation part applies ISM to generate a RIR that corresponds to the reverberant settings and positions of listeners and sources.

In HRTF modification part, the distance and direction between sound source and receiver are firstly calculated by their coordinates in a given rectangular room. Then a measured far-field HRTF of fixed distance, corresponding to the direction between the source and receiver is chosen from the HRTF database. After that, the DVF is computed to describe the change in the transfer function of a sound source from the far-field distance to the distance calculated between source and receiver. Finally, the modified HRTF which is the near-field HRTF here is synthesized by combining far-field HRTF with the computed DVF.

The artificial reverberation module takes advantage of ISM to generate a RIR of a given rectangular room with direct sound subtracted. The RIR is often decomposed into two parts, namely early reflection (direct path together with a series discrete early reflections) and late (diffuse) reflection in acoustic and engineering literature [7]. As the HRTF introduced, the direct-path signal in the early reflection should be subtracted.

The artificial reverberation module here is decomposed into the early reflection module and late reflection module. The early reverberation module obtains the RIR with direct-path signal subtracted. Then a cut-off pointmarked as t_c and an energy decay curve (EDC) by Eric A. Lehmann are calculated for the late reverberation [8]. t_c is the specific transition time between the early reflection and the late reflection. The EDC is to compute an approximate envelope for the synthesis of the late reflection. Then, a feedback delay net (FDN) is applied to implement the late reverberation that corresponding to the EDC.

To generate binaural 3D audio with reverberation, the original audio is first convoluted with the HRTF synthesized by the DVF, then convoluted with the early part of the RIR obtained by the ISM. Finally, the audio is filtered by the FDN that implements the late reverberation. The specific details are discussed in following section:

2.1   HRTF Modification

A HRTF or its time-domain representation the HRIR describes the propagation that the sound wave travels from the source to the listener's ear drum in free field. HRTF varies as the position of source changes. Also, HRTF differs badly between different listeners.

When to generate binaural 3D audio with accurate location of the source, the listener's HRTF at the right position needs to be measured in advance. However, due to the difficulties mentioned in the introduction, it is impractical to get all the needed HRTFs of the listener.

To avoid the difficulties, Alan Kan proposed DVF to synthesis near-field HRTF from far-field HRTF. The DVF represents the ratio of pressure at the surface of a rigid sphere which simulates the human head due to a far-field sound source to the pressure on the rigid sphere due to a near-field sound source. The pressure at a point X on the surface of a rigid sphere of radius a due to a sinusoidal source of sound of angular frequency ω at distance r away from the centre of the sphere, is derived by:

$$p\left(a, \omega, \theta, r\right) = -kr \sum_{m=0}^{\infty} \left(2m + 1\right) \frac{h_m(kr)}{h_m'(ka)} P_m\left(cos\theta\right) e^{-ikr} \tag{1}$$

$h_m\left(kr\right)$ is the spherical Hankel function of the first kind of order m and $h_m'\left(ka\right)$ is its first derivative at radius a, ω/c is the wave number, c is the speed of sound in air, $P_m\left(\lambda\right)$ is the Legendre polynomial of degree m, and θ is the angle between a vector from the centre of the sphere to the point X on the surface of the sphere and a vector from the centre of the sphere to the sound source.

To synthesize a near-field HRTF, the DVF for a sound source at a near-field distance d_n is calculated as the ratio of the pressure arising from a sound source at distance d_n to the pressure arising from a sound source at a distance d_f, where d_f is the distance at which the far-field HRTF was measured:

$$DVF = \frac{p_n\left(a, \omega, \theta, d_n\right)}{p_f\left(a, \omega, \theta, d_f\right)} \tag{2}$$

A separate DVF is calculated for each ear. To synthesize a near-field HRTF from a far-field HRTF, we calculate:

$$HRTF(d_n) = DVF \times HRTF(d_f) \tag{3}$$

To synthesize sounds at multiple locations and distances, a DVF needs to be calculated for every sound source distance and direction and for each ear.

The DVF method presented here can synthesize near-field HRTFs from far-field HRTFs. The DVF approximates the ILD cues of the HRTF as a function of sound source distance. Using this method, the difficulties when measuring near-field HRTFs on human subjects are avoided.

2.2   Artificial Reverberation

The artificial reverberation technique is based on the widely-used model that the RIR can be decomposed into early reverberation and late reverberation. In this part, two modules are included. One is the early reverberation module which takes advantage of ISM to generate the room impulse response between any source and receiver in a given rectangular room. After the RIR is obtained, a cut-off point, marked as t_c, has to be calculated for the decomposition of the early reverberation and late reverberation. And an energy decay curve (EDC) is also calculated for the computation of an approximate envelope for the synthesis of the later reverberation. Then, a feedback delay network (FDN) is applied to implement the late reverberation that meets the requirements of the EDC.

Early Reverberation. In this module, ISM is applied to get a simulation of the room impulse response. The original ISM was first proposed by Allen and Berkley, then soon became a popular technique for simulation of sound propagation in reverberant settings. The image-source method is briefly described as follows. A model of a rectangular room with dimensions $L_x \times L_y \times L_z$ contains a sound source and a receiver, respectively located at $P_s = (x_s, y_s, z_s)^T$ and $P_r = (x_r, y_r, z_r)^T$. The reflection coefficient $\beta = (\beta_{x_1}, \beta_{x_2}, \beta_{y_1}, \beta_{y_2}, \beta_{z_1}, \beta_{z_2})^T$ for each of six enclosure surfaces is used to characterize the acoustic properties of the rectangular room. To get a simulation of the RIR between the source and receiver, the method assumes that the sound waves reflect specularly on the enclosure boundaries. The $RIR(t)$, is computed by taking account of a series of mirrored sources extending in 3 dimensions, and then by summing every image sources contribution at the receiver:

$$r(t) = \sum_{\mu=0}^{1} \sum_{\nu=-\infty}^{\infty} A(\mu,\nu) \cdot \delta(t - \tau(\mu,\nu)) \tag{4}$$

$\mu = (\mu_x, \mu_y, \mu_z)$ and $\nu = (\nu_x, \nu_y, \nu_z)$ are triplet parameters controlling the indexing of the mirrored sources in three dimensions. $A(\cdot)$ is the amplitude factor and $\tau(\cdot)$ is the time delay of the considered mirrored source respectively. In (4) the sum over μ (respectively, ν) is used to represent a triple sum over each of the triples internal indices.

To be more specific the Eq. (4) can be written as follow:

$$r(t) = \sum_{\mu=0}^{1} \sum_{\nu=-\infty}^{\infty} \beta_{x_1}^{|n-q|} \beta_{x_2}^{|n|} \beta_{y_1}^{|l-j|} \beta_{y_2}^{|l|} \beta_{z_1}^{|m-k|} \beta_{z_2}^{|m|} \times \frac{\delta(t - |R_\mu + R_\nu|/c)}{4\pi(|R_\mu + R_\nu|)} \tag{5}$$

R_μ here expresses in terms of the integer three-dimensional vector $\mu = (q, j, k)$ as: $R_\mu = (x_s - x_r + 2qx_r, y_s - y_r + 2jy_r, z_s - z_r + 2kz_r)$ and R_ν here expresses in terms of another integer three-dimensional vector $\nu = (n, l, m)$ as: $R_\nu = 2(nL_x, lL_y, mL_z)$. The βs are the reflection coefficients of the six boundaries of the enclosure, with subscript 1 referring to the boundaries adjacent to the origin of the coordinate, and 2 referring to the opposing one. In Eq. (5), the sum $\sum$ with index μ is to indicate three sum, namely one for each of the three components of $\mu = (q, j, k)$, which is similar to $\nu = (n, l, m)$'s. These sums are over a three-dimensional lattice of points. For μ, there are eight, and for ν, the lattice is infinite. After the $r(t)$ obtained, we subtract the direct sound here by setting the first non-zero value of the RIR to zero to get the RIR with direct signal subtracted.

Late Reverberation. After the room impulse response is computed, t_c, defined as the time for which the overall energy of the signal in the RIR decreased by a certain amount in Δ_c dB:

$$t_c \triangleq E^{-1}(\Delta_c) \tag{6}$$

Once the early reflections, i.e., for $t \leq t_c$, has been computed using the ISM technique, the late reverberation part of the RIR can be simulated by generating a noise signal whose energy decay is determined by the EDC for the given rectangular room. The energy decay curve can then be computed by normalized Schroeder integration method:

$$E(t) = 10 \cdot \log_{10} \left(\frac{\int_t^\infty r^2(\xi)\, d\xi}{\int_0^\infty r^2(\xi)\, d\xi} \right) \tag{7}$$

The EDC is then applied for the computation of an approximate envelope for the synthesis of the later reverberation, and is also limited to the case of small-room acoustics in rectangular enclosures with specular reflections. The approach used for EDC prediction relies on the following observation: the acoustic power $h(t)$ received at the receiver at a given time t corresponds to the addition of the contributions from all the image sources located on a sphere of radius $\rho = c \cdot t$ around the receiver (c denotes the propagation speed of acoustic waves).

After t_c the and EDC is obtained, an FDN is applied to implement the late reverberation. The FDN is built using N delay lines, each having a length in seconds given by $\tau_i = mT$, where $T = 1/f_s$ is the sampling period, and a feedback matrix which is the Householder matrix we used in this paper. The transmission function $H_i(\cdot)$ of the filters in the FDN can be derived by the follow equation:

$$120 \log_{10} \left| H_i\left(e^{j\omega}\right) \right| = -60 \frac{T}{T_r(\omega)} \left(m_i - \frac{\arg\left[e^{j\omega}\right]}{\omega} \right) \tag{8}$$

while the phase delay is usually negligible when compared to m_i, the equation above is then rewritten as:

$$20 \log_{10} \left| H_i\left(e^{j\omega}\right) \right| = -60 \frac{\tau_i}{T_r(\omega)} \tag{9}$$

The $T_r(\omega)$ here describes the time that the energy attenuates r dB compared to the direct signal. $T_r(\omega)$ can be derived from the EDC by equation:

$$T_r(\omega) = E^{-1}(-r, \omega) \tag{10}$$

while $E^{-1}(-r, \omega)$ is the inverse function of $E(t)$.

3 Distance Perception Experiment

A distance perceptual experiment is conducted to compare the perceptual fidelity of binaural 3D audio generated by the DVF-ISM method to that using near-field HRTFs synthesized by DVF.

In the experiment, a piece of music signal of the MPEG's standard test sequences is taken as the original signal and it is shown in the Fig. 2. For convenience, the HRTFs of subject-165 in CIPIC database is taken as the original HRTF to be modified [9].

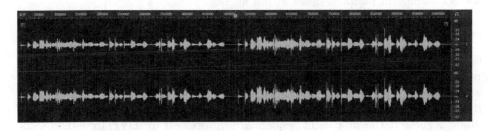

Fig. 2. The figure on the left of the red line shows the wave form of the original signal, and the one on the right is a example of DVF-ISM processed audio when sound source is located 30 cm away from listeners. (Color figure online)

8 listeners aged from 20 to 35 took part in the experiment. All the listeners were first trained in a $6 \times 4 \times 2.8\,\mathrm{m}^3$ room whose reflection coefficients of six surfaces can be considered equal to 0.9. A speaker is used to represent the original signal in different positions to train the listeners. The source positions are shown in the Fig. 3:

First, the training starts from the farthest position in front of the listeners.

Then, the source moves towards the listener step by step. Between each step, there is a 5 s break. After all the positions in the front of the listener trained, a small test is conducted to verify if the listener is able to recognize the target distance between him or herself and the sound source. If the listener can make correct indication of the target distance 3 times in a row. We carry on the training on the two sides and the back of listener. If the listener fails in the test, the training need to be repeated, until the listener succeeds. The training procedure on the sides and the back of the listener is the same as the one in front of the listener.

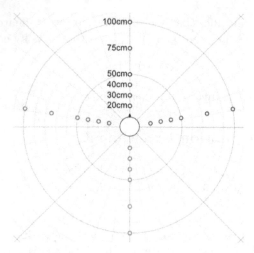

Fig. 3. The listeners are seated in the center of the room. Before they begin the distance perception test, they all have to be trained to distinguish the locations of different sound sources at different positions shown in this figure.

After qualified to recognize the distance, the listener begins listening to 16 sequences of processed audio with in-ear tube phones to compare the distance perception fidelity. The 16 sequences is a mix of 8 DVF-ISM processed audio at 4 different target distances and 8 DVF processed audio at the same 4 target distances. The order of the sequences is randomized that the listeners are not able to guess if the first eight sequences is processed by DVF-ISM. Listeners have to listen to each sequence 3 times, and then give their perceptual distance for each time. The perceptual distance given is asked to be chosen from the distances in Fig. 3.

After given the perceptual distance, the listeners have to do an extra cmos test to compare DVF-ISM processed audio and DVF processed audio to training signal, respectively. The rule of score is shown in Table 1:

Table 1. Rule of score

Score	Feeling of distance
2	Far further
1	Further
0	Same
−1	Closer
−2	Very closer

We summarized the perception result on different directions. As the paper focus on the distance reconstruction, the discussion below is limited to the distance perception performance. In Fig. 4, the distance perception results show that the mean perceptual distance using the DVF-ISM is closer to the real source distance than that using DVF when the target distance is in 75 cm. When target distance is 75 cm or 100 cm away from the listener, the distance perception result of the DVF-ISM is almost the same as the one of the DVF. Besides, the range of perceptual distance using DVF-ISM is obviously smaller than that using DVF, which means the DVF-ISM may produce 3D audio with steadier feeling of distance.

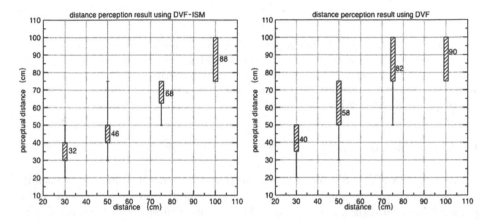

Fig. 4. Here is the distance perception result. Numbers around the boxes is the mean perceptual distance when the target distance is 30 cm, 50 cm, 75 cm or 100 cm. The height of each box means the range of perceptual distance. Lines reach out of the boxes is the outliers in the tests.

The cmos test also indicates that the DVF-ISM method can generate 3D audio with more accurate and steadier feeling of distance (Fig. 5).

The DVF-ISM maintains the intensity and ILD that DVF utilizes to indicate the change in the propagation when sound source moves. In addition, the reverberation added using the DVF-ISM introduces an energy ratio of direct-to-reverberant (D/R) sound which provides an absolute cue to distance perception according to [10]. When the source is near around the listener, the change in distance between source and listener can affect the D/R obviously, which explain why the mean perceptual distance is closer to the target when using DVF-ISM. When the source is not near enough, the change in distance between source and listener may affect the D/R so slightly that the listeners are not able to tell if the D/R is changed. About the reduction of variance of perceptual distance, we now have not get a clear explanation. As we human subjects are always doing distance perception practice unconsciously in rooms, halls, movie theatre or somewhere else that reverberation always exists, we guessed that the artificial reverberation might make the listeners more familiar with reverberant audio. Further study will be made in our following work about the guess. While, there is still a variation for different listeners. The variation for different listeners may blame for non-individual HRTF here.

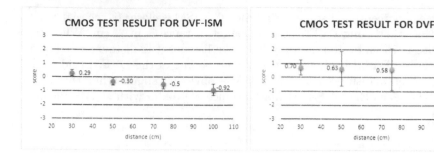

Fig. 5. Here is the comparison between the distance perception performances of DVF-ISM and DVF. The mean cmos scores of DVF-ISM is closer to 0 than the scores of DVF when the distance is 30 cm, 50 cm or 75 cm. When distance is 100 cm, the mean cmos scores is almost the same. Besides that, the variation of DVF-ISM is smaller than that of DVF no matter what the distance is.

4   Conclusion

In this paper, a DVF-ISM method was proposed to generate binaural 3D audio with more accurate feeling of distance. The method combined the DVF with artificial reverberation simulated by ISM to control intensity, interaural differences and reverberation at same time. We applied DVF to indicate the change in intensity and ILD when the position of sound source changes. Also, we applied ISM to simulate the reflection and absorption of sound wave in the reverberant settings to add an D/R factor to help the distance perception. The result of distance perception experiments shows that DVF-ISM can use to synthesize audio with better accuracy and stability of feeling of distance. Further research will be done to reduce the complexity of the DVF-ISM method in our following work.

Acknowledgments. This work is supported by National High Technology Research and Development Program of China (863 Program) No. 2015AA016306; National Nature Science Foundation of China (No. 61231015, 61471271, 61662010).

References

1. Blauert, J.: Spatial Hearing, pp. 372–392 (1997)
2. Algazi, V.R., et al.: The CIPIC HRTF database. In: 2001 IEEE Workshop on the Applications of Signal Processing to Audio and Acoustics. IEEE (2001)
3. Qu, T., et al.: Distance-dependent head-related transfer functions measured with high spatial resolution using a spark gap. IEEE Trans. Audio Speech Lang. Process. **17**(6), 1124–1132 (2009)
4. Kan, A., Jin, C., van Schaik, A.: A psychophysical evaluation of near-field head-related transfer functions synthesized using a distance variation function. J. Acoust. Soc. Am. **125**(4), 2233–2242 (2009)
5. Brungart, D.S., Rabinowitz, W.M.: Auditory localization of nearby sources. Head-related transfer functions. J. Acoust. Soc. Am. **106**(3), 1465–1479 (1999)

6. Berry, J.S., Roberts, D.A., Holliman, N.S.: 3D sound and 3D image interactions: a review of audio-visual depth perception. In: IS&T/SPIE Electronic Imaging. International Society for Optics and Photonics (2014)
7. Habets, E.A.P., Gannot, S., Cohen, I., et al.: Joint dereverberation and residual echo suppression of speech signals in noisy environments. IEEE Trans. Audio Speech Lang. Process. **16**(8), 1433–1451 (2008)
8. Bronkhorst, A.W., Houtgast, T.: Auditory distance perception in rooms. Nature **397**(6719), 517–520 (1999)
9. CIPIC HRTF Database Files, Realase 1.2, 23 September 2004. http://interface.cipic.ucdavis.edu/CIL_html/CIL_HRTF_database.htm
10. Zahorik, P., Anderson, P.W.: The role of amplitude modulation in auditory distance perception. Proc. Meetings Acoust. **21**(1), 050006 (2015). Acoustical Society of America

Color-Introduced Frame-to-Model Registration for 3D Reconstruction

Fei Li$^{(\boxtimes)}$, Yunfan Du, and Rujie Liu

Fujitsu Research and Development Center Co., Ltd., Beijing, China
{lifei,duyunfan,rjliu}@cn.fujitsu.com

Abstract. 3D reconstruction has become an active research topic with the popularity of consumer-grade RGB-D cameras, and registration for model alignment is one of the most important steps. Most typical systems adopt depth-based geometry matching, while the captured color images are totally discarded. Some recent methods further introduce photometric cue for better results, but only frame-to-frame matching is used. In this paper, a novel registration approach is proposed. According to both geometric and photometric consistency, depth and color information are involved in a unified optimization framework. With the available depth maps and color images, a global model with colored surface vertices is maintained. The incoming RGB-D frames are aligned based on frame-to-model matching for more effective camera pose estimation. Both quantitative and qualitative experimental results demonstrate that better reconstruction performance can be obtained by our proposal.

Keywords: 3D reconstruction · Color mapping · Registration · Frame-to-model matching · Optimization

1 Introduction

As one of the most important tasks in computer vision and graphics, high-quality digitization of real-world objects has always been a hot research theme. Generally speaking, digital 3D objects cannot be directly acquired by some equipment, and they are often reconstructed based on the captured raw data, such as depth maps and color images. In the last decades, several 3D reconstruction methods have been explored. With the development of sensing technology, consumer-grade RGB-D cameras have appeared recently. Since they are often with low cost, easy portability and high streaming rate, these cameras have been widely used in various applications, such as augmented reality, computer games, and virtual shopping. Although with many advantages, consumer-grade RGB-D cameras are often with inevitable distortions, and the obtained data are usually not accurate enough, especially for the captured depth maps [1]. Therefore, how to obtain satisfactory 3D reconstruction performance with inaccurate capture devices has been paid more and more attention in recent years [2–12].

3D reconstruction with RGB-D cameras mainly resorts to the obtained depth information, and there are two main steps in the whole process: registration and

© Springer International Publishing AG 2017
L. Amsaleg et al. (Eds.): MMM 2017, Part II, LNCS 10133, pp. 112–123, 2017.
DOI: 10.1007/978-3-319-51814-5_10

integration. The procedures can be simply described as follows. Since multiple depth maps are captured from different positions and directions, they are first aligned into the same coordinate space based on the estimated camera poses. Then the registered models are combined together to get the final reconstruction results.

The accuracy of model alignment has a major influence on the final reconstruction performance, hence most research work mainly focuses on the approach of registration. The most direct idea is to estimate the camera pose of each depth map as accurate as possible, and two frequently adopted ways are frame-to-frame matching [13] and frame-to-model matching [2,3]. Frame-to-frame matching only considers two consecutive frames at a time, and estimates the camera pose of each new depth map by aligning it to its last frame. While frame-to-model matching maintains a model constructed by all the frames coming before, and registers the incoming depth map to the growing model. Since the existing frames are made full use of, frame-to-model matching is often more effective and more robust than frame-to-frame matching [3]. No matter which kind of matching way is adopted, most registration approaches pay more attention to real-time 3D reconstruction, and only consider the frames before the incoming depth map. When the real-time requirement is not necessary, more useful information can be involved for more accurate results. Two-pass registration [4] is such an example. In the first pass, all the available frames are used to construct a whole model; and in the second pass, each frame is revisited again and aligned to the acquired model. As the frames after the incoming depth map are also taken into account, more stable camera pose estimation results can be obtained.

Besides kinds of approaches for camera pose estimation, camera distortions are also addressed in some recent research. Two categories of methods are often adopted. The first category of methods are mainly based on calibration, and try to estimate a specific distortion function with a pre-defined form for the given camera [14,15]. Since specialized calibration sequences are required, the applicable cases of these methods are largely confined. Moreover, the real distortion function is usually irregular and complicated, so the assumption about its form may be not exact. The second category of methods do not explicitly estimate camera distortions, but attempt to correct them by introducing non-rigid deformation, and the most suitable deformation parameters are often calculated by optimization. Elastic registration [5] finds the most appropriate mapping for each 3D point. Because of the lack of prior knowledge, it always leads to unnecessary warping in the final reconstruction results. Moreover, it is quite time-consuming. By factorizing the non-rigid deformation into a rigid localization component and a latent non-rigid calibration component, the method of SLAC [7] effectively conducts localization and calibration at the same time. As both of the two aspects are elaborated considered in one joint optimization framework, it achieves better reconstruction performance. Meanwhile, since the number of parameters to be optimized is much smaller, its overall computational cost is dramatically reduced.

Most representative systems only adopt depth maps for 3D reconstruction, and the methods of geometric alignment have been largely explored. However, as only distance information is recorded, the captured depth map has a certain limitations. Apart from its inaccuracy and missing data, depth-based reconstruction is prone to drift and failure in the presence of smooth surfaces. Therefore, some methods attempt to introduce color images and utilize photometric cue for performance improvement [16–20]. Considering real-time requirements, complicated image processing operations cannot be conducted, and pixel-level dense matching is often adopted. In the existing methods, only frame-to-frame matching is used for photometric consistency based registration. That is to say, for color-involved camera pose estimation, only the incoming image and the previous one are considered. Like the case for depth-based registration, it is hoped that frame-to-model matching will also outperform frame-to-frame matching when color information is taken into account.

In this paper, with the basic idea to utilize color images as well as depth maps for better reconstruction performance, a novel registration approach is proposed. The whole system is with the similar workflow to KinectFusion [2,3]. However, its maintained global model contains not only surface vertices and their normals, but also the corresponding color values. When a new pair of depth map and color image come, the idea of frame-to-model matching is adopted, and both geometric and photometric consistency are considered to develop a unified optimization problem. Since the relationship between the new captured data and the existing RGB-D frames is fully explored, more accurate camera pose estimation results are obtained. Moreover, by introducing non-rigid correction functions, our proposal can be easily extended to involve camera distortions in the same framework.

The rest of the paper is organized as follows. Section 2 describes our proposed color-introduced frame-to-model registration approach in detail. Our experimental results are illustrated in Sect. 3. Some conclusions and analysis of future work follow in Sect. 4.

2   Color-Introduced Frame-to-Model Registration

In this section, first we explain the approach to represent the maintained global model with color information. Then we present the optimization problem based on both geometric and photometric consistency for frame-to-model registration, and talk about how to effectively solve it. Finally, we discuss some extensions of our proposed method.

2.1   Colored Global Model Representation

In KinectFusion [2,3], the whole geometry information of the 3D object to be reconstructed is represented by a volumetric truncated signed distance function (TSDF). The TSDF value of each point is defined as the signed distance between its calculated depth and the value of its projection position in the depth map, and the distance is truncated into a pre-defined interval. For each iteration, when a new depth map comes, after the step of registration for camera pose estimation, its corresponding TSDF volume is calculated and aligned to the maintained global TSDF volume. Then in the step of integration, the calculated TSDF volume and the global one are combined together by a simple running weighted average [21].

Given the integrated TSDF volume, with the idea to find the points with zero-valued TSDF, pixel-level raycast [22] is performed to determine the vertices on the model surface, and the corresponding normals are obtained by gradient extraction. For the global TSDF volume constructed by $(k-1)$ depth maps, the calculated surface vertices and their normals are denoted as $\mathbf{V}_{k-1}^g(\mathbf{u})$ and $\mathbf{N}_{k-1}^g(\mathbf{u})$, respectively, where the superscript "g" indicates that they are defined in the global world space, and $\mathbf{u}$ is the pixel position in the image. These data are used as the "model" for depth-based frame-to-model matching [2,3].

In our proposal, TSDF volume is also adopted for maintaining the overall geometry information, the surface vertices $\mathbf{V}_{k-1}^g(\mathbf{u})$ and their normals $\mathbf{N}_{k-1}^g(\mathbf{u})$ are calculated in the same way as that in KinectFusion [2,3]. To further involve color information, each surface vertex is transformed into the camera space and projected to the available color images. Let the already estimated camera poses for the $(k-1)$ RGB-D frames be $\{\mathbf{T}_{g,1}, \mathbf{T}_{g,2}, \cdots, \mathbf{T}_{g,k-1}\}$, which indicate the transformation matrices from each camera space to the global world space. Thus in the m-th $(m = 1, 2, \cdots, k-1)$ camera space, the surface vertices are calculated as

$$\mathbf{p}_m = [x_m, y_m, z_m, 1]^T = \mathbf{T}_{g,m}^{-1} \mathbf{V}_{k-1}^g(\mathbf{u}) \tag{1}$$

where homogeneous coordinates are adopted, and $\mathbf{T}_{g,m}$ is a 4×4 matrix involving both rotation and translation. Let $H(\cdot)$ denote the projection operation from 3D space to 2D space, f_x and f_y be the focal lengths, and (c_x, c_y) be the coordinates for the principal point, then the projection position is

$$\mathbf{v}_m = H(\mathbf{p}_m) = \left[\frac{x_m f_x}{z_m} + c_x, \frac{y_m f_y}{z_m} + c_y \right]^T \tag{2}$$

Given the $(k-1)$ color images $\{\mathcal{F}_1, \mathcal{F}_2, \cdots, \mathcal{F}_{k-1}\}$, the color of each surface vertex can be calculated by averaging all the color values of the corresponding projection positions, namely

$$C_{k-1}^g(\mathbf{u}) = \frac{1}{k-1} \sum_{m=1}^{k-1} \mathcal{F}_m(\mathbf{v}_m) \tag{3}$$

The above discussion only considers the projection model for each separated point, but does not address its visibility. In fact, the surface vertex $\mathbf{V}_{k-1}^g(\mathbf{u})$ may be unseen from the viewpoint of the m-th camera, thus the color value of the projection position will be meaningless. Therefore, a constraint is introduced, and we only choose the vertices whose calculated depth in the m-th camera space and the depth value of the projection position are close enough. Let the $(k-1)$ depth maps be denoted as $\{\mathcal{D}_1, \mathcal{D}_2, \cdots, \mathcal{D}_{k-1}\}$, the constraint can be formulated as

$$|z_m - \mathcal{D}_m(\mathbf{v}_m)| < \theta \tag{4}$$

where θ is a pre-defined threshold. Since it is more likely that the surface vertex cannot be seen from the viewpoint of images captured a long time before, the color calculation is confined for only considering the previous M images to reduce the computational load. Therefore, the final color of each surface vertex is determined as

$$C_{k-1}^g(\mathbf{u}) = \frac{\sum_{i=1}^{M} \mathcal{F}_{k-i}(\mathbf{v}_{k-i}) \cdot I\left(|z_{k-i} - \mathcal{D}_{k-i}(\mathbf{v}_{k-i})| < \theta\right)}{\sum_{i=1}^{M} I\left(|z_{k-i} - \mathcal{D}_{k-i}(\mathbf{v}_{k-i})| < \theta\right)} \tag{5}$$

where $I(\cdot)$ is the indicator function, namely $I(A) = 1$ when A is true, otherwise $I(A) = 0$.

With the constantly updated TSDF volume, all the surface vertices may be totally changed with the new RGB-D frames, thus the color value of each surface vertex must be recalculated in each iteration. Fortunately, the calculation can be efficiently implemented in parallel by GPU, thus the computational time cost is quite low. In our proposal, the calculated colors $C_{k-1}^g(\mathbf{u})$, as well as the surface vertices $\mathbf{V}_{k-1}^g(\mathbf{u})$ and the corresponding normals $\mathbf{N}_{k-1}^g(\mathbf{u})$, are utilized as the global model in the step of registration.

2.2   Frame-to-Model Matching Based Optimization

When the k-th depth map $\mathcal{D}_k$ comes, its corresponding vertices $\mathbf{V}_k(\mathbf{u})$ and normals $\mathbf{N}_k(\mathbf{u})$ are obtained by the method in KinectFusion [2,3]. As they are expressed in the camera space, the superscript "g" is not used. The incoming color image $\mathcal{F}_k$, as well as the calculated $\mathbf{V}_k(\mathbf{u})$ and $\mathbf{N}_k(\mathbf{u})$, are treated as the "frame" information.

In our proposal, both geometric and photometric consistency are taken into account for registration, and the way of frame-to-model matching is adopted for the two aspects. By fully exploring the available depth maps and color images in a unified optimization framework, the camera pose $\mathbf{T}_{g,k}$ can be estimated more accurately.

For frame-to-model geometry matching, the case is the same as that in KinectFusion [2,3]. By finding the corresponding projection positions and involving point-to-plane distance, the cost term is defined as

$$E_1(\mathbf{T}_{g,k}) = \sum_{(\mathbf{u},\hat{\mathbf{u}}) \in \mathcal{U}} \left\| \left(\mathbf{T}_{g,k}\mathbf{V}_k(\mathbf{u}) - \mathbf{V}_{k-1}^g(\hat{\mathbf{u}})\right)^T \mathbf{N}_{k-1}^g(\hat{\mathbf{u}}) \right\|^2 \tag{6}$$

where $\mathbf{u}$ and $\hat{\mathbf{u}}$ are corresponding projection positions for the same point in 3D space, and $\mathcal{U}$ is the set of corresponding position pairs found by considering both the vertex coordinates and the normal directions.

For frame-to-model color matching, the color of each 3D vertex in the global model is compared with the value of the 2D projection position in the k-th color image, and it is hoped that the two color values should be as close as possible. According to Eqs. (1) and (2), we calculate each surface vertex represented in the k-th camera space $\mathbf{p}_k$ and its corresponding projection position $\mathbf{v}_k$, and the cost term is described as

$$E_2(\mathbf{T}_{g,k}) = \sum_{\mathbf{u} \in \mathcal{W}} \left\| C_{k-1}^g(\mathbf{u}) - \mathcal{F}_k(\mathbf{v}_k) \right\|^2$$

$$= \sum_{\mathbf{u} \in \mathcal{W}} \left\| C_{k-1}^g(\mathbf{u}) - \mathcal{F}_k\left(H\left(\mathbf{T}_{g,k}^{-1}\mathbf{V}_{k-1}^g(\mathbf{u})\right)\right) \right\|^2 \tag{7}$$

where $\mathcal{W}$ is the set of valid projection positions defined by considering a similar constraint as that in Eq. (4)

$$\mathcal{W} = \left\{ \mathbf{u} \mid |z_k - \mathcal{D}_k(\mathbf{v}_k)| < \theta \right\}$$

$$= \left\{ \mathbf{u} \mid \left| z_k - \mathcal{D}_k\left(H\left(\mathbf{T}_{g,k}^{-1}\mathbf{V}_{k-1}^g(\mathbf{u})\right)\right) \right| < \theta \right\} \tag{8}$$

where z_k is the z-coordinate of $\mathbf{p}_k$, indicating the calculated depth in the k-th camera space.

The aforementioned two cost terms are linearly combined in our proposal, and the final cost function involving both geometric and photometric consistency is defined as

$$E(\mathbf{T}_{g,k}) = E_1(\mathbf{T}_{g,k}) + \lambda E_2(\mathbf{T}_{g,k})$$

$$= \sum_{(\mathbf{u},\hat{\mathbf{u}}) \in \mathcal{U}} \left\| \left(\mathbf{T}_{g,k}\mathbf{V}_k(\mathbf{u}) - \mathbf{V}_{k-1}^g(\hat{\mathbf{u}})\right)^T \mathbf{N}_{k-1}^g(\hat{\mathbf{u}}) \right\|^2$$

$$+ \lambda \sum_{\mathbf{u} \in \mathcal{W}} \left\| C_{k-1}^g(\mathbf{u}) - \mathcal{F}_k\left(H\left(\mathbf{T}_{g,k}^{-1}\mathbf{V}_{k-1}^g(\mathbf{u})\right)\right) \right\|^2 \tag{9}$$

where λ is a balanced coefficient for the two terms. It may remain unchanged for all the RGB-D frames or vary with different values. An example for introducing variable coefficient is to consider the blurriness of each color image. λ can be set to a smaller value for more blurry images, as it is likely that the photometric cue obtained from blurry images is inaccurate. The camera pose $\mathbf{T}_{g,k}$ can be calculated by minimizing the overall cost function, which also means maximizing the geometric and photometric consistency.

2.3   Solution to Optimization Problem

As how to minimize the first cost term $E_1(\mathbf{T}_{g,k})$ in the optimization problem has been detailedly explained in KinectFusion [3], we pay more attention to the second term $E_2(\mathbf{T}_{g,k})$ in this section.

The solution of KinectFusion is an iterative approach, and the camera pose to be determined in one iteration is locally linearized around its value obtained in the last iteration. That is to say, in the n-th round of iteration for calculating $\mathbf{T}_{g,k}$, we have

$$\mathbf{T}_{g,k}^{(n)} \approx \Delta\mathbf{T}\ \mathbf{T}_{g,k}^{(n-1)}$$

$$= \begin{bmatrix} 1 & -\gamma & \beta & a \\ \gamma & 1 & -\alpha & b \\ -\beta & \alpha & 1 & c \\ 0 & 0 & 0 & 1 \end{bmatrix} \mathbf{T}_{g,k}^{(n-1)} \tag{10}$$

Here a vector $\mathbf{x} = [\alpha, \beta, \gamma, a, b, c]^T \in \mathbb{R}^6$ is adopted for parameterizing the incremental transformation matrix $\Delta\mathbf{T}$, in which $[\alpha, \beta, \gamma]^T$ and $[a, b, c]^T$ are used for describing the tiny rotation and translation variations, respectively. To deal with $\mathbf{T}_{g,k}^{-1}$ in the second cost term $E_2(\mathbf{T}_{g,k})$, considering that all the six elements in the vector $\mathbf{x}$ are with small values, it can be easily obtained

$$\left(\mathbf{T}_{g,k}^{(n)}\right)^{-1} \approx \left(\mathbf{T}_{g,k}^{(n-1)}\right)^{-1}(\Delta\mathbf{T})^{-1}$$

$$\approx \left(\mathbf{T}_{g,k}^{(n-1)}\right)^{-1}\begin{bmatrix} 1 & \gamma & -\beta & -a \\ -\gamma & 1 & \alpha & -b \\ \beta & -\alpha & 1 & -c \\ 0 & 0 & 0 & 1 \end{bmatrix} \tag{11}$$

In the iterative solution for minimizing the first cost term $E_1(\mathbf{T}_{g,k})$, the problem is transformed into a linear equation in KinectFusion. Since the derivation process is complicated, the final result is simply denoted as $\mathbf{Ax} = \mathbf{b}$, where $\mathbf{A}$ is a 6×6 symmetric matrix, and $\mathbf{b}$ is a 6×1 vector. For more details about the derivation, the readers can be referred to [3]. For minimizing the second cost term $E_2(\mathbf{T}_{g,k})$, since it is with the form of non-linear least squares, the Gauss-Newton method can be adopted. In the n-th round of iteration, we calculate the residual vector $\mathbf{r} = [\mathbf{r_u}]$ where

$$\mathbf{r_u} = C_{k-1}^g(\mathbf{u}) - \mathcal{F}_k\left(H\left(\left(\mathbf{T}_{g,k}^{(n-1)}\right)^{-1}\mathbf{V}_{k-1}^g(\mathbf{u})\right)\right) \tag{12}$$

as well as its Jacobian matrix $\mathbf{J}$ with respect to $\mathbf{x}$, then the parameterized vector $\mathbf{x}$ can be updated by $\mathbf{J}^T\mathbf{J}\mathbf{x} = -\mathbf{J}^T\mathbf{r}$. In our proposal, by taking both of the two cost terms into account, $\mathbf{x}$ is calculated by solving the equation

$$\left(\mathbf{A} + \lambda\mathbf{J}^T\mathbf{J}\right)\mathbf{x} = \left(\mathbf{b} - \lambda\mathbf{J}^T\mathbf{r}\right) \tag{13}$$

Like KinectFusion [3], we can also down-sample the depth maps and color images for multi-scale representations, and conduct a coarse-to-fine framework to effectively solve the optimization problem.

2.4   Extensions

In the above discussion, we only talk about the case when all the depth maps have their corresponding color images. While in the practical applications with RGB-D cameras, depth maps and color images may be captured with different frame rates, and it is more likely that color images are obtained with lower frame rate for larger resolution. To extend our method to manage the situation, our proposed color-introduced frame-to-model registration is only utilized when the pair of depth map and color image exist, otherwise only geometry matching is adopted for camera pose estimation.

Our proposal can also be easily extended for involving camera distortions. Like the method in [23], in order to deal with the optical aberrations, a non-rigid correction function $\mathbf{L}_k$ over the image plane is introduced, and the cost term for photometric consistency is modified as

$$E_2(\mathbf{T}_{g,k}) = \sum_{\mathbf{u} \in \mathcal{W}} \left\| C_{k-1}^g(\mathbf{u}) - \mathcal{F}_k\big(\mathbf{L}_k(\mathbf{v}_k)\big) \right\|^2 \tag{14}$$

where $\mathbf{L}_k$ is directly defined for some pre-given positions, and generalized to other positions in the image plane by bilinear interpolation. The camera pose $\mathbf{T}_{g,k}$ and the parameters of the correction function $\mathbf{L}_k$ can be iteratively calculated by joint optimization.

3   Experimental Results

To evaluate the performance of our proposed approach, some experiments are implemented on two data sets. For quantitative evaluation, camera pose estimation and 3D reconstruction are conducted on the RGB-D SLAM benchmark [24]. Three sequences "fr1/desk", "fr1/room", as well as "fr3/long_office_household" from the benchmark are adopted, and the first 100 pairs of depth maps and color images in each sequence are used. Some captured color images in the three sequences are shown in Fig. 1. Since the benchmark provides ground-truth trajectories obtained from a high-accuracy motion capture system, the estimated camera poses can be compared with the ground-truth data. The absolute translational root mean square error [24] is utilized as the performance measure.

Fig. 1. Some captured color images in the sequences "fr1/desk", "fr1/room", and "fr3/long_office_household" from the RGB-D SLAM benchmark (Color figure online)

Three other registration approaches are used for comparison, including depth-based frame-to-frame matching (D_F2F) [13], depth-based frame-to-model matching (D_F2M) [2,3], as well as D_F2M further involving frame-to-frame color matching (D_F2M+C_F2F) [20]. The absolute translational root mean square errors for the three methods and our proposal are listed in Table 1, it can be seen that our approach produces the best results for all the sequences. D_F2F only adopts two consecutive depth maps for camera pose estimation. As much useful information is not well explored, its performance is the worst of all. Compared with D_F2M, both D_F2M+C_F2F and our proposal introduce photometric consistency in the step of registration. Their superiority demonstrates that it is reasonable to utilize color images as additional cue for obtaining more accurate trajectories. As far as the two color-involved approaches are concerned, like the case of geometric alignment, frame-to-model matching also outperforms frame-to-frame matching for color-introduced registration. Therefore, our proposal is more effective than D_F2M+C_F2F.

Table 1. Absolute translational root mean square errors (in centimeters) on difference sequences from the RGB-D SLAM benchmark

Sequence	D_F2F	D_F2M	D_F2M+C_F2F	Our proposal
fr1/desk	4.53	2.03	1.99	1.83
fr1/room	4.70	4.33	4.14	3.89
fr3/office	4.74	1.90	1.84	1.78

To demonstrate the difference of the estimated camera poses more clearly, as an example, the reconstruction results of D_F2M and our proposal for the sequence "fr1/desk" are placed in the same coordinate space and illustrated in Fig. 2. We can see that there are obvious displacements between the two results. Similar cases can be obtained for other sequences, and the displacements between the reconstruction results of our proposal and other approaches always exist as well. It is known that inaccurate estimated camera poses will hinder subsequent steps such as color mapping, thus effective registration approach is of great importance for 3D processing.

We also conduct experiments on our own data, which consists of 1328 pairs of depth maps and color images captured from various viewpoints for a toy teddy bear. All the depth maps are adopted in the process of depth-based reconstruction. For involving color information, 42 images with low blurriness are chosen to ensure the accuracy of the introduced photometric cue. In order to better compare different approaches, colored reconstruction results are illustrated here, and the color of each surface vertex is simply determined by averaging all the corresponding color values in the images. The reconstructed toy teddy bears by D_F2M and our proposal are shown in Fig. 3. We can see that even only about 3.2% (42/1328) color images are adopted for registration, our result is more clear than that by D_F2M, especially for the area of characters on the box, which indicates that the estimated camera poses by our proposal are more accurate. It should be noted that only simple color mapping is implemented here, thus the performance is not satisfactory. If more elaborate color mapping methods, such as [23], are utilized, better results can be achieved.

Fig. 2. Reconstruction results of D_F2M and our proposal for the sequence "fr1/desk" placed in the same coordinate space

(a) D_F2M   (b) Our proposal

Fig. 3. Reconstructed toy teddy bears by D_F2M and our proposal

4   Conclusions and Future Work

In this paper, by taking both geometric and photometric consistency into account, a novel registration approach for 3D reconstruction is proposed. Since they can provide additional information over depth maps, color images are reasonably introduced and largely explored. To make full use of the existing RGB-D frames, the maintained global model contains not only surface vertices and their normals, but also the corresponding color values. Frame-to-model geometry matching and color matching are simultaneously considered in a unified optimization framework, and an iterative solution is well developed. Furthermore, our method can be easily extended to deal with the case when depth maps and color images are captured with different frame rates, and it is convenient to further involve camera distortions. Experiments demonstrate that our proposal can achieve more accurate camera pose estimation results.

For the next research work, we will mainly focus on how to involve camera distortions more effectively. In the existing methods to address camera distortions by non-rigid transformation, the correction function is usually directly defined on a uniform lattice, and the lattice is kept the same for all the color images. In our proposal, only the projection positions of the surface vertices are useful for camera pose estimation. Generally speaking, they are not evenly distributed in the image plane, and their distributions are different for each image. Therefore, if only the same uniform lattice is adopted, the interpolation results may be inaccurate for the positions not on the lattice, and more complex lattices should be considered for better representation of non-rigid transformations. In the future, we will pay our attention to the problem of how to adaptively determine the most effective lattice for each color image, and attempt to efficiently finish its implementation.

References

1. Smisek, J., Jancosek, M., Pajdla, T.: 3D with kinect. In: Fossati, A., Gall, J., Grabner, H., Ren, X., Konolige, K. (eds.) Consumer Depth Cameras for Computer Vision, pp. 3–25. Springer, Heidelberg (2013)
2. Izadi, S., Kim, D., Hilliges, O., Molyneaux, D., Newcombe, R., Kohli, P., Shotton, J., Hodges, S., Freeman, D., Davison, A., Fitzgibbon, A.: KinectFusion: real-time 3D reconstruction and interaction using a moving depth camera. In: Proceedings of ACM Symposium on User Interface Software and Technology, pp. 559–568 (2011)
3. Newcombe, R.A., Izadi, S., Hilliges, O., Molyneaux, D., Kim, D., Davison, A.J., Kohli, P., Shotton, J., Hodges, S., Fitzgibbon, A.: KinectFusion: real-time dense surface mapping and tracking. In: Proceedings of IEEE International Symposium on Mixed and Augmented Reality, pp. 127–136 (2011)
4. Zhou, Q.Y., Koltun, V.: Dense scene reconstruction with points of interest. ACM Trans. Graph. **32** (2013)
5. Zhou, Q.Y., Miller, S., Koltun, V.: Elastic fragments for dense scene reconstruction. In: Proceedings of IEEE International Conference on Computer Vision, pp. 473–480 (2013)
6. Nießner, M., Zollhöfer, M., Izadi, S., Stamminger, M.: Real-time 3D reconstruction at scale using voxel hashing. ACM Trans. Graph. **32** (2013)
7. Zhou, Q.Y., Koltun, V.: Simultaneous localization and calibration: self-calibration of consumer depth cameras. In: Proceedings of IEEE International Conference on Computer Vision and Pattern Recognition, pp. 454–460 (2014)
8. Choe, G., Park, J., Tai, Y.W., Kweon, I.S.: Exploiting shading cues in kinect IR images for geometry refinement. In: Proceedings of IEEE International Conference on Computer Vision and Pattern Recognition, pp. 3922–3929 (2014)
9. Zollhöfer, M., Nießner, M., Izadi, S., Rhemann, C., Zach, C., Fisher, M., Wu, C., Fitzgibbon, A., Loop, C., Theobalt, C., Stamminger, M.: Real-time non-rigid reconstruction using an RGB-D camera. ACM Trans. Graph. **33** (2014)
10. Newcombe, R.A., Fox, D., Seitz, S.M.: DynamicFusion: reconstruction and tracking of non-rigid scenes in real-time. In: Proceedings of IEEE International Conference on Computer Vision and Pattern Recognition, pp. 343–352 (2015)
11. Dou, M., Taylor, J., Fuchs, H., Fitzgibbon, A., Izadi, S.: 3D scanning deformable objects with a single RGBD sensor. In: Proceedings of IEEE International Conference on Computer Vision and Pattern Recognition, pp. 493–501 (2015)

12. Innmann, M., Zollhöfer, M., Nießner, M., Theobalt, C., Stamminger, M.: VolumeDeform: real-time volumetric non-rigid reconstruction. arXiv preprint arXiv:1603.08161 (2016)
13. Rusinkiewicz, S., Hall-Holt, O., Levoy, M.: Real-time 3D model acquisition. ACM Trans. Graph. **21**, 438–446 (2002)
14. Herrera, D., Kannala, J., Heikkilä, J.: Joint depth and color camera calibration with distortion correction. IEEE Trans. Pattern Anal. Mach. Intell. **34**, 2058–2064 (2012)
15. Teichman, A., Miller, S., Thrun, S.: Unsupervised intrinsic calibration of depth sensors via SLAM. In: Robotics: Science and Systems (2013)
16. Kerl, C., Sturm, J., Cremers, D.: Dense visual SLAM for RGB-D cameras. In: Proceedings of IEEE/RSJ International Conference on Intelligent Robots and Systems, pp. 2100–2106 (2013)
17. Kerl, C., Sturm, J., Cremers, D.: Robust odometry estimation for RGB-D cameras. In: Proceedings of IEEE International Conference on Robotics and Automation, pp. 3748–3754 (2013)
18. Whelan, T., Johannsson, H., Kaess, M., Leonard, J.J., McDonald, J.: Robust real-time visual odometry for dense RGB-D mapping. In: Proceedings of IEEE International Conference on Robotics and Automation, pp. 5724–5731 (2013)
19. Wang, K., Zhang, G., Bao, H.: Robust 3D reconstruction with an RGB-D camera. IEEE Trans. Image Process. **23**, 4893–4906 (2014)
20. Choi, S., Zhou, Q.Y., Miller, S., Koltun, V.: A large dataset of object scans. arXiv preprint arXiv:1602.02481 (2016)
21. Curless, B., Levoy, M.: A volumetric method for building complex models from range images. ACM Trans. Graph. 303–312 (1996)
22. Parker, S., Shirley, P., Livnat, Y., Hansen, C., Sloan, P.P.: Interactive ray tracing for isosurface rendering. In: Proceedings of IEEE Conference on Visualization, pp. 233–238 (1998)
23. Zhou, Q.Y., Koltun, V.: Color map optimization for 3D reconstruction with consumer depth cameras. ACM Trans. Graph. **33** (2014)
24. Sturm, J., Engelhard, N., Endres, F., Burgard, W., Cremers, D.: A benchmark for the evaluation of RGB-D SLAM systems. In: Proceedings of IEEE/RSJ International Conference on Intelligent Robots and Systems, pp. 573–580 (2012)

Compressing Visual Descriptors of Image Sequences

Werner Bailer[✉], Stefanie Wechtitsch, and Marcus Thaler

DIGITAL – Institute for Information and Communication Technologies, Joanneum
Research Forschungsgesellschaft mbH, Steyrergasse 17, 8010 Graz, Austria
{werner.bailer,stefanie.wechtitsch,marcus.thaler}@joanneum.at

Abstract. In recent years, there has been significant progress in developing more compact visual descriptors, typically by aggregating local descriptors. However, all these methods are descriptors for still images, and are typically applied independently to (key) frames when used in tasks such as instance search in video. Thus, they do not make use of the temporal redundancy of the video, which has negative impacts on the descriptor size and the matching complexity. We propose a compressed descriptor for image sequences, which encodes a segment of video using a single descriptor. The proposed approach is a framework that can be used with different local descriptors, including compact descriptors. We describe the extraction and matching process for the descriptor and provide evaluation results on a large video data set.

1 Introduction

Instance search, i.e., finding video clips containing a similar foreground object, background or scene as in the query, is still a challenging problem in large-scale video collections. In contrast to video copy detection, the problem cannot be addressed only by global visual descriptors, due to the variability with which the object of interest may be depicted. In recent years, there has been significant progress in defining more compact visual descriptors, typically by aggregating local descriptors (either sampled from interest points or densely) and applying means such as dimensionality reductions and binarization. Examples of such methods are Fisher Vectors [12], VLAD [8] and its improvements [4], VLAT [13] and CDVS [3]. While these descriptors achieve good matching performance even at small descriptor sizes, they are all descriptors for still images that need to be applied independently to individual frames of the video. Thus, they do not make use of the temporal redundancy of the video. This is not only an issue of the size of the extracted descriptor, but also of the matching complexity, as pairwise matching of the frame descriptors has to be performed.

We propose a descriptor for image sequences, which encodes a set of consecutive and related frames (i.e., a segment such as a shot) as a single descriptor. The descriptor is created from an aggregation of sets of local descriptors from each of the images, and contains an aggregation of global descriptors and a time and location indexed set of the extracted local descriptors. The proposed method can

L. Amsaleg et al. (Eds.): MMM 2017, Part II, LNCS 10133, pp. 124–135, 2017.
DOI: 10.1007/978-3-319-51814-5_11

use compact still image descriptors (such as CDVS) as its basis. The descriptor extraction is based on a method for local descriptor extraction from interest points and a method for aggregation of such descriptors to a global descriptor, but is agnostic of the specific type of descriptor and aggregation method (as long as they fulfill certain properties). The descriptor extraction process can be parametrized for different descriptor bitrates. Depending on the bitrate, temporal subsampling and possibly lossy compression of local descriptors is applied. The matching process is hierarchical, in the sense that matching of details is only performed if some degree of similarity is found on the coarser level.

The rest of this paper is organized as follows. Section 2 defines the proposed descriptor and introduces the notation. The extraction of the descriptor from an image sequence is described in Sect. 3 and the matching of a pair of descriptors in Sect. 4. Section 5 presents evaluation results on a large video data set and Sect. 6 concludes the paper.

2 Descriptor Definition

Let $\mathcal{I} = \{I_1, \ldots, I_N\}$ be the sequence of images of a video, and let further $\mathcal{S} = \{S_1, \ldots, S_K\}$ be the set of segments of the video, with $S_k = \{I_1^k, \ldots, I_{M_k}^k\}$ being the set of images of a segment. A segment is only characterized by visual similarity and continuous changes between subsequent frames, but does not necessarily coincide with a semantic unit of the video. However, in practice, a segment is likely to be a shot or subshot.

Let $P^m = \{p_1^m, \ldots, p_n^m\}$ be a set of interest points extracted from image I_m^k (using a method such as DoG [10], ALP [5] or Hessian Affine [11]), $D^m = \{d_1^m, \ldots, d_n^m\}$ be a corresponding set of descriptors of the surrounding region of each of the interest points, such as SIFT [10], SURF [6] or ORB [14], and G^m be a descriptor of the frame obtained from aggregating the descriptors in D^m using a method such as Fisher Vectors (FV) [7], Scalable Compressed FV, VLAD [8], VLAT [13]. Let G_0^m be an encoded version of G^m, such as after dimension reduction. If the method chosen for descriptor aggregation already yields a binary descriptor, then let $G_0^m = G^m$. The descriptor $\bar{d}_i^m$ is an encoded version of the local descriptor, e.g. as defined by the local descriptor encoding in CDVS [3].

The compact image sequence descriptor for a segment is composed of a global and a local descriptor $\mathcal{D}_{S_k} = (\mathcal{G}_{S_k}, \mathcal{L}_{S_k})$, where the global descriptor is $\mathcal{G}_{S_k} = (G_0^{\tilde{m}}, \{\Delta_G^j | j \in I^k, j \neq \tilde{m}\})$, i.e. consists of one descriptor for frame $\tilde{m}$ (selected as described in Sect. 3), and a set of descriptors for all or a subset of the other frames in the shot. Δ_G^j is calculated as $\mathrm{enc}(G_0^j \oplus G_0^{\tilde{m}})$, with $\mathrm{enc}()$ being the encoding defined in Sect. 3.

The local part of the descriptor is defined as $\mathcal{L}_{S_k} = (T, f_{\tilde{m}}, \bar{\Delta}, \{\lambda_j\})$. It consists of a time map T indicating the presence of descriptors in the frames of the segments, the encoded local descriptors $f_{\tilde{m}}$ of the medoid frame, a (sub)set $\bar{\Delta}$ of local descriptors of other frames, which are encoded as the differences of the feature descriptors, with $\mathrm{enc}()$ being the encoding defined in Sect. 3, and the set of encoded locations of the descriptors λ_j in each of the frames j.

Figure 1 shows an example application of the proposed descriptor for matching two image sequences. The two sequences were shot at nearby locations at different times, and show largely overlapping background scenery. For each of the images, interest point detection and local descriptor extraction is performed, and an aggregated global descriptor per frame is determined. In order to represent the entire sequence efficiently, the descriptors from each of the frames are encoded as the global and local descriptors of a reference frame (determined by the medoid global descriptor of the sequence), and differentially coded global and local descriptors of the other frames. Matching is then done efficiently using only the segment descriptors.

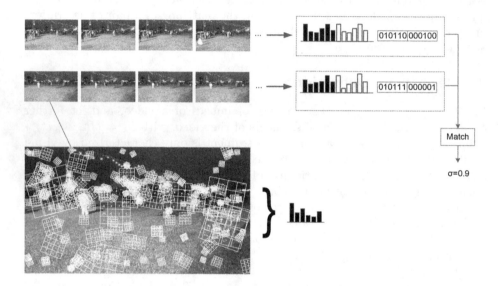

Fig. 1. Example of matching two image sequences.

3   Descriptor Extraction

The extraction process is depicted in Fig. 2. The extraction can be performed from all or a subset of frames of the image sequence, selected by regular subsampling by factor f_s. Clearly, this parameter impacts the size of the resulting descriptor and determines the upper boundary of the temporal localization precision of the descriptor. An additional input parameter is the average upper boundary of the descriptor size S_{max}, specified in kilobytes per second of video.

The first part is the temporal segmentation of the video in visually homogeneous segments. For every frame I^m, interest points P^m are sampled, local descriptors D^m are extracted and aggregated to a global descriptor G^m.

This is done using the similarity of the extracted descriptor, defining a segment as starting from frame $\hat{i}$

$$S_k = \{I_i | \delta_g(G_i, G_{i-1}) \leq \theta_g \wedge I_{i-1} \in S_k, i = \hat{i} \ldots \inf\}, \tag{1}$$

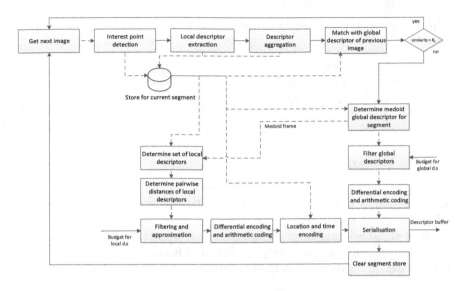

Fig. 2. Segment descriptor extraction.

where δ_g is an appropriate distance function for the chosen global descriptor, and θ_g is a threshold chosen for the desired segmentation properties. Smaller values will yield more homogeneous segments (in terms of visual variations) and shorter duration, but more compact descriptors for these segments.

Once segments are identified, the descriptor for a segment is encoded by aggregating global and local descriptors of the segments.

3.1   Global Descriptor

From the set of global descriptors $G^m, G^n, m, n \in S_k$, the pairwise distances $\delta_g(G^m, G^n)$ are determined, and the medoid frame

$$\tilde{m} = \text{argmin}_j \sum_i \delta_g(G^i, G^j), \tag{2}$$

is selected. The corresponding descriptor is encoded as $G_0^{\tilde{m}}$. For the other sampled frames $i \neq \tilde{m} \in S_k$, $\bar{\Delta}_G^i = G_0^i \oplus G_0^{\tilde{m}}$ is determined, i.e., the bit-wise differences of the binarized global descriptors are calculated. The rationale is to obtain descriptors of the same size, but with a lower number of bits set. Then, adaptive binary arithmetic coding (ABAC) is applied, yielding $\Delta_G^i = \text{enc}(\bar{\Delta}_G^i)$. Depending on the choice of S_{max}, all or only a subset of the descriptors are included in the descriptor for the segment, optionally using a minimum distance of θ_S for descriptors to be included. In case descriptors need to be removed, they are removed by ascending values of $\delta_g(G^i, G^{\tilde{m}})$ i.e. descriptors more similar to the medoid descriptor are removed first. The remaining number of difference descriptors is denoted K_g. In the minimum case $K_g = 0$, i.e. the resulting global descriptor consists only of the medoid descriptor. The descriptors are output in this order

$$[G_0^{\tilde{m}}, \Gamma_0, \ldots, \Gamma_{K_g}], \text{where}$$
$$\Gamma_i = \begin{cases} \Delta_G^{k_0}, k_0 = \text{argmax}_k \delta_g(G^k, G^{\tilde{m}}), & \text{if } i = 0 \\ \Delta_G^{k_i}, k_i = \text{argmax}_k \delta_g(G^k, G^{k_{i-1}}), G^k \notin \{G^{k_0}, \ldots, G^{k^{i-1}}\}, & \text{otherwise} \end{cases} \tag{3}$$

3.2 Local Descriptor

The construction of the local descriptor is done as follows. For each of the frames feature selection is performed as defined in the encoding process for $\bar{d}_i^m$, but encoding is not yet performed.

Each of these selected descriptors $d_i^m = \{x, y, \pi, f\}$ is a tuple of location, selection priority and feature descriptor. Starting from the medoid frame $\tilde{m}$, the sufficiently dissimilar descriptors are collected:

$$L = \{d_i^m | \delta_l(d_i^m, d_j^n) \geq \theta_l; \forall i, j; \\ m = \tilde{m}, \tilde{m}-1, \tilde{m}+1, \ldots; n = \tilde{m}-1, \tilde{m}+1, \ldots\}, \tag{4}$$

where θ_l is a threshold, chosen depending on the intended descriptor size, $\delta_l(\cdot)$ is an appropriate distance function. For descriptors omitted due to high similarity, a reference to the most similar descriptor is kept. This results in a set F_L of feature descriptors. For each $l \in F_L$, the set l^T of frames m_i in which this (or a very similar) descriptor appears, as well as their location is described as:

$$l_i = (f_i, l_i^T) \tag{5}$$

$$l_i^T = \{(\text{t}(m_i), x_{m_i}, y_{m_i})\} \tag{6}$$

The frames are identified by time points $\text{t}(m_i)$ relative to the segment's start.

For the set of descriptors in F_L, the most similar descriptor in F_L is determined, and the feature descriptor is determined as the difference of the encoded descriptors, i.e.

$$\bar{\delta}_i = \bar{f}_{d_i} - \bar{f}_{d_j}, \text{where} j = \text{argmin}_j \delta_l(f_{d_i} - f_{d_j}). \tag{7}$$

Adaptive binary arithmetic encoding is applied to the difference descriptors $\bar{\delta}_i$. A counter records in which of the frames instances of the local descriptors are present:

$$T = (\tau_{\tilde{m}}, \tau_{\tilde{m}-1}, \tau_{\tilde{m}+1}, \ldots),$$
$$\tau_i = \sum_{l_q \in F_L} \begin{cases} 1 & \text{if } i \in l_q^T \\ 0 \end{cases} \tag{8}$$

The differential part of the descriptor is thus composed as

$$\bar{\Delta} = \{(j, \text{enc}(\bar{\delta}_i)), \forall \delta_i\}, \tag{9}$$

with j being the index of the descriptor from use for difference calculation. The encoding of locations is performed as described in the CDVS standard (function `locenc()`).

The local part of the segment descriptor is composed of the set of the time map, the local descriptors appearing in the medoid frame, the set of encoded local difference descriptors, and of the locations of all descriptors:

$$(T, f_{\tilde{m}}, \overline{\Delta}, (\text{locenc}(\overline{L_{\tilde{m}}}), \text{locenc}(\overline{L_{\tilde{m}+1}}), \text{locenc}(\overline{L_{\tilde{m}-1}}), \ldots)), \qquad (10)$$

where $\bar{L}_i$ is the set of locations of local descriptors present in frame i. The global and local segment descriptors thus obtained are combined into a segment descriptor.

4   Descriptor Matching

Given two segment descriptors A and B, with $|A|, |B|$ denoting the respective segment lengths, matching is performed to obtain a matching score σ, which is determined as described below and shown in Fig. 3.

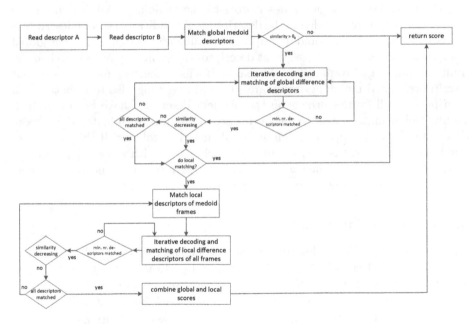

Fig. 3. Segment descriptor matching process.

4.1   Global Descriptor Matching

First, determine the similarity σ_g of the medoid descriptors G_0^A and G_0^B of the two frames, using threshold θ_m. If the matching score is below the threshold, $\sigma_g = 0$, and matching terminates.

Otherwise, the matching process continues with iterative decoding and matching of global descriptors. The similarity σ_g is compared against a second threshold Θ_γ, with $\Theta_\gamma > \Theta_m$ and the match count is determined as

$$c^G = \begin{cases} 0, & \sigma_g \leq \theta_\gamma \\ 1, & \text{otherwise} \end{cases} \tag{11}$$

and score $\sigma_0 = \sigma_g c^G$. The process proceeds to incrementally decode global descriptors $G_1^A, \ldots, G_K^A$ and $G_1^B, \ldots, G_{K'}^B$ and match them against all global descriptors decoded so far, yielding similarities $\rho_1, \ldots, \rho_{KK'/2}$. Then the match count c^G is increased by one for every $\rho_k > \theta_\gamma$, and σ_k is calculated as

$$\sigma_k = \frac{1}{c^G} \begin{cases} \rho_k, & \rho_k \leq \theta_\gamma, \\ 0, & \text{otherwise.} \end{cases} \tag{12}$$

A minimum number of $\min(2 + \lfloor \max(|A|, |B|) s_{min} \rfloor, |A|, |B|)$ descriptors are matched, with s_{min} being a constant ≥ 1. The constant factor two ensures that the most dissimilar global descriptors to the medoid global descriptor are matched (if they were encoded in the descriptor). As additional global descriptors are more similar to the medoid descriptor, decoding and matching further global descriptors from either of the segment descriptors stops after having matched the minimum number of frames if σ_j decreases. If this is the case for both segment descriptors, global matching terminates. Global matching also terminates if all descriptors of all frames present in the segment descriptor have been matched. If only global matching is to be performed, matching terminates. The score σ^G of the global descriptor matching is calculated as follows. If the number of matching frames exceeds $n_{\min} = \lceil m_{\min} \min(|A|, |B|) \rceil$, with a scaling parameter $m_{\min}$ $(0 < m_{\min} \leq 1)$, then σ^G is calculated as median of the $n_{\min}$ highest pairwise similarities, otherwise σ^G is set to 0.

4.2   Local Descriptor Matching

For matching of the local descriptors, we decode the temporal index, the local descriptors and (if encoded) their locations, and perform matching of the local descriptors of the frames corresponding to the two medoid local descriptors, yielding a set of similarities $\sigma_0^L = \{\sigma_{0,0}^L, \ldots, \sigma_{P_{\tilde{m}}^A, P_{\tilde{m}}^B}^L\}$ for the $P_{\tilde{m}}^A P_{\tilde{m}}^B$ pairs of local descriptors in the two frames (using an appropriate similarity metric for the type of local descriptor being used). If relevance information of local descriptors is available, it may be used to match only descriptors with higher relevance. If location information is encoded in the local descriptors, matching may also include spatial verification. As the descriptors are also referenced from other frames, the similarities will be stored for later use. Each of the similarities $\sigma_{p,q}^L$ of the medoid descriptors is compared against a threshold Θ_λ, and the number matching descriptor pairs is counted. A local match count is initialized, $c^L = 0$. If a minimum number of matching descriptor pairs are found (and confirmed by spatial verification, if performed), then the local match count c^L is increased by 1 for each such pair of frames.

The matching of the local descriptors is done in the same sequence as for global descriptors (and with the same number of minimum frames to be matched), and for the corresponding frames, calculating new distances or reusing the already calculated ones. In the same way as for global descriptors, the average similarity is updated from the matching frames, and matching terminates when it is found that the matching score decreases or all descriptors of all frames present in the segment descriptor have been matched. Like for the local descriptors of the medoid frame, the local match count is increased if a minimum number of matching descriptor pairs is found. If the local match count c^L exceeds $n_{\min}$ (as determined above for global descriptor matching), the local matching score σ^L is calculated as median of the $n_{\min}$ highest pairwise similarities.

The global matching score σ^G and the local matching score σ^L are combined into a total matching score σ. The total matching score σ may be determined according to any suitable method, preferably as a weighted sum (e.g., assigning equal weight to both) of the scores σ^G and σ^L, or as the maximum value, $\max(\sigma^G, \sigma^L)$.

5 Evaluation

While the proposed descriptor could be implemented using different local descriptors and aggregation methods, we base the compact image sequence descriptor on the MPEG CDVS descriptor, making use of the global and local parts of the descriptor. A CDVS descriptor contains a set of local SIFT descriptors [10] sampled around ALP interest points [3], which are quantized to a ternary representation. In addition, it contains an aggregated global descriptor, represented as Scalable Compressed Fisher Vector (SCFV) [9] as a binary vector.

MPEG[1] has collected a data set for an activity called Compact Descriptors for Visual Analysis (CDVA) for evaluating technologies for this purpose [1]. We use this data set for our experiments. The dataset contains in total around 23,000 video clips with durations ranging from about one minute to more than an hour. The material contains broadcast and user generated content in different resolutions and frame rates, and with diverse contents. It is divided into a set of reference and query clips, which contain different views of one object or scene, embedded into noise clips. In addition, part of query clips have been modified with transformations (e.g., resolution and frame rate changes, overlays, screen capture). The rest of the set contains distractor material for retrieval experiments.

We perform pairwise matching of the 9,715 queries against the 5,128 reference clips, and report the true positive rate at 1% false positive rate and the temporal localization performance measured as Jaccard index. Further details on

---

[1] http://mpeg.chiariglione.org/.

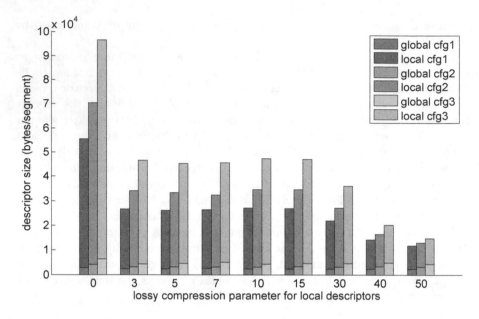

Fig. 4. Mean descriptor sizes of uncompressed sequence descriptor (0), and compressed descriptors with lossy local descriptor coding (θ_l).

the data and the evaluation metrics can be found in [2]. In addition to matching performance metrics, we measure the reduction in descriptor sizes due to the proposed compression.

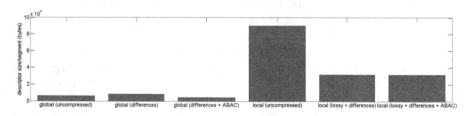

Fig. 5. Size of local and global descriptor components when applying lossy compression ($\theta_l = 30$). Both the numbers for applying lossy compression and differential encoding with and without subsequent adaptive binary arithmetic coding (ABAC) are shown.

As a first step, shot boundary detection using matching of color histograms is performed, and subsequent frames with high similarity discarded. This creates an irregularly sampled set of key frames for each shot. We then extract CDVS descriptors for each of the remaining frames. We compare an uncompressed version of the descriptor (i.e., a set of single frame descriptors) with compressed versions that apply lossless compression to the global descriptors

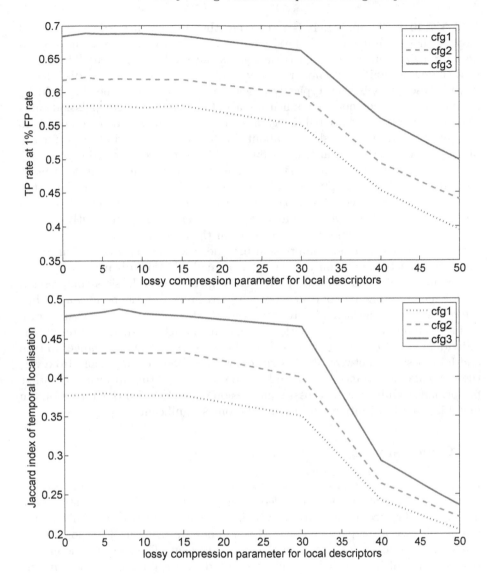

Fig. 6. True positive rate (top) and temporal localization performance (bottom) of uncompressed sequence descriptor (0), and compressed descriptors with lossy local descriptor coding (θ_l) for pairwise matching.

and lossy compression to the local descriptors. We compare three configurations of the descriptor extraction (cfg1, cfg2, cfg3), which differ in terms of the temporal subsampling factor of the input sequence $f_S \in \{4, 3, 2\}$ and the distance threshold for discarding subsequent similar frames $\theta_S \in \{0.7, 0.6, 0.5\}$.

Figure 4 shows the obtained mean descriptor sizes for a segment descriptor. For the compressed descriptor, lossy compression is applied with $\theta_l \in$

$\{0, 3, 5, 7, 10, 15, 30, 40, 50\}$. It is interesting that the smaller value for $\theta_l = 3$ already allows gaining most of the size reduction, resulting in 48% of the original descriptor size, while setting θ_l to 5 only gains one additional percent. This shows that there is a significant number of very similar descriptors, but the distance to further descriptors is then typically larger, so that increasing the threshold by a small margin does not have much impact. Only at higher compression rates ($\theta_l \geq 30$) significant additional size gains can be achieved. The reported numbers are per segment, and translate to about 1.7–3.0 kB per second of video for the uncompressed descriptor, and about 600 B–1 kB for the compressed descriptor.

Both the sizes after differential encoding of the global and local descriptors and after additionally applying adaptive binary arithmetic coding are reported in Fig. 5 for $\theta_l = 30$. For the global descriptor, differential encoding alone does not provide any advantage, but binary arithmetic encoding significantly reduces the size due to the sparser non-zero values in the difference descriptor. For the local descriptor, the lossy compression provides the main size reduction, while binary arithmetic coding only contributes a small additional benefit.

Figure 6 shows the resulting performance metrics. The lossless compression of global descriptors does of course not have any impact on the performance, but the results show that at moderate compression rates, both the true positive rate and the temporal localization performance remain unchanged. Actual results for individual videos do vary in terms of matching scores, with both changes in true and false positives between the descriptors with different compression. However, these results are balanced over the large data set, resulting in nearly constant performance. Only at higher lossy compression rates ($\theta_l > 30$) the impact on the matching and localization performance becomes significant.

6   Conclusion

In this paper, we have proposed a framework to extract and match compressed visual descriptors for segments of video, rather than using independent still image descriptors. The reduction in descriptor size originates from differentially coding the global descriptor part (aggregated frame descriptors) and from lossy compression of the local descriptors of the entire segment. The resulting descriptor is less than half in size than the uncompressed (already compact) descriptors, and can be matched more efficiently due to the structure of the descriptor.

Acknowledgments. The research leading to these results has received funding from the European Union's Seventh Framework Programme (FP7/2007–2013) under grant agreement no 610370, ICoSOLE, and from the Austrian Research Promotion Agency under the KIRAS grant E.V.A.

References

1. Call for proposals for compact descriptors for video analysis (CDVA) - search and retrieval. Technical report ISO/IEC JTC1/SC29/WG11/N15339 (2015)

2. Evaluation framework for compact descriptors for video analysis - search and retrieval - version 2.0. Technical report ISO/IEC JTC1/SC29/WG11/N15729 (2015)
3. ISO/IEC 15938-13: Information technology - multimedia content description interface - part 13: compact descriptors for visual search (2015)
4. Arandjelovic, R., Zisserman, A.: All about VLAD. In: 2013 IEEE Conference Computer Vision and Pattern Recognition (CVPR), pp. 1578–1585, June 2013
5. Balestri, M., Francini, G., Lepsøy, S.: Keypoint identification. Patent application WO 2015/011185 A1 (2013)
6. Bay, H., Ess, A., Tuytelaars, T., Van Gool, L.: Speeded-up robust features (SURF). Comput. Vis. Image Underst. 110(3), 346–359 (2008)
7. Duan, L.-Y., Gao, F., Chen, J., Lin, J., Huang, T.: Compact descriptors for mobile visual search and MPEG CDVS standardization. In: IEEE International Symposium on Circuits and Systems, pp. 885–888 (2013)
8. Jegou, H., Douze, M., Schmid, C., Perez, P.: Aggregating local descriptors into a compact image representation. In: IEEE Conference on Computer Vision and Pattern Recognition, pp. 3304–3311, June 2010
9. Lin, J., Duan, L.-Y., Huang, Y., Luo, S., Huang, T., Gao, W.: Rate-adaptive compact fisher codes for mobile visual search. IEEE Sig. Process. Lett. 21(2), 195–198 (2014)
10. Lowe, D.G.: Distinctive image features from scale-invariant keypoints. Int. J. Comput. Vis. 60(2), 91–110 (2004)
11. Mikolajczyk, K., Schmid, C.: Scale & affine invariant interest point detectors. Int. J. Comput. Vis. 60(1), 63–86 (2004)
12. Perronnin, F., Dance, C.: Fisher kernels on visual vocabularies for image categorization. In: IEEE Conference Computer Vision and Pattern Recognition, June 2007
13. Picard, D., Gosselin, P.-H.: Improving image similarity with vectors of locally aggregated tensors. In: IEEE International Conference on Image Processing, Brussels, BE, September 2011
14. Rublee, E., Rabaud, V., Konolige, K. Bradski, G.: ORB: an efficient alternative to SIFT or SURF. In: 2011 IEEE International Conference on Computer Vision (ICCV), pp. 2564–2571, November 2011

Deep Convolutional Neural Network for Bidirectional Image-Sentence Mapping

Tianyuan Yu[✉], Liang Bai, Jinlin Guo, Zheng Yang,
and Yuxiang Xie

College of Information System and Management,
National University of Defense Technology, Changsha 410073, China
{yutianyuan92, xabpz}@163.com, gjlin99@gmail.com,
yz_nudt@hotmail.com, yxxie@nudt.edu.cn

Abstract. With the rapid development of the Internet and the explosion of data volume, it is important to access the cross-media big data including text, image, audio, and video, etc., efficiently and accurately. However, the content heterogeneity and semantic gap make it challenging to retrieve such cross-media archives. The existing approaches try to learn the connection between multiple modalities by direct utilization of hand-crafted low-level features, and the learned correlations are merely constructed with high-level feature representations without considering semantic information. To further exploit the intrinsic structures of multimodal data representations, it is essential to build up an interpretable correlation between these heterogeneous representations. In this paper, a deep model is proposed to first learn the high-level feature representation shared by different modalities like texts and images, with convolutional neural network (CNN). Moreover, the learned CNN features can reflect the salient objects as well as the details in the images and sentences. Experimental results demonstrate that proposed approach outperforms the current state-of-the-art base methods on public dataset of Flickr8K.

1 Introduction

The ubiquitous adoption of mobile Internet has made multimedia documents available everywhere in daily life in forms of web pages, images, videos, and even mobile services like interactive micro-blogs, social networks, etc., which are usually composed of multimedia formats and content descriptions. Meanwhile, the rapid increase of data volume also makes it more and more difficult for web users to access valuable and customized information for the massive information oceans. The above difficulty has triggered much attention to information retrieval approaches in research communities.

Cross-media information retrieval is challenging because of the so-called *semantic gap* problem, which means the query descriptions and returned results can hardly be corresponded accurately, especially when they belong to different modalities. As a result, one key problem in this task is how to measure the distances or similarities between multiple modalities from the view of semantics. One solution is to align the two feature spaces so that they can be comparable and such semantic mapping has attracted much research interest. However, the detection of saliency such as scenes,

L. Amsaleg et al. (Eds.): MMM 2017, Part II, LNCS 10133, pp. 136–147, 2017.
DOI: 10.1007/978-3-319-51814-5_12

objects, etc. from the visual media is not enough, the task of bidirectional multimedia retrieval also requires the machine to understand the details from images, texts, etc., and more importantly, their semantic connections with each other. As shown in Fig. 1, the detection of "house" and "window" might be noisy when providing meaningful description of the image, though they take up large part of the image. A machine needs to learn the useful correlations (such as "jump" and "trampoline") and neglect unimportant visual and textual information (such as " house" and "up on a").

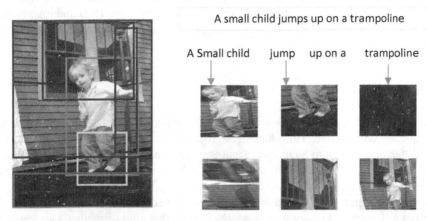

Fig. 1. Representation of the mapped image segments and a relevant sentence. The difficult is to learn the useful correlations and neglect the unimportant visual and textual information.

To deal with the above challenges, modern cross-media information retrieval approaches try to query visual media and texts alternatively, i.e. searching for relevant images with textual query, or vice versa. At the beginning of cross-media research, the task only focused on a limited number of keywords or classification tags [1]. Since one word label cannot fully represent the whole image, more recently researchers started to use long sentences or articles to search for images of interest [2, 3], and even describe a target image with appropriate captions [4]. In a more challenging task as introduced in [5], an answer can be returned through a visual Turing test when the machine is provided with an image and a corresponding textual question.

As a major breakthrough in artificial intelligence, deep learning has been successfully applied in various fields. Among the deep networks, convolutional neural network (CNN) is a typical architecture for visual feature representation [6, 7]. Compared to features extracted by traditional approaches, those derived from CNN are proved to have better performances in various computer vision tasks [8, 9] and multimedia retrieval challenges [10, 11]. Similarly in cross-media information retrieval, a large number of researchers use images labels as the targets in their networks [12, 13] aiming at classifications. Because of the limits of one word label representation, semantic details are neglected during the training process, which can definitely affect the final ranked results.

In this work, a novel deep model is introduced which learns mixed features in a common feature space from visual and textual representations respectively, and the mapped features are used to correctly determine whether the texts and images are relevant or not. Our contributions are three-fold: (1) A deep convolutional neural network which maps cross-media data into a common feature space is introduced. (2) The CNN-like model is used to analyze the textual information and extract features from textual information. (3) The attention model is combined in CNN to extract visual features from images. (4) Comprehensive evaluations in experiment demonstrate that the proposed approach differentiates from previous work in that the mixed features extracted have better representations in the common space between texts and images. In particular, the deep network achieves convincing performance on Flickr8K dataset [14] for cross-media retrieval task.

2   Related Work

Domain difference between queries and retrieval results leads to the difficulty that they are not directly comparable. This challenges cross-media information retrieval and the map of different domains to a common feature space is necessary, so that the distance between them can be measured. In this section, related work on how to model such common feature spaces is presented and discussed.

Original work in this area used low-level feature spaces to represent simple visual descriptors or linguistic keywords, separately. That is, this kind of methods are carried out in a *extract-and-combine* manner, i.e. extracting the highly correlated features in different spaces first, which are then used to construct a correlated representation in a common feature space. Though simple visual and textual features are used in these approaches, they performed well and kept the state-of-the-art results for a long time in the past. Representative methods in this category include cross-media hashing [13], canonical correlation analysis [15] and its extension [16].

The defect of above extract-and-combine approaches is obvious in that simple features cannot represent the semantic meaning correctly, leaving the semantic gap still unbridged between different modalities. As a result, advanced semantic features are proposed and extracted to construct the mid-level feature space so as to improve the performance. The most popular method in this category is multimodal topic model [17]. Similarly in [18], Latent Dirichlet Allocation (LDA) is used to build better mapping between texts and images by Blei and Jordan. However, LDA method only works well when the features are discrete, such as traditional bag-of-words features, and is not flexible enough to be adapted to other advanced features. In [19], a mutual semantic space is proposed by Pereira et al. in which texts and images are mapped to a pre-defined vocabulary of semantic concepts according to probabilities in order to utilize the underlying semantic information more directly. Based on the probabilities representation, the distance between texts and images can be measured. Because this method highly depends on manual annotations for learning the semantic concepts, it is less flexible when a new dataset is given. In such cases, a new vocabulary has to be made manually, which is undoubtedly time consuming and labor intensive.

Recently, deep learning methods are also applied in this area aiming at developing a common feature space with the learned features. In [20], a deep visual-semantic embedding model is introduced to identify visual objects using labeled images as well as semantic information gleaned from unannotated textual corpus. Similarly, Socher et al. propose a dependency tree recursive neural network (DT-RNN) to process textual information [12]. Among these methods, recurrent or recursive neural networks are used to deal with textual information and inner product is employed to strictly measure the correspondence between cross-media features to describe similarity/relevance. Except Karpathy's method, other models reason about objects only on global level. Because the information extracted from images or texts are usually represented at global level, such as background or salient objects in an image, the inner product with global features can cause inevitable mistakes, especially when the extracted keywords may not match the saliency in the image, as discussed in Fig. 1. In [21], Karpathy et al. propose a model which works on a finer level and embeds fragments of images and sentences into a common space. Though the state-of-the-art is achieved, sentence fragments are not always appropriate, especially when it comes to multiple adjectives for one noun or numeral, as they mention in [21]. Furthermore, it is hard to correspond image fragments with words or phrases in the relevant sentence. Instead, our model focuses on both local and global features in images and sentences. The proposed mixed features are demonstrated to be better compared to previous global methods.

3 Two-Stream Deep Network

The aim of this paper is to construct a deep learning model, automatically finding the semantic similar pair of images and sentences close to each other in this common space. For this purpose, a novel two-stream deep model is introduced to extract the mixed features and correctly determine the relevance relationship based on this new representation, as shown in Fig. 2.

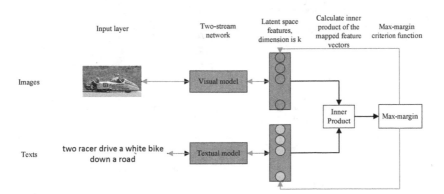

Fig. 2. Paradigm of the proposed two-stream model. Textual and visual features are extracted separately first and combined into a feature space in which max margin is used to optimize the relevance relationship.

The proposed two-stream network consists of three main components: (1) Textual Model (T-model), is responsible for training textual data with CNN and extracting the textual features. (2) Visual Model (P-model), is responsible to map the images into a common space where textual information has already been embedded. (3) Multi-Modal Embedding, involves a criterion function in order to encourage the relevant pair to have a high inner product.

The proposed model is trained on a set of images and sentences with their relationship labeled as relevant or irrelevant. In the training stage, we forward propagate the whole network to map the textual and visual information into a common space. Then inner product and max margin are used in the criterion function to backward propagate the whole network with stochastic gradient descent (SGD) method to force the semantic similar cross-media information to be close to each other in the new space. Three components of the proposed model can be described in details as follows.

3.1 Textual Model

Deep semantic similarity model (DSSM) introduced in [22] has been proved to achieve significant quality improvement on automatic highlighting of keywords and contextual entity search. One advantage of this model is that it can extract local and global features from a sentence. However, the convolutional layer in this model fixed the number of words in the group of input, which limits its function in extracting potentially relevant words. For example, for the phrase "a black and white cat", it is impossible to link the adjective "black" and the noun "cat" if the group number for relevance searching is less than four. To tackle this weakness, we extend this model and relax this constraint by searching phrases with arbitrary length. The overview of our textual model is shown in Fig. 3, which is constructed as a CNN composed of hashing layer, convolutional layer, max-pooling layer, and fully-connected layer.

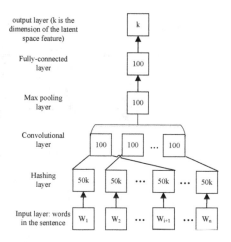

Fig. 3. Illustration of the network architecture and information flow of the textual model. The number in the rectangle represents the dimension in the layer.

As we can see from Fig. 3, the raw input of textual model includes each word in a sentence. In the hashing layer, a vector of 3-grams (tri-letter vector) is built for each word. The prominent advantage of tri-letter vector is that this representation can significantly reduce the total number of dimensions. Though English words is numerous, the number of tri-letter used to represent them can be very small. According to [23], a set of 500K-word vocabulary can boil down to only 30621 tri-letters.

After the tri-letter vectors are inputted to the convolutional layer, the local features of sentence are extracted in this layer. During the process of textual feature extraction, a sliding window is employed to concatenate words within the window to generate a new vector which is used as the input to a linear function and *tanh* activation in the last layers of the textual model. Since each word has a chance to be relevant to any other words in the sentence, the size of window is varied from one to the total number of words in the sentence. In the process, the duplicated words will increase their importance so that the extracted local features are more representative.

In the next layer of a max-pooling, the extracted feature vectors of words in sentence turns to a fixed dimension feature vector representing the sentence with the maximum operation. This is implemented by setting the ith value of the output vector of max-pooling layer as the maximum value of all the ith values in the input vectors. The step is to encourage the network to keep the most useful local features and form the mixed feature for each sentence. The features extracted by convolutional and max-pooling layers mainly represents the keywords and important phrases in the sentence while other useful details are kept and meaningless items are removed.

The final step of in textual model is the fully-connected layer. Like the common CNN models, there are two fully-connected layers to reduce the dimension of extracted mixed features. Going through the whole textual model, the initial sentences can be converted to vectors in a fixed-dimensional space.

3.2   Visual Model

In this section, we use the attention model originated by human visual system to extract feature from images. When people look through a picture, they usually focus on the salient parts rather than the entire image. To imitate the biological phenomenal, the attention modal is proposed to focus on different parts of the input according to different tasks.

In this work, we use the spatial transformer network introduced in [24] to focus on the visual feature in part of an image. The visual model is illustrated in Fig. 4. The input image is separated as several sections through spatial transformer network. Then, we extract the feature of each section of the input image by convolutional neural networks. Finally, the features of image sections are combined by the method of weighting. The modality of the extracted visual feature is the same with that of the extracted textual feature.

The spatial transformer network is utilized as attention model because it imitates the human visual system. The network focuses on the parts of an image, which contains much more information than others so that the useless details can be neglected. With the duplication of the more informative parts in the image, the global feature extracted

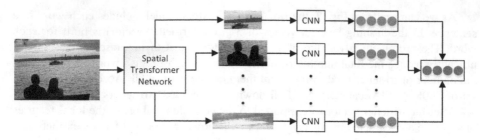

Fig. 4. Illustration of the visual model

in this paper would remain much more information delivered by the image than that extracted directly by CNN.

The spatial transformer network in our work learns four variables $\alpha_1, \alpha_2, \beta_1, \beta_2$, which makes the points (x', y') in the extracted section satisfies the Eq. (1). In the equation, (x, y) is the point in the original image. In this way, the original image can be transformed to the sections which contain much information.

$$\begin{cases} x' = \alpha_1 \bullet x + \beta_1 \\ y' = \alpha_2 \bullet y + \beta_2 \end{cases} \tag{1}$$

3.3   Multi-modal Embedding

The previous two sections of 3.1 and 3.2 have shown how the textual and visual media data can be mapped into the features with the same dimension, which means their features share a common feature space. In this section, a multi-modal objective function is defined in order to learn joint image-sentence representations. The aim of objective function is to force the corresponding pairs of images and sentences to have higher inner products than any other unrelated pairs. Since traditional classification functions such as logistic function cannot be flexibly used here to train the ranking information, we take the measure of max-margin objective function to force the difference between the inner products of correct pairs and other pairs to reach a fixed margin, which can be formalized as:

$$\begin{aligned} loss = &\sum_{(i,j) \in P} \sum_{(i,k) \notin P} \max(0, margin - v_i^T t_j + v_i^T t_k) \\ &+ \sum_{(i,j) \in P} \sum_{(k,j) \notin P} \max(0, margin - v_i^T t_j + v_k^T t_j) \end{aligned} \tag{2}$$

where v_i is a column vector denoting the output of our visual model for the i-th image, t_j is a column vector representing the output of textual model for the j-th sentence. We also define P as the universal set of all the corresponding image-sentence pairs (i,j). It is obviously time-consuming if all the irrelevant cross-media information are used to optimized this model. For the purpose of efficiency, we randomly select 9 false samples for each true sample to restrict the scale of training dataset. The hyper-parameter margin is usually set around 1. However, the range of the variable is wide, for example, it is set to 3 in [12] while 0.1 in [20]. In this paper, the margin is set to 0.5.

4 Experiment and Results

4.1 Dataset and Experiment Setup

Dataset. We use the dataset of Flickr8K [14] which consists of 8000 images, each with 5 sentences as its descriptions. Two exemplar image samples together with its sentences are shown in Fig. 5. In our experiment, we split the data into 6000 images for training, 1000 for validating, and 1000 for testing. Since there are 5 labeled description for each image, we finally obtained 30,000 training sentences and 5000 testing sentences.

1. A child in a pink dress is climbing up a set of stairs in an entry way
2. A girl going into a wooden building
3. A little girl climbing into a wooden playhouse
4. A little girl climbing the stairs to her playhouse
5. A little girl in a pink dress going into a wooden cabin

1. A black dog and a spotted dog are fighting
2. A black dog and a tri-colored dog playing with each other on the road
3. A black dog and a white dog with brown spots are staring at each other in the street
4. Two dogs of different breeds looking at each other on the road
5. Two dogs on pavement moving toward each other

Fig. 5. Two examples in the dataset of Flickr8K.

Baselines. In the comparison to other methods, several state-of-the-art methods are used as baselines including (in italics): In 2013, *Hodosh et al.* [14] introduced the dataset of Flickr8K and propose a method of bidirectional ranking on the dataset. Later, Google achieved the state-of-the-art performance on the 1000-class ImageNet using a deep visual-semantic embedding model *DeViSE* [20]. Although they focused on the potential image labels and zero-shot predictions, their model laid the foundation for the latter models. Socher et al. [12] embedded a full-frame neural network with the sentence representation from a semantic dependency tree recursive neural network (*SDT-RNN*), which has made prominent progress in the indices such as mean rank and recall at position k (*R@k*) compared to *kCCA*. Recently, deep fragment embedding proposed by *Karpathy et al.* [21] achieved a major breakthrough in the available datasets.

Evaluation Metrics. We use the popular indices recall at position k (*R@k*) and median rank scores as evaluation metrics. R@k is the percentage of ground truth among the first k returned results and is a widely used index of performance especially for search engines and ranking systems. The median rank indicates the location of *k* at which the result has a recall of 50%.

Implementation Settings. In the textual model, we directly use the results of tri-letters dictionary released by the open source demo "sent2vec"[1], which includes about 50,000 tri-letters. If a new tri-letters vector occurs which is not included in the dictionary, it is then appended into the dictionary. Using this dictionary, the image captions are mapped into tri-letter vectors after punctuations are removed.

We set the number of pairs in a batch as 10, and use the 10 corresponding pairs to get 90 irrelevant pairs. Before each epoch, we shuffled the dataset in order to force the network to adapt to more irrelevant image-sentence pairs. We set the dimension of the common feature space as 20. Once the training is completed, the network model is evaluated on testing set of images and sentences. The evaluation process scores and sorts the image-sentence pair in the testing dataset. In the meantime, the locations of ground truth results are recorded.

4.2   Feature Extracted by Textual Model

In this section, the feature extracted by textual model is analyzed. Recall that all the sentences in the test dataset have been mapped into the resulted multi-modal space. From this result, we can determine which words or phrases are extracted into the final space by our network model. Typical resulted samples are shown in Fig. 6. From this figure, we can find that the global feature repeat the keywords in order to keep the features, which satisfies our needs and demands. In Fig. 6, the first underlined words (blue) are the main source of the extracted features, followed by the second underlined words (green), then the third lines (red). There are still other words existing in the final global feature, which only take up a low proportion.

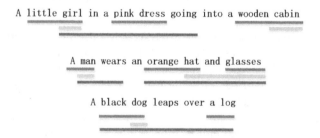

Fig. 6.   Features extracted by the textual model represent keywords and key information in the sentence. (Color figure online)

---

[1] http://research.microsoft.com/en-us/downloads/731572aa-98e4-4c50-b99d-ae3f0c9562b9/default. aspx.

Table 1 shows the results of the average rank of the ground truth in the list. We can find that bag of words (BoW) method can achieve a good performance in the calculation of mapped sentence similarity. In Table 1, the RNN method performs the worst which is also been reported in [3]. One possible reason is the representation of RNN is dominated by the last words, which are usually not the most important words in image captions.

Table 1. Comparison of textual processing to baselines. The rank is expected to be lower because sentences describing the same image should be closer in the common feature space.

Model	Random	BoW	Bigrams	Trigrams	RNN	kCCA	CNN+tri-letter
Med r	998.3	24.4	22.7	21.9	38.1	20.3	19.7

4.3 Image Annotation and Searching

This experiment evaluates the performance of the proposed model finding the desired textual or visual information that is more related to the content of the given image or sentence. The results in this task are shown in Table 2. In the paper, most results listed are based on the results in [21]. When comparing with Hodosh et al. [14], we only use a subset of N sentences out of total 5N so that the two approaches can be comparable. From Table 2, we can find that our model outperforms the state-of-the-art methods on most of criteria. The main reason might lie in that [21] requires the fragments of images and sentences to be matched exactly to each other, which is a very strict constraint especially when the sentences are only focused on a part of contents in the images. Such cases tend to result in wrong matches in evaluation. Instead, in our model, the extracted textual features can effectively represent the key information in the sentence, which is more likely to match the salient objects and details of the corresponding image. Besides, the attention model used in the visual model repeats the key information in the images while the textual model repeats the key word in the texts. In the meanwhile, both of the networks can neglect the useless details in the input. Therefore, the extracted features of semantic similar pair of cross-media information can correspond more closely in our work.

Table 2. Result comparison on Flickr8K data

Flickr8K								
Model	Image Annotation				Image Search			
	R@1	R@5	R@10	Med r	R@1	R@5	R@10	Med r
Random Ranking	0.1	0.5	1.0	635	0.1	0.5	1.0	537
DeViSE [20]	4.8	16.5	27.3	28	5.9	20.1	29.6	29
SDT-RNN [12]	4.5	18.0	28.6	32	6.1	18.5	29.0	29
Karpathy et al. [21]	12.6	32.9	44.0	14	**9.7**	**29.6**	42.5	**15**
Our model	**12.8**	**34.9**	**48.7**	**12**	9.5	29.1	**43.7**	**15**
*Hodosh et al. [14]	8.3	21.6	30.3	34	7.6	20.7	30.1	38
*Karpathy et al. [21]	9.3	24.9	37.4	21	**8.8**	27.9	41.3	**17**
Our model	**9.6**	**25.3**	**37.9**	**19**	8.5	**28.4**	42.8	**17**

5 Conclusion

In this paper, we introduced a novel two-stream network model to fulfill the task of bidirectional cross-media information retrieval. This model first maps the textual and visual media into a common feature space. In the textual model, tri-letter vector is used to duplicate the key words and key phrases, and neglect the meaningless details. In the visual model, attention mechanism is combined in the visual model to focus on the partial salient objects in the images so that the most information can be remained and least information can be filtered. During this procedure, the cross-media pairs are judged and their relevance relationships are optimized in the proposed multi-modal embedding methods, in order to determine whether sentences or images are relevant. Comprehensive experiments on publically available dataset demonstrates that the proposed model outperforms the baselines including the state-of-the-arts and prevailing methods. The mixed features extracted by our model are also shown to be advantageous in representing the semantics in images and sentences.

References

1. Jeon, J., Lavrenko, V., Manmatha, R.: Automatic image annotation and retrieval using cross-media relevance models. In: Proceedings of the 26th Annual International ACM SIGIR Conference on Research and Development in Informaion Retrieval, pp. 119–126. ACM (2003)
2. Srivastava, N., Salakhutdinov, R.R.: Multimodal learning with deep Boltzmann machines. In: Advances in Neural Information Processing Systems, pp. 2222–2230 (2012)
3. Wu, F., Lu, X., Zhang, Z., et al.: Cross-media semantic representation via bi-directional learning to rank. In: Proceedings of the 21st ACM International Conference on Multimedia, pp. 877–886. ACM (2013)
4. Vinyals, O., Toshev, A., Bengio, S., et al.: Show and tell: a neural image caption generator. In: Proceedings of the IEEE Conference on Computer Vision and Pattern Recognition, pp. 3156–3164 (2015)
5. Malinowski, M., Rohrbach, M., Fritz, M.: Ask your neurons: a neural-based approach to answering questions about images. In: IEEE International Conference on Computer Vision, pp. 1–9. IEEE (2015)
6. Krizhevsky, A., Sutskever, I., Hinton, G.E.: Imagenet classification with deep convolutional neural networks. In: Advances in Neural Information Processing Systems, pp. 1097–1105 (2012)
7. Szegedy, C., Liu, W., Jia, Y., et al.: Going deeper with convolutions. In: Proceedings of the IEEE Conference on Computer Vision and Pattern Recognition, pp. 1–9 (2015)
8. Girshick, R.: Fast R-CNN. In: Proceedings of the IEEE International Conference on Computer Vision, pp. 1440–1448 (2015)
9. Xu, Z., Yang, Y., Hauptmann, A.G.: A discriminative CNN video representation for event detection. In: Proceedings of the IEEE Conference on Computer Vision and Pattern Recognition, pp. 1798–1807 (2015)
10. Paulin, M., Douze, M., Harchaoui, Z., et al.: Local convolutional features with unsupervised training for image retrieval. In: Proceedings of the IEEE International Conference on Computer Vision, pp. 91–99 (2015)

11. Matsuo, S., Yanai, K.: CNN-based style vector for style image retrieval. In: Proceedings of the 2016 ACM on International Conference on Multimedia Retrieval, pp. 309–312. ACM (2016)
12. Socher, R., Karpathy, A., Le, Q.V., et al.: Grounded compositional semantics for finding and describing images with sentences. Trans. Assoc. Comput. Linguist. **2**, 207–218 (2014)
13. Zhuang, Y., Yu, Z., Wang, W., et al.: Cross-media hashing with neural networks. In: Proceedings of the ACM International Conference on Multimedia, pp. 901–904. ACM (2014)
14. Hodosh, M., Young, P., Hockenmaier, J.: Framing image description as a ranking task: data, models and evaluation metrics. J. Artif. Intell. Res. **47**, 853–899 (2013)
15. Hardoon, D.R., Szedmak, S., Shawe-Taylor, J.: Canonical correlation analysis: an overview with application to learning methods. Neural Comput. **16**(12), 2639–2664 (2004)
16. Ballan, L., Uricchio, T., Seidenari, L., et al.: A cross-media model for automatic image annotation. In: Proceedings of International Conference on Multimedia Retrieval, p. 73. ACM (2014)
17. Wang, Y., Wu, F., Song, J., et al.: Multi-modal mutual topic reinforce modeling for cross-media retrieval. In: Proceedings of the 22nd ACM International Conference on Multimedia, pp. 307–316. ACM (2014)
18. Blei, D.M., Jordan, M.I.: Modeling annotated data. In: Proceedings of the 26th Annual International ACM SIGIR Conference on Research and Development in Informaion Retrieval, pp. 127–134. ACM (2003)
19. Pereira, J.C., Coviello, E., Doyle, G., et al.: On the role of correlation and abstraction in cross-modal multimedia retrieval. IEEE Trans. Pattern Anal. Mach. Intell. **36**(3), 521–535 (2014)
20. Frome, A., Corrado, G.S., Shlens, J., et al.: Devise: a deep visual-semantic embedding model. In: Advances in Neural Information Processing Systems, pp. 2121–2129 (2013)
21. Karpathy, A., Joulin, A., Li, F.F.F.: Deep fragment embeddings for bidirectional image sentence mapping. In: Advances in Neural Information Processing Systems, pp. 1889–1897 (2014)
22. Gao, J., Deng, L., Gamon, M., et al.: Modeling interestingness with deep neural networks: U. S. Patent 20,150,363,688, 17 December 2015
23. Huang, P.S., He, X., Gao, J., et al.: Learning deep structured semantic models for web search using clickthrough data. In: Proceedings of the 22nd ACM International Conference on Information & Knowledge Management, pp. 2333–2338. ACM (2013)
24. Jaderberg, M., Simonyan, K., Zisserman, A., Kavukcuoglu, K.: Spatial transformer networks. In: Advances in Neural Information Processing Systems 28: Annual Conference on Neural Information Processing Systems, (NIPS 2015), pp. 2017–2025 (2015)

Discovering Geographic Regions in the City Using Social Multimedia and Open Data

Stevan Rudinac$^{(\boxtimes)}$, Jan Zahálka, and Marcel Worring

Informatics Institute, University of Amsterdam, Amsterdam, The Netherlands
{s.rudinac,j.zahalka,m.worring}@uva.nl

Abstract. In this paper we investigate the potential of social multimedia and open data for automatically identifying regions within the city. We conjecture that the regions may be characterized by specific patterns related to their visual appearance, the manner in which the social media users describe them, and the human mobility patterns. Therefore, we collect a dataset of Foursquare venues, their associated images and users, which we further enrich with a collection of city-specific Flickr images, annotations and users. Additionally, we collect a large number of neighbourhood statistics related to e.g., demographics, housing and services. We then represent visual content of the images using a large set of semantic concepts output by a convolutional neural network and extract latent Dirichlet topics from their annotations. User, text and visual information as well as the neighbourhood statistics are further aggregated at the level of postal code regions, which we use as the basis for detecting larger regions in the city. To identify those regions, we perform clustering based on individual modalities as well as their ensemble. The experimental analysis shows that the automatically detected regions are meaningful and have a potential for better understanding dynamics and complexity of a city.

Keywords: Urban computing · Social multimedia · Open data · Human mobility patterns · Semantic concept detection · Topic modelling

1 Introduction

A modern city is a complex organism shaped by the dynamics of various processes, related to e.g., economy, infrastructure and demographics. Administrative divisions, therefore, often do not match the *actual* regions in the city determined by their common functionality, architectural resemblance and ever-changing human flows. Identifying those regions is of critical importance for better understanding and modelling the processes in a city. Recently, open data has been shown invaluable in solving various problems a modern metropolis is faced with. However, open data is often associated with a lack of content connecting various sources and the absence of "factual" information about human flows and their perception of the habitat, so it does not provide the full picture of city dynamics.

L. Amsaleg et al. (Eds.): MMM 2017, Part II, LNCS 10133, pp. 148–159, 2017.
DOI: 10.1007/978-3-319-51814-5_13

While much of social multimedia is spontaneously captured, we conjecture that useful information can still be encoded in it. User-contributed images and their associated metadata in particular may offer valuable insights about the properties of a geographic area. Indeed, users with similar background and interests are intuitively more likely to reside, work or seek entertainment in the same geographic area, which will be reflected in their mobility patterns. Regions of a city associated with a similar functionality, such as business, education or entertainment, are further more likely to share certain visual attributes and be described by the users in similar way. Similarly, regions of the city constructed during a particular epoch may bear resemblance in terms of architectural style, function or demographics. Social multimedia has recently been succesfully deployed in analysing various urban and geographic phenomena [5,6,11,15,17,21,29]. Its full utilization remains challenging due to e.g., a wide variety of user interests, which makes their contributed content extremely topically heterogeneous. Additionally, as compared to professional or curated content, valuable information is only implicitly embedded in social multimedia. Appropriate multimedia analysis techniques are needed to bring out the best from social multimedia.

Fig. 1. A map of districts and postal regions in the Amsterdam Metropolitan area used in our study.

In this paper we investigate the potential of social multimedia and open data for identifying and characterizing functional regions of the city. For this purpose we systematically crawl the location-based social networking platform Foursquare and the content sharing platform Flickr. Additionally, we make use of open data related to various neighbourhood statistics, such as demographics, housing, transportation and services. In our analysis we focus on investigating both the descriptive power of each individual information channel, as well as its usefulness in interpreting results obtained by the other channels. As the test bed for our analysis we chose the Amsterdam Metropolitan Area (cf. Fig. 1), due to its moderate size, multicultural nature and a unique blend of historic and modern architecture. Next to the potential for aiding the analysis of various urban issues, our approach is applicable in a number of multimedia problems, such as location recommendation [26, 28], recognition and exploration [5], region summarization [17] and content placing [12, 13, 24].

The remainder of this paper is organised as follows. Section 2 provides a brief overview of related work. In Sect. 3 we introduce our approach and in Sect. 4 we report on experimental results. Section 5 concludes the paper.

2 Related Work

Recently, a number of approaches to detecting characterizing regions within a city have been proposed. The majority of these methods focus on utilizing information about points of interest (POIs) and human mobility patterns. For example, to discover functional regions of a city Yuan et al. analyse the distribution of POI categories together with information about road network topology and the GPS trajectories generated by taxi vehicles [27], while Toole et al. utilize spatio-temporal information about mobile phone calls [23]. More recently, Andrienko et al. proposed a visual analytics approach, which jointly utilizes mobile phone usage data together with information about land use and points of interest for discovering place semantics [1].

While the above-mentioned approaches have been proven effective in various use cases, they often rely on scarce and difficult to obtain data, which limits their wider adoption. Following a different logic, Thomee and Rae propose a multimodal approach to detecting locally characterizing regions in a collection of geo-referenced Flickr images [22]. The regions are detected at several scales, corresponding to e.g., world, continent, and country level. With regard to data use and research goal, the analysis conducted by Cranshaw et al. [3] is probably most relevant to our study. The authors of the paper utilize information about location and users of a large number of Foursquare venues to identify dynamic areas comprising the city, which they name *livehoods*. While offering several original ideas and an interesting qualitative analysis, the study does not utilize open data and the content (i.e., text and visual) richness of social multimedia.

3 Region Detection

In this section we describe our data collection procedure and then elaborate on our approach for identifying regions within a city based on different modalities. The approach pipeline is illustrated in Fig. 2.

3.1 Data Collection

To ensure reasonable granularity and to allow easier visual interpretation of the results, we use postal code regions as the unit for data collection and analysis. As parameter for querying the Foursquare venue database we use geo-coordinates of the centres of the regions corresponding to street-level postal codes, and set the query radius to 100 m. For each venue we further download images as well as the information about the users who recommended it.

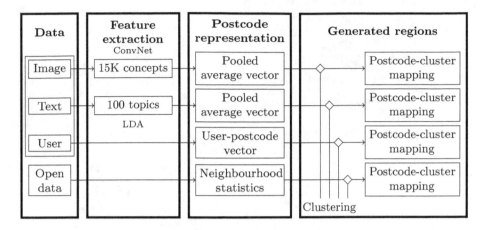

Fig. 2. The pipeline of our region detection approach.

As the Flickr API imposes fewer restrictions on the number of results per request, we query it using geo-coordinates of the centres of neighbourhood-level postal regions (cf. Fig. 1), modularly adjusting the radius to the size of the region. Images are downloaded together with the information about their uploaders as well as the title, tags and description. We restricted our search to the Creative Commons images assigned with the highest available location accuracy.

Due to an irregular shape of postal code regions, the above-mentioned data collection procedure can result in image or venue assignment to multiple postal code regions. To make precise assignments, we perform reverse geo-coding using postal code shapefiles downloaded from Google Maps [8].

Finally, we make use of official neighbourhood statistics, regularly compiled by the local government. The collection includes a large number of variables related to neighbourhood *demographics, companies, housing, energy consumption, motor vehicles, surface area*, and *facilities*. In Sect. 4.3 we will discuss the most discriminative variables from the above-mentioned categories.

3.2   Feature Extraction

Visual Features: Given the wide variety of potentially interesting visual attributes that should be captured and to make interpretation of results easier, we opt for image representation at an intermediate semantic level. For each image we extract 15,293 ImageNet concepts [4] output by a customised Caffe [10] implementation of "Inception" network [20]. A postcode region is then represented with a concept vector generated by applying average pooling [2] on concept vectors of individual images captured within it.

Text Features: We index the *title, description* and *tags* text of each Flickr image. Before computing a bag of words representation, we pre-process the text by removing HTML elements and stopwords. We further apply Online Latent

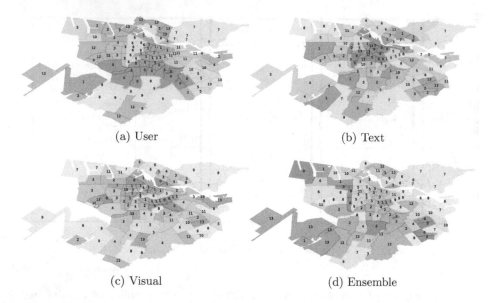

(a) User   (b) Text

(c) Visual   (d) Ensemble

Fig. 3. A map of Amsterdam regions produced by spectral clustering based on (a) user (b) text and (c) visual modalities as well as (d) their ensemble.

Dirichlet Allocation (LDA) [9] and extract 100 latent topics using the gensim framework [16]. Finally, we use average pooling over LDA vectors of individual images to produce a topical representation for a postal code region.

User Features: We represent each postal code region with a binary vector to indicate if a particular Flickr or Foursquare user visited it. We associate Foursquare users with a postal code region if they recommended a venue within it. Flickr users uploading an image captured within a postal code region are associated with it.

Open Data: For each postal code we create representations containing neighbourhood statistics of a particular category (e.g., *demographics* or *housing*), which we further combine in an early fusion fashion. The vectors are scaled to the [0,1] range.

3.3   Clustering

After the representations of postcode regions are computed for each social media modality (i.e., visual, text and user), we proceed by identifying the borders of larger geographic regions within the city. For that purpose, in the visual and text domain we independently compute cosine similarity between the semantic concept and LDA postcode representations. Similarly, for the user modality, we compute Jaccard similarity between binary user vectors representing postal code regions. The similarities are further used as an input into spectral clustering [14]. After the clustering in individual modalities is performed, we deploy an ensemble (consensus) clustering approach [19] to produce a single, reinforced clustering. In case of open data, we resort to k-means clustering, which is more appropriate for relatively short feature vectors.

We set the number of clusters to be equal to the number of administrative units in the city (e.g., 14 city quarters and municipalities visualised in Fig. 1). While performing clustering, unlike most related work, we intentionally chose not to impose a geographic constraint. This choice is motivated by our specific goal to investigate whether meaningful boundaries of geographic regions could be identified using individually information on human mobility patterns, visual properties and users' perception of the city. Since most clustering algorithms attempt to identify larger regions, imposing a geographic constraint would make conclusions of such analysis unreliable. For example, two neighbouring postal code regions would more likely belong to the same cluster, independently of their topical similarity, which could easily lead to misinterpretation of results.

4   Experimental Results

In this section we seek to answer the following questions:

1. What do the social media channels tell us about the regions in a city?
2. To what degree do the computed region boundaries agree with the official administrative division?
3. Which regions can be identified based on neighbourhood statistics?

Applying the procedure described in Sect. 3.1 to Amsterdam, we arrive at a collection consisting of 1,136 verified venues, 4,605 unique users and 34,419 images from Foursquare and 59,417 Flickr images uploaded by 2,342 users. The neighbourhood statistics are provided as open data by Statistics Netherlands (CBS) [18].

4.1   Regions Shaped by Social Multimedia

The maps of Amsterdam regions output by spectral clustering based on user, text and visual modality as well as the ensemble clustering are respectively shown in Figs. 3a–d. Our first observation is that relatively large and coherent regions can be identified in all four visualisations, most notably the one shown in Fig. 3a. The regions revealed by user modality are furthermore particularly easy to interpret. For example, the number "1" region, comprising a postal code in south-west corner and another ten in the central part of the figure corresponds to Amsterdam Airport Schiphol and the centre of the city (i.e., the Amsterdam Canal District). Such association is not surprising, since Amsterdam is a popular tourist destination[1] with more than 15,000,000 visits per year, many of which begin with arrival to Schiphol airport, followed by transfer to Amsterdam Central station. Additional three postcodes at the south of the same region correspond

---

[1] www.amsterdam.info/basics/figures/.

to the area around museum quarter, which is again, very popular among tourists. Another general observation related to Fig. 3a is that the user mobility patterns are normally limited to a particular quarter of the city.

The results of visual clustering reveal interesting patterns as well. For example, a region detected in south-west corner of Fig. 3c comprises visually characteristic large industrial areas. In the central quarter of the city, for example, a distinction is correctly made between the oldest city parts featuring typical canal houses and two postcode regions constructed more recently. Similar observations can be made about large regions in the north, east and west of the city, which were built in coherent architectural styles. Finally, a large cluster encompassing the postcodes in those three districts of the city corresponds to the decaying residential neighbourhoods.

Table 1. Evaluation of the agreement between partitioning into regions output by four different clustering algorithms and the official administrative division; the agreement is reported in terms of Adjusted Mutual Information (AMI).

	GRT	VIS	TXT	USR	ENS	RAN
GRT	•	0.134	0.115	0.452	0.265	0.014
VIS	0.134	•	0.106	0.114	0.367	0.005
TXT	0.115	0.106	•	0.106	0.295	0.035
USR	0.452	0.114	0.106	•	0.524	0.012
ENS	0.265	0.367	0.295	0.524	•	0.052
RAN	0.014	0.005	0.035	0.012	0.052	•

4.2   Agreement with the Official Division

Table 1 summarizes the agreement between the following area partitions: the ground truth official administrative division (GRT) as visualized in Fig. 1; the output of the four different clusterings: visual (VIS), text (TXT), user information (USR), and ensemble (ENS); the results of random clustering (RAN) serving as a reference point. As the measure of agreement we use Adjusted Mutual Information (AMI), which enumerates how the individual partitionings agree with each other [25]. An AMI of 1 means perfect agreement, while 0 corresponds to the average agreement with random partitioning.

Relatively low agreement between text, visual and user partitionings indeed indicates that all modalities provide complementary information. At the same time, all automatically generated partitionings score well above random, showing that the disagreement is topical, not manifesting by chance. The clustering by user modality is the closest to the administrative division, suggesting that the official administrative districts do roughly follow *vox populi*. A lower agreement with the administrative division observed in the case of visual and text modalities may reflect the fact that the large official administrative regions, for example, do not have uniform visual appearances or functionality. Ensemble clustering agrees with the constituent unimodal approaches reasonably, most dominantly with the user-based clustering, which is the least fragmented one (cf. Fig. 3). Ensemble thus does a good job in smoothing out the fragmentation brought to the table by visual and text clusterings, while including their complementary information. Making use of multimodal data and analysis is thus shown to bring rich information about the city, reflecting diverse aspects and showing promise for urban computing.

4.3   Regions by Neighbourhood Statistics

Figures 4a–d show the regions generated based on statistics about population, living, services and the combined neighbourhood statistics. Again, we observe large coherent and meaningful regions. For example, region "5" in Fig. 4d corresponds to some of the most exclusive neighbourhoods of the city, including the parts of Amsterdam Canal District and Old South. Further, large regions in the west, south-east and north of the city, corresponding to the known disadvantaged neighbourhoods are also correctly identified. Finally, region "6" encompasses several municipalities neighbouring Amsterdam. Similar observations can be made in Fig. 4c, where the regions with comparable socio-economic structure are grouped together.

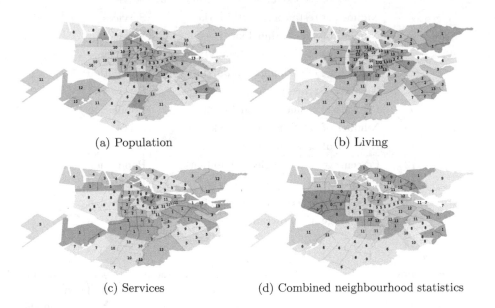

(a) Population

(b) Living

(c) Services

(d) Combined neighbourhood statistics

Fig. 4. A map of Amsterdam regions produced by k-means clustering based on different categories of neighbourhood statistics.

Table 2. Evaluation of the agreement between partitioning into regions based on different categories of official neighbourhood statistics and social multimedia; the agreement is reported in terms of Adjusted Mutual Information (AMI).

	GRT	VIS	TXT	USR	ENS	RAN
ENE	0.145	0.124	0.051	0.090	0.054	0.007
COM	0.260	0.029	0.008	0.093	0.040	−0.004
LIV	0.239	0.120	0.119	0.161	0.148	0.010
SUR	0.149	0.175	0.059	0.125	0.120	0.004
VEH	0.233	0.150	0.043	0.125	0.110	−0.006
SER	0.444	0.132	0.086	0.242	0.143	0.013
POP	0.385	0.127	0.112	0.239	0.182	0.008
ALL	0.462	0.147	0.112	0.260	0.188	0.043

Table 3. A list of top-15 variables comprising neighbourhood statistics, sorted by their importance in discriminating between the regions shown in Fig. 4d.

Rank	Variable	Category
1	Number of department stores within 5 km	Services
2	Number of hotels within 5 km	Services
3	Number of secondary schools within 5 km	Services
4	Number of restaurants within 5 km	Services
5	Number of bars within 1 km	Services
6	Number of attractions within 50 km	Services
7	Percentage of surinamese immigrants	Population
8	Number of grocery stores	Services
9	Percentage of married inhabitants	Population
10	Number of cafeterias within 5 km	Services
11	Number of bars within 5 km	Services
12	Percentage of unmarried inhabitants	Population
13	Percentage of turkish immigrants	Population
14	Percentage of western immigrants	Population
15	Percentage of moroccan immigrants	Population

The results shown in Table 2 suggest a high agreement between the regions yielded by the analysis of user mobility patterns on one side and the statistics about demographics, services and the combined neighbourhood statistics on the other. This agreement is also apparent in Figs. 3a, 4a, c and d, which further confirms the importance of information about human flows for better understanding the city. Another interesting observation in Table 2 is that the visual modality in general does a better job than text in "emulating" official neighbourhood statistics. In particular, visual modality appears to be effective in capturing the characteristics of a neighbourhood related to surface use, motor vehicles, services and population.

Finally, we train an extra-tree classifier [7] to investigate which variables are the most useful for discriminating between the regions generated based on combined neighbourhood statistics. Based on the results from Table 3 we conclude that the demographics and services are important forces shaping up the city.

5 Conclusions

In this paper we have investigated the feasibility of utilizing social multimedia and open data for identifying actual regions of the city. Our analysis involved information about human mobility patterns, visual appearance and user perception of the city as well as the official neighbourhood statistics. The experiments show that the areas proposed by our approach reflect diverse semantic aspects of the city, with each information channel contributing unique information. Region detection based on social multimedia and open data thus shows promise in reflecting the vibrant and ever-changing nature of the city, making it a solid basis for further urban computing research.

References

1. Andrienko, N., Andrienko, G., Fuchs, G., Jankowski, P.: Scalable and privacy-respectful interactive discovery of place semantics from human mobility traces. Inf. Vis. **15**(2), 117–153 (2016)
2. Boureau, Y.-L., Ponce, J., LeCun, Y.: A theoretical analysis of feature pooling in visual recognition. In: Proceedings of the 27th International Conference on Machine Learning (ICML 2010), pp. 111–118 (2010)
3. Cranshaw, J., Schwartz, R., Hong, J., Sadeh, N.: The livehoods project: utilizing social media to understand the dynamics of a city. In: International AAAI Conference on Weblogs and Social Media (2012)
4. Deng, J., Dong, W., Socher, R., Li, L.-J., Li, K., Fei-Fei, L.: Imagenet: a large-scale hierarchical image database. In: IEEE Conference on Computer Vision and Pattern Recognition, CVPR 2009, pp. 248–255, June 2009
5. Fang, Q., Sang, J., Xu, C.: Giant: geo-informative attributes for location recognition and exploration. In: Proceedings of the 21st ACM International Conference on Multimedia, MM 2013, pp. 13–22. ACM, New York (2013)
6. Boonzajer Flaes, J., Rudinac, S., Worring, M.: What multimedia sentiment analysis says about city liveability. In: Ferro, N., Crestani, F., Moens, M.-F., Mothe, J., Silvestri, F., Nunzio, G.M., Hauff, C., Silvello, G. (eds.) ECIR 2016. LNCS, vol. 9626, pp. 824–829. Springer, Heidelberg (2016). doi:10.1007/978-3-319-30671-1_74
7. Geurts, P., Ernst, D., Wehenkel, L.: Extremely randomized trees. Mach. Learn. **63**(1), 3–42 (2006)
8. Google Maps. Postcodes Amsterdam. http://goo.gl/hHoZWi. Accessed Nov 2015
9. Hoffman, M., Bach, F.R., Blei, D.M.: Online learning for latent Dirichlet allocation. In: Advances in Neural Information Processing Systems, NIPS 2010, pp. 856–864 (2010)

10. Jia, Y., Shelhamer, E., Donahue, J., Karayev, S., Long, J., Girshick, R., Guadarrama, S., Darrell, T.: Caffe: convolutional architecture for fast feature embedding. In: Proceedings of the 22nd ACM International Conference on Multimedia, MM 2014, pp. 675–678. ACM New York (2014)

11. Kennedy, L., Naaman, M., Ahern, S., Nair, R., Rattenbury, T.: How flickr helps us make sense of the world: context and content in community-contributed media collections. In: Proceedings of the 15th ACM International Conference on Multimedia, MM 2007, pp. 631–640. ACM, New York (2007)

12. Larson, M., Soleymani, M., Serdyukov, P., Rudinac, S., Wartena, C., Murdock, V., Friedland, G., Ordelman, R., Jones, G.J.F.: Automatic tagging, geotagging in video collections, communities. In: Proceedings of the 1st ACM International Conference on Multimedia Retrieval, ICMR 2011, pp. 51:1–51:8. ACM, New York (2011)

13. Luo, J., Joshi, D., Yu, J., Gallagher, A.: Geotagging in multimedia and computer vision–a survey. Multimedia Tools Appl. 51(1), 187–211 (2011)

14. Ng, A.Y., Jordan, M.I., Weiss, Y.: On spectral clustering: analysis and an algorithm. In: Dietterich, T., Becker, S., Ghahramani, Z. (eds.) Advances in Neural Information Processing Systems 14, pp. 849–856. MIT Press, Cambridge (2002)

15. Porzi, L., Rota Bulò, S., Lepri, B., Ricci, E.: Predicting and understanding urban perception with convolutional neural networks. In: Proceedings of the 23rd ACM International Conference on Multimedia, MM 2015, pp. 139–148. ACM, New York (2015)

16. Řehůřek, R., Sojka, P.: Software framework for topic modelling with large corpora. In: Proceedings of the LREC 2010 Workshop on New Challenges for NLP Frameworks, pp. 45–50. ELRA, Valletta, May 2010

17. Rudinac, S., Hanjalic, A., Larson, M.: Generating visual summaries of geographic areas using community-contributed images. IEEE Trans. Multimedia 15(4), 921–932 (2013)

18. Statistics Netherlands. Neighbourhood statistics. https://www.cbs.nl/nl-nl/maatwerk/2015/48/kerncijfers-wijken-en-buurten-2014. Accessed Nov 2015

19. Strehl, A., Ghosh, J.: Cluster ensembles – a knowledge reuse framework for combining multiple partitions. J. Mach. Learn. Res. 3, 583–617 (2003)

20. Szegedy, C., Liu, W., Jia, Y., Sermanet, P., Reed, S., Anguelov, D., Erhan, D., Vanhoucke, V., Rabinovich, A.: Going deeper with convolutions. In: 2015 IEEE Conference on Computer Vision and Pattern Recognition (CVPR), pp. 1–9, June 2015

21. Thomee, B., Arapakis, I., Shamma, D.A.: Finding social points of interest from georeferenced and oriented online photographs. ACM Trans. Multimedia Comput. Commun. Appl. 12(2), 36:1–36:23 (2016)

22. Thomee, B., Rae, A.: Uncovering locally characterizing regions within geotagged data. In: Proceedings of the 22nd International Conference on World Wide Web, WWW 2013, pp. 1285–1296 (2013)

23. Toole, J.L., Ulm, M., González, M.C., Bauer, D.: Inferring land use from mobile phone activity. In: Proceedings of the ACM SIGKDD International Workshop on Urban Computing, UrbComp 2012, pp. 1–8. ACM, New York (2012)

24. Trevisiol, M., Jégou, H., Delhumeau, J., Gravier, G.: Retrieving geo-location of videos with a divide & conquer hierarchical multimodal approach. In: Proceedings of the 3rd ACM International Conference on Multimedia Retrieval, ICMR 2013, pp. 1–8. ACM, New York (2013)

25. Vinh, N.X., Epps, J., Bailey, J.: Information theoretic measures for clusterings comparison: variants, properties, normalization and correction for chance. J. Mach. Learn. Res. **11**, 2837–2854 (2010)
26. Yin, H., Cui, B., Huang, Z., Wang, W., Wu, X., Zhou, X.: Joint modeling of users' interests and mobility patterns for point-of-interest recommendation. In: Proceedings of the 23rd ACM International Conference on Multimedia, MM 2015, pp. 819–822. ACM, New York (2015)
27. Yuan, J., Zheng, Y., Xie, X.: Discovering regions of different functions in a city using human mobility and POIs. In: Proceedings of the 18th ACM SIGKDD International Conference on Knowledge Discovery and Data Mining, KDD 2012, pp. 186–194. ACM, New York (2012)
28. Zahálka, J., Rudinac, S., Worring, M.: Interactive multimodal learning for venue recommendation. IEEE Trans. Multimedia **17**(12), 2235–2244 (2015)
29. Zheng, Y., Capra, L., Wolfson, O., Yang, H.: Urban computing: concepts, methodologies, and applications. ACM Trans. Intell. Syst. Technol. **5**(3), 38:1–38:55 (2014)

Discovering User Interests from Social Images

Jiangchao Yao[1](✉), Ya Zhang[1], Ivor Tsang[3], and Jun Sun[2]

[1] Cooperative Medianet Innovation Center, Shanghai Jiao Tong University,
Shanghai, China
{Sunarker,ya_zhang}@sjtu.edu.cn

[2] Institute of Image Communication and Network Engineering, Shanghai Jiao Tong
University, Shanghai, China
junsun@sjtu.edu.cn

[3] Centre for Artificial Intelligence, University of Technology Sydney,
Sydney, Australia
Ivor.Tsang@uts.edu.au

Abstract. The last decades have witnessed the boom of social networks. As a result, discovering user interests from social media has gained increasing attention. While the accumulation of social media presents us great opportunities for a better understanding of the users, the challenge lies in how to build a uniform model for the heterogeneous contents. In this article, we propose a hybrid mixture model for user interests discovery which exploits both the textual and visual content associated with social images. By modeling the features of each content source independently at the latent variable level and unifies them as latent interests, the proposed model allows the semantic interpretation of user interests in both the visual and textual perspectives. Qualitative and quantitative experiments on a Flickr dataset with 2.54 million images have demonstrated its promise for user interest analysis compared with existing methods.

Keywords: User interest mining · Multimedia analysis · Coupled learning

1 Introduction

Discovery of user interests is essential for personalized service, market analysis and demographic analysis, which has drawn wide attention from researchers in the past decades. There are a lot of works devoted to mining user interest from photos, texts, links, click data and social information [20]. In general, they fall into three categories, i.e., textual data based user mining [1], visual data based user mining [4,15,20] and hybrid data based user mining [5,6,9]. While different models have been explored and showed promise in a series of applications, they still suffer from some problems, e.g., handcrafted features, unclear interpretation about the interests and simply combinational multi-view methods.

© Springer International Publishing AG 2017
L. Amsaleg et al. (Eds.): MMM 2017, Part II, LNCS 10133, pp. 160–172, 2017.
DOI: 10.1007/978-3-319-51814-5_14

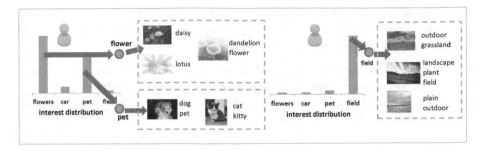

Fig. 1. Two Flickr user examples with different preferences. They take different photos and post the corresponding tags based on their specific interests that are sampled from their interest distributions.

In this paper, we consider these above three problems in discovering user interests from social images i.e., image contents and their tags. The motivation of our model is illustrated in Fig. 1. We make the assumption that each user has a latent interest distribution and they take photos and post tags based on their interests. For example, there are two users in Fig. 1 and the first user prefers "flower" and "pet" while the second user likes "field" more. So they generate different photos and tags according to the specific interests. Besides, since users have different degrees about their interests, they have different number of instances related to the interests, e.g., the first user has more flower-related photos and tags than that of pet-related photos and tags because he/she likes "flower" more than "pet". Based on above thoughts, we propose to simulate the procedure from the user interest distribution to photos and tags with a generative model, i.e., our hybrid mixture model. In the model, we similarly represent each user with a latent user interest distribution and for each (photo, tags) pair, we sample a specific interest assigned to them from the distribution, then based on this interest, the photo and its tags are generated from the visual and textual spaces respectively. Here, the interest bridges the image features and tags features at a latent variable level, which brings the potential to interpret the interest both in the visual and textual perspectives. In addition, to facilitate the generation from the interest (scale) to the photo feature (vector) and its tags feature (vector) in the model, we introduce the Gaussian mixtures to constitute two spaces. A variational EM is used to infer the posteriors of latent variables and learn the model parameters.

The major contribution of our model could be summarized into three points as follows:

- We proposed a hybrid mixture model which discovers user interests from social images, meanwhile considers the relationship of visual features and textual features at a latent variable level.
- The latent user interest vector learnt by our model has a clear interpretation and could be understood both in the visual and textual perspectives.

– We evaluate our model on a collection of 2.54 million Flickr images. Qualitative and quantitative analysis both demonstrate its promise in the real world applications compared with several existing methods.

The rest of the paper is organized as follows. Section 2 reviews the related work. Then we introduce our model and inference algorithm in details in Sect. 3. After that, we present the experimental performance on a large dataset in Sect. 4. Finally, Sect. 5 concludes this paper.

2   Related Work

Visual and textual data on the social media websites provides a comprehensive view about the characteristics of users. Therefore, a large range of research is devoted into using these two domains to improve the real world applications related to user interests. In the following, we give a brief review.

Some researchers build their models based on a single type of data e.g., textual metadata or visual data. For instance, Rosen-Zvi et al. [11] proposed a topic model to learn the author's topic distribution by considering the contribution of all the authors of one document. Yao et al. [22] mined the user profile in a coupled view by considering the effect of both the user and the object to the social tag's generation. With the coupled structure, the profiles of the user and the object are extracted from the non-IID tags at the same time. Xie et al. [20] explored the user-interested themes only from the personal photos with a hierarchical User Image Latent Space model. But the features they used in that work are at the pixel level, which brings a high computational complexity in the inferrence. Guntuku et al. [15] investigated approaches for modeling personality traits based on a collection of images the user tagged as "favorite". They demonstrated that personality modeling needed high level features in the image such as pose, scene and saliency. However, totally, the data in a single domain can not reflect all the preference information about user and to some extent, optimizing exhaustedly in one domain might lead to bias in the user profiling.

Some works on multi-domain data with low level features for user mining are explored and have been demonstrated the promising performance. For example, Pennacchiotti and Popescu et al. [14] implemented Twitter user classification by considering many kinds of data including user metadata, user tweeting behavior and linguistic contents etc and their final results outperformed that of each domain. Ikeda et al. [8] mined the user profile by facilitating both textual data and community information. A hybrid method was proposed to analyze the demographics of Twitter users for market analysis. Tang et al. [17] presented a combination approach to web user profiling with all kinds of user information and a series of feature selection procedures. Some similar works were also conducted by incorporating visual features, textual features and contextual features with late fusion strategies such as [3,6,7]. However, in these works, experts might be required in choosing the suitable features and merging the effect of each domain. It is a trivial work and lack of a generally simply way to deal with the fusion of domain features. Besides, most of above models used low level visual or textual features.

Recently, Deep neural network have shown significant advantages in computer vision area due to their capacity in learning semantic concepts compared with traditional machine learning methods. Some works to facilitate DNN have been explored in user interests mining. Joshi *et al.* [9] built Flickr user profiles on fusion of deep visual features extracted with CNN and textual TFIDF features and validated their model with group membership to demonstrate the performance. Elkahky *et al.* [5] proposed a multi-view deep learning approach for cross domain user modeling which could learn the profile from multiple domains of item information. Experiments on several large-scale real world data sets indicated the method worked much better than other systems by large margin. Niu *et al.* [13] proposed to analyze user sentiment towards a specific topic (e.g., brand) in the multi-view and explored a benchmark with existing methods i.e., early fusion, late fusion and multimodal deep Boltzmann machines [12] on the Twitter dataset. Although these works achieved promising performance in the applications, explicit interpretation about learnt user interest in their models is impossible, which departs from the nature of interest. In fact, latent variable models or some classical topic models like [2,19], are powerful to capture the coupled dependencies among multiple domains of data and give a clear interpretation about latent parameters. But there is still little work combining them with features learnt by deep neural network in user interest mining.

3   User Interest Discovery

3.1   Problem Statement

Given a set of (photo, tags) pairs of users as observations, we assume each user is a mixture of a small number of interests. And the creation of each (photo, tags) pair is attributable to the user's interests. Specifically, Let U, PT be the set of users and (photo, tags) pairs, for which the numbers are M and N_m ($m = 1, \ldots, M$) respectively. Let PT_{mn} be the nth photo and its tags of user $u_m \in U$, whose features are D_1 dimensional vector w_{mn} and D_2 dimensional vector v_{mn} extracted by LDA [2] and GoogLeNet [16] respectively. The interest distribution of u_m is indicated by θ_m and z_{mn} is a sample from the interest classes. $\{N(\mu_k^1, \sigma_k^1 \mathbf{I})\}_{k=1,\ldots,K}$ and $\{N(\mu_k^2, \sigma_k^2 \mathbf{I})\}_{k=1,\ldots,K}$ constitute the textual space as well as visual space. θ_m, μ_k^1 and μ_k^2 are all vectors in $\mathbb{R}^K$, $\mathbb{R}^{D_1}$ and $\mathbb{R}^{D_2}$ respectively.

Given the collection $\{(w_{11}, v_{11}), \ldots, (w_{MN_M}, v_{MN_M})\}$, our goal is to calculate the user interest distribution $\{\theta_m\}_{m=1,\ldots,M}$ and parameters α, $\{N(\mu_k^1, \sigma_k^1 \mathbf{I})\}_{k=1,\ldots,K}$ and $\{N(\mu_k^2, \sigma_k^2 \mathbf{I})\}_{k=1,\ldots,K}$.

3.2   A Hybrid Mixture Model

We transform this problem into a generative process and propose a hybrid mixture model following the idea of the graphical model. Formally, the generative procedure is summarized as follows:

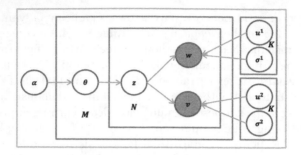

Fig. 2. The plate notation of our model. The shaded nodes w and v means observed variables, θ and z are the latent variables and the other nodes are model parameters.

- Draw the user interest distribution $\theta_m \sim Dirichlet(\alpha)$.
- Draw an interest class $z_{mn} \sim Multinomial(\theta_m)$.
- For each tags feature w_{mn},
  draw $w_{mn} \sim Gaussian(\mu^1_{z_{mn}}, \sigma^1_{z_{mn}} I)$.
- For each photo feature v_{mn},
  draw $v_{mn} \sim Gaussian(\mu^2_{z_{mn}}, \sigma^2_{z_{mn}} I)$.

We present the plate notation of our model in Fig. 2. In Fig. 2, each Gaussian mixtures builds a semantic space, i.e., textual space and visual space. Latent interest z bridges two spaces and is decided by both w and v, which makes the interest distribution interpretable in the visual and textual perspectives. The novelty and main differences from existing models are that we build our model on high-level features and introduce two Gaussian mixtures to model visual features and textual features respectively with combining two domains at the latent variable level.

3.3   Inference and Parameter Estimation

The key inferential problem of our model is to compute the posterior distribution of latent variables θ and z given data, $p(\theta, z|w, v)$. However, it is not easy to achieve the exact inference due to intractable $p(w, v)$ in the Bayes theorem. Therefore, an alternative algorithms, variational inference, could be taken into consideration.

For the inference of latent variables, we use two factorized distributions $q(\theta, z) = q(\theta|\gamma)q(z|\psi)$ of θ, z by introducing two free independent parameters γ, ψ to approach the real distribution $p(\theta, z)$ [21]. In accordance with original generative distribution, we choose $q(\theta|\gamma)$ as the Dirichlet form while the other one $q(z|\psi)$ keeps the multinomial. By minimizing Kullback-Leibler (KL) divergence between $p(\theta, z)$ and $q(\theta, z)$, we obtain the following solutions.

$$\psi_{mnk} \propto \exp(\Psi(\gamma_{mk}))N(w_{mn}|\mu^1_k, \sigma^1_k)N(v_{mn}|\mu^2_k, \sigma^2_k)$$
$$\gamma_{mk} = \alpha_k + \sum_n \psi_{mnk} \tag{1}$$

For the estimation of model parameters μ^1, σ^1, μ^2, σ^2 and α, the Jensen's inequality is first used to gain an adjustable lower bound. Then with previous optimal free parameters γ, ψ, it could be maximized by setting the derivatives to zero with respect to the parameters μ^1, σ^1, μ^2, σ^2 respectively, that is, the following solutions.

$$\mu_k^1 = \frac{\sum_m \sum_n \psi_{mnk} w_{mn}}{\sum_m \sum_n \psi_{mnk}} \ , \ \sigma_k^1 = \frac{\sum_m \sum_n \psi_{mnk} \left(\mu_k^1 - w_{mn}\right)^T \left(\mu_k^1 - w_{mn}\right)}{D_1 \sum_m \sum_n \psi_{mnk}}$$

$$\mu_k^2 = \frac{\sum_m \sum_n \psi_{mnk} v_{mn}}{\sum_m \sum_n \psi_{mnk}} \ , \ \sigma_k^2 = \frac{\sum_m \sum_n \psi_{mnk} \left(\mu_k^2 - v_{mn}\right)^T \left(\mu_k^2 - v_{mn}\right)}{D_2 \sum_m \sum_n \psi_{mnk}} \qquad (2)$$

And for Dirichlet prior α, we use Newton-Raphson method to iteratively update it like LDA [2]. Iterating the inference and parameter estimation steps, the final solution to our model could be achieved gradually.

4 Experiments

4.1 Dataset

The dataset we used to perform our experiments is a subset of Yahoo Flickr Creative Commons 100 Million (YFCC100M) dataset [18]. To form it, we use the WordNet[1] to do tag lemmatization and remove all the photos that had no tags. Then, we randomly select 10,000 users whose photo number ranged from 50 to 1000 from the remaining dataset since too few and too many is not normality. Then after crawling their photos and meta information from the Flikcr website, we acquired the final dataset. In summary, it consists of 10,000 users, 2,451,574 photos and a vocabulary of 20,167 words. For each user, there are mean 245.1 photos and For each photo, there are 3.6 tags in average. In the experiments, we use the GoogLeNet [16] to extract 1024-dimension visual vector for each photo and use LDA [2] to extract 1000-dimension textual feature for the tags of each photo to ensure enough feature granularity.

4.2 Qualitative Analysis

In Fig. 2, we introduced two Gaussian mixtures to cluster the visually and textually similar features independently in each space and claimed that there was a correspondence between them to help us clearly understand the latent interest. To demonstrate it, 6 interests learnt by our model are illustrated in Fig. 3.

From these 6 interests, we find that the image clusters are visually similar and the textual topics clearly interpret the interest. For example, the 1st interest is represented by some church-style architecture photos in the visual domain. And in the textual space, it presents two location-related topics (1st and 3rd topics)

---

[1] http://www.nltk.org.

(a) Visual Domains	(b) Textual Domains

Fig. 3. Learned interests in the visual domains and textual domains. Each row corresponds to an interest represented by 8 representative images and top 3 topics.

and one religious building topic (2nd). As we know, when someone likes religious buildings and travels to see them, they not only talk about the buildings with architectural vocabulary, but also tag them with location words meaning that these places may contain the religious information. In fact, Turkey and Spain are really religion-related. Therefore, textual domain complementarily express this interest along with visual domain. For another example, in the 5th interest, it aggregates some photos about snacks and desserts, and the three topics really gives the words related to food. So this is an eating interest. In fact, there are also a lot of other interests that we do not present in this paper due to the

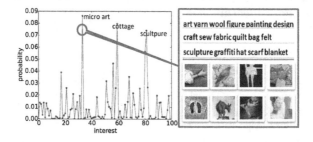

Fig. 4. The interest distribution of 7365168@N03. His significant interests are manually labeled by considering the corresponding textual and visual domain, e.g., "micro art".

limited space, but almost all of them reflect one of human's tastes in a specific perspective in two domains.

In our model, each user is represented by an interest distribution θ. By observing it, we can figure out what the user is interested in. Figure 4 presents such an example. As we can see, 7365168@N03 has high values regarding "mircro art", "cottage" and "sculpture", so he might be an artist-style person, which indicates that the θ learnt by our model helps us clearly understand users.

4.3 Gender Classification

To show its promising usage compared with existing methods in the supervised applications, we would implement a gender classification experiment. Here, the classical SVM[2] is chosen as our classifier with different n-fold cross-validations ($n = 4, 6, 8, 10$) and four baselines are used to compare with our model HM.

- tag based methods: TFIDF [9], Doc2Vec [10].
- photo based methods: Kmeans-V (use deep features extracted by GoogLeNet [16]).
- tag and photo based methods: Multi-UP [9].

Besides, we also investigate the relationship between classification performance and the interest number $K = 30, 50, 100, 200, 500$ in our model. The performance is presented in Tables 1 and 2.

Table 1. Performance of five methods

Methods	n = 4	n = 6	n = 8	n = 10
TFIDF	0.6544	0.6594	0.6634	0.6650
Doc2Vec	0.6623	0.6642	0.6649	0.6654
Kmeans-V	0.6946	0.6932	0.6942	0.6941
Multi-UP	0.7102	0.7114	0.7160	0.7160
HM_{100}	**0.7466**	**0.7482**	**0.7458**	**0.7489**

From Table 1, we find that classification results with the textual feature are usually lower than those with visual features comparing TFIDF, Doc2Vec with Kmeans-V. Methods using multiple sources perform better than those using single source from the results of Multi-UP and our model. For the multi-source methods, our model HM_{100} owns the best performance (around 0.74) compared with Multi-UP (about 0.71). One possible reason could trace back to Eq. (1) where two domain features adaptively fuse together to decide the free parameter ϕ_{ijk} and further feedback to γ_{ik}. This makes the fusion in each photo level. However, Multi-UP mainly considers the fusion in the final classification step which is to find a best linear combination weight.

---

[2] http://www.support-vector-machines.org/.

Table 2. Our model HM with different K

Interests	n = 4	n = 6	n = 8	n = 10
HM$_{30}$	0.6592	0.6592	0.6600	0.6604
HM$_{50}$	0.7339	0.7358	0.7333	0.7347
HM$_{100}$	0.7466	0.7482	0.7458	0.7489
HM$_{200}$	**0.7728**	**0.7724**	**0.7723**	**0.7726**
HM$_{500}$	0.7663	0.7721	0.7712	0.7659

According to Table 2, there is an incline trend in the gender classification along with interest number increasing. When we choose 30 interests, the precision is almost lowest 0.65. It may be because low interest number compresses much different cluster information which is beneficial to distinguish the user's gender. And when we choose 200 interests, it achieves almost best performance 0.77. However, up to 500, the results decrease gradually, which might attribute that more detailed information brings the classifier slightly overfitting.

When we apply the linear kernel in the SVM, it is easy to find the important interests in classifying gender by comparing the SVM weights for each feature dimension. To figure out the difference, we present 5 interests with higher weights for females and males by illustrating the topical word as well as 8 photos of each interest in Fig. 5. From the illustration, we find that females tends to more like taking selfies, while males more like taking photos for "car". In terms of the landscape interest, females are fond of close shot of flowers and trees while males likes panorama view of spacious scene such as mountain, sunset and field. Females like taking more photos of lovely pets or creative arts in the daily life, but males are fond of tall building photos.

4.4   Friend Recommendation

In this section, we evaluate our model on friend recommendation. We randomly divide one user's data into two equal parts and use the user interest distribution

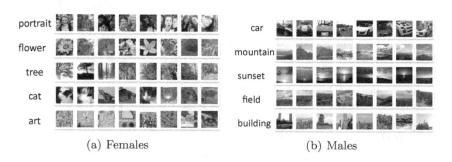

(a) Females   (b) Males

Fig. 5. Five important interests of females and males. Here, we only illustrate one representative tag of the topics and eight photos for one interest to save space.

Table 3. Recommendation performance compared with three baselines which respectively uses textual features, visual features and both features.

Methods	K = 30	K = 50	K = 100	K = 200	K = 500
TFIDF	0.4079	0.4079	0.4079	0.4079	0.4079
Doc2Vec	0.6174	0.6194	0.6143	0.5307	0.4853
Kmeans-V	0.7994	0.8257	0.7528	0.7877	0.6985
Multi-UP	0.8200	0.8370	0.8055	0.8228	0.7450
HM	**0.8289**	**0.8922**	**0.9373**	**0.9476**	**0.9602**

learnt from one part to retrieve that of the other part. Formally, we use L1-norm to compute the distance of two interest distributions and use mean reciprocal rank[3] (MRR) to compute the retrieval results. We set the interest number 30, 50, 100, 200, 500 and choose four methods TFIDF, Doc2Vec [10], Kmeans-V and multi-view user profiling (Multi-UP) [9] as our baselines like previous section. The results are illustrated in Table 3.

From Table 3, our model usually has the best MRR value and achieves the best score about 0.96 when $K = 500$. TFIDF remains the same performance 0.40 with the interest number changing. Doc2Vec achieves 0.6194 when $K = 50$ and Kmeans-V has the best performance at 0.82. But both two methods lose compared with our model because of limited information in each domain. Besides, Multi-UP is also lower than our model when $K = 30, 50, 100, 200, 500$. Therefore, the experimental results indicate that our model extracts more complementary information from two domains and reliably represent user interests than Multi-UP. To further study the performance of users who have different number of photos, we split the users into different groups by partitioning the photo number 50-1000 into 10 intervals, <100, 100–200, ..., 900–1000. Then the MRR of the users in different intervals is computed and plotted in Fig. 6. From it, an interesting phenomenon has been discovered, i.e., users who have few social images will have low MRR. It means the less social images you have, less reliable the recommendation is. This encourages users to share more social images if they want to get better recommendation.

To directly visualize its recommendation performance, we randomly choose one user 8844520@N02 and recommend top three users who have small L1-norm distances with him in Fig. 7. From the figure, the target user 8844520@N02 is very interested in cars, which is indicated by his car photos and some frequent tags such as ford, race, motor and mustang in the tag cloud. Besides, in fact, there is a peak in the 2nd interest of his θ which is car-related from the user interest distribution. From the recommendation results, the top three users really share the similar interests with the target user according to the distributions, photos and the tag clouds. For example, the top one user 10938603@N03 has only 0.14 difference with the target user in the interest distribution. And from

---

[3] https://en.wikipedia.org/wiki/Mean_reciprocal_rank.

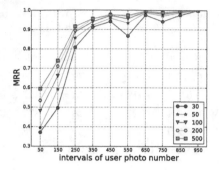

Fig. 6. The Mean Reciprocal Rank of different users whose photo number ranging in different intervals, e.g., less than 100. We plot each MRR in the middle of the corresponding interval with different interest number $K = 30, 50, 100, 200, 500$.

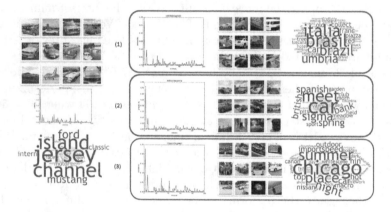

Fig. 7. One example on friend recommendation. We plot the representative photos and tag clouds of the target user at the left. Then we recommend three top similar users to them by calculating L1-norm distance of user interest distributions. Their representative photos and tag clouds are also illustrated correspondingly.

his photos, a lot of cars, a small number of helicopter photos and other kinds of photos could be found. In the tag cloud, some frequent tags, e.g., car, race, driver, auto, all indicate the user's taste.

5   Conclusion

In this paper, we study the problem of user interest discovery from social images. We proposed a hybrid mixture model to learn the user interest distribution from the high level visual features and textual features. Experiments on a 2.54 million Flickr photos showed that the interests learnt by our model had a clear interpretation both in the visual and textual perspectives and had potential usage in analyzing social users. In addition, a gender classification experiment and a friend recommendation application were explored in our paper, which showed the promising performance in analyzing demographical characteristics of users and retrieving the people who shared the similar interests.

Acknowledgements. The work is partially supported by the High Technology Research and Development Program of China 2015AA015801, NSFC 61521062, STCSM 12DZ2272600.

References

1. Abel, F., et al.: Twitter-based user modeling for news recommendations. In: Proceedings of the 23rd International Joint Conference on Artificial Intelligence. AAAI Press (2013)
2. Blei, D.M., et al.: Latent Dirichlet allocation. J. Mach. Learn. Res. (2003)
3. Campbell, W.M., et al.: Content+ context networks for user classification in Twitter. In: In Frontiers of Network Analysis: Methods, Models, and Applications Workshop at Neural Information Processing Systems (2013)
4. Crandall, D., Snavely, N., et al.: Modeling people and places with internet photo collections. Commun. ACM (2012). ACM
5. Elkahky, A.M., et al.: A multi-view deep learning approach for cross domain user modeling in recommendation systems. In: WWW, World Wide Web Consortium (2015)
6. Farseev, A., et al.: Harvesting multiple sources for user profile learning: a big data study. In: Proceedings of the 5th ACM on International Conference on Multimedia Retrieval (2015)
7. Huang, S., et al.: Social friend recommendation based on network correlation and feature co-clustering. In: Proceedings of the 5th ACM on International Conference on Multimedia Retrieval (2015)
8. Ikeda, K., et al.: Twitter user profiling based on text and community mining for market analysis. Knowl.-Based Syst. (2013)
9. Joshi, D., et al.: Building user profiles from shared photos. In: Proceedings of the 2015 Workshop on Community-Organized Multimodal Mining: Opportunities for Novel Solutions. ACM (2015)
10. Le, Q.V., Mikolov, T.: Distributed representations of sentences and documents. Comput. Sci. (2014)
11. Rosen-Zvi, M., et al.: The author-topic model for authors and documents. In: Proceedings of the 21st Conference on Uncertainty in Artificial Intelligence. Morgan Kaufmann Publishers Inc. (2004)
12. Srivastava, N., Salakhutdinov, R.R.: Multimodal learning with deep Boltzmann machines. In: Frontiers of Network Analysis: Methods, Models, and Applications Workshop at Neural Information Processing Systems (2012)
13. Niu, T., Zhu, S., Pang, L., Saddik, A.: Sentiment analysis on multi-view social data. In: Tian, Q., Sebe, N., Qi, G.-J., Huet, B., Hong, R., Liu, X. (eds.) MMM 2016. LNCS, vol. 9517, pp. 15–27. Springer, Heidelberg (2016). doi:10.1007/978-3-319-27674-8_2
14. Pennacchiotti, M., Popescu, A.M.: A machine learning approach to twitter user classification. In: Proceedings of the 21st International Joint Conference on Artificial Intelligence. AAAI Press (2011)

15. Guntuku, S.C., Roy, S., Weisi, L.: Personality modeling based image recommendation. In: He, X., Luo, S., Tao, D., Xu, C., Yang, J., Hasan, M.A. (eds.) MMM 2015. LNCS, vol. 8936, pp. 171–182. Springer, Heidelberg (2015). doi:10.1007/978-3-319-14442-9_15
16. Szegedy, C., et al.: Going deeper with convolutions. In: The IEEE Conference on Computer Vision and Pattern Recognition (2015)
17. Tang, J., et al.: A combination approach to web user profiling. ACM Trans. Knowl. Discov. Data (2010)
18. Thomee, B., et al.: YFCC100M: the new data in multimedia research. Commun. ACM (2016). ACM
19. Wang, Z., et al.: Bilateral correspondence model for words-and-pictures association in multimedia-rich microblogs. ACM Trans. Multimedia Comput. Commun. Appl. (2014)
20. Xie, P., et al.: Mining user interests from personal photos. In: Proceedings of the 29th AAAI Conference on Artificial Intelligence (2015)
21. Xing, E.P., et al.: A generalized mean field algorithm for variational inference in exponential families. In: Proceedings of the 19th Conference on Uncertainty in Artificial Intelligence. Morgan Kaufmann Publishers Inc. (2002)
22. Yao, J., et al.: Joint latent Dirichlet allocation for non-iid social tags. In: IEEE International Conference on Multimedia and Expo (2015)

Effect of Junk Images on Inter-concept Distance Measurement: Positive or Negative?

Yusuke Nagasawa[✉], Kazuaki Nakamura, Naoko Nitta,
and Noboru Babaguchi

Graduate School of Engineering, Osaka University, Suita, Japan
{nagasawa,k-nakamura,naoko,babaguchi}@nanase.comm.eng.osaka-u.ac.jp

Abstract. In this paper, we focus on the problem of inter-concept distance measurement (ICDM), which is a task of computing the distance between two concepts. ICDM is generally achieved by constructing a visual model of each concept and calculating the dissimilarity score between two visual models. The process of visual concept modeling often suffers from the problem of junk images, i.e., the images whose visual content is not related to the given text-tags. Similarly, it is naively expected that junk images also give a negative effect on the performance of ICDM. On the other hand, junk images might be related to its text-tags in a certain (non-visual) sense because the text-tags are given by not automated systems but humans. Hence, the following question arises: Is the effect of junk images on the performance of ICDM positive or negative? In this paper, we aim to answer this non-trivial question from experimental aspects using a unified framework for ICDM and junk image detection. Surprisingly, our experimental result indicates that junk images give a positive effect on the performance of ICDM.

Keywords: Inter-concept distance · Junk image detection · Image reliableness · Iterative calculation

1 Introduction

With the rapid growth of social networking services (SNS) on which users can share their own photos or images with others such as Facebook, Twitter, and Flickr, the amount of images stored in the web space has been increasing recently. Many of the images on SNS are publicly available as well as annotated with text-tags by their owners. This provides a large-scale dataset of tagged images, which accelerates researches in the field of image retrieval and recognition. One example of such researches is visual concept learning (VCL) [1–6], which is a task of constructing visual models of tags, or concepts, using their image instances. VCL is a hot topic today, because it may be able to bridge the *sematic gap* [7] between low-level visual features and high-level semantic concepts. Inter-concept distance measurement (ICDM) [6, 8–13] is another topic driven by the large-scale tagged dataset, which is a task of computing the distance between two concepts.

© Springer International Publishing AG 2017
L. Amsaleg et al. (Eds.): MMM 2017, Part II, LNCS 10133, pp. 173–184, 2017.
DOI: 10.1007/978-3-319-51814-5_15

This can be achieved by calculating the dissimilarity score between two concept models constructed by VCL. ICDM can play an important role in various vision applications including image retrieval, automatic image annotation, indexing, and clustering. This paper focuses on this task, i.e., ICDM.

The general procedure of ICDM for two concepts u and v is as follows: (i) A set of images annotated with u and another annotated with v are gathered, typically from SNS websites. (ii) Visual models of u and v are constructed by using the respective image set. (iii) The dissimilarity score between the two visual models is calculated. Ideally, in the step (i), only "relevant images", i.e., the images whose visual content is truly related to the given tags should be gathered. However, in reality, some of the gathered images would be "junk images", i.e., the images whose content is visually unrelated to the given tags, because the tags of SNS-based images are variously given depending on their owner's background knowledge, perception, and personality. Kennedy et al. [14] indicate that *there is a roughly 50/50 chance that the concept actually appears in the image* in the case of Flickr.

Junk images are the main factor of performance degradation of VCL. Aiming to construct better visual models, several methods for removing junk images from a gathered dataset have been proposed [1,15,16]. Similarly, we can naively guess that junk images also degrade the performance of ICDM. However, this might not be always true. Because tags are given to images by not automated systems but humans in SNS, there might be some semantical (non-visual) relationships between the image content and the given tags even in the case of junk images, and the relationship might be useful for ICDM. For example, in the dataset we gathered from Flickr in our previous study [13], there are some images annotated with both "lake" and "sandwich". This would be because some Flickr users went to lake for fishing or something and had sandwiches for lunch at the lake. Of course "sandwich" itself is not so deeply related to "lake" because people can eat sandwiches not only at lake but also at many other places. Hence, the images annotated with "sandwich" often act as junk images for the concept of "lake". This is also the case with the concept of "river", which also co-occurs with "sandwich" in the same dataset. However, humans would intuitively feel that the concepts of "lake" and "river" are close with each other. Therefore, the images annotated with "sandwich", which can co-occur with both "lake" and "river", could provide a positive effect to a calculation process of the distance between "lake" and "river".

As seen in the above example, not only relevant images but also junk images might reflect a certain property of the target concept and sometimes might provide useful information in ICDM. On the other hand, it is a very reasonable idea that junk images degrade the performance of ICDM. Now the following question arises: Is the effect of junk images on the performance of ICDM positive or negative? In this paper, we aim to answer this question from experimental aspects. To this end, we first propose a unified framework that integrates the processes of ICDM and junk image detection, and then experimentally examine the above question using the proposed framework. The contribution of this paper is summarized as follows: First, we integrate ICDM and junk image detection into a unified framework. Although there are a lot of previous studies focusing on either ICDM or junk image detection, this is the first work integrating these two tasks, to the best of our knowledge. Second, we experimentally clarify how junk images affect the performance of ICDM, which is a non-trivial question from the above reasons.

The remainder of this paper is organized as follows: we first review some previous woks focusing on ICDM and junk image detection in Sect. 2. In Sect. 3, we describe the proposed framework in detail. In Sect. 4, we experimentally evaluate the effect of junk images on the performance of ICDM and answer the above question. Finally, we conclude this paper in Sect. 5.

2   Related Works

2.1   Inter-concept Distance Measurement

As mentioned in Sect. 1, ICDM is generally achieved by calculating the dissimilarity score between two visual concept models constructed by VCL. There are two types of VCL approaches: discriminative one [1–4,6] and generative one [5]. For each concept c, the former tries to learn a classifier which classifies the presence or absence of the concept c in images. On the other hand, the latter tries to estimate a probability distribution $p(\boldsymbol{x}|c)$, where $\boldsymbol{x}$ is a certain feature vector extracted from each image. In VCL, discriminative approach is more widely studied than generative approach due to its high performance. However, it is not a straightforward problem to calculate the dissimilarity score between two classifiers, especially in the case that each classifier has a complex discriminant hypersurface. In contrast, the dissimilarity score between two probability distributions can be directly calculated as information divergence. Hence, in ICDM, it is more common to employ generative approach for constructing visual concept models. For instance, Kawakubo et al. [8] model the distribution $p(\boldsymbol{x}|c)$ as $p(\boldsymbol{x}|c) = \int p(\boldsymbol{x}|h)p(h|c)dh$ using probabilistic latent semantic analysis (pLSA), where h is a hidden state. Then they calculate JS divergence [17] between $p(h|u)$ and $p(h|v)$ for each concept pair (u,v) and use it as the inter-concept distance (ICD) between u and v. Wu et al. [9] also employ pLSA and calculate JS divergence-based ICD as $d(u,v) = \iint p(h|u)p(h'|v)D_{JS}[h||h']dhdh'$, where $D_J S[h||h']$ is JS divergence between $p(\boldsymbol{x}|h)$ and $p(\boldsymbol{x}|h')$. Fan et al. [6], Katsurai et al. [10], and Wang et al. [11] do not employ JS divergence, but their methods are also based on the generative approach of VCL. Most of these methods employ a single kind of visual feature as $\boldsymbol{x}$, whereas Zhang and Liu [12] propose to use multiple kinds of visual features for further performance improvement. Although it is reported that the above methods could achieve a good performance, most of them do not focus on the effect of junk images. Kawakubo et al. [8] consider it, but they only remove junk images from a gathered image dataset. Unlike this, we attempt to experimentally examine the effect of junk images on the performance of ICDM and answer the question described in Sect. 1.

2.2 Junk Image Detection

For the purpose of improving the performance of VCL, several methods for detecting and removing junk images from a gathered image dataset have been proposed. These methods are roughly divided into two types: classifier-based approach and corpus-based approach. Classifier-based approach generally makes the following assumption: a large set of relevant images can be hardly gathered, whereas a relatively small set of relevant images can be gathered manually. Based on the assumption, this type of approach tries to train a binary classifier that can distinguish whether an input image is junk or not for each concept, using such a small set of relevant images. Setz et al. [15] simply uses visual features for training the classifier. To improve the performance, Schroff et al. [16] use not only visual features but also metadata such as image titles and filenames. These methods are quite similar to discriminative VCL approach. One drawback of these methods is insufficient accuracy of the trained classifiers, because it is difficult to train a good classifier with a small dataset, especially in the case of abstract concepts.

On the other hand, corpus-based approach focuses on the relationship between text-tags, which is generally modeled with a certain corpus. Suppose that there is an image I annotated with tag c. Let $T(I)$ be a set of tags given to the image I except for c. If many members of $T(I)$ are semantically related to c, the image I can be considered as relevant for c. Conversely, if many members of $T(I)$ are unrelated to c, the image I seems to be junk for c. Based on this notion, corpus-based approach calculates the degree of relatedness between various pairs of tags in order to distinguish whether an input image is junk or not. The method of Zhu et al. [1], which is a typical example of corpus-based approach, models the degree of relatedness between tags $t \in T$ and c as conditional probability $p(t|c)$, and calculates the following score for each image I:

$$r(I) = \frac{1}{|T(I)|} \sum_{t \in T(I)} p(t|c).$$ (1)

If the score $r(I)$ is lower than some threshold, I is judged as junk for c. The conditional probability $p(t|c)$ is typically computed as a co-occurrence frequency of tag pair (t, c) divided by a term frequency of tag c, using a certain corpus. The probability $p(t|c)$ in Eq. (1) would be strongly related to ICD between t and c; two concepts having smaller ICD would co-occur with each other more frequently.

3   Unified Framework for ICDM and Junk Detection

Ideally, to examine the effect of junk images, we should use a set of tagged images in which ground truth labels of each image, i.e., labels indicating whether the image is junk or not for each given tag, are available. However, currently there are no such datasets, to the best of our knowledge. Moreover, adding such ground truth labels to an existing dataset is time-consuming and labor-expensive. Therefore we employ corpus-based approach of junk image detection to judge whether each image is junk or not, which can be easily scaled to a large dataset than classifier-based approach. As seen in Sect. 2.2, corpus-based approach requires a conditional probability $p(t|c)$ for each concept pair (t, c), which is strongly related to ICD between t and c. Hence, it is considered that we can detect junk images more accurately if we employ a better ICDM method for obtaining the probability $p(t|c)$. Based on this consideration, we propose a unified framework which integrates ICDM and corpus-based junk image detection. In this section, we describe the framework in detail, whose overview is as follows: First, we introduce the index named "image reliableness", which quantifies how likely the image is relevant image. Then we iteratively perform the following two steps: (a) estimating image reliableness for each image using a current result of ICDM and (b) calculating ICD for each concept-pair based on the estimated image reliableness.

3.1   Notation

Before describing the proposed framework in detail, we first summarize the notation used in this paper. Let $\mathcal{C}$ denote a finite set of concepts. For each concept $c \in \mathcal{C}$, let $I_i^c (i = 1, \ldots, N_c)$ denote the i-th image instance of concept c, where N_c is the total number of image instances of concept c. Each image is converted into the same kind of feature vector $\boldsymbol{x}$ by some feature extractor. Let $\boldsymbol{x}_i^c$ denote the actual value of feature vector $\boldsymbol{x}$ extracted from image I_i^c. For each concept pair $(u, v) \in \mathcal{C}^2$, let $d(u, v)$ be the ICD between u and v. For each image I_i^c, let $r(I_i^c)$ denote its reliableness.

3.2   Initial Calculation of ICD

In the first step of the proposed method, we compute ICD for every concepts pair without considering image reliableness. The computed ICD is used as the initial value for the proposed iterative process.

As mentioned in Sect. 2.1, probability distribution $p(\boldsymbol{x}|c)$ is first estimated in a general ICDM framework. In this paper, we model $p(\boldsymbol{x}|c)$ with a K-dimensional categorical distribution, quantizing feature vector $\boldsymbol{x}$ into K-representative values $\{\boldsymbol{y}_k | k = 1, \cdots, K\}$ by using a vector quantization method. Let $\boldsymbol{\theta}^c = (\theta_1^c \cdots \theta_K^c)^T$ denote a model parameter of the categorical distribution $p(\boldsymbol{x}|c)$. Using a set of image instances $\{I_i^c\}$ of concept c, $\boldsymbol{\theta}^c$ is estimated as

$$\theta_k^c = \frac{1}{Z_c} \sum_{i=1}^{N_c} \delta(\boldsymbol{x}_i^c, \boldsymbol{y}_k), \tag{2}$$

where Z_c is a constant for satisfying $\sum_{k=1}^{K} \theta_k^c = 1$, and $\delta(\boldsymbol{x}_i^c, \boldsymbol{y}_k)$ is an assignment function that is defined as

$$\delta(\boldsymbol{x}_i^c, \boldsymbol{y}_k) = \begin{cases} 1 & \text{if } k = \text{argmin}_l ||\boldsymbol{x}_i^c - \boldsymbol{y}_l|| \\ 0 & \text{otherwise} \end{cases}. \tag{3}$$

ICD between concept $u \in C$ and $v \in C$ is calculated as the dissimilarity score between the distribution parameterized by $\boldsymbol{\theta}^u$ and that parameterized by $\boldsymbol{\theta}^v$. To this end, we use JS divergence [17], which is defined as

$$d(u,v) = D_{\mathrm{JS}}\left[\boldsymbol{\theta}^u \parallel \boldsymbol{\theta}^v\right] = \sum_{k=1}^{K} \left\{ \frac{\theta_k^u}{2} \log \frac{2\theta_k^u}{\theta_k^u + \theta_k^v} + \frac{\theta_k^v}{2} \log \frac{2\theta_k^v}{\theta_k^u + \theta_k^v} \right\}, \tag{4}$$

JS divergence is symmetric and nonnegative. Its value becomes 0 if and only if $\boldsymbol{\theta}^u = \boldsymbol{\theta}^v$. Moreover, the square root of JS divergence satisfies the triangle inequality. These facts mean the square root of JS divergence is a strict metric.

3.3   Estimation of Image Reliableness Based on ICD

For estimating the reliableness $r(I_i^c)$ for each image I_i^c, we basically follow the method of Zhu et al. [1], which is formulated as Eq. (1). The difference between Zhu's method and our method is to compute $p(t|c)$ not using a text corpus but using $d(t,c)$ calculated by Eq. (4). It can be naturally assumed that two concepts t and c having smaller $d(t,c)$ would co-occur with each other more frequently. Based on the assumption, we propose to compute $p(t|c)$ as a monotonically decreasing function of $d(t,c)$. In this paper, we focus on the fact that the square root of JS divergence is a strict metric and employ a kind of Gaussian function

$$g(t,c) = \hat{g}(d(t,c)) = \exp\left\{-\frac{d^2(t,c)}{2\sigma}\right\} \tag{5}$$

as the monotonically decreasing function, which is used in place of $p(t|c)$. Let $T(I_i^c)$ be a set of tags given to image $T(I_i^c)$ except for c. Based on Eq. (5), we calculate the reliableness $r(I_i^c)$ as

$$r(I_i^c) = \frac{1}{|T(I_i^c)|} \sum_{t \in T(I_i^c)} g(t,c), \tag{6}$$

where $|\mathcal{S}|$ means the size of set $\mathcal{S}$. Note that other monotonically decreasing functions can be also used instead of Gaussian function.

3.4   Calculation of ICD Based on Image Reliableness

Using the estimated image reliableness $r(I_i^c)$, we re-estimate model parameter $\boldsymbol{\theta}^c$ for each concept $c \in C$ and calculate ICD again for all concept pairs $(u,v) \in C^2$. The idea is quite simple: each image I_i^c (or each feature vector $\boldsymbol{x}_i^c$) is weighted by $r(I_i^c)$ before Eq. (2) is calculated. More specifically, we estimate $\boldsymbol{\theta}^c$ as

$$\theta_k^c = \frac{1}{Z_c} \sum_{i=1}^{N_c} r(I_i^c)\delta(\boldsymbol{x}_i^c, \boldsymbol{y}_k) \tag{7}$$

instead of Eq. (4), where Z_c is a normalization constant for satisfying $\sum_{k=1}^{K} \theta_k^c = 1$. Then we calculate ICD for each concept pair, using Eq. (4). Since the reliableness $r(I_i^c)$ becomes smaller if I_i^c is more likely to be junk, we can decrease the effect of junk images on the calculation process of ICD using Eq. (7). Conversely, we can increase the effect of junk images by using

$$\theta_k^c = \frac{1}{Z_c} \sum_{i=1}^{N_c} \left(1 - r(I_i^c)\right) \delta(\boldsymbol{x}_i^c, \boldsymbol{y}_k), \tag{8}$$

which is another option of the proposed framework. Comparing the effectiveness of ICD calculated with Eq. (7) and that calculated with Eq. (8), we can examine whether the effect of junk images is positive or negative.

The proposed framework iteratively performs the above procedures until the mean difference between current $r(I_i^c)$ and previous $r(I_i^c)$ becomes less than a certain threshold ϵ.

4   Experiment

To examine the effect of junk images on the performance of ICDM, we conducted an experiment, in which we compared the following three types of ICDM methods: one considering junk image detection negatively (called "n-JID" in the remainder), one considering junk image detection positively (called "p-JID"), and one without junk image detection (called "w-JID"). Note that ICD is calculated based on Eqs. (7), (8), and (2) in n-JID, p-JID, and w-JID respectively. The performance of these methods was evaluated on the basis of correlation coefficients with JCN [18] described by Jiang and Conrath, which is an ontology-based ICDM method using WordNet and can provide ICDs close to human perception [19]. (However JCN is disadvantageous in that they can handle only a limited number of concepts, i.e., those that are included in the ontology).

4.1   Experimental Setting

To make a dataset for the experiment, we gathered tagged images from Flickr, using the following 10 words as the queries: "animal", "architecture", "city", "food", "nature", "people", "spring", "summer", "autumn" and "winter". The dataset originally has more than 600,000 unique tags, but most of them are given to a very few number of images. So we only focused on the tags that are given to more than 1,500 images and erased all other tags. After that, we excluded the images that cannot keep two or more tags from the dataset because of the limitation of our proposed framework: it cannot appropriately compute the reliableness of image I_i^c unless the I_i^c has at least one more tag other than c. Through the above pre-process, we finally used 506 tags and 967,485 images for this experiment. The number of images per concept ranged from 1,553 to 79,337 and the average number was 8,062.

To extract visual features from the gathered images, we employed AlexNet CNN model pre-trained on ImageNet and applied it to each image, obtaining a 4096-dimensional vector as visual feature x by choosing the output of the "fc7" layer. As for the parameter ϵ described in Sect. 3.4, we empirically set as $\epsilon = 3.0 \times 10^{-7}$.

4.2 Results and Discussion

Figure 1 shows the performance of the three methods in terms of correlation coefficients with JCN using different settings of parameter K. (K is the dimensionality of the categorical distribution $p(x|c)$.) It is derived from Fig. 1 that larger K has a tendency to lead higher performance. This is because smaller K decrease the descriptive power of categorical distributions. More importantly, we can see that p-JID outperforms w-JID and n-JID, and n-JID results in the worst performance among the three methods. This indicates that junk images provide a positive effect on the performance of ICDM. To deeply discuss the reason, we evaluated this result from several viewpoints.

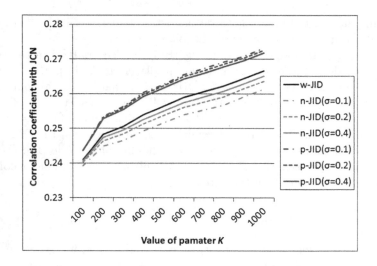

Fig. 1. Performance comparison of n-JID, p-JID, and w-JID

First we evaluated the effectiveness of the methods for junk image detection and visual concept modeling employed in the proposed framework. Figure 2 shows 12 example images of the concept "cake" which have the lowest reliableness. In Fig. 2, there are only a few real photos capturing a cake; the other images are regarded as junk for most humans. Similar tendency can be seen for many other concepts. This result indicates that junk images are effectively detected in the proposed framework. Next, to evaluate the effectiveness of the method for visual concept modeling, we listed top 20 concepts that have the lowest/highest

Fig. 2. 12 example images of concept "cake" that have lowest reliableness

entropy of feature distribution $p(\boldsymbol{x}|c)$ in Table 1/Table 2. Generally speaking, the entropy of $p(\boldsymbol{x}|c)$ becomes high if the concept c is abstract because such concept can have diverse image instances, whereas the entropy of $p(\boldsymbol{x}|c)$ becomes low if c is concrete because image instances of such c are very similar with each other. This kind property could be obtained if each concept c is successfully modeled by the distribution $p(\boldsymbol{x}|c)$. In fact, we can see that most of the concepts listed in Table 1 are concrete objects such as animals and foods, whereas those listed in Table 2 are abstract concepts such as the name of month. These results indicate that a visual model of each concept is successfully constructed in the proposed framework.

Table 1. Top 20 concrete concepts (concepts that have the lowest entropy of $p(\boldsymbol{x}|c)$)

Rodent	Lizard	Kitty	Salad	Reptile
Soup	Puppy	Bacon	Feline	Squirrel
Sandwich	Recipe	Tomato	Insect	Pork
Waterfall	Bug	Pet	Cheese	Beef

Table 2. Top 20 abstract concepts (concepts that have the highest entropy of $p(\boldsymbol{x}|c)$)

Holiday	January	Free	February	Photography
November	May	June	Color	Stock
Photo	July	Picture	August	March
April	September	Image	Explore	Rebel

Based on the above analysis, we next evaluated the relationship between the distance of concept pair (u, v) and their level of abstraction, i.e., the entropies of $p(\boldsymbol{x}|u)$ and $p(\boldsymbol{x}|v)$. To this end, we divided the concept set $\mathcal{C}$ into two subsets

(a) $(u_H, v_H) \in \mathcal{C}_H^2$ (b) $(u_L, v_L) \in \mathcal{C}_L^2$ (c) $(u_H, v_L) \in \mathcal{C}_H \times \mathcal{C}_L$

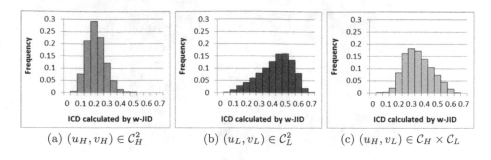

Fig. 3. Histogram of ICD for three types of concept pairs

		Actual similarity	
		similar	dissimilar
Type of concept pair	pair of abstract concepts	A	B
	pair of abstract concept and concrete concept	C	D
	pair of concrete concepts	E	F

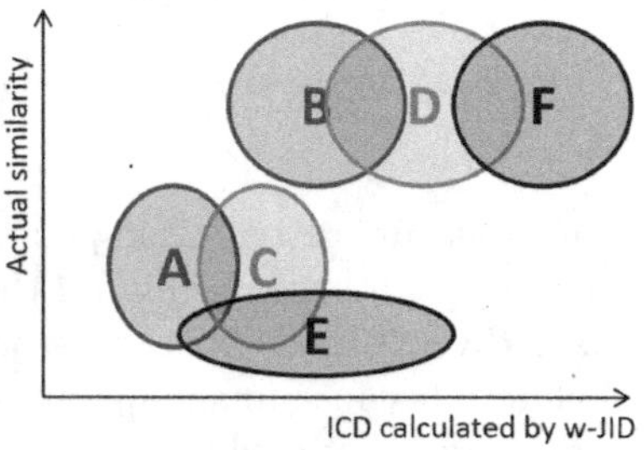

Fig. 4. Relationship between actual similarity of concept pairs and their ICD

$\mathcal{C}_H$ and $\mathcal{C}_L$, where $\mathcal{C}_H$ is a set of concepts that have high entropy of $p(\boldsymbol{x}|c)$ and $\mathcal{C}_L$ is a set of concepts that have low entropy of $p(\boldsymbol{x}|c)$, and computed the histogram of ICD between concept pair $(u_H, v_H) \in \mathcal{C}_H^2$, $(u_L, v_L) \in \mathcal{C}_L^2$, and $(u_H, v_L) \in \mathcal{C}_H \times \mathcal{C}_L$, respectively. Note that we use w-JID for calculating ICD in this evaluation. Figure 3 shows the result. It is derived from the Fig. 3 that ICD tends to become small in the case of abstract concepts and become large in the case of concrete concepts. This is because the forms of high-entropy distributions are close to each other even if their peaks are not so close, whereas this is not true in the case of low-entropy distributions. Therefore, in most ICDM methods including the proposed one, small ICD is often calculated between two abstract concepts that are semantically not close. We briefly illustrate the situation in Fig. 4, which raises a serious problem: ICD between type-B pairs often become smaller than ICD between type-E pairs. Unfortunately, removing junk images is a not suitable solution for this problem; junk image removal generally makes ICD larger for concrete concept pairs, whereas it tends to give only an insignificant effect to ICD between abstract concept pairs. This is because the distribution of visual feature $\boldsymbol{x}$ extracted from junk images spreads widely, so its entropy becomes generally high. As a result, ICD between type-E pairs becomes further larger whereas ICD between type-B pairs does not change so much. This is also demonstrated by Fig. 5 which shows the histogram of the difference between ICD calculated by n-JID and that calculated by w-JID for each type of concept pair $(u_H|v_H) \in \mathcal{C}_H^2$, $(u_L|v_L) \in \mathcal{C}_L^2$, and $(u_H|v_L) \in \mathcal{C}_H \times \mathcal{C}_L$. The difference becomes

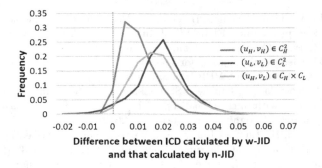

Fig. 5. Relationship between the abstractness of concept pair and the difference in their ICD by junk image removal

negative if n-JID gives smaller ICD than w-JID and becomes positive otherwise. It is derived from Fig. 5 that ICD for concrete concept pairs tends to become larger by n-JID, whereas ICD for abstract concept pairs only changes a little. This means junk image removal does not solve the above problem. Rather, it makes the problem more serious. In contrast, p-JID has properties opposite to n-JID, so it can solve the problem. That is why p-JID outperforms w-JID and n-JID in the result shown in Fig. 1.

5 Conclusion

In this paper, we evaluated the effect of junk images on the performance of ICDM. Junk images are the main factor of performance degradation of VCL. Similarly, it is naively expected that junk images also degrades the performance of ICDM because VCL is a basis of ICDM. On the other hand, junk images might be related to its text-tags in a certain sense because the text-tags are given by not automated systems but humans. Hence, the following question arises: Is the effect of junk images on the performance of ICDM positive or negative? To answer the question, we first proposed a unified framework which integrates ICDM and junk image detection and then conducted an experimental analysis using the proposed framework. Our result indicates that junk images give a positive effect on the performance of ICDM despite of its negative effect on the performance of VCL. The result also indicates the fact that ICD calculated by visual feature-based ICDM methods including our framework strongly depends on the level of abstraction of each concept. Based on the fact, we will attempt to use different methods for different concept pairs depending on their level of abstraction in the future.

This work was supported by JSPS KAKENHI Grant Number 26730090.

References

1. Zhu, S., Wang, G., Ngo, C., Jiang, Y.: On the sampling of web images for learning visual concept classifiers. In: Proceedings of 9th ACM International Conference on Image and Video Retrieval, pp. 50–57 (2010)

2. Yang, J., Li, Y., Tian, Y., Duan, L., Gao, W.: Per-sample multiple kernel approach for visual concept learning. EURASIP J. Image Video Process. **2010**, 1–14 (2010)
3. Qi, G., Aggarwal, C., Rui, Y., Tian, Q., Chang, S., Huang, T.: Towards cross-category knowledge propagation for learning visual concepts. In: Proceedings of 2011 IEEE Conference on CVPR, pp. 897–904 (2011)
4. Sjöberg, M., Koskela, M., Ishikawa, S., Laaksonen, J.: Real-time large-scale visual concept detection with linear classifiers. In: Proceedings of 21st International Conference on Pattern Recognition, pp. 421–424 (2012)
5. Zhuang, L., Gao, H., Luo, J., Lin, Z.: Regularized semi-supervised latent dirichlet allocation for visual concept learning. Neurocomputing **119**, 26–32 (2013)
6. Fan, J., He, X., Zhou, N., Peng, J., Jain, R.: Quantitative characterization of semantic gaps for learning complexity estimation and inference model selection. IEEE Trans. Multimedia **14**(5), 1414–1428 (2012)
7. Gudivada, V., Raghavan, V.: Content-based image retrieval systems. IEEE Comput. **28**(9), 18–22 (1995)
8. Kawakubo, H., Akima, Y., Yanai, K.: Automatic construction of a folksonomy-based visual ontology. In: Proceedings of 6th IEEE International Workshop on Multimedia Information Processing and Retrieval, pp. 330–335 (2010)
9. Wu, L., Hua, X., Yu, N., Ma, W., Li, S.: Flickr distance: a relationship mea-sure for visual concepts. IEEE Trans. PAMI **34**(5), 863–875 (2012)
10. Katsurai, M., Ogawa, T., Haseyama, M.: A cross-modal approach for extracting semantic relationships between concepts using tagged images. IEEE Trans. Multimedia **16**(4), 1059–1074 (2014)
11. Wang, M., Yang, K.: Constructing visual tag dictionary by mining community-contributed media corpus. Neurocomputing **95**, 3–10 (2012)
12. Zhang, X., Liu, C.: Image annotation based on feature fusion and semantic similarity. Neurocomputing **149**, 1658–1671 (2015)
13. Nakamura, K., Babaguchi, N.: Inter-concept distance measurement with adaptively weighted multiple visual features. In: Jawahar, C.V., Shan, S. (eds.) ACCV 2014. LNCS, vol. 9010, pp. 56–70. Springer, Heidelberg (2015). doi:10.1007/978-3-319-16634-6_5
14. Kennedy, L.S., Chang, S., Kozintsev, I.V.: To search or to label?: predicting the performance of search-based automatic image classifiers. In: Proceedings of 8th ACM International Workshop on Multimedia Information Retrieval, pp. 249–258 (2006)
15. Setz, A.T., Snoek, C.G.M.: Can social tagged images aid concept-based video search?. In: Proceedings of IEEE International Conference on Multimedia and Expo 2009, pp. 1460–1463 (2009)
16. Schroff, F., Criminisi, A., Zisserman, A.: Harvesting image databases from the web. IEEE Trans. PAMI **33**(4), 754–766 (2011)
17. Fuglede, B., Topsøe, F.: Jensen-shannon divergence and hilbert space embedding. In: Proceedings of 2004 International Symposium on Information Theory, p. 30 (2004)
18. Pedersen, T., Patwardhan, S., Michelizzi, J.: Wordnet::Similarity - measuring the relatedness of concepts. In: Proceedings of 5th Annual Meeting of the North American Chapter of the Association for Computational Linguistics, pp. 38–41 (2004)
19. Budanitsky, A., Hirst, G.: Evaluating wordnet-based measures of lexical semantic relatedness. Comput. Linguist. **32**(1), 13–47 (2006)

Exploiting Multimodality in Video Hyperlinking to Improve Target Diversity

Rémi Bois[1]([✉]), Vedran Vukotić[2], Anca-Roxana Simon[4], Ronan Sicre[3], Christian Raymond[2], Pascale Sébillot[2], and Guillaume Gravier[1]

[1] CNRS, IRISA and Inria Rennes, Rennes, France
{remi.bois,guillaume.gravier}@irisa.fr
[2] INSA Rennes, IRISA and Inria Rennes, Rennes, France
{vedran.vukotic,christian.raymond,
pascale.sebillot}@irisa.fr
[3] Inria, IRISA and Inria Rennes, Rennes, France
ronan.sicre@irisa.fr
[4] University Rennes 1, IRISA and Inria Rennes, Rennes, France
anca-simon@outlook.com

Abstract. Video hyperlinking is the process of creating links within a collection of videos to help navigation and information seeking. Starting from a given set of video segments, called anchors, a set of related segments, called targets, must be provided. In past years, a number of content-based approaches have been proposed with good results obtained by searching for target segments that are very similar to the anchor in terms of content and information. Unfortunately, relevance has been obtained to the expense of diversity. In this paper, we study multimodal approaches and their ability to provide a set of diverse yet relevant targets. We compare two recently introduced cross-modal approaches, namely, deep auto-encoders and bimodal LDA, and experimentally show that both provide significantly more diverse targets than a state-of-the-art baseline. Bimodal autoencoders offer the best trade-off between relevance and diversity, with bimodal LDA exhibiting slightly more diverse targets at a lower precision.

1 Introduction

The automatic generation of hyperlinks within video collections recently became a major subject, in particular via evaluation benchmarks within MediaEval and TRECVid [8,9,20]. The key idea is to create hyperlinks between video segments within a collection, enriching a set of anchors that represent interesting entry points in the collection. Links can be seen as recommendations for potential viewers, whose intent is not known at the time of linking. The goal of the links is thus to help viewers gain insight on a potentially massive collection of videos so as to find information of interest, following a search and browse paradigm.

Creating video hyperlinks from a given anchor traditionally implements two steps. A segmentation step that aims at determining potential targets is followed

© Springer International Publishing AG 2017
L. Amsaleg et al. (Eds.): MMM 2017, Part II, LNCS 10133, pp. 185–197, 2017.
DOI: 10.1007/978-3-319-51814-5_16

by a selection step in which relevant targets are selected. The vast majority of approaches developed for the selection step rely on direct pairwise content-based similarity, seeking targets whose content is very similar to the anchor. Unsurprisingly, most use textual and/or visual content comparison [1,2,6,7,12,14,17,21]. Maximizing content-based similarity between anchors and targets showed to offer good relevance, as evidenced in [14] where n-gram bag-of-words are used to emphasize segments sharing common sequences of words.

Unfortunately, emphasizing relevance by rewarding highly similar content in terms of words and visual concepts does not offer diversity in the set of targets that are proposed for a given anchor. This lack of diversity is considered as detrimental in many exploration scenarios, in particular when users' intentions and information needs are not known at the time of linking. In this case, providing relevant links that cover a number of possible extensions with respect to the anchor's content is desirable. Clearly, having a set of diverse targets strongly improves the chance for any user to find at least one interesting link to follow, whatever his/her initial intentions. Additionally, target diversity directly improves serendipity, *i.e.*, unexpected yet relevant links, offering the possibility to drift from the initial anchor in terms of information so as to gain a better understanding of what can be found in the collection. This objective of providing diverse results is gaining traction in the hyperlinking community and has been explicitly included in the latest TRECVid evaluation benchmark [10].

In this paper, we investigate cross-modal approaches recently introduced for multimedia content matching, namely, bimodal autoencoders [27] and bimodal LDA [4], as a mean to improve the diversity of targets within a limited set. The intuition is that cross-modality unveils relevant links that would not be captured with standard approaches, and is a good candidate to improve diversity. Indeed, providing links to visual content related to spoken content, and conversely, is bound to reduce the similarity between anchors and targets while maintaining high relevance. For instance, a target could talk about what is shown in the anchor or show things that are discussed in the anchor, and thus bring complementary information. While multimodal approaches have been proposed to increase content similarity between anchors and targets [1,6,17], cross-modality has been seldom considered so far and no evaluation regarding the diversity of targets has been run to this date. In Sects. 2 and 3, we introduce the two cross-modal systems that are used in this study, namely a bidirectional deep neural network, and a cross-modal topic model. Section 4 describes the evaluation protocol used to assess diversity, presents the corpus, and discusses results obtained by a user-centric study, as well as automatic measures.

2   Bidirectional Deep Neural Networks

The first approach that we consider to improve diversity relies on distributional representations of words and their multimodal extensions. Word vector representations, such as *word2vec* [19], have proven to be of interest for information retrieval [16, 18, 29], and were recently experimented for video hyperlinking [21, 27]. Interestingly, this representation of words can easily be used for cross-modal matching, often in conjunction with deep neural networks [11, 27, 30], with a strong potential for diversity. In this study, we rely on bidirectional symmetrical deep neural networks (BiDNN) operating on averaged *word2vec* representations [5] of words, obtained by automatic transcription, and on visual concepts detected in keyframes.

Autoencoders are neural networks, used in unsupervised learning, that are setup to reconstruct their input, while the middle layer is being used as a new representation of the data. In a multimodal setting, autoencoders are used to combine separate input representations to yield a joint multimodal representation in the middle hidden layer.

Typical multimodal autoencoders come in two varieties: (i) extended classical single-modal autoencoders where multimodality is achieved by concatenating the modalities at their inputs and outputs and (ii) truly multimodal autoencoders that have separate inputs and outputs for each modality, as well as one or more separated fully connected layers assigned to each. Both share a common central point: a fully connected layer connecting both modalities and used to obtain a multimodal embedding. However, multimodal autoencoders have some downsides. First, to enable cross-modal translation, one modality is often sporadically removed from the input, while autoencoders are asked to reproduce both modalities at their output. This means that an autoencoder has to learn to represent the same output both when a specific modality is present and when it is zeroed, which is less optimal than having direct cross-modal translation. Secondly, central fully connected layers are influenced by both modalities (either directly or through other fully connected layers). While this is good for multimodal embedding, it does not provide a clean cross-modal translation.

Bidirectional symmetrical deep neural networks tackle these two problems by first creating straight-forward cross-modal translations between modalities and then providing a common representation space where both modalities are projected and a multimodal embedding is formed. Learning is performed in both directions: one modality is presented as input and the other as the expected output while, at the same time, the second modality is presented as input and the first as expected output. This architecture is presented as two networks—one translating from the first modality to the second and the other conversely—where the variables in the central part are tied to enforce symmetry, as illustrated in Fig. 1. Implementation-wise, the variables representing the weights in the hidden layers are shared across the two networks and are in fact the same variables. Learning the two cross-modal mappings is thus performed simultaneously thanks to the symmetric architecture in the middle. The joint representation formed in the middle layer while learning acts as a multimodal pivot representation enabling translation from one modality to the other.

Formally, let $\mathbf{h}_i^{(j)}$ denote (the activation of) the hidden layer at depth j in network i ($i = 1, 2$, one for each modality), $\mathbf{x_i}$ the feature vector for modality i and $\mathbf{y_i}$ the output of the network for modality i. Networks are defined by their

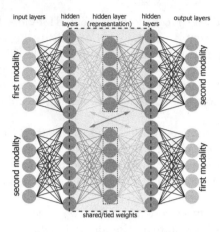

Fig. 1. Architecture of bidirectional symmetrical deep neural networks [27]

weight matrices $\mathbf{W}_i^{(j)}$ and bias vectors $\mathbf{b}_i^{(j)}$, for each layer j, and admit f as activation function. The entire architecture is then defined by:

$$\mathbf{h}_i^{(1)} = f(\mathbf{W}_i^{(1)} \times \mathbf{x}_i + \mathbf{b}_i^{(1)}) \qquad i = 1, 2 \tag{1}$$

$$\mathbf{h}_1^{(2)} = f(\mathbf{W}^{(2)} \times \mathbf{h}_1^{(1)} + \mathbf{b}_1^{(2)}) \tag{2}$$

$$\mathbf{h}_1^{(3)} = f(\mathbf{W}^{(3)} \times \mathbf{h}_1^{(2)} + \mathbf{b}_1^{(3)}) \tag{3}$$

$$\mathbf{h}_2^{(2)} = f(\mathbf{W}^{(3)\mathbf{T}} \times \mathbf{h}_2^{(1)} + \mathbf{b}_2^{(2)}) \tag{4}$$

$$\mathbf{h}_2^{(3)} = f(\mathbf{W}^{(2)\mathbf{T}} \times \mathbf{h}_2^{(2)} + \mathbf{b}_2^{(3)}) \tag{5}$$

$$\mathbf{o}_i = f(\mathbf{W}_i^{(4)} \times \mathbf{h}_i^{(3)} + \mathbf{b}_i^{(4)}) \qquad i = 1, 2 \tag{6}$$

It is important to note that the weight matrices $\mathbf{W}^{(2)}$ and $\mathbf{W}^{(3)}$ are used twice due to weight tying, respectively in Eqs. 2, 5 and Eqs. 3, 5. Training is performed by applying batch gradient descent to minimize the mean squared error of $(\mathbf{o}_1, \mathbf{x}_2)$ and $(\mathbf{o}_2, \mathbf{x}_1)$ thus effectively minimizing the reconstruction error in both directions and creating a joint representation in the middle.

Given such an architecture, cross-modal translation can be done straightforwardly by presenting the first modality as $\mathbf{x}_i$ and obtaining the output in the representation space of the second modality as $\mathbf{y}_i$. However, to improve relevance while preserving diversity, we experimented with the multimodal embedding of the hidden layer. In practice, for a video segment, each modality is projected to the hidden layer with the corresponding network and the two resulting vectors are concatenated. More specifically, if both modalities are present, each one is presented to its respective input of the bidirectional deep neural network and the values are propagated through the network. The values from the central layer, where the common representation space lies, are concatenated and a multimodal embedding is formed. When only one modality is available, it is presented to its respective input of the bidirectional deep neural network and the values are propagated to the network. The values of the central layer are duplicated, as to form an embedding of an equal size as when both modalities are present. This allows for transparent comparison of video segments regardless of modality availability. When one video segment has only one modality while the other has both, a distance computed on such multimodal embedding would automatically compare the one available modality from one video segment with the two modalities of the other video segment. This all happens in the new common representation space where both modalities are projected and transparent comparisons are made possible.

Note that while the embedding is multimodal, it corresponds to a space dedicated to cross-modal matching and thus significantly differs from classical joint multimodal spaces. Figure 2 illustrates the task of video hyperlinking with bidirectional deep neural networks: for all video segments, cross-modal translations between embedded automatic transcripts, embedded visual concepts, and back, are learned. Then, for the specific video segments that are compared, their respective embedded automatic transcripts and embedded visual concepts are presented (regardless of modality availability) and their multimodal embeddings in the new common representation space are formed. Finally, the two multimodal embeddings are compared with a cosine distance to obtain a similarity score.

We implemented bidirectional neural networks in *Lasagne*[1]. All embeddings have a dimension of 100 as larger dimensions did not bring any significant improvement. The architectures used had 200-100-200 hidden layers as other smaller sizes performed worse and larger sizes did not perform better. The networks were trained with stochastic gradient descent (SGD) with Nesterov momentum, dropout of 20%, in mini-batches of 100 samples, for 1000 epochs (although convergence was achieved quite earlier). Since all the methods described belong to unsupervised learning, the learning was performed on the part of the dataset that contains both transcripts and visual concepts and tested on the whole dataset.

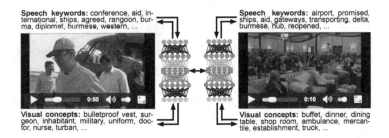

Fig. 2. Video hyperlinking with bidirectional symmetrical deep neural networks [28]

---

[1] https://github.com/Lasagne/Lasagne.

3   Cross-Modal Topic Model

Another potential solution to diversity is the use of topic models, such as latent Dirichlet allocation (LDA), where the similarity between two documents is measured via the similarity of the latent topics they share rather than by direct content comparison [3]. Recently, based on seminal work on multilingual topic modeling [24], multimodal extensions of LDA were proposed for cross-modal video hyperlinking [4], combining the potential for diversity offered by topic models and by multimodality. As for BiDNN, words extracted from the automatic transcripts and the visual concepts from the keyframes are used in the bimodal LDA (BiLDA).

The LDA model is based on the idea that there exist latent variables, *i.e.*, topics, which explain how words in documents have been generated. Fitting such a generative model to a document means finding the best set of such latent variables in order to explain the observed data. As a result, documents are seen as mixtures of latent topics, while topics are probability distributions over words. The multimodal extension in [4] considers that each latent topic is defined by two probability distributions, one over each modality (or language in [24]). The BiLDA model is thus trained on parallel documents, assuming that the underlying topic distribution is common to the two modalities. In the case of videos, parallel documents are straightforwardly obtained by considering the transcripts and the visual concepts of a video segment as two parallel documents sharing the same underlying topic distribution. Training, *i.e.*, determining the topics from a given collection of videos, is achieved by Gibbs sampling, as for standard LDA [25], with the number of latent topics set to 700. Given a set of documents in the text (resp. visual) modality with vocabulary V_1 (resp. V_2), the probability that a word $w_i \in V_1$ (resp. visual concept $c_i \in V_2$) corresponds to topic z_j is estimated as

$$p(w_i|z_j) = \frac{n_{z_j}^{w_i} + \beta}{\sum_{x=1}^{|V_1|} n_{z_j}^{w_x} + \beta|V_1|}, \tag{7}$$

where $n_{z_j}^{w_i}$ is the number of times that topic z_j was assigned to word w_i in the training data and β is a Dirichlet prior.

Table 1. Three multimodal topics represented by their top-5 words and visual concepts

Topic 3	Words	Love, home, feel, life, baby
	Visual concepts	Singer, microphone, sax, concert, flute
Topic 7	Words	Food, bit, chef, cook, kitchen
	Visual concepts	Dig, acorn, pumpkin, guava, zucchini
Topic 25	Words	Years, technology, computer, key, future
	Visual concepts	Tape-player, computer, equipment, machine, appliance

This training step provides a mapping between topics of the two modalities by means of the topics. Table 1 displays examples of this mapping obtained from the corpus used in this study (see corpus description Sect. 4). For each topic, we show the 5 most probable words and visual concepts. Sometimes, words and visual concepts are directly related (e.g., *computer* in topic 25). However, the relation can be more subtle, as in topic 3 where visual concepts describe a stage, and words are utterances frequently encountered in the lyrics of songs.

The interest of topic models lies in the fact that video segments dealing with similar topics will tend to have similar distribution over the latent topics. This enables the indirect comparison of two video segments by comparing the distribution of latent topics, rather than using their multimodal content, thus potentially enabling a diversity of content (within documents from closely related topics). Formally, given a video segment d, the idea is to represent the segment as a vector collecting the topic probabilities

$$p(d|z_j) = \left(\prod_{i=1}^{n_x} p(w_i|z_j) \right)^{1/n_x}, \tag{8}$$

where n_x is the size of the vocabulary in d and w_i is the ith word or visual concept in d. Note that $p(d|z_j)$ is an approximation of the posterior $p(z_j|d)$, considering a uniform distribution of topics, which is a reasonable assumption. The similarity score between any two segments is given by a cosine similarity between their corresponding vectors after $L2$-normalization.

In practice, the probabilities $p(d|z_j)$ can be obtained from either one of the two modalities (using the corresponding distributions $p(\cdot|z_j)$), thus enabling multimodal and cross-modal matching as illustrated in Fig. 3. In this paper, we considered visual to text matching, representing the distribution of topics based on visual concepts for the anchor and on automatic transcripts for the targets.

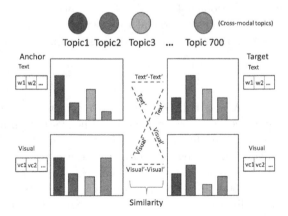

Fig. 3. Illustration of the multimodal and cross-modal matching with the BiLDA model

4   Experimental Results

4.1   Experimental Setup

Experimental evaluation of the different methods and of the diversity of targets is performed using data from the TRECVid 2015 video hyperlinking task [20] and the corresponding annotations. The original data consists of approximately 2,700 h of BBC programs from which 100 anchors were defined, with an average length of 71 s. Anchors were selected by experts as being segments of interest that a user would like to know more about. As a result of the 2015 evaluation, a set of potential targets along with relevance judgments is also provided, these targets being the top-10 targets proposed by each participating team. Relevance assessment of each of those targets was achieved post hoc on Amazon Mechanical Turk (AMT). In total for the 100 anchors, 21,176 targets are available with their relevance judgments, of which 25.4% are actually relevant.

In this work, the content matching methods are evaluated via a reranking task where the set of potential targets are reordered for each anchor, thus getting rid of segmentation issues. For each anchor, reranking operates on an average of 212 targets proposed in 2015. Apart from practical reasons due to the lack of extensive ground-truth on the whole data set, the reranking task is justified by the fact that we want to assess the properties of different methods for the target *selection* step. We should however stress a minor bias in this setting due to the fact that anchors were initially proposed by the 2015 participants. Hence, the targets that we rerank are all somehow related to the anchor and we cannot appreciate the potential of the methods to discard totally irrelevant targets. This, however, does not hinder the potential of the methods compared here, in particular with respect to diversity.

Anchor and target segments are described according to two modalities. On the one hand, automatically generated speech transcripts [13] provide a lexical representation after lemmatization and stopword removal. On the other hand, the automatic detection of 1,537 visual concepts [26], averaged over the keyframes of a segment, provides a visual representation. As some anchors are very short, a context of 30 s around each actual anchor is considered.

4.2   Results

We first compare BiLDA and BiDNN with a transcript-only baseline system, on their ability to find relevant targets. The baseline system implements a bag-of-words representation for each segment with tf-idf weighting [22] along with cosine similarity. Inverse document frequencies were estimated on the set of anchors plus the set of proposed targets. The BiDNN was implemented with an architecture of 200-100-200 hidden layers, dropout in the central part of 50% and trained with batch gradient descent on the video segments that appear on the groundtruth. The BiLDA model was trained on the full TRECVid 2015 dataset.

Results are reported in Table 2, where precision at 10 are given for the whole set of anchors and for the 16 anchors that gave the best results, and which were

Table 2. Precision at rank 10 on target reranking

Anchors	Baseline	BiDNN	BiLDA
All	0.59	0.57	0.24
16	0.80	0.80	0.40

retained for perceptual study of diversity (see below). Results for the baseline and for BiDNN are state of the art while BiLDA matching exhibits weaker results. The good baseline results are partly explained by the fact that, in 2015, participants mainly used the textual modality for target selection. Hence the list of targets to rerank contains a significant number of relevant targets with high lexical similarity and using a bag-of-words representation with cosine similarity is adequate. This also explains why baseline results on the top-16 anchors are very strong and, to some extent, why LDA-based approaches fail to be on par with the baseline. Consistently with results in [27], the BiDNN approach performs as well as the baseline, however using a cross-modal approach, showing the interest of autoencoders for multimodal multimedia retrieval. Finally, we note that the post hoc AMT-based annotation process does not encourage diversity. This is beneficial to the baseline, which ranks high segments very similar to the anchor, but detrimental to LDA-based approaches. This consideration also motivates the specific study on diversity described hereunder.

The diversity of the results returned by either one of the methods can be assessed either intrinsically, e.g., by measuring how diverse are the relevant segments found, or by means of subjective human-based judgments. The latter requires that a limited number of anchors be considered to enable a significant number of votes for a single anchor. Diversity is thus evaluated on the 16 anchors for which the best results were obtained with the baseline. For each anchor, we consider the top-5 relevant targets found and test for diversity among these targets. In addition to the lexical and visual representation used for content comparison, we also extracted 10 key words and 10 key concepts for each segment using a tf-idf ranking—a method commonly used as baseline for key word extraction [15]. This compact representation enables comparing the content of the targets and help user quickly apprehend the content of a target.

Table 3 reports a number of intrinsic indicators of the diversity of a list of targets. n_u ($\in [10, 50]$) is the average number of unique key words/concepts in the top-5 relevant segments of an anchor, where the bigger n_u the better the diversity. A value of 10 indicates that all targets have the same key words/concepts. $\overline{d}_a$ is the average cosine similarity between the anchor and the top-5 relevant targets computed over the transcript or over the set of visual concepts. $\overline{d}_i$ measures the similarity within the top-5 targets of an anchor, computed as the average cosine similarity between any two pairs of targets in the top-5 list. In these last two cases, the larger the value, the less diverse the list of 5 targets. Results in Table 3 clearly demonstrate that the cross-modal approaches offer a significantly greater diversity of relevant targets than the baseline. Diversity shows both from

Table 3. Intrinsic evaluation of the diversity of the top-5 relevant targets

	Transcripts			Concepts		
	n_u	$\overline{d}_a$	$\overline{d}_i$	n_u	$\overline{d}_a$	$\overline{d}_i$
Baseline	29.8	0.51	0.61	35.6	0.61	0.71
BiDNN	40.8	0.20	0.12	46.7	0.42	0.31
BiLDA	40.0	0.25	0.16	38.0	0.48	0.41

the lexical standpoint and from the visual one, where the difference between the baseline and cross-modal methods is stronger at the lexical level. BiDNN appears to be slightly better than BiLDA in terms of average distance from targets to anchor as well as in terms of target dispersion.

Results as measured by the intrinsic indicators above are confirmed by user evaluations, where users were presented with an anchor and three lists of 5 relevant targets, one for each method, and asked to rank those lists from the least diverse (rank 1) to the most diverse (rank 3). In the evaluation interface, the anchor appeared on the top of the page, followed by 3 columns of 5 targets each, in a randomized order. Each segment was represented by a key image from which the video could be played, along with 10 key words and 10 key concepts to facilitate the task, potentially avoiding the need to watch all 16 video segments. A session consisted in ranking the lists for the 16 anchors selected, however not all evaluators completed their session. Since the order of anchors was also randomized per session, we kept all votes to report results on as many judgments as possible. In total, 25 persons, mostly from academia, participated in the evaluation, the vast majority of them not familiar with the video hyperlinking task. A total of 176 votes were recorded, with an average of 11 votes per anchor. The annotation took approximately 16 min to complete (median time), which corresponds to about one minute per anchor. Results are summarized in Fig. 4 where the average rank is plotted (dot) for each method, with an error bar depicting the dispersion of judgments among users—the lowest/highest average rank assigned to the method by a particular user.

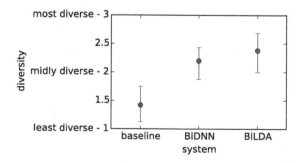

Fig. 4. Average rank of systems with respect to diversity as perceived by evaluators

Perceptive evaluations by users confirm the results obtained with intrinsic evaluations, with a significant difference between the transcript-only baseline (average rank of 1.42) and the two cross-modal methods (average ranks of 2.20 and 2.38 for BiDNN and BiLDA resp.). It is also interesting to note that judgments are rather consistent across evaluators, for instance with average ranks from 1.12 to 1.75 for the baseline, confirming the ability of humans to judge diversity. However, contrary to intrinsic evaluations, the relevant targets found by BiLDA were globally perceived as more diverse than those found by BiDNN (significant at $\alpha = 0.01$ according to a paired one-tailed t-test), even though BiLDA performs less than BiDNN in terms of relevance.

5   Conclusion

The study presented in this paper focuses on cross-modal approaches for target selection in video hyperlinking as a mean to offer a diversity of targets. Intrinsic and perceptive evaluations show that cross-modal approaches are significantly better than a text-only baseline at diversity. Bidirectional symmetrical DNNs offer a very good compromise between relevance and diversity. Bimodal LDA offers better potential for diversity but weak performance in terms of relevance still appears as a limitation for this method. However, recent perceptual studies on LDA-derived targets show that combination of topic models can yield performance equivalent to the baseline [23]. Another interesting outcome of the experiments presented here is the fact that diversity can be assessed not only using perceptual tests, but also using intrinsic dispersion measures. The latter are easy to obtain and yield conclusions similar to the one made with the former, opening the door to large-scale studies on diversity in video hyperlinking.

References

1. Barrios, J.M., Saavedra, J.M., Ramirez, F., Contreras, D.: ORAND at TRECVID 2015: instance search and video hyperlinking tasks. In: Proceedings of TRECVID (2015)
2. Bhatt, C., Pappas, N., Habibi, M., Popescu-Belis, A.: Idiap at MediaEval 2013: search and hyperlinking task. In: Proceedings of the MediaEval Workshop (2013)
3. Blei, D.M., Ng, A.Y., Jordan, M.I.: Latent dirichlet allocation. J. Mach. Learn. Res. 3, 993–1022 (2003)
4. Bois, R., Şimon, A.-R., Sicre, R., Gravier, G., Sébillot, P.: IRISA at TrecVid2015 2015: leveraging multimodal LDA for video hyperlinking. In: Proceedings of TRECVID (2015)
5. Campr, M., Ježek, K.: Comparing semantic models for evaluating automatic document summarization. In: Král, P., Matoušek, V. (eds.) TSD 2015. LNCS (LNAI), vol. 9302, pp. 252–260. Springer, Heidelberg (2015). doi:10.1007/978-3-319-24033-6_29
6. Cheng, Z., Li, X., Shen, J., Hauptmann, A.G.: CMU-SMU@TRECVID 2015: video hyperlinking. In: Proceedings of TRECVID (2015)

7. De Nies, T., De Neve, W., Mannens, E., Van de Walle, R.: Ghent University-iMinds at MediaEval 2013: an unsupervised named entity-based similarity measure for search and hyperlinking. In: Proceedings of the MediaEval Workshop (2013)
8. Eskevich, M., Aly, R., Racca, D.N., Ordelman, R., Chen, S., Jones G.J.F.: The search and hyperlinking task at MediaEval 2014. In: Proceedings of the MediaEval Workshop (2014)
9. Eskevich, M., Jones, G.J., Chen, S., Aly, R., Ordelman, R., Nadeem, D., Guinaudeau, C., Gravier, G., Sébillot, P., Nies, T.D., Debevere, P., de Walle, R.V., Galuščáková, P., Pecina, P., Larson, M.: Multimedia information seeking through search and hyperlinking. In: ACM International Conference on Multimedia Retrieval (2013)
10. Eskevich, M., Larson, M., Aly, R., Sabetghadam, S., Jones, G.J.F., Ordelman, R., Huet, B.: Multimodal video-to-video linking: turning to the crowd for insight and evaluation. In: Proceedings of the 23rd International Conference on Multimedia Modeling (2017)
11. Feng, F., Wang, X., Li, R.: Cross-modal retrieval with correspondence autoencoder. In: ACM International Conference on Multimedia, pp. 7–16 (2014)
12. Galuscáková, P., Krulis, M., Lokoc, J., Pecina, P.: CUNI at MediaEval 2014 search and hyperlinking task: visual and prosodic features in hyperlinking. In: Working Notes Proceedings of the MediaEval Workshop (2014)
13. Gauvain, J.-L., Lamel, L., Adda, G.: The LIMSI broadcast news transcription system. Speech commun. 37(1), 89–108 (2002)
14. Guinaudeau, C., Gravier, G., Sébillot, P.: IRISA at MediaEval 2012: search and hyperlinking task. In: Working Notes Proceedings of the MediaEval Workshop (2012)
15. Hasan, K.S., Ng, V.: Conundrums in unsupervised keyphrase extraction: making sense of the state-of-the-art. In: Proceedings of the 23rd International Conference on Computational Linguistics (2010)
16. Huang, E.H., Socher, R., Manning, C.D., Ng, A.Y.: Improving word representations via global context and multiple word prototypes. In: Proceedings of the 50th Annual Meeting of the Association for Computational Linguistics: Long Papers-Volume 1
17. Le, H.A., Bui, Q., Huet, B., et al.: LinkedTV at MediaEval 2014 search and hyperlinking task. In: Proceedings of the MediaEval Workshop (2014)
18. Le, Q.V., Mikolov, T.: Distributed representations of sentences and documents. In: Proceedings of International Conference on Machine Learning
19. Mikolov, T., Sutskever, I., Chen, K., Corrado, G.S., Dean, J.: Distributed representations of words and phrases and their compositionality. In: Proceedings of Advances in Neural Information Processing Systems (2013)
20. Over, P., Awad, G., Michel, M., Fiscus, J., Kraaij, W., Smeaton, A.F., Quénot, G., Ordelman, R.: TRECVID 2015 – an overview of the goals, tasks, data, evaluation mechanisms and metrics. In: Proceedings of TRECVID (2015)
21. Pang, L., Ngo, C.-W.: VIREO @ TRECVID 2015: video hyperlinking. In: Proceedings of TRECVID (2015)
22. Salton, G., McGill, M.J.: Introduction to Modern Information Retrieval. McGraw-Hill Inc., New York (1986)
23. Simon, A.-R.: Semantic structuring of video collections from speech: segmentation and hyperlinking. Ph.D. thesis, Université de Rennes 1 (2015)
24. Smet, W.D., Moens, M.: Cross-language linking of news stories on the web using interlingual topic modelling. In: ACM Workshop on Social Web Search and Mining (2009)

25. Steyvers, M., Griffiths, T.: Probabilistic topic models. Handb. Latent Semant. Anal. **427**(7), 424–440 (2007)
26. Tommasi, T., Aly, R.B.N., McGuinness, K., Chatfield, K., et al.: Beyond metadata: searching your archive based on its audio-visual content. In: Proceedings of the International Broadcasting Convention (2014)
27. Vukotić, V., Raymond, C., Gravier, G.: Bidirectional joint representation learning with symmetrical deep neural networks for multimodal and crossmodal applications. In: Proceedings of the ACM International Conference on Multimedia Retrieval (2016)
28. Vukotic, V., Raymond, C., Gravier, G.: Multimodal and crossmodal representation learning from textual and visual features with bidirectional deep neural networks for video hyperlinking. In: ACM Multimedia 2016 Workshop: Vision and Language Integration Meets Multimedia Fusion (iV&L-MM 2016), Amsterdam, Netherlands. ACM, October 2016
29. Vulić, I., Moens, M.-F.: Monolingual and cross-lingual information retrieval models based on (bilingual) word embeddings. In: Proceedings of the 38th International ACM SIGIR Conference on Research and Development in Information Retrieval
30. Weston, J., Bengio, S., Usunier, N.: Large scale image annotation: learning to rank with joint word-image embeddings. Mach. Learn. **81**(1), 21–35 (2010)

Exploring Large Movie Collections: Comparing Visual Berrypicking and Traditional Browsing

Thomas Low[1](✉), Christian Hentschel[2], Sebastian Stober[3], Harald Sack[4], and Andreas Nürnberger[1]

[1] Otto von Guericke University Magdeburg, Magdeburg, Germany
{thomas.low,andreas.nuernberger}@ovgu.de
[2] Hasso Plattner Institute for Software Systems Engineering, Potsdam, Germany
christian.hentschel@hpi.de
[3] University of Potsdam, Potsdam, Germany
sebastian.stober@uni-potsdam.de
[4] FIZ Karlsruhe, Karlsruhe, Germany
harald.sack@fiz-karlsruhe.de

Abstract. We compare Visual Berrypicking, an interactive approach allowing users to explore large and highly faceted information spaces using similarity-based two-dimensional maps, with traditional browsing techniques. For large datasets, current projection methods used to generate maplike overviews suffer from increased computational costs and a loss of accuracy resulting in inconsistent visualizations. We propose to interactively align inexpensive small maps, showing local neighborhoods only, which ideally creates the impression of panning a large map. For evaluation, we designed a web-based prototype for movie exploration and compared it to the web interface of *The Movie Database* (TMDb) in an online user study. Results suggest that users are able to effectively explore large movie collections by hopping from one neighborhood to the next. Additionally, due to the projection of movie similarities, interesting links between movies can be found more easily, and thus, compared to browsing serendipitous discoveries are more likely.

Keywords: Exploratory interfaces · Media retrieval · Multidimensional scaling · User study

1 Introduction

Exploring movie collections containing thousands of items described by textual metadata such as title, genre, actors and plot summary is difficult for users, as it takes some effort to judge the relevance of all retrieved objects. In this paper, we present and evaluate a map-based interface in order to support especially exploratory processes when working with highly faceted information spaces. Current projection methods suffer from increased computational costs and a loss of accuracy resulting in inconsistent visualizations. However, previous work has shown that similarity-based two-dimensional maps can be very useful for

© Springer International Publishing AG 2017
L. Amsaleg et al. (Eds.): MMM 2017, Part II, LNCS 10133, pp. 198–208, 2017.
DOI: 10.1007/978-3-319-51814-5_17

exploratory information retrieval [12]. They quickly convey information about the general structure and coverage of a collection and help to identify groups of similar objects. Grouping by similarity helps to assess the relevance of entire clusters once the relevance of a few representative objects is known.

In order to evaluate the effectiveness of our proposed map-based exploration method, we compared it with standard browsing techniques when exploring a large collection of movies. Our exploration model enables a user to quickly identify movies that share common features such as common actors or similar plots. Navigation through the collection is possible by panning towards a specific feature subset.

In [11] the conceptual idea of Visual Berrypicking was demonstrated by extracting visual features from a large image collection. In this paper, we transfer and evaluate this approach for a more complex movie dataset by combining its diverse metadata, e.g., a movie's rating, plot description and more. We follow the approach of [11] and visualize only the set of k-nearest neighbors for a given reference or *seed* object. Selecting a new seed from the currently visualized subset leads to the retrieval of a new set of k-nearest neighbors and the computation of a new map. The old and the new set of visualized neighbors most likely overlap to some extent. This topological similarity is exploited to align the two consecutive maps and create a meaningful link between them. Additionally, animated transitions aim to help keeping track of positional changes for objects contained in both maps. Ideally, with largely consistent transitions, this creates the impression of panning a large (global) map of the collection. Most importantly, establishing links between consecutive maps allows users to transfer knowledge about the content and relevance of individual objects accumulated during the search process from one visualization to the next.

For evaluation, we conducted an online user study with more than 100 participants. Participants were asked to explore the collection of *The Movie Database*[1] (TMDb) and find interesting movies they were eager to watch using both the native website as well as our proposed map-based movie explorer. Details about our study as well as its results are discussed in Sect. 4. First, we briefly review related work in Sect. 2 and present details of the Visual Berrypicking method in Sect. 3.

2 Related Work

Media retrieval systems are based on either or both extracted audiovisual features and other metadata. Approaches that focus on extracted audiovisual features usually try to tag movies or individual scenes and provide faceted search interfaces. As an example, the authors in [6] present a semantic video search engine that enables shot-accurate exploration of a cultural heritage video archive. While the interface allows for filtering retrieval results using content-based facets, a query refinement needs to be performed by altering the initial query string.

---

[1] The Movie Database Website: https://www.themoviedb.org/.

A dense visualization of a large collection of videos is presented in [1]. Browsing is supported by tree-based navigation, each level representing a two-dimensional projection of similar sequences. While this extremely compact layout maximizes use of available display space, identifying clusters of similar elements becomes difficult and impossible when groups are located in different hierarchy levels. Furthermore, exploration is limited by navigating up and down branches of the collection tree. Different from the aforementioned approaches, we do not further analyze movies. Our approach solely relies on metadata provided by TMDb and therefore can be easily applied to other media collections, e.g., images, music or simple text documents.

Rubner et al. [13] were among the first to propose multidimensional scaling (MDS) for iterated search in large image collections – a technique that we also follow in this paper. The authors propose a local MDS on the nearest neighbors of a query image. In contrast to our approach, consecutive maps are not aligned to each other. In [12] image arrangements are constructed using MDS and layouted continuously as well as in a structure preserving grid layout (in order to reduce overlap). The authors conclude that arranging a set of thumbnails according to their similarity is useful and helps to divide the set into simple clusters. Several other map-based approaches for information retrieval have been described, e.g. for large document collections using self-organizing maps [8], graph-based approches [5] or adaptive multi-view systems [10].

Recent work [14] reviewed and compared different dimensionality reduction algorithms for the visualization of large music collections. Based on a user study, MDS was favored as best layout algorithm when the collection undergoes changes due to newly added items. In contrast to this paper, [14] pursued a global map approach for MDS. It was suggested to use Procrustes analysis [4] to better align newly generated maps with their respective predecessors, which we will also adopt here as it leads to a reduced confusion of users when navigating with aligned maps. Similar to our approach, Kleiman et al. [7] arrange images on a fully populated similarity based grid layout that can be interactively panned and zoomed. Due to the dense layout, similarities and clusters cannot be inferred from the layout alone. In case of images, humans are able to quickly assess similarities based on the visual information. When using other media, map-based approaches need to illustrate similarities and clusters in order to support the user.

Evaluating map-based retrieval and exploration systems is very challenging due to the complexity of the exploration process. Often, systems are evaluated in controlled lab experiments in order to get a better understanding of how users interact with them and whether there are able to utilize the two-dimensional arrangement. In [12] random and similarity based organization of images in browsing scenarios are compared. Strong et al. [15] evaluate the user's effectiveness in finding images with specific properties. These evaluation scenarios do not conform to a typical exploration task, where a user's information need might change during the exploration process. In this paper we aim to evaluate a user's exploration effectiveness while looking for movies that are considered to be worth watching tonight. To our knowledge, there is no online study comparing a map-based exploration approach to an established, traditional browsing-based system that evaluates the user's effectiveness in an open exploration scenario.

3 Visual Berrypicking

In document-based information retrieval, *berrypicking* describes the user's behavior during the search process [2]. Instead of a single query, the user performs a series of evolving queries in order to find relevant information. While inspecting individual documents, the user gets a better understanding of his or her own information need, which is then used to modify the query. At any time during this process, useful information can be identified, which will all contribute to satisfying the user's information need. When applying the idea of berrypicking to map-based visualizations, choosing a neighboring information or movie object as a new seed corresponds to modifying a search query, which we call Visual Berrypicking. We hypothesize that being able to iteratively inspect parts of the information space by Visual Berrypicking stimulates exploration, which helps to learn about information objects and their relations, and thus, enhances overall user experience.

Visualization of similarities on a two-dimensional map requires dimensionality reduction of the typically much higher input feature space. Multidimensional scaling (MDS, [9]) is a popular distance-preserving technique that can be used for this purpose. Naturally, any projection into lower dimensional spaces will cause projection errors that increase with the number of dimensions to reduce and the size of the collection. As a result, neighboring objects may turn out to be not that similar after all (degrading *trustworthiness*[2]) and similar objects may not be visualized as neighbors but far apart from each other (degrading *continuity*[2]). Such problems may disturb the process of berrypicking, since users might get confused about information objects that are visualized far apart though being perceived as similar. By limiting the number of items used to compute the projection, we try to reduce the impact of these problems. Large-scale collections with millions of items cannot be reasonably handled in their entirety anyway and users will therefore always focus on a small subset. For each view of the dataset, our prototype presents a map of the k most similar items only, given a user defined seed item. By clicking on any of the presented items, a new map is created using the selected item as a new seed.

Transitions between consecutive maps are animated with the aim of giving the user the impression of panning on a large map representing the collection as a whole. In order to make these transitions as consistent as possible, we use the overlap between any two consecutive maps to align them based on their common neighbors, see Fig. 1. We use Procrustes analysis [4] to reduce the sum of the squared differences between the two sets of items that remain visible by translation, scaling, rotation and reflection. The alignment error corresponds to the

---

[2] The measures of trustworthiness and continuity are introduced in [16] together with a discussion of common problems that arise from using map visualizations for information retrieval.

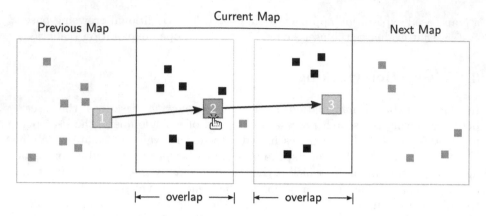

Fig. 1. Similarity-based projection of nearest neighbors (squares) using three seed items (colored squares 1, 2, 3). Common neighbors (black squares) overlap between consecutive maps and are used for alignment when navigating from one item to the next. (Color figure online)

difference between the two-dimensional positions of common nearest neighbors in consecutive maps after alignment by Procrustes analysis. Because of small relative position changes of common neighbors and the alignment of subsequent maps by Procrustes analysis this transition is ideally perceived as panning a structurally stable map. As a result, the user benefits from continuity that allows to transfer knowledge from one map to the next and more stable navigation directions (items with certain properties can be found in the same corner during multiple interactions). Thus, the user is less likely to get lost, which supports the process of exploration.

4   Evaluation

For evaluation we implemented a web-based prototype for movie exploration called NEMP (Neighborhood Exploration using MDS and Procrustes analysis) and compared it to the web interface of *The Movie Database* (TMDb), see Figs. 2 and 3. We used the 10,000 most popular movies from TMDb, including a movie's title, cover image, rating, genres, actors, directors and plot description. Because the proposed technique only considers neighborhoods of a constant size for map generation, it does not depend on the collection size and thus is easily scalable – given that the k-nearest neighbors can be retrieved efficiently.

For computing movie similarities, we linearly combined five individual measures: linear difference in release date, jaccard similarity of genres, directors and actors, and plot similarity using tf-idf. Optimal weights were determined based on preliminary test trials. The number of k-nearest neighbors was fixed to 30. All pairwise similarities were calculated before the experiment, such that the k-nearest neighbors can be retrieved in $\mathcal{O}(n)$ time. However, for larger datasets

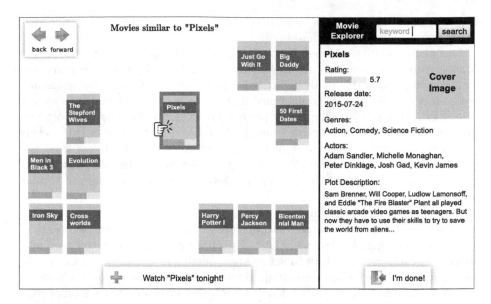

Fig. 2. NEMP User interface and study overlay (buttons with shadow): a projection of movies visualizes similarities and local clusters with respect to, e.g., genre (left bottom cluster), actors (top right cluster) and director (bottom right cluster); back and forward buttons (top left) as well as buttons to add a movie or quit are used as an overlay in both study interfaces. Cover images (grey areas) are omitted for legal reasons.

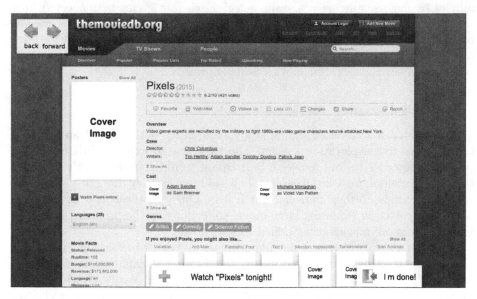

Fig. 3. TMDb user interface and study overlay: movies can be explored by inspecting lists of popular movies or following recommendations on movie pages. Screenshot adapted to fit page, cover images omitted. Original version at https://web.archive.org/web/20151003235059/https://www.themoviedb.org/movie/257344-pixels.

there are more efficient ways to retrieve the k-nearest neighbors, e.g. approximately via locality-sensitive hashing [3].

Our map-based layout of movie items is constructed by applying classic MDS to the given seed item and all its k-nearest neighbors. Consecutive maps are aligned using Procrustes analysis. In order to make use of all available display space, item coordinates are transformed linearly to better fit the screen's aspect ratio. Also, overlaps of similar movies are prevented by dividing the viewport into grid cells of fixed size and assigning movies to their closest free cell. Due to this transformation, the arrangement in the user interface slightly differs from the coordinates provided by MDS. We believe that the gain in usability outweighs the loss in precision of relative distances. Most importantly, clusters of similar movies are usually preserved.

The user interface is composed of a simple search bar, a sidebar showing detail information, and a large viewport presenting movie covers arranged in our two-dimensional layout, see Fig. 2. By clicking on a movie cover, a new similarity search is started and its k-nearest neighbors are shown. Changes in the position of movies are animated. Hovering over a movie cover allows to inspect its information inside the details panel. Users are able to click on genres, actors and directors, which will result in a corresponding search query. Also, genres, actors, directors and plot terms that are common to both – the currently selected and hovered movie – are highlighted.

In contrast, the TMDb web interface follows a traditional browsing approach, see Fig. 3. Users may browse through several lists of popular movies, top rated movies or movies starring a specific actor. Each movie can be inspected in a separate details page, which also provides a list of related movies as recommendations. In comparison to NEMP, more information are presented for each movie, including, e.g., its runtime, budget, additional pictures and reviews. Both interfaces allow to search for movies using keywords.

4.1   Study Design

We conducted a web-based online study comparing both NEMP and TMDb in an interactive exploration session. The task was to explore TMDb's movie collection in order to find movies a participant would consider worth watching tonight. Therefore, we asked participants to add movies they thought to be worth watching tonight to their personal watch lists. Our hypothesis is that users are better supported in exploring large movie collections by the proposed interface, resulting in more movies being added to watch lists.

In order to effectively test our hypotheses, we tried to avoid that users will add movies they have not found during the process of exploration. Therefore, we asked them to write down movie titles they could easily remember and would consider worth watching tonight before using any of the two websites. We assumed that participants would prefer to explore movies given that we asked them not to add movies they were aware of beforehand. The study interface itself was designed as consecutive pages and forms asking participants for 20–30 min of their time, their age, gender, profession and a self-assessment of

Table 1. Evaluation statements that were rated according to a 5-point Likert scale by all participants immediately after using an interface.

helpful	I have the impression I was able to find interesting movies
easy	I thought the interface was easy to used
complex	I found the interface unnecessarily complex
intuitive	I found navigation using this interface intuitive
inconsistent	I thought there was too much inconsistency when new movies were shown
interesting	I was able to find interesting links between different movies
random	The movies presented seemed random

their own "movie knowledge". After a number of instructions, e.g., to use the fullscreen mode of their browser window, they were presented both user interfaces in random order (later denoted as *first* and *second* interface). During both sessions we asked participants to add relevant movies to their personal watch lists. Therefore, we overlayed both interfaces by additional buttons that allowed to add and remove movies, as well as to proceed with the study once finished, see Figs. 2 and 3. Immediately after each interface session, participants were asked to rate statements listed in Table 1 using a 5-point Likert scale.

4.2   Study Results

A total of 110 participants (48 female, 62 male, 31.8 years old on average) completed our study. Since participants performed our study online without any human supervision, results are unsurprisingly diverse. Although we implemented an automatic check for browser size, which shows a large green tick or red cross indicating whether the browser size is appropriate, a total of 67 participants chose to proceed with the study even though their window was rated to be too small. However, a large window size is important in order to provide enough display space to be able to identify clusters using the NEMP interface. Therefore, it would be interesting to compare our results with a controlled lab experiment.

In 9 cases, participants spent less than one minute in one or both of the two interfaces. Also, results suggest a strong bias towards the interface presented first, e.g., the average time spent on the first interface is 7:09 (min:sec) compared to 4:56 for the second interface. Due to the design of our online study, it was not possible to even out the number of trials starting with a particular interface. In total, 47 participants started with NEMP, 63 with TMDb. In order to do meaningful statistical comparisons, we randomly removed 16 results for participants that started with TMDb.

Figure 4 shows aggregated ratings for all statements of Table 1. Although results do not show a clear winner, some important aspects clearly stand out. The most prominent difference was expressed for the statement that interesting links can be found more easily using NEMP, which is the fundamental goal of our exploration interface and supports our initial hypothesis that users will be supported in learning about information objects and their relations.

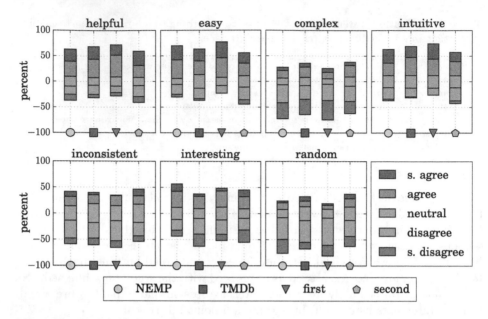

Fig. 4. Evaluation results presented in diverging stacked bar charts showing percentage of participants rating both interfaces according to all seven statements, see Table 1. Results for the first and second interface are shown for comparison.

Presumably due to the inherent differences in both interfaces, e.g., TMDb contains a lot more information, images, etc. about each movie, NEMP is rated slightly less helpful in finding interesting movies, but also more easier to use and less complex. Also, we found strong correlations between the participants' browser size and perceived complexity. TMDb was rated more complex with increasing browser size (Pearson correlation coefficient $pcc = 0.25$), while NEMP was rated less complex ($pcc = -0.12$), see Fig. 5. Similarly, we found a positive correlation between NEMP's helpfulness and browser size ($pcc = 0.16$), which might be due to NEMP's two-dimensional presentation, which benefits from large screens. Finally, there was a strong correlation between the time participants spent using NEMP and their perceived randomness of presented movies ($pcc = -0.21$), suggesting that participants needed to get familiar with our approach.

On average 5.08 (standard deviation $std = 4.45$) movies were added to watch lists during exploration sessions using NEMP, and 5.77 ($std = 5.17$) were added during sessions using TMDb. Therefore, our hypothesis that more movies will be added while using the proposed exploration interface can not be validated. However, since NEMP is considered to be an unconventional user interface, it could still be clearly demonstrated that participants were able to effectively use it for exploration of a large movie dataset even though they used it for the first time.

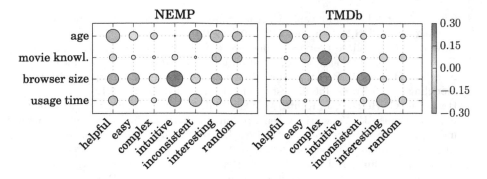

Fig. 5. Pearson correlation coefficients for study parameters and statement ratings of both interfaces, see Table 1. Circle radius and color intensity corresponds to correlation strength, hue to orientation. (Color figure online)

5   Conclusions

We have compared Visual Berrypicking to traditional browsing techniques using a large-scale movie collection. Local maps are aligned during navigation from one neighborhood to the next, which ideally creates the impression of panning a large global map. At the same time, small maps can be computed much quicker and have higher visual accuracy. We have presented a web-based prototype for movie exploration using multidimensional scaling for map generation and Procrustes analysis for alignment. For evaluation, we have compared our prototype with the website of *The Movie Database* in an interactive online study. Results indicate that users can effectively explore large movie collections, even though our approach was unfamiliar to them. Additionally, they find more interesting links between different movies.

Although our study was conducted using the example of movie exploration, the proposed approach is not restricted to a specific application. Given a meaningful similarity measure and thumbnails (e.g. images or music cover art) our approach is generally applicable to any kind of media. Future work will focus on training feature weights during the user interaction process in order to further adapt maps to the user's retrieval focus. As local neighborhood maps are easily computed on the fly, adaptations of the underlying similarity space can be visualized immediately.

References

1. Barthel, K.U., Hezel, N., Mackowiak, R.: Graph-based browsing for large video collections. In: He, X., Luo, S., Tao, D., Xu, C., Yang, J., Hasan, M.A. (eds.) MMM 2015. LNCS, vol. 8936, pp. 237–242. Springer, Heidelberg (2015). doi:10. 1007/978-3-319-14442-9_21
2. Bates, M.J.: The design of browsing and berrypicking techniques for the online search interface. Online Inf. Rev. **13**(5), 407–424 (1989)

3. Gionis, A., Indyk, P., Motwani, R., et al.: Similarity search in high dimensions via hashing. In: VLDB, vol. 99, pp. 518–529 (1999)
4. Gower, J.: Generalized procrustes analysis. Psychometrika **40**(1), 33–51 (1975)
5. Gretarsson, B., O'donovan, J., Bostandjiev, S., Höllerer, T., Asuncion, A., Newman, D., Smyth, P.: Topicnets: visual analysis of large text corpora with topic modeling. ACM Trans. Intell. Syst. Technol. (TIST) **3**(2), 23 (2012)
6. Hentschel, C., Hercher, J., Knuth, M., Osterhoff, J., Quehl, B., Sack, H., Steinmetz, N., Waitelonis, J., Yang, H.: Open up cultural heritage in video archives with mediaglobe. In: Proceedings of the 12th International Conference on Innovative Internet Community Services, I2CS (2012)
7. Kleiman, Y., Lanir, J., Danon, D., Felberbaum, Y., Cohen-Or, D.: Dynamicmaps: similarity-based browsing through a massive set of images. In: Proceedings of the 33rd Annual ACM Conference on Human Factors in Computing Systems, CHI 2015 pp. 995–1004. ACM (2015)
8. Kohonen, T., Kaski, S., Lagus, K., Salojarvi, J., Honkela, J., Paatero, V., Saarela, A.: Self organization of a massive document collection. IEEE Trans. Neural Netw. **11**(3), 574–585 (2000)
9. Kruskal, J.: Multidimensional scaling by optimizing goodness of fit to a nonmetric hypothesis. Psychometrika **29**, 1–27 (1964)
10. Lee, H., Kihm, J., Choo, J., Stasko, J., Park, H.: iVisClustering: an interactive visual document clustering via topic modeling. Comput. Graph. Forum **31**, 1155–1164 (2012). Wiley Online Library
11. Low, T., Hentschel, C., Stober, S., Sack, H., Nürnberger, A.: Visual berrypicking in large image collections. In: Proceedings of the 8th Nordic Conference on HCI, NordiCHI 2014, pp. 1043–1046. ACM (2014)
12. Rodden, K., Basalaj, W., Sinclair, D., Wood, K.: Does organisation by similarity assist image browsing? In: Proceedings of the SIGCHI Conference on Human Factors in Computing Systems, CHI 2001, pp. 190–197. ACM (2001)
13. Rubner, Y., Guibas, L., Tomasi, C.: The earth mover's distance, multi-dimensional scaling, and color-based image retrieval. In: Proceedings of the ARPA Image Understanding Workshop, pp. 661–668 (1997)
14. Stober, S., Low, T., Gossen, T., Nürnberger, A.: Incremental visualization of growing music collections. In: 14th International Conference on Music Information Retrieval, ISMIR 2013, pp. 433–438 (2013)
15. Strong, G., Hoeber, O., Gong, M.: Visual image browsing and exploration (vibe): user evaluations of image search tasks. In: An, A., Lingras, P., Petty, S., Huang, R. (eds.) AMT 2010. LNCS, vol. 6335, pp. 424–435. Springer, Heidelberg (2010). doi:10.1007/978-3-642-15470-6_44
16. Venna, J., Kaski, S.: Local multidimensional scaling. Neural Netw. **19**, 889–899 (2006)

Facial Expression Recognition by Fusing Gabor and Local Binary Pattern Features

Yuechuan Sun[1] and Jun Yu[1,2(✉)]

[1] Department of Automation, University of Science and Technology of China,
Hefei 230027, China
ycsun@mail.ustc.edu.cn
[2] State Key Laboratory for Novel Software Technology, Nanjing University,
Nanjing, People's Republic of China
harryjun@ustc.edu.cn

Abstract. Obtaining effective and discriminative facial appearance descriptors is a challenging task for facial expression recognition (FER). In this paper, a new FER method which combines two of the most successful facial appearance descriptors, namely Gabor filters and Local Binary Patterns (LBPs), is proposed considering that the former one can represent facial shape and appearance over a broader range of scales and orientations while the latter one can capture subtle appearance details. Firstly, feature vectors of Gabor and LBP representations are generated from the preprocessed face images respectively. Secondly, feature fusion is applied to combine these two vectors and dimensionality reduction is conducted. Finally, the Support Vector Machine (SVM) is adopted to classify prototypical facial expressions using still images. The experimental results on the CK+ database demonstrate that the proposed method promotes the performance compared with that using Gabor or LBP descriptor alone, and outperforms several other methods.

Keywords: Facial expression recognition · Gabor wavelet · Local binary patterns · Feature fusion

1 Introduction

Computers and other electronic devices are penetrating our daily lives at ever increasing rates, bringing us with more and more user-friendly interactions. However, current human-computer interactions (HCIs) mostly rely on traditional interactive modes, such as text input and screen touch. It is believed that to truly achieve human-centered HCI, computers are required to understand and react naturally to users' implicit information [1], including affective state. Therefore, automatic recognition of facial expressions is crucial to HCI for its potential applications in everyday life.

In terms of feature representation of expression analysis, the methods reported in literature can be classified mainly into geometric feature-based ones [2,3] and appearance-based ones [4,5]. The former track some predefined facial

© Springer International Publishing AG 2017
L. Amsaleg et al. (Eds.): MMM 2017, Part II, LNCS 10133, pp. 209–220, 2017.
DOI: 10.1007/978-3-319-51814-5_18

points to infer the shape and locations of facial components such as the mouth or the eyebrows to classify the facial expression. The second approaches describe the appearance changes caused by expression by applying image filters to extract feature vectors.

One of the most extensively employed appearance descriptors is the Gabor filter, which proves highly powerful for its superior multi-scale and multi-direction representation [6,7]. LBP [5,8] is another successful descriptor which captures small texture details. LBP features show decent performance for its invariance against monotonic transformation and computational simplicity [8]. Being motivated by this, we attempt to combine the advantages of these two features to attain better FER performance.

Recently, deep convolutional neural networks (CNNs) have yielded excellent performance in vision-related applications [9,10] and boosted the performance of FER [11–13]. CNNs can learn multiple levels of representations which enable algorithms to find complex expressive patterns from image data. In particular, Liu et al. [11] proposed a 3D CNN with deformable action parts constraints to learn part-based representations for expression recognition. Jung et al. [12] extracted facial landmarks based shape features and image based shape features through a combined CNN. Finally, the authors of [13] presented a CNN-based deep architecture which modeled facial expressions by utilizing a set of local action unit features. However, training a robust CNN model requires a lot of data, which is difficult to collect.

In this paper, we propose a new approach (See Fig. 1) for recognizing six basic expressions (i.e., Anger, Disgust, Fear, Happiness, Sadness and Surprise) along with a neutral state by combining Gabor and LBP descriptors. In particular, our method first preprocesses the raw expression images for normalization, then extracts facial features by means of Gabor and LBP respectively, followed by Principal Component Analysis (PCA) implementation for dimensionality reduction. Feature vectors are then fused and fed to Linear Discriminant Analysis (LDA) for feature optimization.

The rest of this paper is organized as follows: Sect. 2 describes the preprocessing procedure. Feature extraction is introduced in Sect. 3, followed by feature

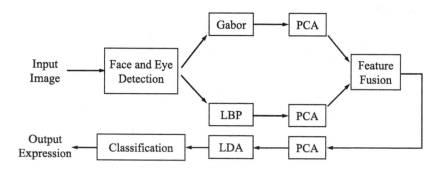

Fig. 1. Flowchart of the proposed method.

fusion method. We evaluate the performance of the proposed method and give a detailed analysis of the results attained in Sect. 5. Section 6 concludes the paper.

2 Preprocessing

Preprocessing is a vital step in facial expression recognition. An ideal preprocessing is supposed to eliminate the irrelevant information (e.g., illumination, background, rotation) for pure expression images, which have uniform size and normalized intensity. For a given image, we first localize the centers of eyes with Adaboost learning algorithm [14] for reference points, followed by an image rotation to live up to eye coordinates. Images are then cropped and resized to 108×120 pixels to remove the background based on the face model shown in Fig. 2(a). Histogram equalization is finally adopted for illumination compensation.

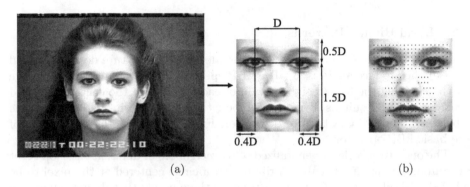

(a)   (b)

Fig. 2. (a) Original image and cropped face region. (b) Inhomogeneous sampling.

3 Feature Extraction

3.1 Gabor Feature Representation

Gabor filters are widely used to extract local features and present the best simultaneous localization of spatial and frequency information. The Gabor function can be defined as [15]:

$$\psi_{u,v}(z) = \frac{\|k_{u,v}\|}{\sigma^2} e^{(-\|k_{u,v}\|^2 \|z\|^2 / 2\sigma^2)} \left[e^{ik_{u,v}z} - e^{-\sigma^2/2} \right]. \tag{1}$$

$z = (x, y)$ gives the pixel position in the spatial domain, and frequency vector $k_{u,v}$ is defined as follows:

$$k_v = \frac{k_{\max}}{f_v} e^{i\phi_u}, \tag{2}$$

where $\phi_u = u\pi/u_{\max}, \phi_u \in [0, \pi)$, and u and v denote the orientation and scale factors of Gabor filters respectively. In our system, we adopt the Gabor filters of five scales ($v = 0, 1, \cdots, 4$) and eight orientations ($u = 0, 1, \cdots, 7$), with $\sigma = 2\pi$, $k_{max} = \pi/2$ and $f = \sqrt{2}$. The same parameters were also chosen in [7,15].

The Gabor wavelet feature representation $O_{u,v}(z)$ is obtained by convolving a given face image $I(z)$ with Gabor filters $\psi_{u,v}(z)$:

$$O_{u,v}(z) = I(z) * \psi_{u,v}(z), \tag{3}$$

and the magnitude of the convolution output is usually used for facial expression recognition for its invariance for displacement [16]. For each pixel of the face image, totally 40 Gabor features are obtained when 40 Gabor filters are used. Thus, the feature vector can be obtained by concatenating the outputs of all the pixels. In practice, inhomogeneous sampling is applied to extract distinctive expression information mainly located at eyes, mouth and nose. 269 fiducial points are selected in the experiments (as shown in Fig. 2(b)), as it achieves a good trade-off between the performance and the feature length.

3.2   Local Binary Patterns

LBP was originally introduced by Ojala et al. in [17] for texture description and later applied to face analysis [5]. The LBP operator labels the pixels of an image by thresholding the 3×3 neighborhood of each pixel with the center value and considering the result as a binary number, then the 256-bin histogram of the labels can be used as a texture descriptor. Figure 3(a) gives the illustration of the basic LBP operator.

The operator was later generalized to use neighbors of different sizes [8]. A set of sampling points are evenly distributed on a circle centered at the pixel to be labeled, and bilinear interpolation is used for these points that do not fall within the pixels. Figure 3(b) shows an example of the circular neighborhood, where the notation (P, R) represents sampling P points on a circle of radius of R.

A successful extension to the original LBP is so called uniform patterns. A local binary pattern is called uniform if it contains at most two bitwise transitions from 0 to 1 or vice versa when the corresponding binary string is considered circular. It is found that the uniform patterns account for nearly 90% of all the patterns in a (8, 1) neighborhood and for 70% in a (16, 2) one in texture image [8]. We use the following notation for the LBP operator: $LBP_{P,R}^{u2}$,

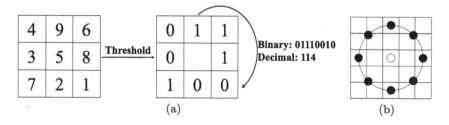

(a)   (b)

Fig. 3. (a) The basic LBP operator. (b) The circular (8,2) neighborhood.

where the subscript means using the operator in a (P, R) neighborhood and the superscript $u2$ represents using uniform patterns and labelling all remaining patterns with a single label. Hence, the number of different output labels with the neighbor of P pixels is $P \times (P - 1) + 3$ instead of 2^P for the standard LBP.

An early stage experiment is conducted to find the optimal parameters for this application, resulting in $P = 8$, and $R = 1$. Hence, $LBP_{8,1}^{u2}$ is adopted in our experiments. A histogram of the labeled image can be defined as:

$$H_i = \sum_{x,y} I\{f_l(x, y) = i\}, i = 0, \cdots, n - 1, \tag{4}$$

where n is the number of different labels produced by LBP operator and

$$I\{A\} = \begin{cases} 1, & \textit{if } A \textit{ is true} \\ 0, & \textit{otherwise} \end{cases}. \tag{5}$$

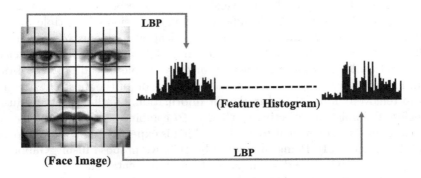

Fig. 4. LBP-based face description.

The LBP histograms contain the information of the distribution of local micropatterns and thus can be used to effectively describe face images. However, spatial information should also be considered to represent the shape information of faces for efficient face representation. For this purpose, face images are divided into small regions to extract LBP features (See Fig. 4 for an illustration). The LBP features extracted from each sub-region are subsequently concatenated into a single, spatially-enhanced feature histogram. In our work, the face images are divided into 72 (8×9) regions, as this is empirically found to give the best feature representation.

4 Feature Fusion

4.1 Dimensionality Reduction by Principal Component Analysis

Before fusing the Gabor and LBP features, we first reduce their dimensionality to remove some of the redundant information with PCA. PCA is a useful technique used to reduce the dimensionality of a feature vector and has been successfully used in face analysis [18,19]. It seeks for a linear projection that maps the original high-dimensional data space to a lower dimensional feature subspace. Using PCA reduces the length of the feature vector while retains the variance of the raw data. Principal components $v_i, i = 1, 2, \cdots, k$ corresponding to the k largest eigenvalues λ_i in descending order are first calculated, and k is determined according to $\sum_{i=1}^{k} \lambda_i \Big/ \sum_{all} \lambda_j = \eta$, where η is the contribution rate. The linear transformation matrix W_{pca} is then obtained by combining the k eigenvectors.

Let $a \in R^m$ and $b \in R^n$ be the Gabor and LBP feature vectors of a face image respectively, then the corresponding lower-dimensional vectors $\alpha = W_{Gabor}^T a$ and $\beta = W_{LBP}^T b$ can be obtained for feature fusion.

4.2   Fusion Method of Gabor and LBP Features

Most of the existing approaches to facial expression recognition use features of just one type. However, research indicates that the fusion of multiple features can improve the performance of pattern classification problems [20]. There are mainly two types of data fusion strategies, namely feature fusion and decision fusion. Feature level fusion plays a very important role in the process of data fusion, for it focuses on the selection and combination of the most effective features to remove redundant and irrelevant information. Of all the feature fusion strategies, a simple but effective method is to combine several sources of raw features to produce a new feature vector, which is expected to be more informative and synthetic [21]. Being motivated by this, we propose fusing Gabor and LBP features to give more discriminative facial representations.

Suppose that we have PCA-reduced feature vectors α and β of the original Gabor and LBP features defined in Sect. 4.1. In order to eliminate the numerical unbalance between these two features and gain satisfactory fusion performance, we first adopt feature normalization to turn α and β into unit vectors as suggested in [20]:

$$\begin{cases} \bar{\alpha} = \alpha/\|\alpha\| \\ \bar{\beta} = \beta/\|\beta\| \end{cases} . \tag{6}$$

Considering that the dimensions of $\bar{\alpha}$ and $\bar{\beta}$ are usually unequal, the higher-dimensional one is more powerful than the lower-dimensional one as it plays a more important role in the scatter matrices in terms of the linear feature extraction after combination. Thus, in order to remove the unfavorable effect resulting from unequal dimensions, we adopt the weighted combination strategy introduced in [20]. The serial combination is formed by $\gamma = [\alpha, \theta\beta]$, where θ is the weight for LBP features. Suppose that the dimensions of α and β are m and n respectively, then let θ be m^2/n^2, and the combined feature vector can be defined as $\gamma = [\bar{\alpha}, \theta\bar{\beta}]$, here the dimension of γ is $m + n$.

The combination weight θ is determined for the following reason. When the lengths of two feature vectors are unequal, since the size of scatter matrices generated by feature vectors α and β are $m \times m$ and $n \times n$, the combination weight θ is considered to be the square of m/n.

4.3 Feature Optimization by Linear Discriminant Analysis

The next stage of the process is to extract optimally discriminative nonlinear features from the combined feature vector γ using LDA. LDA seeks for an optimal projection that best discriminates data among classes, the goal is to maximizing the ratio of the between-class scatter and the within-class scatter [19]. The between-class scatter matrix S_B and within-class scatter matrix S_W can be defined as:

$$S_B = \sum_{i=1}^{c} N_i \left(\mu_i - \mu\right) \left(\mu_i - \mu\right)^T, \tag{7}$$

$$S_W = \sum_{i=1}^{c} \sum_{j=1}^{N_i} (X_{ij} - \mu_i)(X_{ij} - \mu_i)^T, \tag{8}$$

where μ_i and μ are the mean images of class i and all classes, X_{ij} denotes the jth sample of class i, c defines the number of classes and N_i is the number of samples of class i. The optimal projection can be obtained by maximizing the ratio $\det |S_B|/\det |S_W|$. To overcome the singularity problem of matrix S_W, PCA is implemented on γ as proposed in [19].

5 Experiments and Analysis

5.1 Experiment

Most facial expression recognition systems recognize six basic emotions proposed by Ekman [22]. In this work, 6-class as well as 7-class (including neutral faces) expression classification experiments are conducted. The performance of the proposed method is evaluated on the Cohn-Kanade (CK+) Facial Expression Database [23]. This well-known dataset consists of 593 image sequences from 123 subjects from 18 to 30 years old. The image sequences vary in duration from neutral faces to apexes, the peak information of the prototypical facial expressions. In our study, 309 sequences from 106 subjects are selected which meet the criteria for one of the six basic expressions, where the last frames are used for 6-class classification. In addition, neutral faces of the first frames are selected to conduct a 7-class classification experiment. Table 1 presents the detailed statistics for the portion of the dataset that is used.

We adopt a 10-fold cross-validation scheme for performance evaluation. The dataset is randomly divided into 10 groups of roughly equal numbers of subjects and each prototypical expression. SVM is chosen to classify different facial expressions and three dominant kernels of SVM (i.e. linear, RBF and polynomial) are used.

Table 1. Overview of the dataset.

Expression	AN	DI	FE	HA	SA	SU	NE	Total	
No. of image	45	59	25	69	28	83	106	309 (6-class)	415 (7-class)

AN = Anger,   DI = Disgust,   FE = Fear,   HA = Happiness,   SA = Sadness, SU = Surprise, NE = Neutral.

5.2   Evaluation and Comparison

We first compare the performance of our proposed method with original Gabor and LBP descriptors using the SVM classifier. There are no significant differences when using different SVM kernels. The results in Table 2 show that the proposed method achieves higher recognition rates than Gabor or LBP applied alone in both 6-class and 7-class experiments, reaching 97.42% and 95.45% respectively.

The confusion matrix for 6-class classification in Table 3 shows that Happiness and Surprise are easiest to recognize. It is an intuitive result as these expressions cause many more facial movements mainly located around mouth and thus are relatively easy to recognize. Other expressions, such as Anger and Sadness, attain lower classification accuracies due to the lack of facial deformation and training samples.

The case of 7-class classification turns out to be less satisfying as shown in Table 4. The confusion matrix shows that Neutral expression causes considerable amount of confusion with subtle expressions including Anger and Sadness and makes it hard to discriminate among them. It is noticeable that the recognition rate for Sadness drops dramatically in this case, probably caused by not having enough training images.

To further evaluate our proposed method, we compare our performance with the results obtained with other approaches which adopted similar experimental protocols on the CK+ dataset. Lucey et al. [23] proposed two methods based on Active Appearance Model (AAM) when they proposed their database, namely similarity normalized shape (SPTS) and canonical appearance features (CAPP). Chew et al. [2] and Jeni et al. [3] proposed constrained local model (CLM) based

Table 2. Comparison of recognition accuracy of different methods using SVM on the CK+ database.

Method		Kernel		
		Linear	RBF	Polynomial
Gabor	6-class	91.92	91.92	92.57
	7-class	89.17	89.18	89.16
LBP	6-class	94.75	95.06	94.75
	7-class	91.55	91.80	91.55
Gabor+LBP	6-class	97.10	**97.42**	96.45
	7-class	**95.45**	**95.45**	94.45

Table 3. Confusion matrix for the 6-class classification on the CK+ dataset

	AN	DI	FE	HA	SA	SU
AN	**93.4**	2.2	2.2	0	2.2	0
DI	0	**98.3**	1.7	0	0	0
FE	4.0	0	**92.0**	0	0	4.0
HA	0	0	0	**100.0**	0	0
SA	3.6	0	0	0	**96.4**	0
SU	0	0	1.2	0	0	**98.8**

Table 4. Confusion matrix for the 7-class classification on the CK+ dataset

	AN	DI	FE	HA	NE	SA	SU
AN	**93.4**	2.2	2.2	0	2.2	0	0
DI	0	**98.3**	0	0	1.7	0	0
FE	8.0	0	**80.0**	0	0	0	12.0
HA	0	0	0	**100.0**	0	0	0
NE	1.9	0	0	0	**97.2**	0.9	0
SA	10.7	0	0	0	10.7	**78.6**	0
SU	0	0	1.2	0	0	0	**98.8**

methods to model the changes of shape information from neutral faces to apexes. The results in Table 5 show that our method outperforms theirs' probably due to that they are all geometric-based, which have poorer performance in person independent scenarios and are insensitive to subtle expressions. In terms of 7-class classification experiment, Shan et al. [5] trained traditional LBP features using SVMs. Jain et al. [24] modeled temporal variations within facial shapes from video sequences using Latent-Dynamic Conditional Random Fields (LDCRFs). Among the methods [25–27] which adopted similar feature fusion strategies to ours, only [27] reported its performance on FER task. However, it selected fewer

Table 5. Recognition rates on CK+ database (6-class)

Method	AN	DI	FE	HA	SA	SU	Ave
AAM [23]	75.0	94.7	65.2	**100.0**	68.0	96.0	83.2
CLM [2]	70.1	92.5	72.1	94.2	45.9	93.6	78.1
3D CLM [3]	77.8	91.5	80.0	98.6	67.9	97.6	85.6
CNN [11]	91.1	94.0	83.3	98.6	85.7	96.4	91.5
3D CNN [12]	**100.0**	**100.0**	84.0	**100.0**	89.3	**98.8**	95.4
Proposed	93.4	98.3	**92.0**	**100.0**	**96.4**	**98.8**	**96.5**

Table 6. Recognition rates on CK+ database (7-class)

Method	AN	DI	FE	HA	NE	SA	SU	Ave
LBP [5]	85.0	97.5	68.0	94.7	90.0	69.5	98.2	86.1
LDCRFs [24]	76.7	81.5	**94.4**	98.6	73.5	77.2	**99.1**	85.9
Gabor+LBP [27]	53.6	75.4	79.7	96.6	96.2	**79.1**	96.9	82.5
AUDN [13]	81.8	95.5	82.7	99.6	95.4	71.4	97.6	89.1
Proposed	**93.4**	**98.3**	80.0	**100.0**	**97.2**	78.6	98.8	**92.3**

fiducial points and did not perform feature normalization before fusion, which leads to a lower recognition rate. The results show that the proposed method shows equal superiority on the performance of 7-class classification. Specifically, the CNN-based methods [11–13] introduced in Sect. 1 improved the recognition performance compared with traditional approaches according to Tables 5 and 6, however, are less competitive to ours.

6   Conclusion

In this paper, we propose a framework of fusing Gabor and LBP features to recognize facial expressions. The feature fusion method fully utilizes the local feature and the texture information to extract expression features. Extensive experiments show that the proposed method yields superior performance compared to Gabor or LBP descriptor alone and outperforms several other methods on the CK+ database. However, the results indicate that the proposed method fails to achieve equivalent recognition rates for subtle expressions, such as Anger and Sadness.

In the future, this work will be extended in three aspects. Firstly, more feature descriptors will be used to give more comprehensive facial representations. Secondly, the fusion strategy will be improved to increase the recognition rate. Finally, we expect to extend this framework to handle other recognition problems, such as facial action unit detection.

Acknowledgement. This work is supported by the National Natural Science Foundation of China (No. 61572450 and No. 61303150), the Open Project Program of the State Key Lab of CAD&CG, Zhejiang University (No. A1501), the Fundamental Research Funds for the Central Universities (WK2350000002), the Open Funding Project of State Key Laboratory of Virtual Reality Technology and Systems, Beihang University (No. BUAA-VR-16KF-12).

References

1. Zeng, Z., Pantic, M., Roisman, G.I., Huang, T.S.: A survey of affect recognition methods: audio, visual, and spontaneous expressions. IEEE Trans. Pattern Anal. Mach. Intell. **31**(1), 39–58 (2009)

2. Chew, S.W., Lucey, P., Lucey, S., Saragih, J., Cohn, J.F., Sridharan, S.: Person-independent facial expression detection using constrained local models. In: IEEE International Conference on Automatic Face and Gesture Recognition and Workshops, pp. 915–920 (2011)
3. Jeni, L.A., Lőrincz, A., Nagy, T., Palotai, Z., Sebők, J., Szabó, Z., Takács, D.: 3D shape estimation in video sequences provides high precision evaluation of facial expressions. Image Vis. Comput. **30**(10), 785–795 (2012)
4. Rivera, A.R., Castillo, J.R., Chae, O.O.: Local directional number pattern for face analysis: face and expression recognition. IEEE Trans. Image Process. **22**(5), 1740–1752 (2013). A Publication of the IEEE Signal Processing Society
5. Shan, C., Gong, S., McOwan, P.W.: Facial expression recognition based on local binary patterns: a comprehensive study. Image Vis. Comput. **27**(6), 803–816 (2009)
6. Donato, G., Bartlett, M.S., Hager, J.C., Ekman, P., Sejnowski, T.J.: Classifying facial actions. IEEE Trans. Pattern Anal. Mach. Intell. **21**(10), 974–989 (1999)
7. Liu, C., Wechsler, H.: Gabor feature based classification using the enhanced fisher linear discriminant model for face recognition. IEEE Trans. Image Process. **11**(4), 467–476 (2002)
8. Ojala, T., Pietikäinen, M., Mäenpää, T.: Multiresolution gray-scale and rotation invariant texture classification with local binary patterns. IEEE Trans. Pattern Anal. Mach. Intell. **24**(7), 971–987 (2002)
9. Krizhevsky, A., Sutskever, I., Hinton, G.E.: Imagenet classification with deep convolutional neural networks. In: Advances in Neural Information Processing Systems, pp. 1097–1105 (2012)
10. Ciregan, D., Meier, U., Schmidhuber, J.: Multi-column deep neural networks for image classification. In: 2012 IEEE Conference on Computer Vision and Pattern Recognition (CVPR), pp. 3642–3649 (2012)
11. Liu, M., Li, S., Shan, S., Wang, R., Chen, X.: Deeply learning deformable facial action parts model for dynamic expression analysis. In: Asian Conference on Computer Vision, pp. 143–157 (2015)
12. Jung, H., Lee, S., Yim, J., Park, S., Kim, J.: Joint fine-tuning in deep neural networks for facial expression recognition. In: The IEEE International Conference on Computer Vision (ICCV), pp. 2983–2991 (2015)
13. Liu, M., Li, S., Shan, S., Chen, X.: Au-aware deep networks for facial expression recognition. In: 2013 10th IEEE International Conference and Workshops on Automatic Face and Gesture Recognition (FG), pp. 1–6 (2013)
14. Viola, P., Jones, M.: Rapid object detection using a boosted cascade of simple features. In: IEEE Computer Society Conference on Computer Vision and Pattern Recognition, vol. 1, p. I-511 (2001)
15. Yang, P., Shan, S., Gao, W., Li, S.Z., Zhang, D.: Face recognition using ada-boosted gabor features. In: IEEE International Conference on Automatic Face and Gesture Recognition, pp. 356–361 (2004)
16. Abdulrahman, M., Gwadabe, T.R., Abdu, F.J., Eleyan, A.: Gabor wavelet transform based facial expression recognition using PCA and LBP. In: 2014 22nd Signal Processing and Communications Applications Conference (SIU), pp. 2265–2268 (2014)
17. Ojala, T., Pietikäinen, M., Harwood, D.: A comparative study of texture measures with classification based on featured distributions. Pattern Recogn. **29**(1), 51–59 (1996)
18. Turk, M.A., Pentland, A.P.: Face recognition using eigenfaces. In: IEEE Computer Society Conference on Computer Vision and Pattern Recognition, pp. 586–591 (1991)

19. Belhumeur, P.N., Hespanha, J.P., Kriegman, D.J.: Eigenfaces vs. fisherfaces: recognition using class specific linear projection. IEEE Trans. Pattern Anal. Mach. Intell. **19**(7), 711–720 (1997)

20. Yang, J., Yang, J.Y., Zhang, D., Lu, J.F.: Feature fusion: parallel strategy vs. serial strategy. Pattern Recogn. **36**(6), 1369–1381 (2003)

21. Mangai, U.G., Samanta, S., Das, S., Chowdhury, P.R.: A survey of decision fusion and feature fusion strategies for pattern classification. IETE Tech. Rev. **27**(4), 293–307 (2010)

22. Ekman, P., Friesen, W.: Facial Action Coding System: A Technique for the Measurement of Facial Movement. Consulting Psychologists, San Francisco (1978)

23. Lucey, P., Cohn, J.F., Kanade, T., Saragih, J., Ambadar, Z., Matthews, I.: The extended Cohn-Kanade dataset (CK+): a complete dataset for action unit and emotion-specified expression. In: IEEE Computer Society Conference on Computer Vision and Pattern Recognition Workshops, pp. 94–101 (2010)

24. Jain, S., Hu, C., Aggarwal, J.K.: Facial expression recognition with temporal modeling of shapes. In: IEEE International Conference on Computer Vision Workshops, pp. 1642–1649 (2011)

25. Zhang, W., Shan, S., Gao, W., Chen, X.: Local Gabor binary pattern histogram sequence (LGBPHS): a novel non-statistical model for face representation and recognition. In: Tenth IEEE International Conference on Computer Vision, pp. 786–791 (2005)

26. Tan, X., Triggs, B.: Fusing Gabor and LBP feature sets for kernel-based face recognition. In: Zhou, S.K., Zhao, W., Tang, X., Gong, S. (eds.) AMFG 2007. LNCS, vol. 4778, pp. 235–249. Springer, Heidelberg (2007). doi:10.1007/978-3-540-75690-3_18

27. Zavaschi, T.H.H., Britto, A.S., Oliveira, L.E.S., Koerich, A.L.: Fusion of feature sets and classifiers for facial expression recognition. Expert Syst. Appl. **40**(2), 646–655 (2013)

Frame-Independent and Parallel Method for 3D Audio Real-Time Rendering on Mobile Devices

Yucheng Song[1,2], Xiaochen Wang[2,3(✉)], Cheng Yang[2,4], Ge Gao[2], Wei Chen[1,2], and Weiping Tu[2]

[1] State Key Laboratory of Software Engineering, Wuhan University, Wuhan, China
yuchengsong_youki@foxmail.com, erhuchen@163.com
[2] National Engineering Research Center for Multimedia Software,
Computer School of Wuhan University, Wuhan, China
clowang@163.com, yangcheng41506@126.com, {gaoge,tuweiping}@whu.edu.cn
[3] Hubei Provincial Key Laboratory of Multimedia and Network Communication
Engineering, Wuhan University, Wuhan, China
[4] School of Physics and Electronic Science, Guizhou Normal University,
Guiyang, China

Abstract. As 3D audio is a fundamental medium of virtual reality (VR), 3D audio real-time rendering technique is essential for the implementation of VR, especially on the mobile devices. While constrained by the limited computational power, the computation load is too high to implement 3D audio real-time rendering on the mobile devices. To solve this problem, we propose a frame-independent and parallel method of framing convolution, to parallelize process of 3D audio rendering using head-related transfer function (HRTF). In order to refrain from the dependency of overlap-add convolution over the adjacent frames, the data of convolution result is added on the final results of the two adjacent frames. We found our method could reduce the calculation time of 3D audio rendering significantly. The results were 0.74 times, 0.5 times and 0.36 times the play duration of si03.wav (length of 27 s), with Snapdragon 801, Kirin 935 and Helio X10 Turbo, respectively.

Keywords: Virtual reality · 3D audio rendering · Mobile devices · Parallel · Framing convolution

1 Introduction

In recent years, the realization of virtual reality (VR) is becoming more practical with the rapid development of multimedia technology and hardware devices. The real-time technology of rendering virtual scene is essential, and the virtual scene is usually constructed from 3D video, 3D audio and other media. As for 3D audio

---

Y. Song—This work is supported by National High Technology Research and Development Program of China (863 Program) No. 2015AA016306; National Nature Science Foundation of China (No. 61231015, 61471271, 61662010).

L. Amsaleg et al. (Eds.): MMM 2017, Part II, LNCS 10133, pp. 221–232, 2017.
DOI: 10.1007/978-3-319-51814-5_19

rendering, due to high computation load and restricted time requirement, there are still some gaps between conventional methods and practical use. Though in common life, mobile services are regarded as the best carriers to implement VR technology, real-time applications are always constrained by the low computational power of the mobile services. For this reason, 3D audio real-time rendering technique on mobile devices is almost becoming one of the bottlenecks of VR implementation.

The best-known technologies for 3D audio field reconstruction are Ambisonics [1], wave field synthesis (WFS) [2,3], amplitude panning (AP) [4] and head-related transfer function (HRTF) [5]. Binaural sound synthesis based on HRTF simulates the procedure of the sound propagation from source location to the binaural eardrums. HRTF technology is able to implement the perceptive reconstruction of 3D audio image with just a two-channel earphone, other than the three 3D audio technologies above requiring for the playback environment of multichannel loudspeakers. However, it would be impractical to implement the playback environment of multichannel loudspeakers on mobile devices for the high hardware cost. For mobile devices, 3D audio technology based on HRTF using two-channel headphones is considered as feasible and efficient solution. So, its low equipment requirement makes HRTF technology a good choice for mobile devices and personal entertainment.

In the last decades, 3D audio technology of HRTF has been receiving much attention from researchers. In 2000, Lee et al. designed a method of 3D sound orchestra for two-channel headphones in the cyber space [6]. In their method, sounds from different instruments are convolved with HRTF and the results are mixed into two channels.

To establish a complete HRTF database publicly available, in 2001, Algazi et al. measured the HRTFs for 45 subjects, and in addition, the database contains anthropometric measurements for each subject [7]. The anthropometric data also provides experimental materials for future research on HRTF individualization. Besides, in 2009, the Peking University (PKU) and Institute of Acoustics (IOA), Chinese Academy of Sciences, together published the PKU&IOA HRTF database [8], which is one of the most completed near-field and far-field HRTF database with high spatial resolution. In their work, a spark gap was used as the acoustic sound source to solve the difficulty of HRTF measurement in the proximal region.

In order to make 3D audio generation based on HRTF more practical, many researchers have contributed to the technology of efficient 3D audio rendering. In 2004, Zotkin et al. presented a set of algorithms (including HRTF interpolation, room impulse response creation, audio scene presentation), for creating virtual auditory space rendering systems using HRTF [9]. With these algorithms, a prototype system was created that ran in real-time on a typical office PC. In 2013, Fu et al. pointed out the computation load of 3D audio rendering using HRTFs is high for real-time applications, and proposed equalized and relative HRTFs to speed up 3D audio image rendering by reducing more than 50% computation load [10]. Furthermore, in 2013, Zhang and Xie constructed a personal computer (PC) platform of the virtual auditory environment (VAE) real-time rendering system [11]. They proposed a scheme including PCA-based (principal components analysis) near-field virtual source synthesis. In their test, the system is able to simultaneously render up to 280 virtual sources using conventional scheme, and 4500 virtual sources using the using the PCA-based scheme.

From an overview of the development of 3D audio generation based on HRTF, the majority of the experimental measurements and the technological implementation were either conducted on the PC computers or high-performance servers. There is little consideration of the problems on mobile applications. In 2015, Iwaya Yukio and Otani Makoto considered the next generation virtual auditory display (VAD) are required to be available on the poor computation power equipment, such as tablet or smart phone. So they proposed a VAD system on smartphone by remote rendering [12]. As for mobile devices, low computation power increases the difficulty of real-time 3D audio rendering. Therefore, how to increase the efficiency of 3D audio synthesis becomes the key problem of 3D audio real-time rendering on mobile devices.

In this paper, we propose a frame-independent and parallel method based on framing convolution. In every parallel thread, the audio frame data is separately convolved with the corresponding HRTF. And in order to refrain from the dependency of overlap-add convolution over the adjacent frames, the data of convolution result is added on the memory regions of the two adjacent frames, so as to realize the convolved calculation with multiple frames parallel together. Parallel calculation was performed on the test machines with Snapdragon 801, Kirin 935, and Helio X10 Turbo. The results demonstrated that our method reduced the 3D audio generation time separately below 0.8 times, 0.6 times and 0.4 times play duration of the audio file, reaching the target of real-time rendering.

The rest of the paper is organized as follows. Section 2 introduces the HRTF and the implementation of 3D audio rendering. Section 3 presents the frame-independent and parallel method of framing convolution. Section 4 analyzes the acceleration experiment with different number of parallel threads and size growth experiment of different data lengths, and the subjective experiment between offline rendered and the real-time rendered audio. And Sect. 5 concludes this paper.

2   3D Audio Rendering Using HRTF

2.1   Head-Related Transfer Function

HRTF consisting of filters for the left and right channels, serves as the key role of implementation for the 3D audio rendering. Sound propagates from source to the eardrum of listener, during the path of which it interacts with the listeners torso, shoulders, pinnae head and so on, so as to have sound quality change. And the listener would determine the location of the sound source according to the interaural time difference (ITD), interaural level difference (ILD) and interaural quality difference, etc. The procedure of sound propagation from source to the eardrum, can be described as a transfer function, termed as HRTF [5]. There are essential perceptive localization cues of human included in the HRTFs. In the free-field situation, the left-ear and right-ear HRTF is defined as

$$\begin{cases} H_L = H_L\left(r, \theta, \varphi, f\right) = P_L\left(r, \theta, \varphi, f\right)/P_0\left(r, f\right) \\ H_R = H_R\left(r, \theta, \varphi, f\right) = P_R\left(r, \theta, \varphi, f\right)/P_0\left(r, f\right) \end{cases}, \tag{1}$$

respectively, where P_L and P_R denote the acoustic pressures that sound source causes at the listener's left and right ear separately; P_0 refers to the acoustic pressure at the corresponding location of the listener's head center (without the listener existing there). As seen by the formula, H_L and H_R are the functions of the azimuth θ, elevation φ, distance r (between source and head center), and frequency f [13].

When a source signal passes through a set of HRTFs, the synthesized signals contain the necessary binaural cues for the sound source localization, which is called binaural sound synthesis, also the basic procedure of 3D audio rendering. By processing the dry signal with the HRTF, the obtained dual-ear signals are

$$\begin{cases} X_L = H_L S \\ X_R = H_R S \end{cases}, \tag{2}$$

where H_L and H_R are the transfer functions for the left and right channels, respectively.

On account of the essentiality of HRTF, many research institutes and laboratories had worked on measurement of HRTF, and exhibited the related experimental results, laying the foundation of future research. Some of them are open to the public so as to have great influence on others, such as MIT HRTF database [14], CIPIC HRTF database [7] and PKU&IOA HRTF database [8].

2.2 Implementation of 3D Audio Rendering

As for implementation, framing convolution is the essential method, considering the long length of audio data. In the conventional process of 3D audio generation, it oughts to utilize head-related impulse response (HRIR) of the specific localization to convolve with the single frame of raw audio data, in order to get the frame data with spatial feeling. Finally, the final results of all the frames are combined end to end, to obtain the entire audio with 3D effect.

For HRTF databases commonly accessed [7, 8, 14], data of filters is provided in the form of HRIR in the time domain, other than HRTF in the spectral domain. So the filtering need to be implemented by the time-domain convolution. Typically, an overlap-add convolution technique is used to segment the signal into appropriate frames [15]. Firstly, the dry audio sequence $x\left(n\right)$ is divided into several frames with the length of N, and the i-th frame is marked as $x_i\left(n\right)$. $h_L\left(n\right)$ and $h_R\left(n\right)$ are the left-ear and right-ear HRIRs, respectively, both of M samples. And the frame convolution results of the i-th frame can be generated as follows,

$$\begin{cases} y'_{L,i} = h_L * x_i \\ y'_{R,i} = h_R * x_i \end{cases}. \tag{3}$$

From the property of linear convolution, it can be seen that either $y'_{L,i}$ or $y'_{R,i}$ is $M + N - 1$ samples long. The overlapping section of the $(i - 1)$-th frame .convolution result ought to be added to the final results of the i-th frame:

$$\begin{cases} y_{L,i}(n) = y'_{L,i}(n) + y'_{L,i-1}(n+N) \\ y_{R,i}(n) = y'_{R,i}(n) + y'_{R,i-1}(n+N) \end{cases}, \tag{4}$$

as $y_{L,i}$ and $y_{R,i}$, whose length both is N. From the formula above, we can see the final result of i-th frame is related to the convolution results of i-th frame and $(i - 1)$-th frame, which reveals the fact that overlap-add calculation is dependent on the adjacent frames. This dependency hinders the parallel for the implementation of framing convolution. In the conventional serial method of convolution, extra buffer is needed to keep the convolution result of the previous frame [16,17]. To solve the problem of the dependency, we propose the frame-independent and parallel method of framing convolution in the following section.

3   Frame-Independent and Parallel Method of Framing Convolution

As for mobile devices, it's a future trend that the central processing unit (CPU) consists of multiple cores. To make better use of that, we consider multithreading method to speed up the calculation of framing convolution, opening parallel threads to process multiple frames. While in the previous section, it has been concluded that overlap-add convolution is dependent on adjacent frames, so that general multi-threading method could not be applied to framing convolution naturally.

Algorithm 1. Pseudocode for one of the parallel threads, with d frames under processing

---

Input: d, h_L, x_i
Output: $y_{L,i}, y_{L,i+1}$
1: **in parallel do**
2:   **lock**(A)
3:     $d \leftarrow d + 1$
4:     $i \leftarrow d$
5:   **unlock** (A)
6:   computing $h_L * x_i \rightarrow y'_{L,i}$
7:   **for** $i \leftarrow 1$ to N **do**
8:     $y_{L,i}(i) \leftarrow y_{L,i}(i) + y'_{L,i}(i)$
9:   **end for**
10: **for** $i \leftarrow 1$ to $M - 1$ **do**
11:     $y_{L,i+1}(i) \leftarrow y_{L,i+1}(i) + y'_{L,i}(i+N)$
12: **end for**
13: output $y_{L,i}, y_{L,i+1}$

In order to solve the problem and increase the efficiency, we propose the frame-independent and parallel method of framing convolution. Because of the additivity of the linear system, we assumed the convolution result of the current . frame and the overlapping section of the previous frame convolution result are two different ways of signals added on the linear system. These two ways of signals together constitute the final result of the current frame, and are able to be calculated in different parallel threads. Here is the calculation process of left-ear data, as an example, to describe the general idea of our proposed method, as the following steps:

1. During initialization, zero setting y_L entirely;
2. Allocating i-th frame of raw audio data x_i to the k-th thread;
3. Convolving x_i with HRIR h_L to obtain the convolution result $y'_{L,i}$;
4. Adding the first N points of $y'_{L,i}$ to the whole N points of $y_{L,i}$, and the last $M-1$ points of $y'_{L,i}$ to the first $M-1$ points of $y_{L,i+1}$, on their corresponding memory regions.
5. Repeating above steps until all the frames have been processed.

The calculation process of the right-ear data is similar to the left-ear. The procedure of the adding operations in the method is shown in Fig. 1. Algorithm 1 shows pseudocode of the implementation for one of the parallel threads (using left-ear data as an example).

The method helps us refrain from the dependency of overlap-add convolution over the adjacent frames, the data of convolution result is added on the memory regions of the two adjacent frames, so as to realize the convolved calculation with

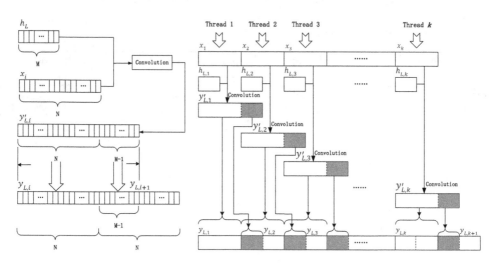

Fig. 1. The procedure of the adding operations on adjacent frames (left-ear)

Fig. 2. The procedure of paralleled calculation of multiple frames with k threads parallel (left-ear)

multiple frames parallel together, thus increasing the performance of rendering from frame level. Additionally, in our method, there's no need for extra buffer to keep the convolution results of the previous frame. Figure 2 demonstrates the overall process of the paralleled calculation.

4   Experiment

To evaluate the performance of our proposed frame-independent and parallel method of framing convolution, we set up three kinds of experiments, the first one for the acceleration evaluation of parallel method with different number of parallel threads compared to the basic convolution method, the second one for the size growth evaluation of different data lengths, the third one for the subjective evaluation between offline rendered and the real-time rendered audio. All the experiments were conducted on the Mi Note LTE cellphone with Qualcomm Snapdragon 801 processor, the Huawei P8 cellphone with Huawei Kirin 935 and the Meizu MX5 cellphone with MediaTek (MTK) Helio X10 Turbo, and the explicit testing environment is described as Table 1. The HRIRs used through the two experiments are from the PKU&IOA HRTF database [8], whose length is 1024 (which is much longer than the HRIR length of 200, from the CIPIC HRTF database [7]). The sources of input signals are monaural audio data, sampled at 48000 Hz and digitized at 16bit. For a statistical significance, all the results shown below have been averaged.

Table 1. The explicit testing environment of mobile devices

Mobile device	Mi Note LTE	Huawei P8	Meizu MX5
Processor	Snapdragon 801	Kirin 935	Helio X10 Turbo
Number of Cores	4*Krait 400	4*Cortex A53 enhanced + 4*Cortex A53	8*Cortex A53
Clock of Processor	2.5 GHz	2.0 GHz (A53 enhanced) & 1.5 GHz (A53)	2.2 GHz
RAM	3 GB	3 GB	3 GB
Minimum Available RAM	1 GB	1 GB	1 GB
System Version	Android 6.0	Android 5.0	Android 5.1

4.1   Acceleration Experiment with Different Number of Parallel Threads

In the acceleration experiment of the parallel method, we measured the 3D audio rendering time of audio data with different number of parallel threads from two to ten threads, compared to the rendering time of the general convolution method

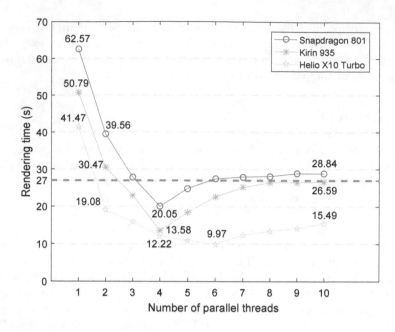

Fig. 3. The 3D audio rendering time of si03.wav (length of 27 s) with Snapdragon 801, Kirin 935 and Helio X10 Turbo

described in Sect. 2.2, in other words, the serial method. The rendering time refers to the total time of the rendering procedure from input mono audio of pulse-code modulation (PCM) transferred into two-channel 3D audio data of PCM (both stored in the random-access memory (RAM) of the test machine). The source signal is si03.wav of 27 s, which is one of standard audio sequences from Moving Picture Experts Group (MPEG). Then we compared 3D audio data of the final results in parallel method with the final result in serial method, and found there were no difference, which confirmed that the calculations were correct.

The results of rendering time are illustrated in Fig. 3. Firstly, we ought to declare that an implementation of 3D audio rendering can be called real-time, only if the rendering time is less than the play duration of the original audio (that is, in Fig. 3, the point is below the red imaginary line). In the serial method, the rendering time is 62.57 s, 50.79 s and 41.47 s with Snapdragon 801, Kirin 935 and Helio X10 Turbo, respectively, which are not real-time. As the number of parallel threads increases, the rendering times with the three CPUs decrease at the beginning, until reaching the lowest points, then increase slowly, and finally flatten out. At the points of 4 threads with Snapdragon 801, 4 threads with Kirin 935 and 6 threads with Helio X10 Turbo, they get the shortest rendering times, which are separately 3.12 times, 3,74 times and 4.16 times faster than the serial method (in Fig. 3, $62.57/20.05 \approx 3.12$, $50.79/13.58 \approx 3.74$ and $41.47/9.97 \approx 4.16$), and are 0.74 times, 0.5 times and 0.36 times the play duration of si03.wav (in Fig. 3, $20.05/27 \approx 0.74$, $13.58/27 \approx 0.5$ and $9.97/27 \approx 0.36$).

To explore whether audio content would impact the rendering efficiency, we also conducted the acceleration experiment with files of different contents, including vocal and instruments. The test was carried out upon Helio X10 Turbo with 6 threads. Detailed information of the test sequences and the experiment results are illustrated with Table 2. The average of the speed-up (the ratio of the serial convolution method to our parallel method) is 3.88, and the rendering time of our parallel method is 0.38 times medially the duration of the four different sequences. These results demonstrate that our parallel method can adapt well to different audio contents.

Table 2. Details of the sequences and test results using Helio X10 Turbo with 6 threads

Sequences	Content	Duration (s)	Serial convolution method (s)	Our parallel method (s)	Speed-up (serial/ our)	Rendering efficiency (duration/ our)
es01	Female vocal	10	14.57	3.9	3.74	0.39
sc01	Trumpet & orchestral music	10	15.27	4.09	3.73	0.41
si03	Pitch pipe	27	41.47	9.97	4.16	0.36
sm02	Carillon	10	14.12	3.62	3.89	0.36
Average	—	—	—	—	3.88	0.38

From the above results, it can be concluded that our parallel method is able to reduce the calculation time of 3D audio rendering substantially. Though it is acknowledged that the computation load of is so high for real-time 3D audio rendering applications [18], we implemented real-time generation of 3D audio on mobile devices, using 4 threads with Snapdragon 801, 4 threads with Kirin 935 and 6 threads with Helio X10 Turbo. The results also reveal the fact that the parallel method takes full advantage of the multicore of mobile processor.

4.2   Size Growth Experiment of Different Data Lengths

To better observe the properties of the parallel method with size growth of the input data, we generated 3D audio with source data of different lengths using 4 threads with Snapdragon 801, 4 threads with Kirin 935 and 6 threads with Helio X10 Turbo, (as they got the shortest rendering times with these threads). The source signals vary from 10 s to 70 s, and are also sampled at 48000 Hz and digitized at 16bit.

The correlation between the duration of audio signal and rendering time is shown in Fig. 4. As we can see, with the audio duration growing, the rendering time is increasing approximately linearly, which reflects the stable performance of our parallel method. In practical applications, the stable performance will

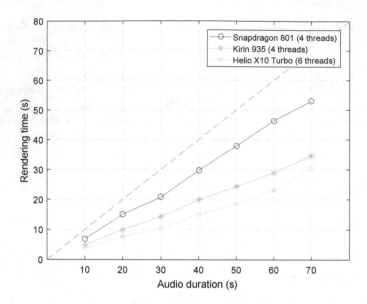

Fig. 4. The 3D audio rendering time of source audios in different lengths with Snapdragon 801 (4 threads), Kirin 935 (4 threads) and Helio X10 Turbo (6 threads)

adapt to strict time constraints and contribute to better estimation of runtime, even though real-time application of 3D audio is very computation-intensive, such as VR audio rendering.

4.3   Subjective Experiment Between Offline Rendered and the Real-Time Rendered Audio

For 3D game, VR or any application related to 3D audio, audio quality plays an important role. Here we also conducted the subjective experiment based on MUSHRA method with 3.5 kHz low-pass filtered anchor to evaluation the sound quality of real-time rendering using our parallel method. The MPEG standard series es01, sc01 and sm02 were used in the test. Figure 5 displays the interface of the test program on Android. There were 10 subjects (4 females and 6 males) from 20 to 30 years old taking part in subjective test. Firstly, offline rendered 3D audio with serial convolution method would be played as the reference sound. Then they would listen to the sounds rendered in three versions in random order, offline rendered 3D audio with serial convolution method (as the hidden reference), real-time 3D audio rendering using our parallel method and offline audio filtered below 3.5 kHz. The subjective evaluation results with 95% confidence interval are shown in Fig. 6. It can be concluded from the results that, there exists a little sound quality decline in real-time parallel 3D audio rendering, although our parallel method accelerate the 3D audio rendering substantially. And conventional serial method performed better than parallel method in sound quality fidelity.

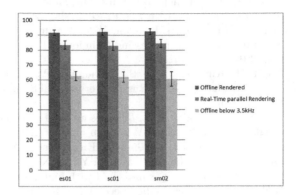

Fig. 5. The interface of the test program on Android

Fig. 6. The subjective evaluation results of sound quality based on MUSHRA method

5   Conclusion

In this paper, we present a frame-independent and parallel method for 3D audio real-time rendering on mobile devices. To refrain from the dependency of overlap-add convolution over the adjacent frames, we assumed that the convolution result of the current frame and the overlapping section of the previous frame convolution result are two different ways of signals added on the linear system, and are able to be calculated in the parallel threads. Then we proposed using parallel threads to process multiple frames, and parallelize the process of 3D audio rendering. Finally, we conducted the acceleration and the size growth experiments of the parallel method, and found our method could reduce the calculation time of 3D audio rendering significantly. The results were 0.74 times, 0.5 times and 0.36 times the play duration of si03.wav (length of 27 s), with Snapdragon 801, Kirin 935 and Helio X10 Turbo, respectively, achieving the target of 3D audio real-time rendering on the mobile devices. However, in subjective evaluation, conventional serial method performed better than our parallel method in sound quality fidelity. The results also reveal the fact that our parallel method takes full advantage of the multicore of mobile processor.

References

1. Gerzon, M.A.: Ambisonics. Part two: studio techniques. Studio sound **17**(8), 24–26 (1975)
2. Berkhout, A.J.: A holographic approach to acoustic control. J. Audio Eng. Soc. **36**(12), 977–995 (1988). Audio Engineering Soc, New York
3. Berkhout, A.J., de Vries, D., Vogel, P.: Acoustic control by wave field synthesis. J. Acoust. Soc. Am. **93**(5), 2764–2778 (1993)

4. Pulkki, V., Karjalainen, M.: Multichannel audio rendering using amplitude panning [DSP applications]. IEEE Sig. Process. Mag. **25**(3), 118–122 (2008). IEEE Press
5. Jianjun, H.E., Tan, E.L., Gan, W.S.: Natural sound rendering for headphones: integration of signal processing techniques. IEEE Sig. Process. Mag. **32**(2), 100–113 (2015). IEEE Press
6. Lee, D.H., Kim, K.N., Lee, S.D., Chong, U.P.: A 3-D sound orchestra in the cyber space. In: The 4th Korea-Russia International Symposium, vol. 2, pp. 12–16. IEEE Press (2000)
7. Algazi, V.R., Duda, R.O., Thompson, D.M., Avendano, C.: The CIPIC HRTF database. In: 2001 IEEE Workshop on the Applications of Signal Processing to Audio and Acoustics, pp. 99–102. IEEE Press (2001)
8. Qu, T., Xiao, Z., Gong, M., Huang, Y., Li, X., Wu, X.: Distance-dependent head-related transfer functions measured with high spatial resolution using a spark gap. IEEE Trans. Actions Audio Speech Lang. Process. **17**(6), 1124–1132 (2009). IEEE Press
9. Zotkin, D.N., Duraiswami, R., Davis, L.S.: Rendering localized spatial audio in a virtual auditory space. IEEE Trans. Multimedia **6**(4), 553–564 (2004). IEEE Press
10. Fu, Z.H., Xie, L., Jiang, D.M., Zhang, Y.N.: Fast 3D audio image rendering using equalized and relative HRTFs. In: 2013 International Conference on Orange Technologies (ICOT), pp. 47–50. IEEE Press (2013)
11. Zhang, C., Xie, B.: Platform for dynamic virtual auditory environment real-time rendering system. Chin. Sci. Bull. **58**(3), 316–327 (2013)
12. Iwaya, Y., Otani, M., Tsuchiya, T., Li, J.: Virtual auditory display on a smartphone for high-resolution acoustic space by remote rendering. In: 2015 International Conference on Intelligent Information Hiding and Multimedia Signal Processing (IIH-MSP), pp. 368–371. IEEE Press, September 2015
13. Xie, B., Zhong, X., Rao, D., Liang, Z.: Head-related transfer function database and its analyses. Sci. China Ser. G Phys. Mech. Astron. **50**(3), 267–280 (2007). Springer
14. Gardner, W.G., Martin, K.D.: HRTF measurements of a KEMAR. J. Acoust. Soc. Am. **97**(6), 3907–3908 (1995). Acoustical Society of America
15. Oppenheim, A.V., Schafer, R.W.: Discrete-Time Signal Processing. Pearson Higher Education, London (2010)
16. Carty, B.: Movements in binaural space: issues in HRTF interpolation and reverberation, with applications to computer music. Doctoral dissertation, National University of Ireland Maynooth (2010)
17. Snchez, I., Bescs, J.: Software modules for HRTF based dynamic spatialisation. Grupo de Tratamiento de imgenes, Universidad Politcnica de Madrid. Script and Documentation can be found on the CD-ROM respectively in Source C++ Bescos and Documentation Class LibrariesXBescos (2007)
18. Zhang, J., Xu, C., Xia, R., Li, J., Yan, Y.: Dependency of the finite-impulse-response-based head-related impulse response model on filter order (2014). https://depositonce.tu-berlin.de/bitstream/11303/181/1/23.pdf

Illumination-Preserving Embroidery Simulation for Non-photorealistic Rendering

Qiqi Shen, Dele Cui, Yun Sheng$^{(\boxtimes)}$, and Guixu Zhang

The School of Computer Science and Software Engineering,
East China Normal University, Shanghai 200062, China
ysheng@cs.ecnu.edu.cn

Abstract. We present an illumination-preserving embroidery simulation method for Non-photorealistic Rendering (NPR). Our method turns an image into the embroidery style with its illumination preserved by intrinsic decomposition. This illumination-preserving feature makes our method distinctive from the previous papers, eliminating their problem of inconsistent illumination. In our method a two-dimensional stitch model is developed with some most commonly used stitch patterns, and the input image is intrinsically decomposed into a reflectance image and its corresponding shading image. The Chan-Vese active contour is adopted to segment the input image into regions, from which parameters are derived for stitch patterns. Appropriate stitch patterns are applied back onto the base material region-by-region and rendered with the intrinsic shading of the input image. Experimental results show that our method is capable of performing fine embroidery simulations, preserving the illumination of the input image.

Keywords: Non-photorealistic rendering · Embroidery simulation · Intrinsic shading · Image processing · Multimedia signal processing

1 Introduction

Non-photorealistic Rendering (NPR) aiming to simulate a wide range of artistic styles with computers has attracted tremendous interest from the research community. Although many efforts have been made in NPR, involving oil painting, penciling drawing, stippling, *etc*, there have been a small number of research papers reported for embroidery simulation in NPR thus far. Chen *et al.* [1] presented a line-drawing-based method of traditional embroidery modeling and rendering, where the rendered embroidery was subsequently texture-mapped onto a deformable 3D object. Yang *et al.* [2] proposed an image-based method to simulate the Chinese Irregular Needling Embroidery (CINE) by constructing a stitch dictionary with a multilayer rendering technique to produce vivid and colorful results. Cui *et al.* [3] proposed a 3D stitch model with the Phong lighting model to create embroidery-like images.

Since real embroidery glints under the light and is sensitive to illumination, a proper simulation to illumination can shade the computer-generated embroidery

© Springer International Publishing AG 2017
L. Amsaleg et al. (Eds.): MMM 2017, Part II, LNCS 10133, pp. 233–244, 2017.
DOI: 10.1007/978-3-319-51814-5_20

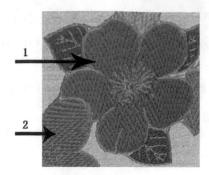

Fig. 1. Incoherent illumination of Chen's result [1].

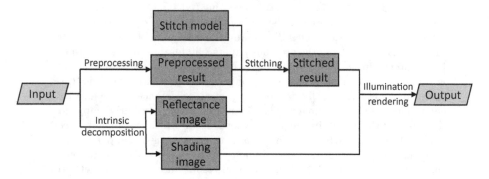

Fig. 2. The flowchart of our embroidery simulation system.

nicely, and give it a 3D illusion and artistic sense; otherwise, incoherent illumination will give rise to unnecessary chaos to the simulation result. For example, both [1] and [2] employed a similar planar stitch model. In order to give their planar stitches a 3D illusion, Chen *et al.* [1] delicately performed tangent modification on the two ends of each stitch and utilized alpha blending to render the stitches, while Yang *et al.* [2] simply performed a linear tuning of intensity over each stitch from one end to the other. Nevertheless, both [1] and [2] failed to pay enough attention to embroidery shading during stitch rendering. The illumination in [1] appears globally inconsistent and chaotic. Figure 1 shows such an example taken from [1], where the direction of shading in the central flower (see where Arrow 1 points to) is from top to bottom, while the shading in the bottom left corner (see where Arrow 2 points to) is from right to left, leading to an incoherent lighting effect. The stitching in [2] was carried out by locally averaging the color of the input image, and thus caused an unavoidable decay in intensity contrast and color saturation, making the embroidery appearance visually dim (as shown in Fig. 12(b) and (e)). Although the global Phong lighting model was used in [3], the paper resorted to a manual process to align the artificial illumination with the shading direction in the original input image, which is, however, uneasy because estimation of the lighting direction from a single image is still an ill-posed problem.

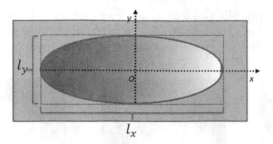

Fig. 3. The stitch model is in red, and $1.15*l_x$ and $1.2*l_y$ are, respectively, the length and width of its bounding box. (Color figure online)

In order to resolve the above problems of inconsistent illumination in the previous methods, we propose an illumination-preserving embroidery simulation method in this paper. Our method transfers the illumination of the input image to the simulation result through intrinsic decomposition so that the illumination captured by physical apparatus will offer embroidery simulation results a more appealing illumination effect. The flowchart of our method is shown in Fig. 2. We first derive some most commonly used stitch patterns based on a self-designed two-dimensional (2D) stitch model, and then preprocess the input image by segmenting it into regions from which the parameters of different stitch patterns can be figured out. Meanwhile, the input image is intrinsically decomposed into its shading and reflectance images. The color of each stitch is determined by the reflectance image during stitching. Finally, the intrinsic shading of the input image is used to render the stitched result.

Our work is also subject to stroke-based rendering (SBR), an NPR technology that renders an image using some predefined primitives, such as the virtual brush stroke, pencil stroke, stipple, and stitch, *etc.* There exist a large number of research papers with SBR, which, according to the stroke layer number, can be categorized into the single-layer [4,5] and multi-layer methods [6,7]. We implement both in this paper.

The rest of this paper is organized as follows: Sect. 2 describes stitch modeling. Section 3 is dedicated to preprocessing as well as determination of stitch parameters. Section 4 introduces intrinsic decomposition and rendering. Section 5 demonstrates the experimental results. Section 6 draws the conclusions.

2   Embroidery Modeling

Since embroidery is created with numerous stitches, a proper stitch model plays a key role in embroidery simulation. It is observed that compared with the central part of a stitch, its two ends are visually thinner and curvier when the stitch goes into and out of the base material. This is caused by interaction forces between the stitch and base material. Meanwhile, we also desire that the stitch shape is easy enough to compute. To this end, we design a 2D ellipse model, as shown in Fig. 3. In this paper, the real length and width of a stitch are respectively set to $1.15 * l_x$, and $1.2 * l_y$ by default. The calculation of other stitch parameters, such as the orientation and color, will be discussed in detail in Sect. 3.

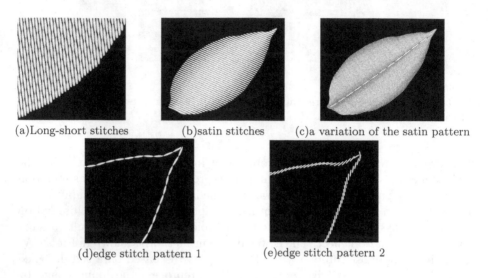

(a)Long-short stitches   (b)satin stitches   (c)a variation of the satin pattern

(d)edge stitch pattern 1   (e)edge stitch pattern 2

Fig. 4. The stitch patterns.

In the following, the stitch patterns namely long-short, satin, and edge stitches are designed and implemented as independent modules. Given a region to stitch, our method is able to perform corresponding stitching by calling one or more of these specific modules.

Long-short stitch good at showing intensity graduation is a common stitch pattern used in various kinds of embroidery. By alternating long and short stitches and changing the color parameter, this type of stitch is capable of displaying a smooth color change. Figure 4(a) shows an example of long-short stitch pattern. In practice, we find this fixed-size pattern fails to clearly depict the area with rich detail and color change. If we decreased the stitch size, the computational complexity would increase hugely and be unnecessary for areas in constant color. Thus, we add an adaptive length varying strategy for the sake of efficiency. Before applying stitches, we calculate a map of local color variances. If the variance value at a stitching point is greater than a predefined threshold, short stitches are used to show more details; otherwise long stitches are applied for computational efficiency. Moreover, we find if randomizing the starting positions and length of stitches to follow a preset normal distribution with an appropriate expectation and derivation, the resulting embroidery image will appear more appealing; otherwise the arrangement of stitches would appear too regular to resemble embroidery.

Satin stitch is the quickest stitch pattern to sew up an area. It consists of a series of long and parallel stitches stretching over the whole area.

Traditionally, satin stitch is used to fill the areas with less visual salience, such as the background. Figure 4(b) shows the implementation of such a stitch pattern. Moreover, when a region is approximately symmetrical, we suggest a variation of satin stitch shown in Fig. 4(c), where we first separate the original region into two halves with a symmetric line, and then apply satin stitches from the symmetric line to both sides of the region. This produces an arrangement of stitches like leaf veins, and makes the region visually more appealing. The calculation of the symmetrical line will be discussed in Sect. 3.

Edge stitch is designed to emphasize the edge of a region. We implement two general patterns of edge stitch, as shown in Fig. 4(d) and (e). The first pattern is easy to implement because it consists of a series of long stitches joining end to end. Since it is likely to infinitely approximate a curve by joining short segments, we do the same here. We sample the edge with respect to the length of the stitch at regular intervals. As for the second edge stitch pattern, all the stitches are placed clockwise, intersecting with the edge of the region at their midpoints with a predefined gap along the edge. We also add a small random offset to stitch orientations in order to avoid over-regularity. It is seen that edge stitch pattern 2 has a better visual effect than the former, but the former is generated more quickly owing to its simplicity.

3   Preprocessing and Selection of Parameters

Before applying stitches to the base material, the input image has to be pre-processed by image segmentation and edge smoothing. In this paper, we segment the input image for three reasons. First, our stitch modules are region-wise, and each region is stitched with a specific stitch module. Second, we need to smooth out region edges to make them visually coherent, and this is of particular importance when applying edge stitches. Third, stitch parameters, such as the length, width, and color, are all decided by properties of the segmented regions.

3.1   Preprocessing

In most cases, embroidery simulation focuses on the visually dominant objects in the foreground and treats the background in a relatively loose manner. In this paper, we use the well-known Chan-Vese active contour [8,9] to segment out the objects of interest. To improve the precision of the Chan-Vese algorithm on low-contrast areas, we use L_0 smoothing [10] to preprocess the grey image before segmentation. Our experiments show that a combination of Chan-Vese segmentation and L_0 smoothing works well in most situations.

Edge smoothing is also required in our method. The regions segmented by the above methods usually end up with rugged edges. If we directly apply stitches along the rugged edges, an abrupt change of edge direction will cause a total chaos of stitch orientation. Here we smooth out the edge with the Fourier descriptors by reconstructing the edge with the low-frequency Fourier coefficients. An experimental illustration of the results produced by the above mentioned algorithms is shown in Fig. 5(b)–(d).

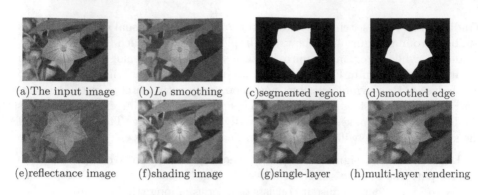

(a)The input image   (b)L_0 smoothing   (c)segmented region   (d)smoothed edge

(e)reflectance image   (f)shading image   (g)single-layer   (h)multi-layer rendering

Fig. 5. Illustration of the embroidery simulation procedure.

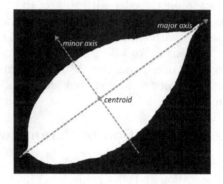

Fig. 6. The major and minor axes.

3.2 Selection of Stitch Parameters

We also calculate the area, centroid, and orientation of major axis of the region. For those approximately symmetrical regions, we calculate their symmetrical lines by morphologically thinning the smoothed regions and then fitting them with the quadratic polynomial. The centroid and orientation of the major axis are obtained by respectively calculating the first and second moments of the region as shown in Fig. 6. The orientation of the minor axis is orthogonal to that of the major one.

In our method, the stitch pattern parameters, such as the color, length, width, and orientation, are dynamically decided according to the input image. The stitch color is set to be the average color of the area the stitch covers in the intrinsic reflectance image. Unlike [2], the shading image decomposed from the input image will be superimposed back onto the stitched result during illumination rendering, thus avoiding the problem of contrast decay caused by the color averaging in [2].

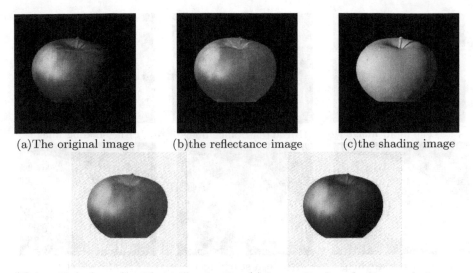

(a)The original image   (b)the reflectance image   (c)the shading image

(d)the stitched result without illumination (e)the result after shading rendering

Fig. 7. Intrinsic decomposition and rendering.

To calculate the stitch length, we adopt the formula developed in [2]:

$$length = \omega(1 + \frac{A(region) - MinA(input)}{MaxA(input) - MinA(input)}) \qquad (1)$$

where $\omega = max(Width_{input}, Height_{input})/50$, $A(region)$ denotes the area of the region, $MaxA(input)$ and $MinA(input)$ denote the maximum and minimum areas of all the segmented regions in the input image respectively. The stitch width is set to $0.2 * length$ empirically. The stitch orientation is set between the major and minor axes of the region.

$$orientation = \varphi * \theta_{major} + (1 - \varphi) * \theta_{minor} \qquad (2)$$

where θ_{major} and θ_{minor} are the orientations of the major and minor axes respectively. φ is the weight, set to 0.25 by default.

4   Intrinsic Decomposition and Rendering

As for intrinsic decomposition, most researchers are primarily interested in two characteristics, namely reflectance and shading. In this context, Land and McCann's Retinex Theory [11] in the early 70 s already made an equivalent representation. An image can be written as a product of reflectance and shading, i.e. $I = sR$, where I is the original image, s is the shading image, and R is the reflectance image.

We use the method introduced in [12] to intrinsically decompose an input image into a shading image and its corresponding reflectance image.

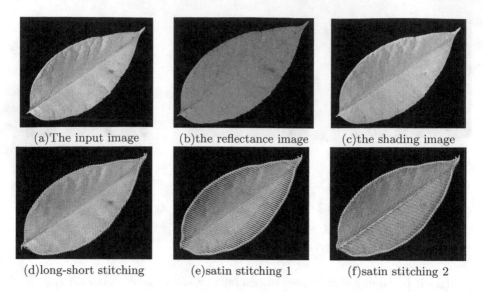

(a)The input image   (b)the reflectance image   (c)the shading image

(d)long-short stitching   (e)satin stitching 1   (f)satin stitching 2

Fig. 8. Intrinsic decomposition and rendering with the long-short and two satin stitch patterns.

Rother *et al.* [12] introduced the global sparsity prior on reflectance by assuming that reflectance values were drawn from a sparse set of basis colors, and built a Conditional Random Field model to specify a probabilistic distribution over the reflectance and shading components of a given image. This intrinsic decomposition method only needs one single image, and is able to generate highly accurate decomposition results. We show an example of this intrinsic decomposition method in Fig. 7.

In the rendering process, we render each stitch with the average color of the area it covers in the intrinsic reflectance image and then illuminate all the colored stitches with the intrinsic shading image. Figure 7(d) is the intermediate result without illumination, while Fig. 7(e) is the result rendered with the shading image. The stitch size in the foreground is smaller than that in the background so that the foreground is strengthened with more details preserved. Moreover, in order to visually distinguish each stitch in the plain areas, e.g. the background in constant color, we add linear intensity graduation to each stitch. Since this intrinsic rendering is achieved by mathematical multiplication, it is speedy to transfer the shading of the input image into the simulation result.

5   Experimental Results

The whole simulation system introduced in this paper is implemented with Matlab R2014a on a PC configured with Core i7 2.2GHz CPU and 4GB memory. In average, it takes approximately 0.4 seconds to finish one stitch on the base material.

(a)The input image   (b)its reflectance   (c)its shading   (d)the output image

Fig. 9. Long-short stitching of a puppy.

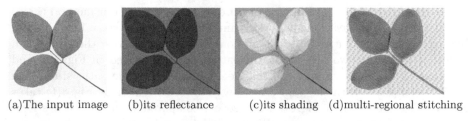

(a)The input image   (b)its reflectance   (c)its shading   (d)multi-regional stitching

Fig. 10. Multi-regional stitching with variant stitch patterns.

(a)Input image 1   (b)input image 2   (c)the result of (a) (d)after shading transfer

Fig. 11. Embroidery simulation with shading transfer.

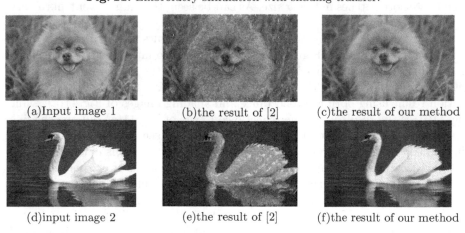

(a)Input image 1   (b)the result of [2]   (c)the result of our method

(d)input image 2   (e)the result of [2]   (f)the result of our method

Fig. 12. Comparison of our method and [2].

   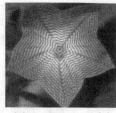   

(a)The input image   (b)the result of [3]   (c)the result of our method

Fig. 13. Comparison of our method and [3].

A long-short stitching result is shown in Fig. 8(d) with the shading image (Fig. 8(c)) of Fig. 8(a). This illumination-preserving nature of our method is also demonstrated by another long-short stitching example as shown in Fig. 9, whose shading image (Fig. 9(c)) is even more complicated. Moreover, the simulation results of using the two different satin patterns are displayed in Fig. 8(e) and (f), where Fig. 8(f) is visually more appealing thanks to the satin stitches in Fig. 8(f) compliant with the vein directions. However, rendering Fig. 8(f) takes almost twice longer than Fig. 8(e). Moreover, compared with the satin stitch, the long-short stitch retains more details of the input image.

Figure 10 shows the multi-regional stitching. All the three leaves are rendered with long-short stitches, their edges with edge stitch pattern 2, and the stem with satin stitches edged with edge stitch pattern 1.

We also implement the multi-layer embroidery technology in this paper. We find multi-layer stitching works better in color and intensity graduation. For example, the purple flower in Fig. 5(g) is stitched with a single layer of long-short stitches, while the same region in Fig. 5(h) is stitched with 4 layers with the length and width of the stitches of each layer decreasing exponentially from bottom to top. The backgrounds of both Fig. 5(g) and (h) are simply rendered with long-short stitches in larger size. As can be seen, the multi-layer embroidery gives a more vivid result.

In addition, the use of intrinsic decomposition enables us to transfer shadings between different images. Figure 11(a) and (b) are two input images with different shadings due to the moved light source. Note that the shadow distribution in Fig. 11(a) is different from that of Fig. 11(b). Figure 11(c) is the embroidery simulation result using both the reflectance and shading images of Fig. 11(a), while Fig. 11(d) is the result using the reflectance image of Fig. 11(a) but the shading image of Fig. 11(b). There are some artifacts clearly seen on the down-right of the apple in Fig. 11(d). These artifacts come from the red stains in the same area of the reflectance image of Fig. 11(a), which are uncovered after shading transfer. This shading-transferring ability enables us to produce various embroidery simulations of the same scenery but with different shadings in a speedy manner, saving us from tedious illumination rendering from scratch.

Compared with the extant embroidery simulation methods [2,3], our method can retain the physical illumination of the input image. As shown in Fig. 12, our method preserves the physical illumination of the input image, eliminating the contrast decay [2]. Compared with [3] in Fig. 13, our method inherits the illumination of the input image, resolving the problem of illumination incoherence caused by the artificial illumination [3]. As can be seen in Fig. 13, the illumination of Fig. 13(c) is closer to the of the input images.

We also conduct objective comparisons to quantitatively demonstrate the preservation of illumination made by our method. Note that there is no general quantitative criterion to measure NPR results. Since the purpose of this paper is to preserve the input illumination, we find the Peak Signal to Noise Ratio (PSNR) and Root Mean Square Error (RMSE) are eligible enough to assess the simulation results. For the PSNR, the larger the better, while for the RMSE, the closer to 0 the better. We take the input images as ground-truth, the PSNR values calculated for Fig. 12(b), (c), (e), and (f) are 17.5729, 27.1063, 14.4869, and 25.4821, respectively, while the corresponding RMSE values are 33.7206, 11.2519, 48.1058, and 13.5654. The PSNR values for Fig. 13(b) and (c) are 16.0065 and 22.8227, respectively, while the corresponding RMSE values are 40.3844 and 18.4249.

6   Conclusions

In order to resolve the problem of inconsistent illumination existing in the previous embroidery simulation methods, we proposed to inherit the physical illumination of the original input for embroidery simulation by intrinsic decomposition. Our work is subject to SBR, and has been experimentally demonstrated feasible using both the single-layer and multi-layer rendering technologies. Hence, we believe that our method may inspire many other extant stroke-based NPR applications, where illumination preservation is desired. Moreover, our method can also work as stylization filter in those multimedia applications, such as Photoshop and Instagram.

It is worth noting that our work is different from woven cloth simulation [13,14] in several aspects. First, our method requires a digital image as input, and all the parameters of stitches are obtained by preprocessing the input, whereas most the woven cloth simulation methods need no input images. Second, unlike woven cloth simulation that emphasizes the realistic sense, our method aims to create a non-photorealistic illusion from the input image. Third, in stitch modeling, those woven cloth simulation methods mainly use coarse and long interlacing knits of weft and warp, while ours uses much smaller stitches with more stitch patterns.

References

1. Chen, X., McCool, M., Kitamoto, A., Mann, S.: Embroidery modeling and rendering. In: Proceedings of Graphics Interface (GI) 2012, pp. 131–139. Canadian Information Processing Society (2012)

2. Yang, K., Zhou, J., Sun, Z., Li, Y.: Image-based irregular needling embroidery rendering. In: Proceedings of the 5th International Symposium on Visual Information Communication and Interaction (VINCI), pp. 87–94. ACM (2012)
3. Cui, D., Sheng, Y., Zhang, G.: Image-based embroidery modeling and rendering. In: Computer Animation and Virtual Worlds (CAVW) (2016)
4. Hertzmann, A.: Fast paint texture. In: Proceedings of the 2nd International Symposium on Non-photorealistic Animation and Rendering (NPAR), pp. 91–97. ACM (2002)
5. Lu, J., Barnes, C., DiVerdi, S., Finkelstein, A.: RealBrush: painting with examples of physical media. ACM Trans. Graph. (TOG) **32**(4), 117:1–117:12 (2013)
6. Hertzmann, A.: Painterly rendering with curved brush strokes of multiple sizes. In: Proceedings of the 25th Annual Conference on Computer Graphics and Interactive Techniques (SIGGRAPH), pp. 453–460. ACM (1998)
7. Shiraishi, M., Yamaguchi, Y.: An algorithm for automatic painterly rendering based on local source image approximation. In: Proceedings of the 1st International Symposium on Non-photorealistic Animation and Rendering (NPAR), pp. 53–58. ACM (2000)
8. Chan, T.F., Vese, L.A.: Active contours without edges. IEEE Trans. Image Process. (TIP) **10**(2), 266–277 (2001)
9. Chan, T., Vese, L.: An active contour model without edges. In: Nielsen, M., Johansen, P., Olsen, O.F., Weickert, J. (eds.) Scale-Space 1999. LNCS, vol. 1682, pp. 141–151. Springer, Heidelberg (1999). doi:10.1007/3-540-48236-9_13
10. Xu, L., Lu, C., Xu, Y., Jia, J.: Image smoothing via L_0 gradient minimization. ACM Trans. Graph. (TOG) **30**(6), 174:1–174:12 (2011). ACM
11. Land, E.H., McCann, J.J.: Lightness and retinex theory. J. Opt. Soc. Am. (JOSA) **61**(1), 1–11 (1971)
12. Rother, C., Kiefel, M., Zhang, L., Schölkopf, B., Gehler, P.V.: Recovering intrinsic images with a global sparsity prior on reflectance. In: Advances in Neural Information Processing Systems (NIPS), pp. 765–773 (2011)
13. Irawan, P., Marschner, S.: Specular reflection from woven cloth. ACM Trans. Graph. (TOG) **11**, 1–20 (2012)
14. Cirio, G., Lopez-Moreno, J., Miraut, D., Otaduy, M.A.: Yarn-level simulation of woven cloth. ACM Trans. Graph. (TOG) **33**(6), 207:1–207:11 (2014)

Improving the Discriminative Power of Bag of Visual Words Model

Achref Ouni[1], Thierry Urruty[1(✉)], and Muriel Visani[2]

[1] XLIM, UMR CNRS 7252, University of Poitiers, Poitiers, France
{achref.ouni,thierry.urruty}@xlim.fr
[2] Laboratory L3i, University of La Rochelle, La Rochelle, France
muriel.visani@univ-lr.fr

Abstract. With the exponential increase of image database, Content Based Image Retrieval research field has started a race to always propose more effective and efficient tools to manage massive amount of data. In this paper, we focus on improving the discriminative power of the well-known bag of visual words model. To do so, we present n-BoVW, an approach that combines visual phrase model effectiveness keeping the efficiency of visual words model with a binary based compression algorithm. Experimental results on widely used datasets (UKB, INRIA Holidays, Corel1000 and PASCAL 2012) show the effectiveness of the proposed approach.

Keywords: Bag of visual words · Visual phrases · Image retrieval

1 Introduction

Content Based Image Retrieval (CBIR) has been an active field in the last decades. The massive amount of data available today has highlighted the needs for efficient and effective tools to manage this data. One important topic from CBIR field is the construction of the *image signature*. Indeed, image signature is at the core of any CBIR system. An accurate and discriminative signature will improve the precision of the retrieval process. It will also help to bridge the well-known *semantic gap* issue between low-level features and the semantic concepts a user perceived in the image.

Among the numerous state-of-the-art approaches that have tried to "narrow down" the semantic gap, some of them have improved the descriptive power of visual features [4,16], improving gradually the existing local and global image descriptors, while others have proposed effective ways to use, mix and optimize the use of these features. Among them, the bag of visual words model (BoVW) [3,15] has become a reference in CBIR. The BoVW model represents images as histograms of visual words, enhancing the retrieval efficiency without losing much accuracy.

More recently, some researchers have stated that the BoVW discriminative power was not enough and have proposed to construct visual phrases or bags

© Springer International Publishing AG 2017
L. Amsaleg et al. (Eds.): MMM 2017, Part II, LNCS 10133, pp. 245–256, 2017.
DOI: 10.1007/978-3-319-51814-5_21

of bags of words by structuring visual words together using different means. However, Bag of visual phrases models [11,14,18] are computationally expensive. In this paper, we present a novel framework, called n-BoVW, to increase the discriminative power of the BoVW model. n-BoVW uses the idea of visual phrases by selecting multiple visual words to represent each key-point but keeps the efficiency of BoVW model with a binary based compressing algorithm. Two methodologies are proposed and combined with the BoVW model to obtain our final image representation. Our experimental results on different datasets highlight the potential of our proposal.

The remainder of this article is structured as follows: we provide a brief overview of bag of visual words and phrases related works in Sect. 2. Then, we explain our different proposals in Sect. 3. We present the experiments on 3 different datasets and discuss the findings of our study in Sect. 4. Section 5 concludes and gives some perspectives to our work.

2   State of the Art

We present in this section a brief overview of the literature of CBIR field that is linked to the BoVW model proposed by Csurka et al. [3]. Its inspiration comes from the Bag of Words model [5] of the Information Retrieval domain. BoVW model contains four main parts in its retrieval framework. For all images, feature detection and extraction has to be done. These two steps detect a list of key-points with rich visual and local information and convert this information into a vector. Many visual descriptors have been created, among them the Scale Invariant Feature Transform (SIFT) [9] and Speeded-up Robust Features (SURF) [2] became two of the most popular descriptors. Then, an off-line process extracts the visual vocabulary, a set of visual words, using a clustering algorithm on the set of visual features. Finally, each key-point of each image is assigned to the closest visual word of the vocabulary. Thus, each image is represented by a histogram of visual word frequencies, i.e. the image signature.

Inspired by the BoVW model, Fisher Kernel [12] or Vector of Locally Aggregated Descriptors (VLAD) [7] have met with great success. The first approach proposed by Perronnin and Dance [12] applies Fisher Kernels to visual vocabularies represented by means of a Gaussian Mixture Model (GMM). VLAD has been introduced by Jégou et al. [7] and can be seen as a simplification of the Fisher kernel. The idea of VLAD is to assign each key-point to its closest visual word and accumulate this difference for each visual word.

Recently, some researchers have focused on improving the discriminative power of the BoVW model. Thus, they have proposed to construct visual phrases or groups/bags of bags of words. Among them, we can cite the work of Yang and Newsam [18] or Alqasrawi et al. [1] who have used the spatial pyramid representation [8] to construct visual phrases from words spatially close or co-occurring in the same sub-region. They obtained good results for classification purposes. Ren et al. [14] have extended the BoVW model into Bag of Bags of Visual Words. They have proposed an Irregular Pyramid Matching with the

Normalized Cut methodology to subdivide the image into a connected graph. Other researchers have chosen to mix several vocabularies with different image resolutions as Yeganli et al. [19].

Most of those methodologies, by considering more meaningful words combinations, reach a better effectiveness than the original BoVW model. However, this improved performance can be reached only at the cost of a lower efficiency, as the processes for extracting/matching word combinations are generally quite costly.

3   Approach

In this section, we first describe our global framework before we detail our contributions. Our main objective is to improve the BoVW model discriminative power without losing much efficiency in the retrieval process. Thus, we use a common CBIR framework without any filtering on image or refining process on the used visual features nor the constructed vocabularies. It insures the reproducibility of our results. As most CBIR systems using the BoVW model are similar, we have a standard off-line learning process to construct the visual vocabulary on a separate dataset.

Figure 1 presents the different steps of our global framework. In the top part of Fig. 1, we find the detection and extraction steps for each image of the dataset. Then, using the visual vocabulary constructed previously, we proposed three different ways to construct the image signature. First "line" is the standard BoVW model that gives for each image a histogram of visual word frequencies as signature (which will be binarized to be combined). The second and third "lines" is our first contribution, an approach we denote n-Bag of Visual Words (n-BoVW). n-BoVW selects n visual words from the vocabulary to represent each detected key-point by a visual phrase. Two different methodologies are studied: (i) selecting the n closest visual words from a key-point (second "line" in the image) and (ii) clustering n nearest key-points together in the visual feature space to obtain a list of n visual words, one word by key-point inside the small cluster. For both proposals, our second contribution is a binary based compression process used to ensure an efficient retrieval. Thus, both methodologies also represent each image by a histogram of frequencies. A final combining step is also proposed to construct the final signature of the image. This step mixes the three obtained histograms to improve the discriminative power of the image signature. The following subsections detail these proposals.

3.1 n-Bag of Visual Words Methodologies

As visual phrases group visual words together to be more discriminative, the first contribution of this paper presents two different methodologies to better describe or represent each key-point by n visual words. Note that visual phrase models from the literature take usually n words from different key-points with the objective to better represent the near sub-region. Our approach differs as

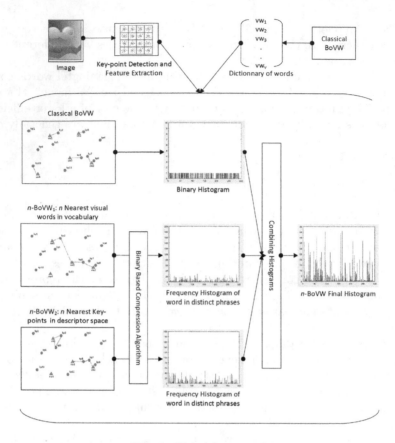

Fig. 1. Global framework

we aim at providing a more precise description of each key-point using a small vocabulary size.

The first methodology we propose is to select n visual words from the same visual vocabulary to represent each key-point of an image, referred as n-BoVW$_1$. Let W denote the vocabulary of v visual words $vw_1, \ldots, vw_v$ constructed using an offline process on a separate dataset. Let KP_i be the set of key-points extracted by the detection step for image i, with kp_{ip} the p-th key-point of image i. For each kp_{ip}, we compute the Euclidean distance ($dL2$) between the key-point and each visual word vw_j from the vocabulary W.

$$dL2(kp_{ip}, w_j) = \sqrt{\sum_{d=1}^{dim} (f_{ip_d} - vw_{j_d})^2}, \tag{1}$$

where f_{ip_d} is the d-th value of the extracted visual feature f of dimension dim for kp_{ip}.

Then, W is sorted according to these distances in order to pick the n nearest visual words from kp_{ip}. Thus, for each key-point kp_{ip}, we obtain a visual phrase vpl_{ip}, i.e. a set of n distinct visual words $vw1_{1_{ip}}, \ldots, vw1_{n_{ip}}$. An example is given Fig. 2(a) with $n = 2$. We can see $kp2$ two nearest visual words are $vw1$ and $vw3$, thus $kp2$ is represented by the visual phrase $(vw1, vw3)$, similarly, $kp8$ is represented by $(vw2, vw3)$.

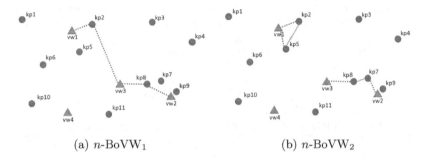

(a) n-BoVW$_1$ (b) n-BoVW$_2$

Fig. 2. Examples for n-BoVW$_1$ and n-BoVW$_2$

With this first methodology, we ensure the description of a key-point by a visual phrase of n distinct words. However, it never takes into account the possibility that the key-point could be better represented by only one and unique word.

Our second proposal, referred as n-BoVW$_2$, is based on this non-possibility we mention and also the fact that a key-point description could be a bit noisy, thus it is interesting to look at his surrounding directly in the descriptor (or visual feature) space. A bit as strong clustering algorithm works, we gather the nearest key-points in the visual feature space to form a strong choice of visual words. Each of those selected key-points is then linked to only one visual word. Note that, the probability to have nearest visual features represented by similar visual words is high. So, it allows the possibility to have only one representative visual word for the small cluster of key-points.

For each key-point kp_{ip}, we compute the Euclidean distances with KP_i, i.e. the other key-points of image i. These distances are then sorted in order to retrieve the n nearest key-points in the visual feature spaces (including the current key-point itself). This set of nearest key-points NKP_{ip} is then used to select the representative visual words. For each key-point of NKP_{ip}, its nearest visual word is calculated using the L2-distance. At the end, for each key-point kp_{ip} we also obtain a visual phrase $vp2_{ip}$, i.e. a set of n visual words $vw2_{1_{ip}}, \ldots, vw2_{n_{ip}}$ with a high probability of duplicates. An example is given in Fig. 2(b). We can see $kp2$ nearest key-point is $kp5$, both key-points are represented by $vw1$, so we have only $vw1$ to represent $kp2$ which is different from first method. However, $kp8$ (with the link to its nearest neighbor $kp7$) is still represented by the same visual phrase $(vw2, vw3)$.

Algorithm 1. Global approach

1: **procedure** CREATEIMAGESIGNATURE
2:     $hisBoVW_i \leftarrow init()$;   ▷ BoVW histogram
3:     $hisNBoVW1_i \leftarrow init()$;   ▷ n-BoVW$_1$ histogram
4:     $hisNBoVW2_i \leftarrow init()$;   ▷ n-BoVW$_2$ histogram
5:     $FinalHisNBoVW_i \leftarrow init()$;   ▷ image i signature
6:     $vp1_i \leftarrow init()$ ▷ set of visual phrases from n-BoVW$_1$
7:     $vp2_i \leftarrow init()$ ▷ set of visual phrases from n-BoVW$_2$
8:     **for** kp_{ip} in KP_i **do**   ▷ for all key-points of an image
9:         $hisBoVW_i \leftarrow hisBoVW_i + computeNearestVisualWords(kp_{ip}, W, 1)$;
10:         $vp1_i \leftarrow vp1_i + computeNearestVisualWords(kp_{ip}, W, n)$;
11:         $NKP_{ip} \leftarrow computeNearestKeypoints(kp_{ip}, KP_i, n)$;
12:         **for** nkp_{ip} in NKP_{ip} **do**
13:             $vp2_i \leftarrow vp2_i + computeNearestVisualWords(nkp_{ip}, W, 1)$;
14:     $hisNBoVW1_i \leftarrow BinaryBasedCompression(vp1_i)$;
15:     $hisNBoVW2_i \leftarrow BinaryBasedCompression(vp2_i)$;
16:     $FinalHisNBoVW_i \leftarrow CombHis(hisNBoVW1_i, hisNBoVW2_i, hisBoVW_i)$;

3.2   Binary Based Compression

The main disadvantage of having such visual phrases from our two proposed methodologies is the number of phrase possibilities, i.e. $\frac{v!}{(v-n)!n!}$ which will be computationally too high for a retrieval system. To deal with this phenomenon, we propose a binary based compression algorithm that is used for both proposals.

We first noticed in literature approaches that visual phrases of only 2 words give better performance [1,18]. So, for each key-point visual phrase vp_{ip} of n words, we construct all possible combinations of 2 visual words. Then, we also observed that for BoVP model approaches with a high number of phrases, only the presence or the absence of the visual phrase is enough to be discriminative. Thus, we decide to binarize the presence of visual phrases in one image. The final step of our compression methodology sums the presence of a word in distinct visual phrases.

The results of our proposal is an histogram of v bins as image signature which is similar to the BoVW model. However, in our approach the histogram contains the frequencies of a word appearing in distinct visual phrases extracted at each key-point.

Algorithms 1 and 2 give the global approach with the binary based compression method. Some part of those algorithms are more detailed to be easier to understand but are obviously optimized in our real code. Of course, the information gathered from both methodologies described previously is different, even from the standard BoVW model. Thus, it is relevant to try and combine the histograms from BoVW model and both n-BoVW methodologies. Out of the different solutions we have tried to combine these histograms, adding the occurrences of visual phrases together before going through the binary based compression process has given the best results. Our exhaustive experimental results are discussed in the next section.

Algorithm 2. Binary Based Compression Algorithm

function BINARYBASEDCOMPRESSION(VP)

2:    $hisVP \leftarrow init()$;   ▷ histogram of presences of words in phrases

    $binaryVP \leftarrow init()$;   ▷ binarized visual phrases of 2 words

4:    **for** vw_j in VP **do**

        for vw_k in VP, $k >= j$ **do**

6:            $tempVP \leftarrow (vw_j, vw_k)$;

            if $tempVP$ is not in $binaryVP$ **then**

8:                $binaryVP \leftarrow binaryVP + tempVP$;

    for $v1$ in W **do**

10:        **for** $v2$ in W, $v1 >= v2$ **do**

            $tempVP \leftarrow (v1, v2)$;

12:            **if** $tempVP$ in $binaryVP$ **then**

                $hisVPv1++$

14:                $hisVPv2++$

    return $hisVP$

---

4 Experimental Results

In this section, we present the experiments done to highlight the potential of our approach. To evaluate our different propositions, 2 low level visual features, Speeded-up Robust Features (SURF) and Color Moment Invariant (CMI), and 3 datasets were considered:

University of Kentucky Benchmark which has been proposed by Nistér and Stewénius [10] is referred as UKB to simplify the reading. UKB contains of 10200 images divided into 2550 groups, each group consists of 4 images of the same object with different conditions (rotated, blurred...). The score is the mean precision over all images for 4 nearest neighbors.

INRIA Holidays [6], referred as Holidays, is a collection of 1491 images, 500 of them are query images, and the remaining 991 images are the corresponding relevant images. The evaluation on Holidays is based on mean average precision score (mAP) [13].

Corel1000 or Wang [17], referred as Wang, is a collection of 1000 images of 10 categories. The evaluation is the average precision of all images for first 100 nearest neighbors.

To construct the initial visual vocabulary, we used the *PASCAL VOC 2012* [4] containing 17225 heterogeneous images categorized into 20 object classes. We use a visual vocabulary of 500 words for each descriptor.

4.1 Performance of n-BoVW

First, we study the effect of the parameter n. Figure 3 shows the performance of the retrieval on the 3 datasets with SURF descriptor. It clearly indicates that $n > 2$ has very little interest ($n = 3$ is better only on Holidays for SURF). Similar observations have also been noticed with CMI descriptor for both methodologies: sometimes, for $n > 2$, the precision is stable, sometimes it drops little by little.

This results is similar to literature visual phrases results where most approaches construct visual phrases of 2 words [1,18]. Thus, we decide to focus the following experiments with $n = 2$ even if $n = 3$ could give small improvement with a specific dataset and descriptor.

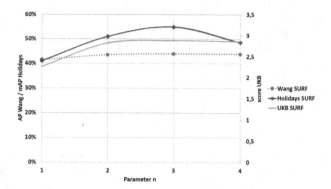

Fig. 3. Study of the effect of parameter n, number of visual words in phrases, for SURF

The next experiments evaluate the performance of the two proposed methodologies. Table 1 presents the performance of using the 2 nearest visual words to represent one key-point, referred as 2-BoVW$_1$. Note that CMI.SURF denotes the concatenation at the end of the process of the two final histograms for retrieval, and 2-BoVW$_1$ + BoVW denotes the addition of BoVW frequencies before the binary based compression step. As one can observe, 2-BoVW$_1$ methodology outperforms the BoVW model in almost all scenarios and when we add the BoVW histogram, the performance is even higher. For example, a score of 3.50 (out of 4) on UKB using 2-BoVW$_1$ + BoVW with the concatenation of both descriptor histograms is very high compared to BoVW (+37%).

Table 1. 2-BoVW$_1$ performance

Dataset	Descriptor	BoVW	2-BoVW$_1$	2-BoVW$_1$ + BoVW
Wang	CMI	40%	48%	48%
	SURF	42%	43%	45%
	CMI.SURF	48%	55%	57%
UKB	CMI	2.52	3.01	3.18
	SURF	2.26	2.82	2.92
	CMI.SURF	2.55	3.41	3.50
Holidays	CMI	41%	51%	52%
	SURF	53%	56%	56%
	CMI.SURF	44%	64%	64%

Table 2 presents the good performance of using the nearest neighbor key-points to obtain visual phrases of $n = 2$ words, referred as 2-BoVW$_2$. We observe that 2-BoVW$_2$ has almost similar results than 2-BoVW$_1$, with only a small decrease on UKB dataset. These two tables clearly highlight the interest of our proposals.

Table 2. 2-BoVW$_2$ performance

Dataset	Descriptor	BoVW	2-BoVW$_2$	2-BoVW$_2$ + BoVW
Wang	CMI	40%	46%	47%
	SURF	42%	42%	44%
	CMI.SURF	48%	54%	56%
UKB	CMI	2.52	3.08	3.04
	SURF	2.26	2.73	2.81
	CMI.SURF	2.55	3.30	3.41
Holidays	CMI	41%	52%	53%
	SURF	53%	56%	56%
	CMI.SURF	44%	65%	65%

Performance Combining Methodologies. As the two proposed method-ologies present good performance but similar, we try to mix both obtained his-tograms together in order to check if the performance of the system could benefit from this combination. On Table 3, we observe that on UKB with single descrip-tor, the precision has increased. Note that we highlight in bold, the precision scores that are strictly above n-BoVW$_1$ or n-BoVW$_2$. However, it is important to notice that combining histograms never decreases the results.

Table 3. Performance combining 2-BoVW$_2$, 2-BoVW$_1$ and BoVW histograms

Dataset	Descriptor	BoVW	2-BoVW	2-BoVW + BoVW
Wang	CMI	40%	48%	48%
	SURF	42%	44%	45%
	CMI.SURF	48%	55%	57%
UKB	CMI	2.52	**3.14**	**3.20**
	SURF	2.26	**2.90**	**3.02**
	CMI.SURF	2.55	3.41	3.50
Holidays	CMI	41%	52%	53%
	SURF	53%	**57%**	**57%**
	CMI.SURF	44%	65%	65%

4.2   Discussion

The observed results show the interest of n-BoVW with the 2 methodologies we have proposed combined with the BoVW model. The precision of the retrieval is clearly higher than the BoVW alone. Most of literature approaches have indeed improved the BoVW model but needed some indexing structure to decrease the loss in efficiency for the retrieval. Constructing the image signature with our framework is obviously more complex than the BoVW model: for one image, 3 histograms are created and combined. The most complex one is the second methodologies n-BoVW$_2$ because it needs to sort all image key-points to pick n nearest ones. Constructing this histogram takes 5 times more longer than BoVW histogram. However, as we obtain an image signature of the same size (vocabulary size) than the BoVW one, the increase in complexity has little effect in the global retrieval process. Extracting the descriptor, and searching for nearest neighbors in the dataset are still preponderant processes.

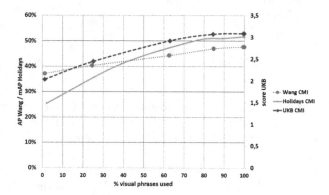

Fig. 4. Performance of n-BoVW with respect to the % of visual phrases used for CMI

Table 4. n-BoVW vs. other methods

Method	UKB score	Wang AP	Holidays mAP
BoVW [3]	2.95	48%	53%
Fisher [12]	3.07	na	69.9%
VLAD [7]	3.17	na	53%
n-Grams [11]	na	34%	na
n-BoVW	3.50	57%	65%

Another point of discussion we highlight is the possibility that for more than one word to describe a key-point could add noise in the image description. Thus, we have tried to put a distance threshold in our algorithm. The visual phrases constructed with words too "far" should not be taken into consideration, replacing the visual phrase by only one word. Figure 4 presents the observed results on the 3 datasets with respect to the percentage of visual phrases used for CMI descriptor. Note that SURF results are similar. The results are a bit surprising because best results are achieved with a percentage of visual phrases close to 100%. Thus, we may conclude all visual phrases are needed in n-BoVW even if results still improve when combining with BoVW. Finally, we compare our approach against few state-of-the-art methods in Table 4. We give here results given by authors when available (na when not available). We observe easily that our proposed approach mostly outperforms other recent methods.

5   Conclusion

This paper presents a more discriminative BoVW framework called n-BoVW. Two different methodologies based on visual phrases model were proposed with for both, results outperforming the BoVW model on all test datasets and with two different descriptors. Mixing these methods together with the BoVW model also improves greatly the performance. Another contribution of this paper is the proposed binary based compression method. It allows the proposed framework to have a similar computational cost than the BoVW model for retrieval. Our perspective will focus first on the notion of distance from a key-point to a visual word discussed in the previous section. We believe it could be useful to adapt automatically the parameter n for each key-point. Thus, different lengths of visual phrases could represent each key-point of the image. A study of the effect of number of visual words in the starting vocabulary would also be interesting even if increasing this number will decrease the efficiency of the retrieval framework.

Acknowledgments. This research is supported by the Poitou-Charentes Regional Founds for Research activities and the European Regional Development Founds (ERDF) inside the e-Patrimoine project from the axe 1 of the NUMERIC Program.

References

1. Alqasrawi, Y., Neagu, D., Cowling, P.I.: Fusing integrated visual vocabularies-based bag of visual words and weighted colour moments on spatial pyramid layout for natural scene image classification. Sig. Image Video Process. **7**(4), 759–775 (2013)
2. Bay, H., Tuytelaars, T., Gool, L.: SURF: speeded up robust features. In: Leonardis, A., Bischof, H., Pinz, A. (eds.) ECCV 2006. LNCS, vol. 3951, pp. 404–417. Springer, Heidelberg (2006). doi:10.1007/11744023_32
3. Csurka, G., Bray, C., Dance, C., Fan, L.: Visual categorization with bags of key-points. In: Workshop on Statistical Learning in Computer Vision, ECCV, pp. 1–22 (2004)
4. Everingham, M., Van Gool, L., Williams, C.K.I., Winn, J., Zisserman, A.: The PASCAL Visual Object Classes Challenge 2012 (VOC2012) Results (2012). http://www.pascal-network.org/challenges/-VOC/voc2012/workshop/index.html

5. Harris, Z.: Distributional structure. Word 10(23), 146–162 (1954)
6. Jegou, H., Douze, M., Schmid, C.: Hamming embedding and weak geometric consistency for large scale image search. In: Forsyth, D., Torr, P., Zisserman, A. (eds.) ECCV 2008. LNCS, vol. 5302, pp. 304–317. Springer, Heidelberg (2008). doi:10.1007/978-3-540-88682-2_24
7. Jégou, H., Douze, M., Schmid, C., Pérez, P.: Aggregating local descriptors into a compact image representation. In: 23rd IEEE Conference on Computer Vision and Pattern Recognition (CVPR 2010), pp. 3304–3311, San Francisco, United States. IEEE Computer Society (2010)
8. Lazebnik, S., Schmid, C., Ponce, J.: Beyond bags of features: spatial pyramid matching for recognizing natural scene categories. In: 2006 IEEE Computer Society Conference on Computer Vision and Pattern Recognition (CVPR 2006), New York, NY, USA, 17–22 June 2006, pp. 2169–2178 (2006)
9. Lowe, D.G.: Object recognition from local scale-invariant features. Int. Conf. Comput. Vis. 2, 1150–1157 (1999)
10. Nistér, D., Stewénius, H.: Scalable recognition with a vocabulary tree. IEEE Conf. Comput. Vis. Pattern Recogn. (CVPR) 2, 2161–2168 (2006)
11. Pedrosa, G., Traina, A.: From bag-of-visual-words to bag-of-visual-phrases using n-grams. In: 2013 26th SIBGRAPI - Conference on Graphics, Patterns and Images (SIBGRAPI), pp. 304–311, August 2013
12. Perronnin, F., Dance, C.R.: Fisher kernels on visual vocabularies for image categorization. In: 2007 IEEE Computer Society Conference on Computer Vision and Pattern Recognition (CVPR 2007), Minneapolis, Minnesota, USA, 18–23 June 2007. IEEE Computer Society (2007)
13. Philbin, J., Chum, O., Isard, M., Sivic, J., Zisserman, A.: Object retrieval with large vocabularies and fast spatial matching. In: Proceedings of the IEEE Conference on Computer Vision and Pattern Recognition (2007)
14. Ren, Y., Bugeau, A., Benois-Pineau, J.: Bag-of-bags of words irregular graph pyramids vs spatial pyramid matching for image retrieval. In: 2014 4th International Conference on Image Processing Theory, Tools and Applications (IPTA), pp. 1–6, October 2014
15. Sivic, J., Zisserman, A.: Video Google: a text retrieval approach to object matching in videos. In: Proceedings of the International Conference on Computer Vision, pp. 1470–1477, October 2003
16. van de Sande, K.E.A., Gevers, T., Snoek, C.G.M.: Evaluating color descriptors for object and scene recognition. IEEE Trans. Pattern Anal. Mach. Intell. 32(9), 1582–1596 (2010)
17. Wang, J.Z., Li, J., Wiederhold, G.: Simplicity: semantics-sensitive integrated matching for picture libraries. IEEE Trans. Pattern Anal. Mach. Intell. 23(9), 947–963 (2001)
18. Yang, Y., Newsam, S.D.: Spatial pyramid co-occurrence for image classification. In: Metaxas, D.N., Quan, L., Sanfeliu, A., Gool, L.J.V. (eds.) IEEE International Conference on Computer Vision, ICCV 2011, Barcelona, Spain, 6–13 November 2011, pp. 1465–1472. IEEE Computer Society (2011)
19. Yeganli, F., Nazzal, M., Özkaramanli, H.: Image super-resolution via sparse representation over multiple learned dictionaries based on edge sharpness and gradient phase angle. Sig. Image Video Process. 9, 285–293 (2015)

M-SBIR: An Improved Sketch-Based Image Retrieval Method Using Visual Word Mapping

Jianwei Niu[1](✉), Jun Ma[1], Jie Lu[1], Xuefeng Liu[2], and Zeyu Zhu[3]

[1] State Key Laboratory of Virtual Reality Technology and Systems,
School of Computer Science and Engineering, Beihang University,
Beijing 100191, China
{niujianwei,melee,lujie}@buaa.edu.cn
[2] Hong Kong Polytechnic University, Hung Hom, Hong Kong
csxfliu@gmail.com
[3] School of Electronics and Information, Xi'an Jiaotong University,
Xi'an 710049, China
asxxzzy@qq.com

Abstract. Sketch-based image retrieval (SBIR) systems, which interactively search photo collections using free-hand sketches depicting shapes, have attracted much attention recently. In most existing SBIR techniques, the color images stored in a database are first transformed into corresponding sketches. Then, features of the sketches are extracted to generate the sketch visual words for later retrieval. However, transforming color images to sketches will normally incur loss of information, thus decreasing the final performance of SBIR methods. To address this problem, we propose a new method called M-SBIR. In M-SBIR, besides sketch visual words, we also generate a set of visual words from the original color images. Then, we leverage the mapping between the two sets to identify and remove sketch visual words that cannot describe the original color images well. We demonstrate the performance of M-SBIR on a public data set. We show that depending on the number of different visual words adopted, our method can achieve $9.8 \sim 13.6\%$ performance improvement compared to the classic SBIR techniques. In addition, we show that for a database containing multiple color images of the same objects, the performance of M-SBIR can be further improved via some simple techniques like co-segmentation.

Keywords: SBIR · Visual word · Mapping · Co-segmentation · M-SBIR

1 Introduction

In order to achieve image retrieval from large-scale image data effectively, digital image repositories are commonly indexed using manually annotated keyword tags that indicate the presence of salient objects or concepts. According to [5,16, 19],there are two types of approaches used for image retrieval: Text-Based Image Retrieval (TBIR) and Content-Based Image Retrieval (CBIR). In this paper, we

© Springer International Publishing AG 2017
L. Amsaleg et al. (Eds.): MMM 2017, Part II, LNCS 10133, pp. 257–268, 2017.
DOI: 10.1007/978-3-319-51814-5_22

are focus on CBIR methods. However, the limitation of traditional CBIR is that, when searching for a particular image, users have to input an "example image", which is almost impossible. Thus, an increasing number of researchers has begun to focus on sketch-based image retrieval. Visual queries such as free-hand sketches are an intuitive way to depict an object's appearance. Compared with traditional CBIR, sketch-based image retrieval (SBIR) has emerged as a more expressive and interactive way to perform image searching.

In most existing SBIR techniques, the bag-of-feature (BoF) model [12,18,21] is used as a framework. In the original method, the color images in a database are first transformed into corresponding sketches. Then, features of sketches are extracted to generate the sketch's visual words for later retrieval. However, transforming color images to sketches will normally suffer from inaccuracy and incur a loss of information, thus decreasing the overall performance of SBIR. The biggest problem in classic SBIR is that the sketches corresponding to the color image cannot reflect the object in the color image perfectly, that is to say, it is difficult to create a sketch that describes the picture accurately.

In this work, to address this problem, we propose a method called M-SBIR, where the M means the mapping relation in our method. In this method, we first extract sketches from color images using an edge extraction algorithm. Then, in parallel with generating sketch visual words, we also generate a set of visual words for the original color image. Next, we construct a mapping from the color image visual words to the sketch visual words. We leverage the mapping between the two sets to identify and remove sketch visual words that cannot describe the original color images well. In addition, we show that using a reference database containing multiple color images for the same objects, the performance of M-SBIR can be further improved via co-segmentation. Even if in some cases it is difficult for the computer to extract accurate sketches, our method can still work well through the use of co-segmentation.

The main contributions of this paper are summarized as follows:

- we present the mapping from the color image visual words to the sketch visual words to identify and remove inaccurate sketch visual words, which can make the sketch index more accurate and improve the performance.
- we use co-segmentation in M-SBIR to extract the edge contour of the object in the image, which could significantly improve the performance.

The remaining sections of this paper are organized as follows. Section 2 surveys the existing literature in this area. Section 3 presents an overview of our method. In Sect. 4, we demonstrate experimental results and in Sect. 5, we conclude the paper.

2   Related Work

Sketch-based image retrieval plays an important role in various applications. In [13,17], the presented methods retrieve trademarks using sketches. [9] presented a new approach to classify, index and retrieve technical drawings based on their content using spatial relationships, shape geometry and high- dimensional indexing mechanisms. Wang et al. [20] described the Sketch2Cartoon system, which creates cartoons automatically using a novel sketch-based clipart image search engine. [3] presented Sketch2photo, which can compose a realistic picture by seamlessly stitching several photographs based on a freehand sketch annotated with text labels.

An image retrieval system must meet the requirement of having a high response speed, and many methods focusing on this aspect have been proposed recently. Among these, many have made use of the BoF model. This model extracts retrieval information during a pre-processing step and stores it in an inverted structure for fast access. Based on the BoF model [18], Eitz [6] designed a sketch interface for image retrieval with a database containing millions of images. The biggest advantage of this model is that the image feature descriptor is obtained in the pretreatment process, and then it is quantified as a vector of visual dictionary. This vector is suitable for the use of inverted index structure storage, which is good to meet the requirements of the retrieval speed. In the BoF framework, the local descriptor is crucial for identifying the visual words within an image. In [4], the three-parameter Generalized Gamma Density (GGD) is adopted for modeling wavelet detail subband histograms and for image texture retrieval. Many methods [2,15,17] extract descriptors based on the image edge map. In [15], strongest edges of images are used to compare with the sketch query in curvature and direction histograms. Meanwhile, some typical descriptors are also used here. Due to the inaccuracy of edges, it will produce errors to extract these descriptors on the edge map directly. Then, researchers proposed some variations of these techniques. SIFT-HOG (SHOG) [7] stores only the most dominant sketch feature lines in the histogram of oriented gradients (HOG), whereas Gradient Field HOG (GFHOG) [11] adapted form of this method, which computes the HOG after constructing the gradient field.

3   M-SBIR: An Improved Sketch-Based Image Retrieval Method

In this paper, we propose a mapping to improve the traditional SBIR and use co-segmentation to ensure that the proposed M-SBIR method will still work in very poor quality cases where we cannot accurately generate a corresponding sketch. Our method is specifically designed to improve the accuracy of retrieval. We will first introduce the construction of M-SBIR and then describe in detail how the mapping and co-segmentation in M-SBIR improves the original SBIR.

3.1   The Overview of M-SBIR

The M-SBIR is based on the BoF model. The BoF model essentially is a text retrieval model, which cannot be directly used for image retrieval. Humans use language, which is composed of words, but images cannot be decomposed in this manner. Thus, the following steps are necessary.

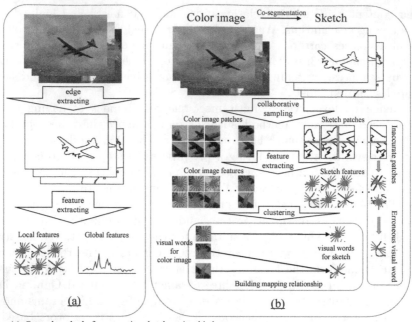

(a). General method of constructing sketch retrieval index
(b). M-SBIR

Fig. 1. The difference between traditional method and our proposed M-SBIR

Define Independent Units. In fact, an image is comprised of pixels. However, a single pixel has no real meaning; only a group organized pixels can represent a specific meaning. In our method, the image is treated as a collection of many independent patches. In other words, each patch is considered as a word of the image. However, a patch is not the original structure in the image, but a geometric structure used to sample, i.e., it is a rectangular sampling window. Therefore, we need to design a reasonable strategy for sampling. In [14], the authors performed a random sampling in the image space. The biggest defect of this method is the randomness of the sampling method. This random sampling generates a lot of redundant information which are many patches with no obvious lines. These patches have no effect on the construction of the sketch index. Therefore, the sampling efficiency of this method is not high. In our approach, we only sample the detected contour.

In this paper, we mainly use two kinds of sampling operators. One of them is SHOG, and another is GFHOG, which are two improved forms of the HOG operator. For the images in database, after edge detection, we take the detected edge as the guide and sample the edge of the color image. In the case of sketches hand-drawn by the user, the line is certainly quite obvious, which means we can do it directly. We use the SHOG and GFHOG operator to extract the features of these patches. Experiments were performed using the two operators in order to verify the universality of our method.

The cluster centers of features extracted from these patches are regarded as visual words.

Build the Inverted Index for the Image Retrieve. After the initial step described in Sect. 3.1, we map all patches of the image into the visual dictionary. In other words, for every patch we find the most similar visual word in the visual dictionary. Then, each image can be regarded as a combination of these visual words with different distribution. In addition, the Term Frequency and Inverse Document Frequency (TF-IDF) algorithm is generally used to adjust the weight which indicates the importance of a word in a image. For the TF, the term frequency will be affected by the size of the image, which requires us normalize it in order to eliminate such effects. For the IDF, the inverse document frequency indicates the degree of occurrence of a visual word and measures its general importance.

Finally, we can obtain the TF-IDF weight by multiplying the Term Frequency with the Inverse Document Frequency.

Then, we construct a standard inverted index data structure, which will be used to compute the similarity between the sketch drawn by the user and the images in the database.

3.2 The Key Contribution 1: Mapping in M-SBIR

The visual vocabulary is a crucial component in the BoF model. In existing work, sketch and image are mapped into the same vocabulary. In those methods, feature is extracted only on the edge of the color image and these feature vectors used as the sketch retrieval index. Based on our survey, we found that many effective feature extractors operate on the edge map of the color images and utilize the similarity between the edge map and the sketch. However, a hand-drawn sketch is usually different from the edge map extracted automatically by the algorithm, especially in characteristics such as thickness and smoothness. Based on existing feature extractors, we propose a new method to build the visual vocabulary so as to further bridge the gap between sketch and color images.

In our method, we build two visual vocabularies, one for the sketch and one for the color image, respectively. We then determine the relationship between the two visual vocabularies. First, we briefly introduce the train data set (BSDS500 Berkeley Segmentation Data Set 500 [1]), which will be used to build the visual vocabularies. In the BSDS500 data set, the public benchmark based on this data consists of all of the grayscale and color segmentations.

We first sampled two patch sets, as shown in Fig. 1. The two sets have a one-to-one correspondence. More specifically, patch P_C^i and P_S^j have corresponding center points in the color image and the sketch. We obtain all patches using the sketch edge pixels as a guide. After this, we obtain the corresponding sets of feature descriptors F_C and F_S. An unsupervised clustering algorithm based on k-means clustering is used to cluster sets F_C and F_S to V_C, V_S, which are the respective sets of visual words.

We cluster the two patch sets based on their features using the following procedure, and obtain two vocabularies, V_C and V_S.

(1) Randomly select K samples from the sample set as the initial cluster centroids: $v_1^u, v_2^u, \cdots, v_k^u \in R^n$

(2) Repeat the following process until convergence is achieved:

    (a) Calculate which cluster centroids the sample feature belongs to for each sample feature $F^{(i)}$

$$v^{(i)} := \arg\min_g \left\| F^{(i)} - v_j^u \right\|^2 \tag{1}$$

    (b) Recalculate the centroid v_j^u of each class j

$$v_j^u := \frac{\sum_{i=1}^{n} 1\left\{ v^{(i)} = j \right\} v_i^u}{\sum_{i=1}^{n} 1\left\{ v^{(i)} = j \right\}} \tag{2}$$

(3) Minimize J:

$$\begin{cases} J = \sum_{i=1}^{n}\sum_{k=1}^{K} r_{ik} \left\| F^i - v_k^u \right\|^2 \\ r_{ik} = \begin{cases} 1, i \text{ belongs to class } k \\ 0, i \text{ does not belong to class } k \end{cases} \end{cases} \tag{3}$$

Finally, we build the mapping from V_C into V_S using Eq. 5, which gives us the degree of consistency for V_C^a and V_S^b. Based on the consistency score, we map every colored visual word into a certain number of sketch words.

$$\begin{cases} \{P_C | P_C^i; i = 1, 2, \ldots, n\} \\ \{P_S | P_S^j; j = 1, 2, \ldots, n\} \end{cases} \tag{4}$$

P_C and P_S represent the color image and sketch image patch sets, respectively, and P_C^i corresponds to P_S^j if and only if $i = j$.

$$C(V_C^a, V_S^b) = \frac{|P_S^{ba}|}{|P_C^a|} \tag{5}$$

P_C^a represents the set of patches which are clustered into the color word V_C^a, while P_S^{ba} represents the set of patches which are clustered into the sketch word V_S^b and have corresponding patches in set P_C^a. In other words, for the example words V_C^a and V_S^b, the degree of their 'consistency' is proportional to the number of patches clustered into V_S^b having corresponding patches clustered into V_C^a.

The degree of 'consistency' determines the mapping from V_C into V_S, which is the core part of our algorithm. This mapping allows us to not simply rely on the edges extracted from the image to build a sketch retrieval index, but to find the best matching sketch word for every part of the image contour.

As shown in Fig. 1, in our M-SBIR, we first map all color image patches into V_C. Then, using the above mapping, the patches are mapped to V_S.

For example, when we build a sketch word index of the plane shown in Fig. 1 after sampling the color image into patches based on the corresponding sketch, we need to use a sketch word to describe each patch. However, any sketch will be flawed, whether it has been generated by computer or a person. If we just use the flawed sketch patch to describe the corresponding color image patch, it will cause the index error. Therefore, for each color patch, we first find the closest cluster center, that is, the corresponding color visual word, and through the mapping relationship we construct, we can find a sketch word which can best describe this color patch.

During the retrieval process, the hand-drawn sketch will be directly mapped into V_S. Combined with an effective feature extractor, our strategy of visual vocabulary construction can achieve better results in enhancing the comparability between two types of images.

The following is a brief discussion about the algorithm complexity of constructing the visual dictionary. We assume that the sketch has n pixels. The algorithm can be divided into three stages: (1) Sampling and extracting features. Because the size of the sample block is fixed, we can say that the time required for feature extraction is a constant time C. The complexity of stage 1 is $\mathcal{O}(Cn)$. (2) Clustering. The clustering algorithm used in this article is based on the k-means, and its complexity is $\mathcal{O}(nkt)$, in which the t is the number of iterative layers. (3) Matching. We should find the best matching sketch visual word (k) for every one of color image visual words (n). The complexity of stage 3 is $\mathcal{O}(nk)$. We can see that the most time-consuming process is the second stage, the time complexity of constructing the visual dictionary in M-SBIR for one image is $\mathcal{O}(nkt)$.

3.3   The Key Contribution 2: Co-segmentation

In some cases where the object's contour in an image is complex, it is difficult for the computer to extract an accurate sketch. Most of the sketch words generated by this inaccurate sketch are wrong, so it is difficult to achieve an improvement using our method. Even if we remove the most obvious mistakes of these inaccurate sketch words by M-SBIR, the rest of the sketch words will still contain significant errors. Therefore, we use co-segmentation to ensure that the generated sketches have sufficient accuracy. It has been proved that co-segmentation gives better segmentation of objects.

The M-SBIR builds retrieval information through counting visual words which reflect the characteristics of local contours in the image. In addition, the boundary is not always identical to the edges of an image. We can consider this difference as follows: (1) the edge is a concept in the field of computer vision, and it is located on those pixels in which gray values change significantly; (2) the boundary represents the joining of background and foreground, which is closely related to our everyday experience; (3) the edge appears near the boundary, however illumination effects may lead to false edges (Fig. 2).

Humans can naturally recognize images through their life experience. In addition, we recognize the most valuable boundaries by analyzing the overall image. In a sketch-based image retrieval system, users are interested in those valuable boundaries rather than the "edges". However, existing edge detection algorithms focus mainly on the edges rather than the boundaries. In other words, it is difficult to extract meaningful boundaries. In this section, we strengthen the characteristics of boundaries, and weaken the impact of false edges. Based on the above consideration, we achieve this target through object detection and segmentation techniques developed in recent years.

The field of computer vision has developed greatly in recent years, and some novel techniques have been proposed. In our method, we utilize a co-segmentation algorithm (Co-segmentation By Composition [8]) which detects common objects from a couple of similar images. Their contour can be considered consistent with sketches drawn by users. Also, compared to object detection from a single image, this class of algorithms yields better results. In order to implement it, for every image in the database we have to obtain similar images. Fortunately, example-based image retrieval systems have matured and are currently widely used, and such techniques help us achieve the above task.

Fig. 2. Some results of M-SBIR

4   Experiments

In this section we first present the data sets used in our experiments, and in Sect. 4.2 we discuss the evaluation measures for image annotation and retrieval. In this section we present a quantitative evaluation of M-SBIR and compare to previous work, qualitative results can be found in Figs. 3 and 4.

4.1   Data Sets and Experimental Setup

We performed experiments using the MATLAB environment on an Intel Core 2 2.83 GHz CPU equipped with 4 GB of RAM. The data set contained 72,924 images, including around 15 K from public dataset [11] and around 55 K images crawled from the Internet, divided into 33 distinct types of shapes (e.g. heart, duck). Then, 10 non-expert sketchers (5 male, 5 female) were asked to query these objects using free-hand sketches. In total, 330 sketches were used to evaluate the performance of the algorithm. Then, we matched similar images using "shitu.baidu.com", which provided enough similar images for the co-segmentation. With this data set, we conducted a comparative performance evaluation.

4.2 Evaluation Measures

In this paper, the aim of our proposed methods is to evaluate the performance of the retrieval system as a whole. We perform a comparative performance evaluation by considering the following aspects: (1) the effectiveness of the mapping; (2) whether co-segmentation can achieve the desired purpose.

For a retrieval system, search results are sorted according to the similarity between all images in the database and the sketches. Higher similarity values result in a higher display priority. In the experiment, we use recall-precision and Mean Average Precision (MAP) to evaluate retrieval results. These quantities are interdependent and need to be considered at the same time. Therefore, we use the P-R curve to evaluate the system. Considering that this is a ranking of search results, MAP is a an alternative measure which complements the limitations of the P-R curve.

Visual Dictionary. We believe that the improved mapping for visual dictionary proposed in this paper has some versatility and robustness, which is reflected in the fact that it can deal with different feature extraction operators. As previously mentioned, Eitz used the BoF model to carry out research on sketch retrieval and achieved good results. He proposed the SHOG feature [7] in the related articles, and Hu [10] proposed the GFHOG feature in another article. The experimental results are shown in Fig. 3.

The K shown in subplot (a) and subplot (c) of Fig. 3 is the visual dictionary size. Through Fig. 3, it can be seen that using the same BoF model framework, the results of using our visual dictionary construction algorithm are better than the original method and our method can adapt to different feature extraction operators.

Through subplots (b) and (d) of Fig. 3, we see that when the visual dictionary size is relatively small ($K < 1300$), an increase in the number of visual words improves the results obtained by our algorithm. However, we also see that the effect is not consistent when we continue to increase the scale of the dictionary. This could be attributed to the fact that, for a fixed image retrieval database, a certain number of visual words can have a relatively wide range of representation. Even if, like the human language, the number of words is limited, perhaps we can find a complete set which contains all the representative visual words.

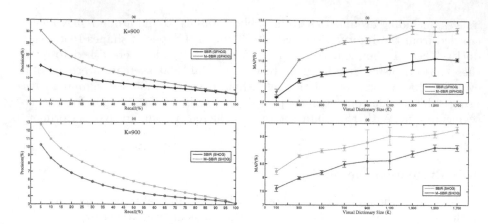

Fig. 3. Results obtained using M-SBIR compared to SBIR. Error bars indicate one standard deviation.

Co-segmentation. In addition, we discuss the impact of using co-segmentation on performance. It is important to note that the number of similar images will influence the effect of co-segmentation. In this experiment, as shown in Fig. 4, SN denotes the number of similar images, that is to say, $SN = 0$ means we do not use co-segmentation. The result shown in subplot (a) of Fig. 4 was carried out with the visual dictionary method using M-SBIR(GFHOG), that is, when $SN = 0$, the experimental results correspond to the Curve M-SBIR(GFHOG) in subplot (a) of Fig. 3. Here we have considered cases up to $SN = 3$. The experimental results shown in subplot (b) of Fig. 4 indicate using co-segmentation yields significantly better results than not using it (the curve of $SN = 0$), and more similar images give higher average retrieval accuracy values. However, the effect is not always obvious (the curves corresponding to $SN = 2$ and $SN = 3$ only have marginal differences).

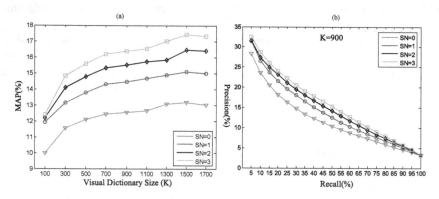

Fig. 4. Results obtained using M-SBIR with different numbers of similar images (SN).

Table 1 shows the maximum F_1 scores under different combinations of SN and K. We can see that there is a significant difference between using ($SN \neq 0$) or not

Table 1. The maximum F_1 scores under different numbers of similar images (SN)

SN	K								
	100	300	500	700	900	1100	1300	1500	1700
0	16.73	18.81	19.47	19.72	19.81	20.03	20.54	20.49	20.38
1	17.92	20.96	21.90	22.45	22.57	22.69	22.59	23.35	23.30
2	19.20	22.25	23.13	23.74	23.86	24.12	23.92	24.81	24.80
3	20.07	23.05	24.08	24.69	24.76	25.02	25.64	25.74	25.70

using co-segmentation ($SN = 0$). Furthermore, when the visual dictionary size is relatively small ($K < 900$), with the increase of the number of visual words, the algorithm performance is getting better obviously. But, it also shows that the effect is not obvious when we continue to increase the scale of the dictionary.

5 Conclusion

In this paper, we introduce a new method for Sketch Based Image Retrieval (SBIR). This method constructs the mapping from color image visual words to sketch visual words. This mapping can help us eliminate the errors in the visual sketch words. In addition, through the use of co-segmentation, our proposed M-SBIR method still gives good results in very poor cases where we can not generate accurate corresponding sketches. From these experimental results we conclude that the combination of a visual dictionary and co-segmentation gives good results, as it combines good recall with high precision over public data sets.

In future work, we will consider extending the Co-segmentation to find objects in the image more delicately and precisely. And if we could be able to use a larger training set like BSDS, the effect of the algorithm will be better. In addition, with the development of object extraction and image fusion technology, we hope that we could use sketch to determine objects and their layout to provide a more convenient image synthesis system.

Acknowledgment. This work was supported by the National Natural Science Foundation of China (Grant Nos. 61572060, 61190125, 61472024) and CERNET Innovation Project 2015 (Grant No. NGII20151004).

References

1. Arbelaez, P., Maire, M., Fowlkes, C., Malik, J.: Contour detection and hierarchical image segmentation. IEEE Trans. Pattern Anal. Mach. Intell. **33**(5), 898–916 (2011)
2. Cao, Y., Wang, H., Wang, C., Li, Z., Zhang, L., Zhang, L.: Mindfinder: interactive sketch-based image search on millions of images. In: Proceedings of the International Conference on Multimedia, pp. 1605–1608. ACM (2010)

3. Chen, T., Cheng, M.M., Tan, P., Shamir, A., Hu, S.M.: Sketch2Photo: internet image montage. ACM Trans. Graph. (TOG) **28**(5), 124 (2009)
4. Choy, S.K., Tong, C.S.: Statistical wavelet subband characterization based on generalized gamma density and its application in texture retrieval. IEEE Trans. Image Process. **19**(2), 281–289 (2010)
5. Datta, R., Joshi, D., Li, J., Wang, J.Z.: Image retrieval: ideas, influences, and trends of the new age. ACM Comput. Surv. (CSUR) **40**(2), 5 (2008)
6. Eitz, M., Hildebrand, K., Boubekeur, T., Alexa, M.: Photosketch: a sketch based image query and compositing system. In: SIGGRAPH: Talks, p. 60. ACM (2009)
7. Eitz, M., Hildebrand, K., Boubekeur, T., Alexa, M.: Sketch-based image retrieval: benchmark and bag-of-features descriptors. IEEE Trans. Vis. Comput. Graph. **17**(11), 1624–1636 (2011)
8. Faktor, A., Irani, M.: Co-segmentation by composition. In: Proceedings of the IEEE International Conference on Computer Vision, pp. 1297–1304 (2013)
9. Fonseca, M.J., Ferreira, A., Jorge, J.A.: Content-based retrieval of technical drawings. Int. J. Comput. Appl. Technol. **23**(2–4), 86–100 (2005)
10. Hu, R., Barnard, M., Collomosse, J.: Gradient field descriptor for sketch based retrieval and localization. In: 17th IEEE International Conference on Image Processing (ICIP), pp. 1025–1028. IEEE (2010)
11. Hu, R., Collomosse, J.: A performance evaluation of gradient field HOG descriptor for sketch based image retrieval. Comput. Vis. Image Underst. **117**(7), 790–806 (2013)
12. Krapac, J., Verbeek, J., Jurie, F.: Modeling spatial layout with fisher vectors for image categorization. In: 2011 IEEE International Conference on Computer Vision (ICCV), pp. 1487–1494. IEEE (2011)
13. Leung, W.H., Chen, T.: Trademark retrieval using contour-skeleton stroke classification. In: 2002 IEEE International Conference on Multimedia and Expo, vol. 2, pp. 517–520. IEEE (2002)
14. Nowak, E., Jurie, F., Triggs, B.: Sampling strategies for bag-of-features image classification. In: Leonardis, A., Bischof, H., Pinz, A. (eds.) ECCV 2006. LNCS, vol. 3954, pp. 490–503. Springer, Heidelberg (2006). doi:10.1007/11744085_38
15. Rajendran, R., Chang, S.F.: Image retrieval with sketches and compositions. In: IEEE International Conference on Multimedia and Expo, vol. 2, pp. 717–720. IEEE (2000)
16. Rui, Y., Huang, T.S., Ortega, M., Mehrotra, S.: Relevance feedback: a power tool for interactive content-based image retrieval. IEEE Trans. Circ. Syst. Video Technol. **8**(5), 644–655 (1998)
17. Shih, J.L., Chen, L.H.: A new system for trademark segmentation and retrieval. Image Vis. Comput. **19**(13), 1011–1018 (2001)
18. Sivic, J., Zisserman, A.: Video Google: a text retrieval approach to object matching in videos. In: Ninth IEEE International Conference on Computer Vision, Proceedings, pp. 1470–1477. IEEE (2003)
19. Smeulders, A.W., Worring, M., Santini, S., Gupta, A., Jain, R.: Content-based image retrieval at the end of the early years. IEEE Trans. Pattern Anal. Mach. Intell. **22**(12), 1349–1380 (2000)
20. Wang, C., Zhang, J., Yang, B., Zhang, L.: Sketch2Cartoon: composing cartoon images by sketching. In: Proceedings of the 19th ACM International Conference on Multimedia, pp. 789–790. ACM (2011)
21. Wang, J.J.Y., Bensmail, H., Gao, X.: Joint learning and weighting of visual vocabulary for bag-of-feature based tissue classification. Pattern Recogn. **46**(12), 3249–3255 (2013)

Movie Recommendation via BLSTM

Song Tang[1]([✉]), Zhiyong Wu[2], and Kang Chen[1]

[1] Department of Computer Science and Technology, Tsinghua National Laboratory
for Information Science and Technology (TNLIST), Tsinghua University,
Beijing 100084, China
tang-s14@mails.tsinghua.edu.cn, chenkang@tsinghua.edu.cn
[2] Graduate School at Shenzhen, Tsinghua University,
Building F, Tsinghua Campus, University Town, Shenzhen, China
zywu@sz.tsinghua.edu.cn

Abstract. Traditional recommender systems have achieved remarkable
success. However, they only consider users' long-term interests, ignoring
the situation when new users don't have any profile or user delete their
tracking information. In order to solve this problem, the session-based
recommendations based on Recurrent Neural Networks (RNN) is proposed to make recommendations taking only the behavior of users into
account in a period time. The model showed promising improvements
over traditional recommendation approaches.

In this paper, We apply bidirectional long short-term memory
(BLSTM) on movie recommender systems to deal with the above
problems. Experiments on the MovieLens dataset demonstrate relative
improvements over previously reported results on the Recall@N metrics
respectively and generate more reliable and personalized movie recommendations when compared with the existing methods.

Keywords: Movie recommendation · Recommendation system ·
BLSTM · RNN

1 Introduction

Movie plays an important role in our daily lives, having become one of the most
popular entertainment forms. With the rapid development of the Internet, there
are many online movie platforms providing massive amounts of resources, which
brings convenience for the users. However, given the rapid growth of network
information resources, people have to spend plenty of time in searching movies
that they are interested. Helping users to find resources that they want rapidly
has become an important requirement. Recommendation systems [2,9] have been
seemed as effective solutions to help the users to find their interesting movie and
filter useless information.

Currently, the most common methods used in movie recommendation are
content-based method (CB) [3,4], collaborative filtering (CF). The former generates recommendations based on the movie or user features extracted from

© Springer International Publishing AG 2017
L. Amsaleg et al. (Eds.): MMM 2017, Part II, LNCS 10133, pp. 269–279, 2017.
DOI: 10.1007/978-3-319-51814-5_23

domain knowledge in advance, while collaborative filtering uses historical data on user preferences to predict movies that a user might like. Generally, there are two successful approaches of CF, namely memory-based [10] and model-based algorithms [8,12]. Memory-based model predicts movie preference of the users based on capturing the relationships between users or items, while Model-based CF uses the observed movie ratings to estimate and learn a model to make predictions.

Crucially, these approaches require the user to be identified when making recommendations. This avoids the cold-start issue when new users didn't have any profile, or they may have deleted their information. The session-based recommendations [7] is proposed to make recommendations based only on the behavior of users in the current session. In [5,6], Recurrent Neural Networks (RNNs) were recently proposed for the session-based recommendation task. The authors showed significant improvements over traditional session-based recommendation models using an RNN model.

However, RNN model used in [5,6] ignore a key point that users' interest can be affected by the previous movie watch record, but also can be related to the behind In this paper, we propose to further study the application of RNNs and use BLSTM model overcome the above problem. We consider the ordering of user actions in a period time based on BLSTM model. More precisely, we consider ordered sequences of movies corresponding to sequences of user actions. Given a sequence of the movie watched in the past, this allows us to infer the future sequence and then to recommend an ordered list of movies to a user.

We start with a discussion of related work in Sect. 2. Then, we present the details of our BLSTM models in Sect. 3 and our experiments on the models in Sect. 4.

2   Related Work

Traditional collaborative filtering techniques utilize a history of item preferences by a set of users in order to recommend movies of interest to a given user. Matrix factorization and neighborhood-based methods are widely utilized for recommender systems in the literature. Neighborhood-based methods involve computing a measure of similarity between pairs of users or items, and obtaining predictions by taking a weighted average of ratings of similar users. Matrix factorization methods are based on the sparse user-item interaction matrix, where the recommendation problem is formulated as a matrix completion task. After decomposing the matrix, each user and item are represented by a latent factor vector. The missing value of the user-item matrix can then be filled by multiplying the appropriate user and item vectors.

The use of deep learning method for recommender system has been of growing interest for researchers. In [14] the authors use Restricted Boltzmann Machines (RBM) to model user-movie interaction and presented efficient learning and inference procedures. Van den Oord et al. [4] use Convolutional neural networks (CNNs) to extract the feature from music audio. Wang et al. [15] introduced a more generic approach called collaborative deep learning (CDL) which extracts content-features and use them to improve the performance of collaborative filtering. In [5], RNNs were proposed for session-based recommendations. The authors compared RNNs with existing methods for session-based predictions and found that RNN-based models performed better than the baselines. Our work is closely related, and we study extensions to their RNN models. In [6], the authors also use RNNs for click sequence prediction; they consider historical user behaviors as well as hand engineered features for each user and item.

Users' interest can be affected by the previous movie play record, but also can be related to the back of the movie play record. In this paper, we utilize BLSTM deep network to model the movie information.

Many approaches have been proposed to improve the predictive performance of trained deep neural networks. Popular approaches include data augmentation, drop-out, batch normalization and residual connections. We seek to apply some of these methods to enhance the training of our recommendation.

3   Our Approach

3.1   Recurrent Neural Networks for Movie Recommendation

A recurrent neural network (RNN) is an extension of a conventional feedforward neural network, which is able to handle a variable-length sequence input. The RNN handles the variable-length sequence by having a recurrent hidden state whose activation at each time is dependent on that of the previous time.

More formally, given a collection $M = \{m_1, \ldots, m_{|M|}\}$ of movie m_i and a movie list $p = \{p_1, \ldots, p_n\}$. Each element p_i of a movie list refers to one movie from M. RNNs update their hidden state h using the following update function:

$$h_t = \mathcal{H}(Wp_t + Uh_{t-1}) \tag{1}$$

Our goal is to estimate a generative model of coherent movie list by modeling the sequential watching behavior. we would like to estimate the distribution $Pr(p)$ of coherent movie list $p = \{p_1, \ldots, p_n\}$.

The probability of $p = \{p_1, \ldots, p_n\}$ can be decomposed into

$$P(p) = \sum_{t=1}^{n} P(p_t|p_1, \ldots, p_{t-1}) \tag{2}$$

We model each conditional probability distribution with

$$P(p_t|p_1, \ldots, p_{t-1}) = P(p_t|h_t) \tag{3}$$

We need some kind distribution of $P(p_t|h_t)$. We use the Softmax distribution over all possible outputs. This means we have to learn a vector for each item. The probability $P(p_t|h_t)$ is now proportional to $exp(h_t^T a_t)$

$$P(p_t|h_t) = \frac{exp(h_t^T a_t)}{\sum_k exp(h_t^T a_k)} \tag{4}$$

where a_t represent the transformation matrix from hidden unit to the output unit.

We can maximize the log-likelihood

$$L(P|H) = \sum_{i=0}^{t-1} log P(p_i|h_i) \tag{5}$$

where P is the movie list $\{p_1, \ldots, p_n\}$, H is the hidden unit or the movie play history.

3.2   Structure of Model

We used the BLSTM in our models for movie recommendation. For a movie list $p = \{p_1, \ldots, p_n\}$, we use $1 - of - N$ encoding to represent the movie in the list p_i as the input of the model. There are many choices in the embedding layer. We use skip-gram model [17] to getting the movie embedding. The core of the network is the BLSTM layer(s) which are added between the input layer and the output layer. The output is the softmax function to calculate the probability of each movie. The movie recommendation based on BLSTM are shown in Fig. 1.

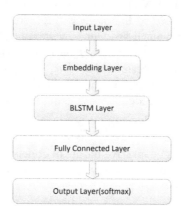

Fig. 1. Generic structure of the network used in our models.

Our model process each sequence $[p_1, p_2, ..., p_r]$ separately and are trained to predict the next movie p_{r+1} in that sequence. The training process is illustrated in Fig. 2.

3.3   BLSTM Mode

Recurrent neural networks (RNNs) are able to process input sequences of arbitrary length via the recursive application of a transition function on a hidden state vector h. Given an input sequence $x = (x_1, \ldots, x_T)$, RNNs compute the

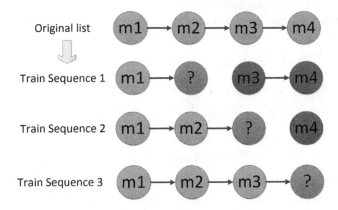

Fig. 2. The train process.

hidden vector sequence $h = (h_1, \ldots, h_T)$ by iterating the following equations from $t = 1$ to T.

$$h_t = \mathcal{H}(W_{ih}x_t + W_{hh}h_{t-1} + b_h) \tag{6}$$

$$y_t = W_{hy}h_t + b_y \tag{7}$$

where $y = (y_1, \ldots, y_T)$ is the output vector sequence. $\mathcal{H}$ is the activation function for hidden state. W represent the weight matrices. b denote the bias vectors.

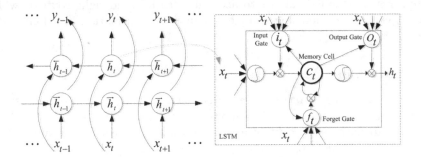

Fig. 3. Bidirectional long short-term memory (BLSTM)

However, it has been observed that it is difficult to train RNNs to capture long-term dependencies since the backpropagated error will tend to either vanish or explode. Long short term memory (LSTM) recurrent neural network is designed to tackle with long time lags. An LSTM layer consists of memory blocks which are a set of connected blocks. The single memory block is shown in Fig. 3. Each block contains four types of units: one or more recurrently connected memory cells, input gate, output gate and forget gate. These gates allow the memory cell to store information over long periods of time and can avoid the vanishing gradient problem. The feedforward process of LSTM is:

$$i_t = \sigma(W_{xi}x_t + W_{hi}h_{t-1} + W_{ci}c_{t-1} + b_i) \tag{8}$$

$$f_t = \sigma(W_{xf}x_t + W_{hf}h_{t-1} + W_{cf}c_{t-1} + b_f) \tag{9}$$

$$c_t = f_t c_{t-1} + i_t tanh(W_{xc}x_t + W_{hc}h_{t-1} + b_c) \tag{10}$$

$$o_t = \sigma(W_{xo}x_t + W_{ho}h_{t-1} + W_{co}c_{t-1} + b_o) \tag{11}$$

$$h_t = o_t tanh(c_t) \tag{12}$$

where σ is the logistic sigmoid function, and i, f, o and c are respectively the input gate, forget gate, output gate, and cell memory. However, the disadvantage of LSTM is that it can only access the previous inputs. Bidirectional LSTM (LSTM) can access both previous and future inputs utilizing the bidirectional architecture. As illustrated in Fig. 3, Bidirectional LSTM (BLSTM) is trained on input sequence in both forward and backward hidden layers, $\overrightarrow{h}$ and $\overleftarrow{h}$. The output sequence y is the combination of $\overrightarrow{h}$ and $\overleftarrow{h}$:

$$\overrightarrow{h} = \mathcal{H}(W_{x\overrightarrow{h}}x_t + W_{\overrightarrow{h}\overrightarrow{h}}\overrightarrow{h}_{t-1} + b_{\overrightarrow{h}}) \tag{13}$$

$$\overleftarrow{h} = H(W_{x\overleftarrow{h}}x_t + W_{\overleftarrow{h}\overleftarrow{h}}\overleftarrow{h}_{t-1} + b_{\overleftarrow{h}}) \tag{14}$$

$$y_t = W_{\overrightarrow{h}y}\overrightarrow{h}_t + W_{\overleftarrow{h}y}\overleftarrow{h}_t + b_y \tag{15}$$

As a result, BLSTM takes into account both left and right context, which makes it capable of modeling complete sequential information in a longer distance.

3.4 Pre-training

We first train a model on the entire dataset. The trained model is then used to initialize a new model using only a more recent subset of the data. This allows the model to have the benefit of a good initialization using large amounts of data, and yet is focused on more recent movie watch record.

3.5 Neural Network Training

During the training process, since the output of softmax layer is a distribution over movies, we utilize the categorical cross-entropy as our objective function

$$H(p, q) = -\sum_x p(x)log(q(x)) \tag{16}$$

where p is the true distribution of movies and q is the predicted distribution. Regularization is vital for good performance with neural networks, as their flexibility makes them prone to overfitting. Here we use Adagrad as our optimizer and introduce dropout technique to avoid overfitting.

4 Experiments

4.1 Datasets

MovieLens. We validate our approach using the data from MovieLens [1]. The MovieLens dataset contains 1,000,209 ratings collected from 6,040 users on 3,900 movies. In this dataset, about 4% of the user-movie dyads are observed. The ratings are integers ranging from 1 (bad) to 5 (good). This dataset also contains additional information about the users and movies, such as age, gender, career and movie types.

4.2 Preprocess

We firstly sort the data of every user according to the time order. Then we use the front 80% of each user as the training set, followed by 20% as the test set. Some detailed statistics of the MovieLens datasets are summarized in Table 1.

Table 1. MovieLens dataset

Name	Users	Movies	Interactions	TrainSet	TestSet
MovieLens	6,040	3,900	1,000,209	808,593	191,616

4.3 Baselines

Item-kNN: An item-based neighborhood [11] approach predicts the rating r_{ui} of a user u for a new item i, using the rating of the user u gave to the items which are similar to i.

SVD: Matrix Factorization methods [8] have demonstrated superior performance vs. neighborhood based models. Each item i and user u is associated with a D-dimensional latent feature vector q_i and p_u respectively. Thus predicted rating is computed by: $\hat{r}_{ui} = \mu + b_i + b_u + q_i^T p_u$.

SVD++: SVD++ [16] is an extension of SVD. For each item i, we add an additional latent factor y_i. Thus, the latent factor vector of each user u can be characterized by the set of items the user have rated. The exact model is as follows: $\hat{r}_{ui} = \mu + b_i + b_u + q_i^T(p_u + |R_u|^{-1/2} \sum_{j \in R_u} y_i)$.

GRU: The GRU-based RNN model [5] was used in session-based recommendations. They modified the basic GRU in order to fit the task better by introducing session-parallel mini-batches, mini-batch based output sampling and ranking loss function.

4.4   Metrics

To compare the performance across different approaches, we use the Recall@N and Precision@N [13] to demonstrate an algorithm's ability to make precise recommendations. For each user, the definition of recall@N is

$$Recall@N = \frac{number\ of\ movies\ the\ user\ likes\ in\ top\ N}{total\ number\ of\ movies\ the\ user\ likes} \tag{17}$$

The recall for the entire system can be summarized using the average recall from all users. Similarly, for each user, the definition of Precision@N is

$$Precision@N = \frac{number\ of\ movies\ the\ user\ likes\ in\ top\ N}{N} \tag{18}$$

4.5   Parameter and Structure Optimization

In this section, we analyze how the changes of the parameters affect the performance of our model. We aim to find the optimal model by applying different parameter configurations on the dataset.

Initialization: As with most latent factor models, we need to initialize the parameter to random noise. The parameters are tuned by cross validation through the grid $\{0.01, 0.05, 0.1, 1\}$. Typically small Gaussian noise like $N(0, 0.1)$ works well.

Dropout: We added dropout to the hidden values since it seemed to help improve the predictions of the model. After the each h_t is calculated, we set half of them to zero. This also seems to help with exploding gradients.

Number of Iterations: With the increasing of the number of iterations, iteration computation is time-consuming because each iteration needs to calculate gradient and update parameter. Lots of iterations indeed can improve performance. Too many times of iterations may argue the overfitting problem appears. So it's important to choose a suitable iteration.

Batch Size: For a large batch, you can take advantage of matrix, linear algebra libraries to accelerate computing. It may be not obvious when the batch is small. However, too bigger batch size may result in the weight will be updated slowly and the training process is too long. So we should choose the batch size which give the model the most rapid improvement in performance.

Hidden Layer: LSTM Layer. As described above, we have adopted a multi-layer LSTM network to implement the model. So LSTM layer is an important hyperparameter for our model. Too many LSTM layers may cause a large amount of calculation.

Table 2. Parameters used for each method

Method	Parameters
Item-kNN	Number of neighbors 200
SVD	$\alpha = 0.05$, $\lambda = 0.01$ and $\theta = 0.0001$
SVD++	$\gamma1 = \gamma2 = 0.007$, $\gamma3 = 0.001$, $\lambda1 = 0.005$, $\lambda2 = \lambda3 = 0.015$
BLSTM & GRU	Hidden layer $= 500$, Embedding layer $= 100$

4.6 Results and Discussion

Quantitative Results. In order to evaluate the effectiveness of our BLSTM model for recommendation system, The parameters for each method could be seen in Table 2. As seen in Fig. 4(a) and 4(b), it's apparent that BLSTM model performs better over all other methods in Recall and Precision. The SVD method performs well on movie dataset and performs much better than Item-kNN.

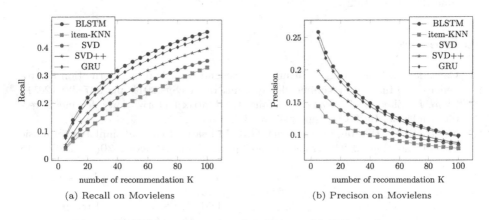

(a) Recall on Movielens　　　　(b) Precison on Movielens

Fig. 4. BLSTM model against baselines on MovieLens dataset.

With the increase of recall, The precision will decrease. Over the MovieLens datasets, BLSTM model is the most significant algorithm in terms of recall as well as precision, outperforming other compared methods. As we expected before, given its naive assumption and simple model design, BLSTM model makes a significant improve to other state-of-the-art models. In summary, our model has multiple advantages. First, it can avoid the aforementioned cold-start issue. Secondly, because it considers both left and right time information, it can predict the user's interest and recommend the movie to users more precisely. Finally, it's very stable and can be pre-trained. We can train the model based on a very big dataset and then predict the new data.

5   Conclusions and Future Work

In this paper, we propose BLSTM models to tackle the task of predicting general users' interest for movies. We chose the task of movie recommendation because it is a practically important area. We modified the BLSTM mode in order to fit the task better. Specifically, we utilize skip-gram embedding model [17] to generate movie-level representations as an input of the model. We showed that our method can significantly outperform popular baselines that are used for this task.

As for the future work, we would like to extend our model by combining other deep neural networks such as CNN or RBM. In addition, we also would like to apply our model in the industrial field to help to make the movie recommendation for more people. It is better for us to conduct online tests.

Acknowledgements. This Work is supported by Natural Science Foundation of China (61433008, 61373145, 61572280, U1435216), National Key Research & Development Program of China (2016YFB1000500), National Basic Research (973) Program of China (2014CB340402).

References

1. Cantador, I., Brusilovsky, P., Kuflik, T.: Second workshop on information heterogeneity and fusion in recommender systems. In: RecSys 2011, pp. 387–388 (2011)
2. Zhao, L., Zhongqi, L., Pan, S.J., Yang, Q.: Matrix factorization+ for movie recommendation. In: IJCAI, pp. 3945–3951 (2016)
3. McFee, B., Barrington, L., Lanckriet, G.R.G.: Learning content similarity for music recommendation. IEEE Trans. Audio Speech Lang. Process. **20**(8), 2207–2218 (2012)
4. van den Oord, A., Dieleman, S., Schrauwen, B.: Deep content-based music recommendation. In: NIPS 2013, pp. 2643–2651 (2013)
5. Hidasi, B., Karatzoglou, A., Baltrunas, L., Tikk, D.: Session-based recommendations with recurrent neural networks. CoRR abs/1511.06939 (2015)
6. Tan, Y.K., Xinxing, X., Liu, Y.: Improved recurrent neural networks for session-based recommendations. CoRR abs/1606.08117 (2016)
7. Dias, R., Fonseca, M.J.: Improving music recommendation in session-based collaborative filtering by using temporal context. In: ICTAI 2013, pp. 783–788 (2013)
8. Koren, Y., Bell, R.M., Volinsky, C.: Matrix factorization techniques for recommender systems. IEEE Comput. **42**(8), 30–37 (2009)
9. Shani, G., Gunawardana, A.: Evaluating recommendation systems. In: Ricci, F., Rokach, L., Shapira, B., Kantor, P.B. (eds.) Recommender Systems Handbook 2011, pp. 257–297. Springer, New York (2011)
10. Takćs, G., Pilśzy, I., Németh, B., Tikk, D.: Major components of the gravity recommendation system. SIGKDD Explor. **9**(2), 80–83 (2007)
11. Papagelis, M., Plexousakis, D.: Qualitative analysis of user-based and item-based prediction algorithms for recommendation agents. Eng. Appl. AI **18**(7), 781–789 (2005)
12. Bell, R.M., Koren, Y., Volinsky, C.: Modeling relationships at multiple scales to improve accuracy of large recommender systems. In: KDD 2007, pp. 95–104 (2007)

13. Cremonesi, P., Koren, Y., Turrin, R.: Performance of recommender algorithms on top-n recommendation tasks. In: RecSys 2010, pp. 39–46 (2010)
14. Salakhutdinov, R., Mnih, A., Hinton, G.E.: Restricted Boltzmann machines for collaborative filtering. In: ICML, pp. 791–798 (2007)
15. Wang, H., Wang, N., Yeung, D.-Y.: Collaborative deep learning for recommender systems. In: KDD, pp. 1235–1244 (2015)
16. Koren, Y.: Factorization meets the neighborhood: a multifaceted collaborative filtering model. In: KDD 2008, pp. 426–434 (2008)
17. Barkan, O., Koenigstein, N.: Item2vec: neural item embedding for collaborative filtering. CoRR abs/1603.04259 (2016)

Multimodal Video-to-Video Linking: Turning to the Crowd for Insight and Evaluation

Maria Eskevich[1]([✉]), Martha Larson[1,2], Robin Aly[3], Serwah Sabetghadam[4],
Gareth J.F. Jones[5], Roeland Ordelman[3], and Benoit Huet[6]

[1] CLS, Radboud University, Nijmegen, Netherlands
m.eskevich@let.ru.nl
[2] TU Delft, Delft, Netherlands
[3] University of Twente, Enschede, Netherlands
[4] TU Vienna, Vienna, Austria
[5] ADAPT Centre, School of Computing, Dublin City University, Dublin, Ireland
[6] EURECOM, Sophia Antipolis, France

Abstract. Video-to-video linking systems allow users to explore and exploit the content of a large-scale multimedia collection interactively and without the need to formulate specific queries. We present a short introduction to video-to-video linking (also called 'video hyperlinking'), and describe the latest edition of the Video Hyperlinking (LNK) task at TRECVid 2016. The emphasis of the LNK task in 2016 is on multi-modality as used by videomakers to communicate their intended message. Crowdsourcing makes three critical contributions to the LNK task. First, it allows us to verify the multimodal nature of the anchors (queries) used in the task. Second, it enables us to evaluate the performance of video-to-video linking systems at large scale. Third, it gives us insights into how people understand the relevance relationship between two linked video segments. These insights are valuable since the relationship between video segments can manifest itself at different levels of abstraction.

Keywords: Crowdsourcing · Video-to-video linking · Link evaluation · Verbal-visual information

1 Introduction

Conventional multimedia information retrieval (MIR) research focuses on addressing an information finding scenario that starts with an ad hoc query, i.e., a query that is freely formulated by the user. The MIR system (or multimedia search engine) then returns results that are potentially relevant to the *information need* underlying the user's query, i.e., what the user had in mind to find. In this scenario the users have complete freedom in what they search for, since the MIR system is able to respond to any query. However, it has a significant disadvantage, if the users do not already have a clear goal in mind and are able to describe it, along with a reasonable understanding of the contents of the collection they are searching, then the 'freedom' to formulate their own query

© Springer International Publishing AG 2017
L. Amsaleg et al. (Eds.): MMM 2017, Part II, LNCS 10133, pp. 280–292, 2017.
DOI: 10.1007/978-3-319-51814-5_24

is of little value. Effectively, the users run the risk of missing content that they might find relevant had they been shown it.

Video-to-video linking (also referred to as 'video hyperlinking') is a MIR scenario that has the potential to address the limitations of search based on ad hoc user queries. In the video-to-video setting, the users first enter a video collection by identifying an initially interesting video or video segment, but from there they can interactively explore the content in the collection by using video-to-video links to navigate between segments. Manual creation of all such links would be prohibitively time consuming, and thus automated methods of video-to-video linking are required. Such systems take an input video segment, which can be seen as analogous to a query, and which we refer to as an *anchor*, and then return a list of video segments, which we refer to as *targets*, to which it might be linked. A user exploring the collection will thus use a link to navigate from an anchor to a target.

A key challenge of video-to-video linking is to appropriately understand and model the relevance relationship between an anchor and its targets. The objective of a conventional MIR system is to seek items relevant to the information need behind the user query. Without a formal query, a video-to-video linking system lacks a well-defined information need. Further, without a way of characterizing the relevance relationship between anchors and targets, it is difficult to design video-to-video linking algorithms, and it is impossible to properly evaluate them.

To advance research in the area of video-to-video linking, we thus need to better understand this relevance relationship between anchors and targets that are automatically identified by a linking system. To address this issue in our work we adopt crowdsourcing methods. Consulting the crowd enables us to gather information from a large and diverse group representative of potential system users. This can also provide provide us with more insights into how users perceive video-to-video relevance. This paper describes our use of crowdsourcing in designing and evaluating the TRECVid 2016 Video Hyperlinking task [1], and the insights we gained on user interpretations of anchors.

The rest of the paper is structured as follows: in Sect. 2 we discuss the history of the video-to-video linking task. We then introduce an innovative new version of the task that focuses specifically on *verbal-visual information*: segments of video in which the videomaker is evidently exploiting, simultaneously, both the visual and the audio channel in order to communicate a message. In Sect. 3, we outline the specifics of the TRECVid 2016 Video Hyperlinking task, which is based on the concept of verbal-visual information. Section 4 explains our multistage crowdsourcing process for evaluating multimodal video-to-video linking systems. This combines an Anchor Verification stage, with evaluation of automatically identified targets. Then in Sect. 5, we discuss the results of the Anchor Verification carried out on Mechanical Turk (MTurk)[1]. In addition to verifying anchors, this task yielded insights about how people interpret video segments. Finally, in Sect. 6, we conclude and outline future work directions.

---

[1] www.mturk.com

2   Video-to-Video Linking

As identified in the Introduction, a key challenge in video-to-video linking is understanding and modeling the relevance relationship between two video segments in the absence of an existing information need formulated independently in the mind of the user. Rather we are seeking to determine whether to create a link on the basis that there might be such an information need in the future, when the user follows the link, although the user may follow the link for a more unfocused exploratory reason.

As a starting point we can observe that for two given video segments, it is possible to define an arguably infinite number of ways in which they can be related. For example, imagine two video segments of people sitting at a table: one from a video of a séance and the other from a talk show. The presence of the table in both segments allows us to assert that the segments are related, but this relationship is not necessarily interesting to users. In this case, the table is essential to the subject of neither video (both a séance and a talk show can happen without a table), nor does it serve to differentiate these segments from other segments. It is certainly possible that a user with great interest in tables would like to navigate from one table to another within a video collection. However, in focusing on providing paths through the collection for this particular user, we risk inadvertently inundating other users with too many possible relevance paths. In other words, rather then assisting them in locating information of interest in another video, we overload them with options to pursue.

It is important to note that comparing videos on the basis of detailed or exhaustive descriptions does not completely solve the problem. Continuing with our example of the video segments of the séance and the talk show: A detailed description would include information such as the names of the people talking, the transcript of what they are saying, a description of their clothing and appearance, and the details of the whole scene including the colour of the table, objects on it, and the background. If video segments are compared at this level of specificity, then we face a high chance that there are no related segments within the entire video collection. Although highly specific comparisons may exclude relationships between video segments exclusively based on the occurrence of incidental elements such as the table, they will also close off paths. The paths closed will be those that involve 'looser', somewhat unexpected relationships between anchor segments and target segments. It is exactly these links that connect video segments that are 'neither too loosely nor too tightly related' that will allow users to explore, and exploit, a video collection, without prior knowledge of what they will find in this collection.

2.1   Past Perspectives on Video-to-Video Linking

After work on automatically creating links in text [9,10] and automatically linking video anchors to text targets [8] and vice versa [2], at MediaEval 2012 we introduced the 'Search and Hyperlinking' (S&H) task [4]. The 'Search' part of the task was devoted to finding anchors in a collection of interest to users, which then provided entry points for the 'Hyperlinking' part. The collection used was a large collection of semi-professional videos from blip.tv which represented a series of channels or 'shows' containing a series of episodes, adopting the style and structure of typical TV broadcasting (discussed further in Sect. 3.1).

S&H 2012 addressed the problem of making the links between anchor and target video segments neither 'too loose' nor 'too tight', by focusing the task on video segments that were likely to have interest to relatively many people in the user population. The anchors for S&H 2012 were selected by crowd workers, using an MTurk Human Intelligence Task (HIT) entitled, 'Find interesting things people say in videos'. Here, 'things' refers to the video segments that were used as anchors in the task. Workers were asked to consider the situation of sharing the video with another person, in order to nudge them to consider 'interesting' with respect to what other viewers would want to have in a video, and not exclusively with respect to their own personal interests.

Subsequently, in 2013–2014 the S&H task moved to professional broadcast content, and in 2015 became the TRECVid Video Hyperlinking (LNK) task. Now, in 2016, we return to both the blip.tv dataset, and also to the question of how to create links that are neither 'too loose' nor 'too tight' and also fairly evaluate them[2].

2.2 Verbal-Visual Information in Video-to-Video Linking

The new version of the video-to-video linking task was inspired by our realization that the 2012 instructions 'Find interesting things people say in videos' missed an important issue. People create video, and people watch video, not just because of what is *said* in video, but also because of what is *seen*. In other words, the intersection of the verbal and the visual modalities is the place to start looking for what users will perceive to be *important* about a video in situations where no independent information need can be assumed. In turn, this importance will provide the basis for solid judgments of video-to-video relevance.

A seemingly trivial observation that we can make about the users who created the videos in the blip.tv collection is that they chose to make videos and not audio podcasts (audio only) or to create images or silent video (video only). This observation is more important than it initially appears since it supports the assumption that people who make video choose video above other media because of the opportunity to exploit *multimodality*, i.e., both the audio and the visual channels. Further, it suggests that messages that videomakers find important to communicate to their audiences, can be found by simultaneously considering both the audio and the visual channels. If the videomaker puts effort into communicating a message to viewers exploiting both multimedia channels, we expect that viewers will have reasonable agreement on what this message is. Their perspectives will not completely converge, but they will tend to have stable interpretations of the aspects of a video message that are most important when

---

[2] For all HITs details, see: https://github.com/meskevich/Crowdsourcing4Video2Vide oHyperlinking/

assessing the relevance relationship between two video segments. On the basis of this line of reasoning, in 2016, we decided to focus the LNK task on segments in the video in which viewers perceive that the *intent* of the videomaker was to leverage both the audio and video modalities to bring across a message to the audience. We refer to such segments as containing *verbal-visual information.*

Intent is a goal or purpose that motivates action or behaviour. In our context, we are interested in the goal the videomaker was trying to achieve in creating the video. The motivation for considering videomaker intent to be important derives from the investigation of uploader intent on YouTube that was carried out in [6]. Among the intent classes identified by the study are: 'convey knowledge', 'teach practice', and 'illustrate'. In [6], some initial evidence for a symmetry between the intent of users uploading video and the intent of users searching for (and viewing) video is uncovered. Our definition of verbal-visual information is agnostic to specific intent classes, such as 'convey knowledge', although addressing such classes might be of interest in future work. Here, our focus is on video segments in which the videomaker is presumably intentionally using both the verbal and visual channel. The driving assumption is that people viewing such segments will have relatively high consensus about what the videomaker is attempting to get across. It is not necessary, or even desirable, that there is complete consensus: rather our goal is to constraint possible relationships between the anchor and the target segments that viewers generally agree on, thereby supporting meaningful evaluation of video-to-video linking systems.

For concreteness, we return to the example of the two video segments above. By focusing on the presumed intent of the videomaker to communicate verbal-visual information, we have moved away from considering the table alone as the basis for linking, which would make the relationship 'too loose'. The reason is that the table is isolated in the visual channel, and is not part of the verbal-visual information that it was the videomaker's intent to bring across. At the same time, we do not insist on links based on similarity of the detailed descriptions of the two video segments, which would make the relationship 'too tight'. Instead, we focus on what is communicated by the videomaker through the interaction of words and visual content. In this way, the visual-verbal information of the video can take the place of the information need in the conventional MIR scenario. It will not be a unique source of relevance, but it will be universal enough to serve a large variety of users, and stable enough to be judged.

3   TRECVid 2016 Video Hyperlinking Task (LNK)

Next we overview the TRECVid 2016 Video Hyperlinking Task (LNK) [1].

3.1   Blip10000 Collection

For LNK at TRECVid 2016, we use the Blip10000 dataset which consists of 14,838 semi-professionally created videos [11]. In addition to the original Blip10000 collection, we used a new set automatic speech recognition (ASR) transcripts provided by LIMSI [7] which uses the latest version of their neural network acoustic models. Using these transcripts and also detected shot boundaries [5], we indexed the entire collection to facilitate the process of defining anchors.

3.2 Defining Verbal-Visual Anchors

The task required us to define a set of anchors with respect to which video-to-video linking systems will be evaluated and compared. Since the set could not include all verbal-visual video segments in the collection, we sought to identify a good-sized subset. Because it is impractical to find these verbal-visual segments by randomly jumping into the collection, we used a search heuristic. Specifically, we searched the collection for 'linguistic cues' in the ASR transcripts. We defined 'linguistic cues' as short phrases that people typically use to signal that something seen rather than said is important to their overall message. We compiled the list of cues by reflecting on what people say when they are showing something, and arrived at the following list: 'can see', 'seeing here', 'this looks','looks like', 'showing', and 'want to show'.

Two human assessors (multimedia researchers) who were familiar with the video collection defined anchor segments by searching the blip.tv collection for occurrences of the cues in the ASR transcripts. They checked a 5 min window around each cue occurrence for verbal-visual information. Cases in which the cue did not lead to video containing verbal-visual information (i.e., someone says 'I can see what you are saying', but nothing is actually being shown in the visual channel) were skipped. For the other cases, an anchor was defined, by making a reasonable choice of a shorter anchor segment (10–60 s) within the five-minute window. The assessors also skipped cases that were very similar to previously chosen anchors, in order to ensure anchors diversity.

For each defined anchor, assessors wrote a short description of the anchor. In developing this summary, assessors attempted to abstract away from literal description of the content of the video. For example, a description would be 'Videos of people explaining where and how they live.' rather than 'Videos of a young man walking through the hall of an apartment and then explaining the contents of the bathroom.' The assessors controlled the level of abstraction by leveraging their familiarity with the collection to write a description for which they found it likely that more than one video segment in the collection could potentially be considered relevant.

3.3 Development and Test Set Anchors: Audio vs. Verbal-Visual

This section provides more details of the anchor sets released for LNK. Two anchor sets were released: 28 development anchors and 94 test anchors. The development anchors were those originally created for use in S&H 2012 [4]. They were defined by a crowdsourcing task carried out on MTurk. The wording of the HIT that was used to collect the anchor was given above in Sect. 2.1. Recall that these anchors were focused on what people 'say' in videos. The Anchor Verification stage of our evaluation process (see Sect. 4.1) determined that a number of these indeed involved verbal-visual information, although they were not created explicitly to do so. The verification stage assured us of the appropriateness of the 2012 anchor set for development purposes in 2016.

4   Crowdsourcing Evaluation

Our evaluation takes the form of three stages (Anchor Verification, Target Vetting, Video-to-Video Relevance Analysis), each realized as a HIT on MTurk. We discuss each in turn. The principles informing the design of our HITs are: (1) provide descriptions that avoid technical terms, but rather allow workers to identify with context of use, and, (2) user quality control mechanisms that are fair and also an integral part of the HIT.

4.1   Stage 1: Anchor Verification

The Anchor Verification stage verifies the verbal-visual nature of the anchors defined by the assessors. Specifically, it checks whether the perceptions of viewers (i.e., potential users of the video-to-video linking system represented by workers on MTurk) align with those of the assessors. We related the HIT to a context of use with which we hope the crowd workers can identify by entitling it 'Watch the video segment and describe why would someone share it'. Further, we avoided reference to 'videomaker intent' or 'verbal-visual information' and instead provided the workers with the following instructions. *We know that people upload those videos for certain reasons. We ask you to think about what the person who made the video was trying to communicate to viewers during this short piece. In other words, what are viewers supposed to understand by watching this particular short piece of the video.* The HIT then asked them to provide a description of the anchor. We asked 'How would you describe the content you see to another person?' We collected a description from three different workers for each video. We then went through the descriptions that the workers provided, and checked whether they contain reference to an element present in the visual channel of the video. The anchors that were not verified to be of verbal-visual nature were not used in the subsequent crowdsourcing stages.

In designing the HIT, we also tried to guide the workers away from highly specific descriptions, since we are not interested in 'tight' relationships between anchors and targets, as discussed above in Sect. 3.2. We used two mechanisms to encourage workers to be more abstract. First, we included three example descriptions (two taken from the originally collected development anchor set and one from the test anchor set). These examples were intended to convey to the workers the diversity of the videos in the collection, and the level of abstraction of description that we were targeting. The examples included anchors bearing the descriptions 'We are looking for videos of people explaining where and how they live.' and 'This video is about features of a computer software application that supports communication.' Second, we stated, 'Please keep in mind that the this description that you write will be shown to someone else (working on Amazon Mechanical Turk), who will be asked to find other videos that are related to this video segment.' As discussed further in Sect. 5, the workers' answers gave us insight into how people abstract from literal descriptions of video clips, to more abstract descriptions.

4.2   Stage 2: Target Vetting

Target Vetting is the first step in assessing the output of video-to-video linking systems. The systems participating in the benchmark returned a list of targets, i.e., video segments, for each anchor in the test set. Target Vetting compares potential targets to the textual description of the anchors originally generated by the assessors. We use Target Vetting because the potential number of targets is very large. A crowdsourcing task to make a video-to-video judgment between anchor and target for every generated target would be prohibitively large. Further, the density of relevant anchor/target pairs could be low, making the task unfulfilling, tedious, and potentially impacting worker performance.

The output of the Target Vetting task is a set of binary decisions about the relevance relationship between the anchor description, and each target that each system has generated for that anchor. We do not ask workers to make the decision directly, since it is hard to enforce consistency in the criteria that they use to make the comparison. Rather we used multiple choice questions which are a good way of creating consistency, since the list of choices conveys information about the answer space and the types of distinctions that are relevant for answering the question. For this reason, the Target Vetting HIT asked workers to watch the target video segment, and then to choose among five anchor descriptions. If the target is relevant to the anchor, the worker should be able to pick the correct anchor description. We force the workers to choose between the five answers, and then ask them whether they are happy with the choice that they made, and whether it was easy to make the decision. If the target is not relevant to the anchor, the worker should find it difficult to pick a correct anchor description, and should express dissatisfaction or discomfort with the pick. Note that asking about the level of discomfort plays the same role as including a 'none of the above' option. The 'forced-choice' question that we use is, however, superior to a 'none of the above' option since the workers need to judge their own reaction (which is familiar territory) rather than contemplate whether or not there might exist an answer that is more fitting than any in the list (which is open ended). The four out of five descriptions are chosen randomly from the dataset, and sometimes could fit the target better than the ones of the original anchors. These cases might help to extend the ground truth for the other anchors, while the current anchor in question earns a non-relevance judgment.

4.3   Stage 3: Video-to-Video Relevance Analysis

Video-to-Video Relevance Analysis is the second step in assessing the output of video-to-video linking systems. Because targets that were not potentially relevant to anchors were eliminated in the previous stage, the crowdsourcing task at this stage can be smaller, and its output can be manually analyzed. This stage consists of a HIT that presents workers with two video segments, the anchor and the corresponding target, and asks them to describe the way in which the two are related. We do not provide choices or suggestions about how the anchor and target could be related (i.e., we do not suggest 'person'/'object'/'speech'), or otherwise constrain answers. Instead, we collected unstructured information in the form of 2–3 natural language sentences. The first stage, Anchor Verification, checked that we were focused on verbal-visual information. We believe that the Video-to-Video Relevance Analysis stage gives us comprehensive qualitative information about the types of relevance between video segments that people perceive. We hope to find relevance aspects that we could not have envisaged beforehand, and also to better understand diversity in relevance relationships, in particular, with respect to the different levels of abstraction at which relevance relationships are perceived.

5   Insights from the Crowd

In this section, we dive more deeply into the insights that are gained from Stage 1: Anchor Verification, described in Sect. 4.1. We ran the HIT on MTurk, with a 0.15 USD reward and restricted to workers with overall 90% HIT Approval Rate. We ran the HIT on 119 anchors (we reduced the original set of 28 development and 94 test sets by taking three anchors as examples) and collected 357 worker judgments. The quality of submitted work confirmed that our HIT design was clear and easy to grasp. There were only two cases in which the submissions had to be rejected for unserious work. A number of workers left comments with thanks, and one reported having enjoyed the task. For each anchor, we compared the anchor description originally provided by the assessors with the workers descriptions both individually, and as a set of three (since three responses were collected per anchor). In total, 51 unique workers carried out the task, with an average of seven judgments per worker, with a maximum of 59 HITs in case of one worker. The rest of this section reports our findings.

5.1   Verbal-Visual Information

The primary purpose of Stage 1: Anchor Verification was to eliminate anchors that were not truly verbal-visual in nature from the test set. Table 1 provides an example of a case where we judged that none of the three workers' descriptions mentioned a visual element to the anchor. As mentioned in Sect. 4.1, such anchors were dropped. Three of the original anchor test set defined by the annotators were affected, leaving us with a final test set of 90 anchors.

We also used Stage 1 to verify whether any of the anchors in the development set could be considered to contain verbal-visual information, although they were

Table 1. Example of an anchor that failed crowd Anchor Verification: it did not contain the visual component necessary to be verbal-visual information, and was dropped.

Anchor ID	Original assessor description	Description by MTurk worker
89	We are looking for videos of a modern, enthusiastic preacher, pastor or spiritual leader	In this video a speaker is talking about how lack of movement and lack of socialization in a social media and computer generation is causing obesity
		A young man is trying to convey to his audience that too much time spent on the computer is leading to bad health and that people need to spend more time meeting each other face to face rather than with social media online
		The video explains how the overuse of social networking sites cause health problems. This is a very important point to note

created by asking about spoken content. It turned out to be the case that some development anchors were indeed verbal-visual. An example of such a case is Anchor ID 25, whose description is 'The launch of a new villain for the Avengers to fight with.' The worker descriptions suggest the critical contribution of an image of Red Skull to the message conveyed by the anchor.

5.2   Observations on Abstraction

Stage 1: Anchor Verification also yielded some insights on the level of abstraction at which people describe video. Recall that with our HIT design, which included examples, we tried to encourage the workers to provide us with abstract anchor descriptions. In general, however, the workers descriptions were much more specific than the descriptions provided by the assessors who originally defined the anchors. Table 2 illustrates such cases. In a given collection, it would be difficult to find two video segments that both fit this description, unless the collection had a special nature or the video segments were near duplicates. We did, however, find cases in which the workers descriptions reached the same level of generality or abstractness as the original descriptions. Table 3 illustrates two examples.

Finally we would like to mention that when analyzing the worker description, we discovered that workers make reference to what we call the *conventional purpose* of the video. We define conventional purpose categories as the categories into which people conceptualize videos. Any category that is widely acknowledged, i.e., conventionally accepted, can be considered. However, we point out the our use of the word *purpose* is motivated by the fact that these categories involve typical situations in which videos are watched and reasons why people watch them. For example, some workers made comments about the genre (that it

Table 2. Examples of anchors for which workers provided detailed descriptions of the literal video content.

Anchor ID	Original assessor description	Description by MTurk worker
106	We are looking for videos that show a collection of people's opinions on the street	This video shows that there is a question as to whether there is a youth movement for voting. There are quick interviews with people in their late teens or early 20s on the topic
74	We are looking for videos of someone fumbling or dropping something by accident	The man on the left is clearly someone of higher status who is ordering his butler (or similar position) to give him his phone. The butler (man on the right) points out that the phone is in the man's hand, and refuses to pick it up when the man clumsily drops it on the floor

Table 3. Examples of anchors where the level of abstraction is matched between the original description and the description provided by the worker.

Anchor ID	Original assessor description	Description by MTurk worker
77	We are looking for videos that give ideas of different kinds of purses and handbags	This video goes over some of the different variations in women's handbags
107	We are looking for videos with tips about using social networking sites such as Facebook	Video about posting to social media. Showing how to physically post information

was a news report or a comedy) and others pointed to other categories reflecting what the videos are used for (that it is for selling something). We observed them mentioning in the description that the video was part of a lecture, commercial, or blooper reel, and that the video was a performance. These references to what the video is used for are also a form of abstraction. We believe that ultimately these sorts of purposes are related to the intent categories mentioned in Sect. 2.2, and are eager to explore user use of these categories further.

6　Summary and Outlook

This paper provided a brief review of the history of video-to-video linking, and introduced a new multimodal video-to-video linking scenario. This new scenario exploits the assumption that users will have more stable judgments about the relevance relationship between anchors and targets, if we focus on cases in which the videomaker is intentionally using both audio and visual information stream to convey a message. We introduced a crowdsourcing strategy for evaluating systems that address this scenario, and also presented findings from the first stage of this evaluation 'Anchor Verification'.

Our next step is carrying out Stage 2 and Stage 3 of the evaluation of the 2016 benchmark, once the participants have submitted their system output. We point out that we anticipate on the basis of general human behaviour that we will find a mismatch between generation and validation: we expect that although workers tend to generate descriptions at a literal level, that they will not insist on this level when they are choosing a description that fits a video segment (i.e., in Stage 2: Target Vetting). Further, we expect to discover that workers will also more naturally move to higher levels of abstraction when comparing anchors and targets (i.e., in Stage 3: Video-to-Video Relevance Analysis).

Additionally, we would like to pursue further insights already revealed by Stage 1: Anchor Verification. We are particularly interested in the relationship between the level of abstractness at which people describe video segments, and their knowledge of the collection from which the video segments are drawn. Here we have seen that the assessors who are familiar with the collection describe videos with a higher level of abstraction. If collection familiarity impacts that level of abstraction that is most appropriate for the video-to-video relevance relationship in the linking scenario, the implications for systems that attempt to hyperlink large collections could be profound.

Acknowledgments. This work has been partially supported by: ESF Research Networking Programme ELIAS (Serwah Sabetghadam, Maria Eskevich); the EU FP7 CrowdRec project (610594); BpiFrance within the NexGenTV project, grant no. F1504054U; Science Foundation Ireland (SFI) as a part of the ADAPT Centre at DCU (13/RC/2106); EC FP7 project FP7-ICT 269980 (AXES); Dutch National Research Programme COMMIT/.

References

1. Awad, G., Fiscus, J., Michel, M., Joy, D., Kraaij, W., Smeaton, A.F., Quénot, G., Eskevich, M., Aly, R., Jones, G.J.F., Ordelman, R., Huet, B., Larson, M.: TRECVID 2016: Evaluating video search, video event detection, localization, and hyperlinking. In: Proceedings of TRECVID 2016, NIST, USA (2016)
2. Bron, M., Huurnink, B., Rijke, M.: Linking archives using document enrichment and term selection. In: Gradmann, S., Borri, F., Meghini, C., Schuldt, H. (eds.) TPDL 2011. LNCS, vol. 6966, pp. 360–371. Springer, Heidelberg (2011). doi:10. 1007/978-3-642-24469-8_37
3. Eskevich, M., Jones, G.J.F., Larson, M., Ordelman, R.: Creating a data collection for evaluating rich speech retrieval. In: Eighth International Conference on Language Resources and Evaluation (LREC), Istanbul, Turkey, pp. 1736–1743 (2012)
4. Eskevich, M., Jones, G.J.F., Chen, S., Aly, R., Ordelman, R.J.F., Larson, M.: Search and hyperlinking task at mediaeval 2012. In: MediaEval CEUR Workshop Proceedings, vol. 927, CEUR-WS.org (2012)

5. Kelm, P., Schmiedeke, S., Sikora, T.: Feature-based video key frame extraction for low quality video sequences. In: 10th Workshop on Image Analysis for Multimedia Interactive Services (2009)
6. Kofler, C., Larson, M., Hanjalic, A.: User intent in multimedia search: a survey of the state of the art and future challenges. ACM Comput. Surv. **49**(2), 1–37 (2016)
7. Lamel, L.: Multilingual speech processing activities in Quaero: application to multimedia search in unstructured data. In: The Fifth International Conference Human Language Technologies - The Baltic Perspective Tartu, Estonia, 4–5 October 2012
8. Larson, M., Newman, E., Jones, G.J.F.: Overview of videoCLEF 2009: new perspectives on speech-based multimedia content enrichment. In: Proceedings of the 10th International Conference on Cross-language Evaluation Forum: Multimedia Experiments (CLEF 2009), Corfu, Greece, pp. 354–368 (2009)
9. Mihalcea, R., Csomai, A.: Wikify!: Linking documents to encyclopedic knowledge. In: Proceedings of the Sixteenth ACM Conference on Conference on Information and Knowledge Management (CIKM 2007), Lisbon, Portugal, pp. 233–242 (2007)
10. Milne, D., Witten, I.H.: Learning to link with Wikipedia. In: Proceedings of the 17th ACM Conference on Information and Knowledge Management (CIKM 2008), Napa Valley, California, USA, pp. 509–518 (2008)
11. Schmiedeke, S., Xu, P., Ferrané, I., Eskevich, M., Kofler, C., Larson, M., Estève, Y., Lamel, L., Jones, G.J.F., Sikora, T.: Blip10000: a social video dataset containing SPUG content for tagging and retrieval. In: Dataset Track. ACM Multimedia Systems, Oslo, Norway (2013)

Online User Modeling for Interactive Streaming Image Classification

Jiagao Hu, Zhengxing Sun$^{(\boxtimes)}$, Bo Li, Kewei Yang, and Dongyang Li

State Key Laboratory for Novel Software Technology, Nanjing University, Nanjing,
People's Republic of China
szx@nju.edu.cn

Abstract. Regarding of the explosive growth of personal images, this paper proposes an online user modeling method for the categorization of the streaming images. In the proposed framework, user interaction is brought in after an automatic classification by the learned classifier, and several strategies have been used for online user modeling. Firstly, to cover diverse personalized taxonomy, we describe images from multiple views. Secondly, to train the classifier gradually, we use an incremental variant of the nearest class mean classifier and update the class means incrementally. Finally, to learn diverse interests of different users, we propose an online learning strategy to learn weights of different feature views. Using the proposed method, user can categorize streaming images flexibly and freely without any pre-labeled images or pre-trained classifiers. And with the classification going on, the efficiency will keep increasing which could ease user's interaction burden significantly. The experimental results and a user study demonstrated the effectiveness of our approach.

Keywords: Image classification · Online user modeling · Streaming images · Nearest class mean classifier · Online metric learning

1 Introduction

Nowadays, almost everybody has a portable photographic device such as smartphones and digital cameras. With the portable camera, we can capture images anytime and anywhere. As a result, new images will flow into our gallery one by one like a stream. Imagine the categorization of this type of streaming images during your daily taking pictures. As you have taken some photos, you can categorize them according to your preferred taxonomy by a few simple drag-and-drop operations and learn an initial classification model. As you take some pictures next time, new images begin to be automatically categorized into categories following your criteria. If the inferred labels are incorrect, you can interactively both correct the labels and improve the learned model. When you take pictures which can be described by none of existing categories, the model will be updated incrementally with new categories to cover them. As you continue taking and

© Springer International Publishing AG 2017
L. Amsaleg et al. (Eds.): MMM 2017, Part II, LNCS 10133, pp. 293–305, 2017.
DOI: 10.1007/978-3-319-51814-5_25

categorizing images, the learned model is refined with these new samples, resulting more and more accurate inferred labels for following images. At the end, you can categorize your huge growing image collection freely. Although simple and intuitive from the user perspective, the user modeling in this type of image categorization scenario remains an open problem.

From our knowledge, there should be at least three requirements for user modeling in this classification scenario. **Firstly, images representation must include multiple views to cover diverse user demands**. Research on image representation have made a lot of great achievements, from the famous BoW methods [7,11] to the impressive deep Convolutional Neural Networks (CNNs) [8,19]. All these methods are describing images from a single viewpoint. But in fact, different people may focus on different points in image categorization. Thus, the problem is *how to combine multiple clues to meet the diverse demands for a user centered classification tool*. **Secondly, the user's classification model must be trained/updated incrementally real-timely**. To ease user's burden on interaction, a customized classification model should be trained. And as new images will flow in anytime, the model must be updated incrementally over time. For image classification, SVM is the most famous model [13,17]. Although there are many incremental extension of SVM [10], they suffer from multiple drawbacks, for example, the extremely expensive update. To ensure the real-time feedback, it should be considered that *how to incrementally update user's classification model in a straightforward way in real time*. **Thirdly, the user's categorization interests must be learned online**. In real-life scenario, user's classification preference cannot be one time setted at the beginning, because a user may concern different along with the increasing of new images. Online learning of user interests has been widely studied in the field of information retrieval by using relevance feedback [16]. But to the best of our knowledge, there is no such research in image classification. Thus, *how to online model user interests in the streaming image classification* still remains unopened.

This article aims to address these issues by an online framework. It incorporate user intervention with online user modeling to classify streaming images. To cover diverse personalized taxonomy, we propose to describe images based on the visual appearance view, the semantic view, and a deep view. The user's preference is modeled based on a variant of the Nearest Class Mean classifier (NCM), i.e., a set of class centroids coupled with a weighted distance measurement. Images are classified as the same class with the nearest centroid measured by the weighted metric. To update user's taxonomy, we design an incremental strategy for the class centroids. And to learn the categorization interests of a particular user, we propose an online learning strategy for the weighted distance metric. Using our framework, people can classify images and train classifiers progressively along with their manipulation and image gathering with simple drag-and-drop operations. The advantages of the our method are threefold:

(1) The streaming images can be classified efficiently without pre-training or pre-labeling, and the newly captured images can be categorized accumulatively.

(2) The classification model can be learned online to handle the dynamic categories, and the more use the more accurate the classification could be.
(3) The uncertainties of images can be solved with user intervention in, and the diversity of taxonomies can be met flexibly in accord with user's preferences.

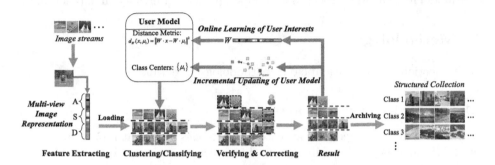

Fig. 1. Overview of proposed method

2   Overview

To classify streaming images with user in the loop, our method incorporate user intervention with online user modeling as shown in Fig. 1. The proposed framework categories images round by round with the following steps.

Firstly, the system will extract image features in multiple views, and then load images in turn for classification. Secondly, the loaded images will be classified into several groups by the user classification model. Thirdly, the result will be shown to user for verifying and correcting. Finally after the correction, images in each classes will be used to update the user model incrementally, and online learn user's interests. In addition, all the loaded images are classified correctly according to the user, so they will be archived to the categorized image collection. A new round of this classification process will start if there are enough images or the user want to. Figure 2 shows an example of the whole classification workflow.

As new images could flow in anytime, the system will put them in a pending pool and pre-extract their feature vectors. To cover diverse user intentions, we proposed to extract features from the visual appearance view, the semantic view, and a deep view (Sect. 3.1). When loading images, it will control the number of loaded images since human users can hardly handle a large number of images at one time. After images loaded, the system will group them into several groups. If the user model has not been initialized, images will be clustered into groups by AP clustering [6]. Otherwise, the user model will be used to classify images into existing categories.

Once images grouped, the user will be asked to verify if they are classified correctly. User can correct misclassified images by simple drag-and-drop operations. There are two cases in which the user need to operate. If (I) an image of a particular category is grouped into another category, the user only need to drag it to the correct category. If (II) an image of a new category which does not exist is grouped into an existing category, the user can drag it to the blank space which indicates the construction a new category, and it will be contained in that category automatically. After that, all these images will be used to update the user model incrementally (Sect. 3.2), and learn user's interests online (Sect. 3.3).

3   Methodology

In this section, we will describe our methodology in details.

3.1   Multi-view Image Representation

In practice, different people may categorize images from different views, such as the visual appearance of images or the objects in images. To cover different user's categorization preferences, we propose to extract image descriptor from multiple views. In this paper, descriptors from the visual appearance view and the semantic view are extracted. Besides the appearance and semantic, user could consider more when categorizing images. Considering the impressive performance of deep learning in recent years, we also extract a descriptor from the deep view. After extracting these descriptors, we then combine them linearly to get the final multi-view representation.

Appearance View. The visual appearance attributes, such as the texture, color, shape, etc. form user's first impression of an image. Thus, describing the appearance of the whole image is helpful for the personalized categorization. The most general way to describe the appearance is to gather the statistics of particular visual characters, such as the famous BoW methods [7,11]. In recent years, there are some higher-level methods [1,18] which use classifiers to get the visual attribute of images. As these methods utilize classifiers instead of statistical histogram, their descriptors are typically more compact.

To describe images from appearance view, we adopt the PiCoDes [1] which pre-trained 2K classifiers to extract the visual attributes in this paper. For a given image, we can get a 2K-dimensional vector describing its appearance.

Semantic View. The semantic attribute describes the semantic content of an image by perceiving the objects it contains. Since most people could classify an image mainly based on the objects it contains, the semantic attribute is also important for the personalized categorization. In this paper, we pre-define a list of objects as Torresani et al. [20] does, and train a set of corresponding binary classifiers to extract the semantic attribute of images. To get more accurate judgment of the existence of objects by simple classifiers, we propose to classify subimages instead of the whole image. Obviously, the classifiers can get more accurate judgment for subimages than the entire image. There are a lot of research about generating region proposals which can be regard as subimages here. Here, we use BING [3] as it can extract accurate region proposals very fast.

Firstly, a set of subimages are generated. Then the object classifiers are used to classify the subimages. Finally, a maximum pooling is applied to get the semantic attribute descriptor with the i-th element indicate the probability that given image contains the i-th objects.

Deep View. In fact, people could consider more which can hardly be summarized besides the appearance and semantic when categorizing images. We think the deep learning-based features can cover some of these unnamable user preference considering that deep learning is a block box. In this paper, we choose the well-known AlexNet architecture from [8], which is a CNN trained on the 1.3-million-image ILSVRC 2012 ImageNet dataset. Specifically, we adopt the already-trained AlexNet provided by the Caffe software package, and use the output from the fc7 layer as the feature descriptor of deep view.

3.2 Incremental Updating of User Model

When categorizing images by individual, the taxonomy cannot be known in advance, and the categories will be change over time. Thus the classification must be modeled online and updated incrementally to be conformed to the user intention. In this paper, we proposed to model the user's taxonomy by an incremental variant of the Nearest Class Mean classifier (NCM) since it have shown promising performance in image classification tasks [14,15].

With each image being represented by a d-dimensional feature vector $x \in \Re^{d \times 1}$, the standard NCM maintains a set of class centroids $\{\mu_c \in \Re^{d \times 1} | c \in [1, C]\}$ with each vector represents a class. Then an image is classified to the class c^* with the closest mean:

$$c^* = \arg\min_{c \in \{1,\dots,C\}} d(x, \mu_c),$$

$$\mu_c = \frac{1}{N_c} \sum_{i:y_i=c} x_i,$$

(1)

where $d(x, \mu_c)$ measure the distance between an image x and the class mean μ_c, y_i is the label of image x_i, and N_c is the number of images in class c. In previous research, class centroids are learned offline with plenty of training images.

As the NCM classifies images merely by the class centroids, the classifier can be updated easily by reforming the class mean vectors. In each classification round, the number of existing images in each class $\{n_c | c \in [1, C]\}$ is recorded and will be updated. Then the class mean vectors can be updated incrementally as follows:

$$\mu_c = \frac{1}{n_c + \sum_i [\![y_i=c]\!]} (n_c \mu_c + \sum_{i:y_i=c} x_i),$$

$$n_c = n_c + \sum_i [\![y_i = c]\!],$$

(2)

where $\sum_i [\![y_i = c]\!]$ computes the number of images belongs to class c in current round. If there are images belongs to a new class, we can add a new class centroid $\{\mu_{new}, n_{new}\}$ and initialize it as $\mu_{new} = \overrightarrow{0}$ and $n_{new} = 0$, then update it with Eq. 2. With the incremental updated class centroids, the NCM can classify images of that new class subsequently.

3.3 Online Learning of User Interests

Different people may concern different focuses when classifying images, it can be reflected as different combination weights of different feature views. Thus, a uniform distance measurement is not an optimal choice to predict the categories of images by NCM. In this paper, we propose to formulate the distance between two images as a weighted Euclidean distance, i.e., $d_W (x_p, x_q) = ||W \cdot x_p - W \cdot x_q||^2$, where $x_p(x_q)$ is the feature vector of the p-th(q-th) image, $W \in \Re^{d \times 1}$ is the weight vector with each element indicates the relative importance of each dimension in feature vectors for distance measurement. Intuitively, it is hard to set the weights online during user's categorization. In this paper, we try to transform this problem to metric learning problem [9], and use online strategy [2] to learn W.

In fact, we can regard W as a diagonal matrix, then the weighted distance can be rewritten as follows:

$$d_W (x_p, x_q) = ||diag\,(W)\,x_p - diag\,(W)\,x_q||^2 = (x_p - x_q)^T D^T D(x_p - x_q), \quad (3)$$

where $D = diag(W)$. Clearly, this is a standard Mahalanobis distance metric. Following the online similarity metric formulation in OASIS [2], we can learn W online.

After each user verifying and correcting, all the classified images in the current round can be used as training data. We sample the training instances as a set of triplets, i.e., $\{(x, x^+, x^-)\}$ where x and x^+ belong to a same class and x^- belongs to a different class. Then the goal is to learn the distance function such that all triplets satisfies the following constraints:

$$d_W (x, x^+) \leq d_W (x, x^-) - \lambda, \quad (4)$$

where $\lambda > 0$ is a safety margin to ensure a sufficiently large distance difference. Then a hinge loss function can be defined as follows:

$$\mathcal{L}_W (x, x^+, x^-) = \max\{0, d_W (x, x^+) - d_W (x, x^-) + \lambda\}. \quad (5)$$

In order to minimize the accumulated loss, we apply the Passive-Aggressive algorithm [4] to optimize W iteratively over triplets:

$$W^{(t)} = \arg\min_W \frac{1}{2}\left\|W - W^{(t-1)}\right\|_{Fro}^2 + \eta \mathcal{L}_W (x, x^+, x^-), \quad (6)$$

where $\|\cdot\|_{Fro}^2$ is the Frobenius norm, η is a parameter that controls a tradeoff between remaining close to the previous $W^{(t-1)}$ and minimizing the loss $\mathcal{L}_W (x, x^+, x^-)$ on the current triplet. W is initialized as $W^{(0)} = ones(d, 1)$.

When $\mathcal{L}_W(x, x^+, x^-) = 0$, it is clear that $W^{(t)} = W^{(t-1)}$ satisfies Eq. 6 directly. Otherwise, we take the derivative Eq. 6 and set it to 0, it is easy to derive that W can be online optimized by the following closed-form solution:

$$W^{(t)} = W^{(t-1)} - \eta \frac{\partial \mathcal{L}_W(x, x^+, x^-)}{\partial W}. \tag{7}$$

As W is a d-dimensional vector, we can compute the derivative of $\mathcal{L}_W$ with respect to W as follows:

$$\frac{\partial \mathcal{L}_W(x, x^+, x^-)}{\partial W} = 2W \cdot \left[(x - x^+) \cdot (x - x^+) - (x - x^-) \cdot (x - x^-) \right]. \tag{8}$$

4 Experiments

4.1 Datasets and Settings

The proposed method is evaluated on three challenging image data sets: UIUI-Sports event dataset [12] (1,579 sport images classified to 8 categories according to the event type), MIT Scene 15 Dataset [11] (4,485 images classified to 15 categories according to the scene type), Caltech 101 [5] (8,677 images classified to 101 categories according to the object type).

4.2 Classification of Streaming Images

Here we will demonstrate the workflow of the proposed streaming image classification. In Fig. 2, the left parts shows the data preparing module which limit the number of loaded images not greater than K (K = 10 here for demonstration purposes, in other experiments K = 20). The middle parts shows the initial result by the user classification model or initial clustering and the categorized collection. Images with yellow border indicate the initial result of loaded images, and the red crosses indicate the misclassified images which is determined by the user. The right parts shows the user's drag-and-drop operation on the misclassified images and the final result. The red borders indicate images corrected by the user, and the red arrows indicate user's dragging paths. For demonstration purposes, only a small part of images and processing are shown.

More specifically, Fig. 2(a) demonstrates the initialization of our tool from (i) to (iii). As the classification model has not been initialized, loaded images are clustered by AP cluster and the initial result is as in Fig. 2(a)(ii). Then user need to drag the misclassified images to correct categories as shown in Fig. 2(a)(iii). The final result will be used to initialize the classification model. Figure 2(b) from (i) to (iii) illustrate the process of adding new category: firstly, the loaded images are classified to three existing categories, then the user drags the misclassified images to the blank space to create a new category, and drag the rest into this new category. With the classification goes on, there will be more and more images classified according to the user's preference, and the classification model will be online train/update time and again. That makes the model fit the user's interests better and better. As in Fig. 2(c) from (i) to (iii) shows that with the categorization going on, there may be more categories and more accurate initial result which might even be all correct without any correction.

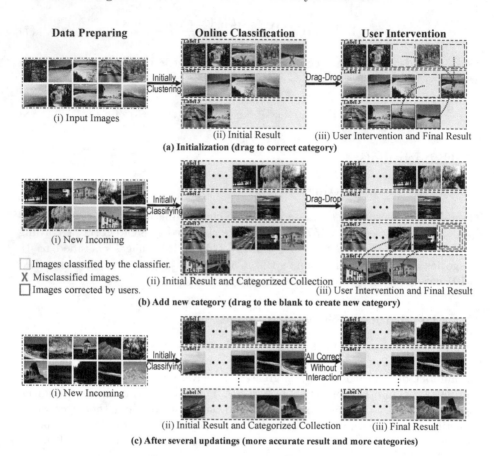

Fig. 2. Illustration of the streaming image classification workflow.

4.3 The Effectiveness of Online Learning

To confirm that the online learning is sure to converge to a high accuracy with the classification continue, and to verify the effectiveness of proposed online learned weighted distance, we carry out this experiment.

We randomly split the dataset into two parts, with 80% as pending set for gradually loading of images and 20% as validation set for testing the performance of the proposed classifier, and regard the groundtruth as user's labels. Images in the pending set are loaded successively to update the class centroids incrementally and online learn the weights, which simulates the image gathering in real-life scenario. After each 4% of all pending images loaded, the current classifier is used to classify the validation images and the accuracy is recorded. The

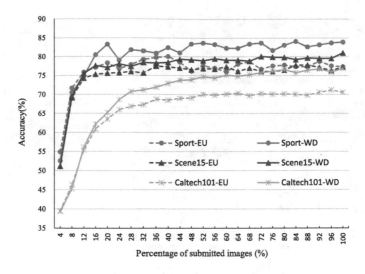

Fig. 3. Accuracies when submitting images gradually. Solid lines are results using weighted distance, and dashed lines are results using Euclidean distance.

solid lines in Fig. 3 illustrates the resulting accuracies of three dataset. It shows that the accuracies can increase with the accumulation of submitted images and will keep stable when reach a maximum. To sum up, we can believe that the online learning strategy of the classifier is effective.

In addition, to verify that the online learned weighted distance (WD) outperform the native Euclidean distance (EU), we run the above experiments using a standard Euclidean distance in the NCM classifier. The results are dashed lines as shown in Fig. 4. It can be seen that the curves of weighted distance lies above the curves of Euclidean distance remarkably. That is to say, the learned weighted distance can improve the classifier obviously. One important thing to note is that the accuracy is not as high as some state-of-the-art, this is because our online setting have restrict the performance of classifiers.

4.4 User Study

To testify that if proposed method can ease user's interactive burden and adapt to diverse classification criterions to meet various demands of real users, we performed a user study. We asked 12 participants to categorize 400 images selected from the original 8 scenes of the Scene 15 dataset. They had not used our system before, and started to group the images directly after our brief description about the function and usage. We did not tell them the component of the image collections, so they didn't know the scope and number of categories. During the classification, they could only browse the categorized images and the loaded images. The whole classification process need 20 rounds to classify all the images (20 images in each round). All the images were loaded in random order.

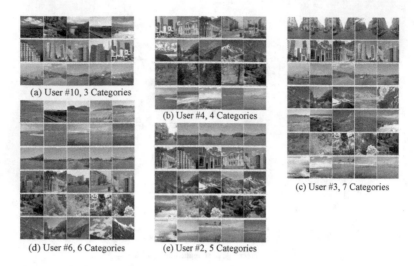

Fig. 4. The representative images of each category from different users.

Personalized Classification. After their classification, we find that the results are different from each other. The class number ranges from 3 to 7, it shows significant diverse classification criterions, and they are obviously different from the standard taxonomies defined in the groundtruth. For each class, we pick out 5 images nearest to its centroid as the representative of that class. Figure 4 shows representative images of each class from 5 users. It can be seen that the results are not very consistent. User #10's classification is the simplest which only contains 3 classes, and user #3's classification is the most complex which covers 7 classes. The consistent of all participants is that they all separate the nature scene from man-made scene. These just prove that proposed method can fit the diverse needs of different users, and they can classify the streaming images flexibly and freely.

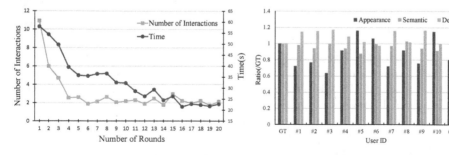

Fig. 5. The interaction number and time cost of each round in the user study.

Fig. 6. The relative ratios of weights of each view for each user to groundtruth.

Efficiency and User Burden. We record the number of interactions and time cost of each round for every user, and average them of all participants. Figure 5 shows the average curves. The average total number of interaction is about 59.2, and the average total time is about 656.3 seconds which is relatively less to classify 400 images one by one by hand. For comparison, we have tried to categorize all the 400 images manually without using our tool, and it takes about 1000 seconds. That is to say, the proposed method can save time and ease user's burden significantly for interactive classification. In addition, from Fig. 5 we can see that the number of interactions is becoming less and less as also the time cost. It demonstrates that with the accumulation of classified images, the system will become more and more accurate to categorize new images which can reduce user's burden greatly. As the participants said, our tool is getting "smarter and smarter" with the classification going on.

Diverse User Intentions. After the categorization, we get a set of weight vectors with each indicates the relative importance of each dimension in feature vectors for every user. We think these weight vectors can reflect user's preference in categorization. To uncover the diverse user intentions, we try to analyze these vectors. Firstly, we learn a weight vector using the groundtruth labels. Then, for each user (including the groundtruth), we compute the weights of each individual feature view by averaging the specific dimensions in the weight vectors corresponding to each view. Finally, we divide the view weights of each user by those of the groundtruth to calculate the relative ratios of each view. Figure 6 shows the result. We can see that the deep view fits better for most of the users compared to groundtruth. Besides the deep view, semantic fits better than appearance view for most participants. But for user #5, #6 and #10, the appearance view fits better, that means they prefer to classify images by appearance more, which agrees with the result in Fig. 4. It demonstrates that the proposed method are effective for diverse user intentions in real-world applications.

5   Conclusion

In this paper, we propose an online image classification framework incorporate online user modeling with user intervention to classify streaming images. To cover diverse personalized taxonomy, we propose to describe images in multiple views. The classification model is based on an incremental variant of the nearest class mean classifier which can be trained incrementally in real time. And to learn diverse interests of different users, an online learning strategy is used to learn weights of different feature views.

The features of our method lies in three aspects. Firstly, the streaming images can be categorized without pre-training or pre-labeling, and the newly captured images can be categorized accumulatively. Secondly, the classifier can be learned online, and the more use the more accurate the classification could be. Thirdly, the diversity of taxonomies can be met flexibly in accord with user's interests. Experimental results and a user study demonstrated the above advantages.

Acknowledgments. This work is supported by National High Technology Research and Development Program of China (No. 2007AA01Z334), National Natural Science Foundation of China (No. 61321491, 61272219), Innovation Fund of State Key Laboratory for Novel Software Technology (No. ZZKT2013A12, ZZKT2016A11), Program for New Century Excellent Talents in University of China (NCET-04-04605).

References

1. Bergamo, A., Torresani, L., Fitzgibbon, A.W.: Picodes: learning a compact code for novel-category recognition. In: NIPS, pp. 2088–2096 (2011)
2. Chechik, G., Sharma, V., et al.: Large scale online learning of image similarity through ranking. J. Mach. Learn. Res. **11**, 1109–1135 (2010)
3. Cheng, M., Zhang, Z., Lin, W., Torr, P.H.S.: BING: binarized normed gradients for objectness estimation at 300fps. In: CVPR, pp. 3286–3293. IEEE (2014)
4. Crammer, K., Dekel, O., Keshet, J., Shalev-Shwartz, S., Singer, Y.: Online passive-aggressive algorithms. J. Mach. Learn. Res. **7**, 551–585 (2006)
5. Fei-Fei, L., Fergus, R., Perona, P.: Learning generative visual models from few training examples: an incremental Bayesian approach tested on 101 object categories. Comput. Vis. Image Underst. **106**(1), 59–70 (2007)
6. Frey, B.J., Dueck, D.: Clustering by passing messages between data points. Science **315**(5814), 972–976 (2007)
7. Csurka, G., Dance, C.R., Fan, L., Willamowski, J., Bray, C.: Visual categorization with bags of keypoints. In: ECCV Workshop, pp. 1–22 (2004)
8. Krizhevsky, A., Sutskever, I., Hinton, G.E.: Imagenet classification with deep convolutional neural networks. In: NIPS, pp. 1106–1114 (2012)
9. Kulis, B.: Metric learning: a survey. Found. Trends Mach. Learn. **5**(4), 287–364 (2013)
10. Laskov, P., Gehl, C., Krueger, S., Müller, K.R.: Incremental support vector learning: analysis, implementation and applications. J. Mach. Learn. Res. **7**, 1909–1936 (2006)
11. Lazebnik, S., Schmid, C., Ponce, J.: Beyond bags of features: spatial pyramid matching for recognizing natural scene categories. In: CVPR, pp. 2169–2178. IEEE (2006)
12. Li, L.J., Fei-Fei, L.: What, where and who? Classifying events by scene and object recognition. In: ICCV, pp. 1–8. IEEE (2007)
13. Lin, Y., Lv, F., Zhu, S., et al.: Large-scale image classification: fast feature extraction and SVM training. In: CVPR 2011, pp. 1689–1696. IEEE (2011)
14. Mensink, T., Verbeek, J.J., Perronnin, F., Csurka, G.: Distance-based image classification: generalizing to new classes at near-zero cost. IEEE Trans. Pattern Anal. Mach. Intell. **35**(11), 2624–2637 (2013)
15. Ristin, M., Guillaumin, M., Gall, J., Gool, L.J.V.: Incremental learning of random forests for large-scale image classification. IEEE Trans. Pattern Anal. Mach. Intell. **38**(3), 490–503 (2016)
16. Rui, Y., Huang, T., Ortega, M., Mehrotra, S.: Relevance feedback: a power tool for interactive content-based image retrieval. IEEE Trans. Circ. Syst. Video Technol. **8**(5), 644–655 (1998)
17. Sanchez, J., Perronnin, F.: High-dimensional signature compression for large-scale image classification. In: CVPR 2011, pp. 1665–1672. IEEE (2011)

18. Su, Y., Jurie, F.: Learning compact visual attributes for large-scale image classification. In: Fusiello, A., Murino, V., Cucchiara, R. (eds.) ECCV 2012. LNCS, vol. 7585, pp. 51–60. Springer, Heidelberg (2012). doi:10.1007/978-3-642-33885-4_6
19. Szegedy, C., Liu, W., Jia, Y., et al.: Going deeper with convolutions. In: CVPR, pp. 1–9 (2015)
20. Torresani, L., Szummer, M., Fitzgibbon, A.: Efficient object category recognition using classemes. In: Daniilidis, K., Maragos, P., Paragios, N. (eds.) ECCV 2010. LNCS, vol. 6311, pp. 776–789. Springer, Heidelberg (2010). doi:10.1007/978-3-642-15549-9_56

Recognizing Emotions Based on Human Actions in Videos

Guolong Wang, Zheng Qin$^{(\boxtimes)}$, and Kaiping Xu

School of Software, Tsinghua University, Beijing, China
{wanggl6,xkpl3}@mails.tsinghua.edu.cn,
qingzh@tsinghua.edu.cn

Abstract. Systems for automatic analysis of videos are in high demands as videos are expanding rapidly on the Internet and understanding of the emotions carried by the videos (e.g. "anger", "happiness") are becoming a hot topic. While existing affective computing model mainly focusing on facial expression recognition, little attempts have been made to explore the relationship between emotion and human action. In this paper, we propose a comprehensive emotion classification framework based on spatio-temporal volumes built with human actions. To each action unit we get before, we use Dense-SIFT as descriptor and K-means to form histograms. Finally, the histograms are sent to the mRVM and recognizing the human emotion. The experiment results show that our method performs well on FABO dataset.

Keywords: Emotion · Action · Spatio-temporal volumes · mRVM

1 Introduction

With the development of the computational technology, people can access a variety of information resources. Video becomes increasingly important content in the database because of its intuitiveness and comprehensiveness. And much interest of researchers is shifted slowly from images to videos. How to understand and organize the video data effectively by user's preference is becoming a hot topic in the field of content-based analysis of video. The study on the video content is very useful to realize the individual multimedia service [1].

At present, many researchers are engaged in the study of video content analysis and classification. The common goal of video content analysis and classification research is human active content and affective content. Research on human active content has grown dramatically, corresponding to demands to construct various important applications including surveillance, human-computer interfaces. Most research focuses on human action recognition. Researchers are now graduating from recognizing simple human actions such as walking and running, and gradually moving towards recognition of complex realistic human activities involving multiple persons and objects. On the other hand, recently some research effort is invested on analyzing affective content. Since 1990 s, many scholars are engaged in the study of video affective content in order to identify emotion type in a video [2, 3]. Nowadays most researchers in cognitive sciences indicate the dynamics of facial and vocal behavior is crucial for their

© Springer International Publishing AG 2017
L. Amsaleg et al. (Eds.): MMM 2017, Part II, LNCS 10133, pp. 306–317, 2017.
DOI: 10.1007/978-3-319-51814-5_26

interpretation. However, sounds and faces are hard to be captured distinctly in many cases especially in videos for reconnaissance and surveillance [4, 5]. In those cases, body movements are the only effective evidence for emotion analysis of human [6]. We can conclude from the preliminary investigation that emotion analysis with human actions in the video is one of the key elements of video affective content.

In this paper, the definition of video affective content is improved with a method establishing the map between body actions and emotion types of human which is active in the video. Our method is divided into three parts: spatio-temporal volume modeling part, action units recognition part, emotion classification part.

2 Related Work

2.1 Emotion Theory

The progress of human emotion cognition has shaped the way emotions being treated in scientific terms. Researchers in psychological science assume that people have 'internal mechanisms for a small set of reactions (typically happiness, anger, sadness, fear, disgust, and interest) that, once triggered, can be measured in a clear and objective fashion'. As such, emotions like anger, sadness and fear are treated as entities that scientists can make discoveries about [7]. To this end, research in psychology has distinguished three major approaches for emotion model construction as followed: categorical approach, dimensional approach and appraisal-based approach [8, 9].

We choose categorical approach to model emotions based on distinct emotion classes. Among the various distinct classes are Ekman's basic emotions of happiness, sadness, fear, surprise, anger and disgust [10] and Plutchik's eight basic emotions of joy, sadness, anger, fear, surprise, disgust, anticipation, and acceptance [11]. A more recent theory is that of Hatice Gunes and Massimo Piccardi with an excellent dataset FABO. It suggests nine basic emotion categories of anger, surprise, fear, anxiety, happiness, disgust, boredom, sadness, uncertainty [12]. The distinct classes are each treated as a category. A category is defined as 'a class of things that are treated as equivalent'. To apply this theory, emotion is viewed as a concept that can be used as a representation for a category (e.g. happiness, surprise and fear). Hence, the categorical approach tends to assign text to a specific emotion category [7]. The assignment of text to a specific category can be achieved using learning based approaches or manually crafted hand rules [13].

2.2 Spatio-Temporal Volumes

The volume data structure mentioned earlier is to emphasize the temporal continuity in an input stream of a video data. The use of spatial-temporal volumes was first introduced in 1985 by Aldelson and Bergen [14], who built motion models based on "image intensity energy" and the impulse response to various filters. There are a number of widely deployed methods for analyzing the STV. A spatio-temporal segmentation method [15] has been proposed to extend segmentation from single image to video. In [16], Alper and Shah proposed a method for action recognition using spatio-temporal action volumes. Ryoo took advantage of spatio-temporal features to predict human activities efficiently [17].

As illustrated in Fig. 1, the STV defines a 3D volume space in a 3D coordinates system denoted by x, y and t (time) axes. In a more observant manner, a STV model is composed of a stack of video frames formed by array of pixels in the time order. To integrate the spatial (coordinates) and temporal (time) information in a single data structure, each smallest element inside of the STV "box" is called a voxel, which holds the pixel and the time information together [18]. The STV data structure enables the video event detection process to distinguish from a conventional frame-based mechanism such as optical flow becomes a real 3D analytical process. Through this transformation, dynamic information can be defined, extracted, and processed as global features rather than the most frame-based empirical local features. Conventional 2D image pattern recognition methods, shape analysis and matching algorithms are anticipated to be developed to adapting the 3D and volumetric natures of the video events. The significant advantage from using the STV model for categorical recognition is rooted in its distinctive ability to provide 3D geometric descriptions for dynamic video content features recorded in footage, which providing a theoretical foundation for event template matching.

Fig. 1. Definition of spatio-temporal volume

3   Proposed Approach

3.1   Part 1: Spatio-Temporal Volume Modeling

Spatio-temporal volume modeling is the first part for our representation of activity. It is used to find the frame which can segment a sequence of human video clips into multiple action units. This part includes two steps: (1) use Harris corner detector to find key points for 3D interpretation; (2) accumulate the trajectories at key points found in the first step to get a coordinate offset.

In the first step of this part, we use the modified Harris corner detector after comparing two eigenvalues of the Harris matrix. We defined $E(u, v)$ as a normal quantitative representation of gray change when one key point shifted from (x, y) to $(x+u, y+v)$.

$$E(u, v) = \sum\nolimits_{x,y} w(x, y)[I(x+u, y+v) - I(x, y)]^2 \qquad (1)$$

After developing a Taylor series expansion of Eq. (1) and ignoring the higher-order terms, the above equation can be changed into Eq. (2). Equation (3) is the Harris matrix.

$$E(u, v) = \sum\nolimits_{x,y} w(x, y)(u, v) \begin{pmatrix} I_x^2 & I_x I_y \\ I_x I_y & I_y^2 \end{pmatrix} \begin{pmatrix} u \\ v \end{pmatrix} \qquad (2)$$

$$M = \sum\nolimits_{x,y} w(x, y) \begin{pmatrix} I_x^2 & I_x I_y \\ I_x I_y & I_y^2 \end{pmatrix} \qquad (3)$$

Different from those traditional methods like computing response function, we directly get the eigenvalues e_1, e_2 of matrix in Eq. (3). Through locating pixels which satisfied the condition in Eq. (4), we coordinate key points $p_1, p_2, \cdots, p_n$ in each frame of action units.

$$L = \max(\min(e_1, e_2)) \qquad (4)$$

Instead of using non-maximum suppression which was used by many researchers [19], we also indicate a tolerance threshold T to ensure L being a global maximum value rather than local one. If p_i, p_{i+1} have been defined as two key points but the distance between p_i and p_{i+1} is less than threshold T, p_{i+1} is eliminated.

By coordinating key points, we can also calculating displacements of every pixel between two adjacent frames. We assume all the collected video clips have same frame rate, and the time between those two frames is Δt, which is always set to one.

The second step is segmenting video clips into short image sequences which we called action units. Video segmentation is not a new research topic. However, it was not until several years ago that segmentation was thought to be important for long video computing because most of the available datasets only consist of short clips containing single action. Some developed algorithms segmented videos by detecting the discontinuity of some video parameter such as intensity, RGB color, and motion vectors. In the work at recent years, velocity magnitude was proposed. All in all, some video phenomena can make the segmentation result a low accuracy rate and it's hard to tune the parameters for all videos. For example, in some sports videos, the parameters change a lot among different videos because players always move fast to a random direction.

Based on preliminary study of motion theory, we conclude that when an action comes to the end, the total shift of key points is short enough to be ignored. On observing different of movie clips, news clips and sports clips, the conclusion is supported by our experimental findings:

$$D_t = \sqrt[n]{\prod_{p(x_{i,t}, y_{i,t})}^{n} \left(\left(x_{i,t} - x_{i,t-1} \right)^2 + \left(y_{i,t} - y_{i,t-1} \right)^2 \right)} \tag{5}$$

where D_t represents the global motion shift at frame F_t, p_i is the ith key point found in frame F_t, $(x_{i,t}, y_{i,t})$ is the position of p_i in the frame. Figure 2 shows the example of video segmentation. For each action unit, the first frame and the last frame have short moving offset. The moving offset reaches the peak in the middle stage of the action unit.

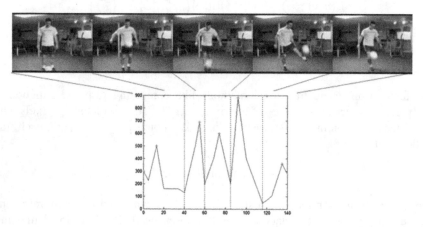

Fig. 2. Action unit segmented by moving offset. The image sequence on the first line contains five images which represent the ends of five actions. The time series on the second line shows moving offset over time. The vertical axis is the moving offset value computed according to Eq. (11) and the horizontal axis is the frame index. The red plus signs on the peak indicate the middle stage of action units, the light blue dots on the bottom mean the segmentation points (Color figure online)

3.2 Part 2: Action Unit Recognition

The spatio-temporal volumes are built in this part for recognizing action units detected in part one. Descriptors is generated which represent local motions occurring in image sequences. In every patch, a descriptor is computed by gradients statistics inside the patch.

After extracting features from spatio-temporal volume, we use 'visual words' which are clusters of 3-D features to represent types of action. K-means is used to form visual words from features extracted from training videos. We use Youtube Action Dataset in [20] as our training dataset. At last, each extracted feature vector in the videos belongs to one of the k visual words.

In modeling the situation, we make the assumption that the same pixels in adjacent frames differ only in position (and with no rotation).

We use a flat windowing function to compute descriptors, the kernels and the descriptors can be written as following equation. J_t is written as temporal transformation of gradient vector field.

$$k(z) = \frac{1}{\sigma_{win}} w(\frac{z}{m\sigma}) \tag{6}$$

$$des(t,i,j) = (k(x)k(y) * J_t)(T + m\sigma \begin{bmatrix} x_i \\ y_j \end{bmatrix}) \tag{7}$$

These modified Dense-SIFT features can reflect image characters and smooth the image according to the scale of key points. We model each action by constructing its histogram. An action histogram of a video is defined as a sequence of feature histograms:

$$H_i(V_l) = [His_i^1(V_l), His_i^2(V_l), \cdots, His_i^{|L|}(V_l)] \tag{8}$$

where |L| is the number of frames in the action unit V_l.and w_j denotes the jth visual word. The value of the jth histogram bin of histogram $His_i^k(V_l)$ written as follows. Figure 3 shows an example feature histogram.

$$His_i^k(V_l)[j] = |\{f|f \in w_j\}| \tag{9}$$

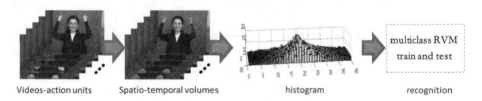

Videos-action units   Spatio-temporal volumes   histogram   recognition

Fig. 3. Flow chart of the proposed emotion recognition method

3.3 Part 3: Emotion Classification

To do emotion classification, our approach aims to establish the mapping between $His(V_l)$ and emotion vectors t. We use sparse Bayesian supervised learning method which is known as mRVM [21]. Assume a training set $H = \{H_1, H_2, \cdots, H_N\} \subset \mathbb{R}^D$ and the target value is $t = \{t_1, t_2, \cdots, t_N\}$, where X_n is a $D^{|L|}$- dimensional sample and $t_n \in \{1, \cdots, C\}, n = 1, \cdots, N$. The corresponding multinomial likelihood function is written as:

$$p(t = i|W, k_n) = \int p(t_n = i|y_n)p(y_n|W, k_n)dy_n \tag{10}$$

where w_i from W follows a zero-mean Gaussian distribution and k_n describes a similarity measure between human action histograms based on spatio-temporal volumes.

$$k_i = d(H_i - H_{i+1}) = \sum_{k=0}^{|L|} |His_i^k - His_{i+1}^k| \tag{11}$$

For each w_i, a hyper-parameter a_i is given and the maximization of the marginal likelihood is:

$$p(t|K, a) = \int p(t|K, W)p(W|a)dW \tag{12}$$

The logarithm counterpart can be expanded as:

$$L(a) = \sum_{c=1}^{C} -\frac{1}{2}[N \log 2\pi + \log|C| + t^T C^{-1} t] \tag{13}$$

where $C = I + Ka^{-1}K^T$ and the determinant and inverse of C can be written as:

$$|C| = |C_{-i}|(1 + a_i^{-1} \varphi_i^T C_{-i}^{-1} \varphi_i) \tag{14}$$

$$C^{-1} = C_{-i}^{-1} - \frac{(C_{-i}^{-1} \varphi_i \varphi_i^T C_{-i}^{-1})}{a_i + \varphi_i^T C_{-i}^{-1} \varphi_i} \tag{15}$$

Then the logarithm can be rewritten as:

$$L(a) = L(a_{-i}) + \sum_{c=1}^{C} \frac{1}{2}[\log a_i - \log(a_i + s_i) + \frac{q_i^2}{a_i + s_i}] \tag{16}$$

where s_i defines the measure of overlap between a sample k_i and the ones already in the model and q_i defines the measure of contribution to a specific emotion class of the sample. We can obtain a_i by derivation at the stagnation point:

$$a_i = \begin{cases} \frac{Cs_i^2}{\sum_{c=1}^{C} q_i + Cs_i^2} & \sum_{c=1}^{C} q_i^2 > s_i \\ \infty & \sum_{c=1}^{C} q_i^2 \le s_i \end{cases} \tag{17}$$

The training procedure involves consecutive updates of the model parameters. The hyper-parameters a, W are updated during each iteration, and the process is repeated until the preset number of iterations has been reached or hyper-parameter values are convergence. A detailed description of the training process of the mRVM classifier can be found in [20].

In the classification procedure, the given test video O can be classified to the emotion class for which $p(t|K, a)$ is maximized:

$$t = arg \ max_i \, p(t|K, a) \tag{18}$$

At the end of this section, the proposed emotion recognition method is summarized. At first, the videos are segmented into action units. Then spatio-temporal volumes are built from action units and the histograms are constructed. The histograms are sent to mRVM for recognition. The flow chart can be illustrated in Fig. 3.

4 Experiments

In this section, the experiment results on both FABO dataset and our self-collected one are reported. FABO dataset which contains approximately 1900 videos of facial expressions recorded by the facial camera, and face and body expressions recorded by the face and body cameras [12]. We also reconstruct a dataset based on datasets proposed by Jiang et al. [22]. The same nine emotion categories are considered according to FABO for the comparison, including "anger", "surprise", "fear", "anxiety", "happiness", "disgust", "boredom", "sadness", "uncertainty". Some videos in Jiang's dataset have low resolution or blocked human bodies. To get more videos in which persons have identifiable expressions and gestures, we download the largest allowed number of videos from each search of 8 YouTube and 8 Baidu searches. These videos are manually classified by 10 annotators (5 males and 5 females). After careful annotations of all the videos by each annotator separately and a group meeting to finalize the emotions carried by the videos, the final dataset contains 1440 videos. Each action unit is labeled one emotion type and we can observed that emotions do share high correlations with certain actions such that "anger" always connects to "punching" and "flinging".

4.1 Experiments on Self-collected Dataset

In each category of our self-collected dataset, ten train-test splits are randomly generated and each of them using 3/4 of the samples for training and 1/4 for testing. The histograms and mRVM classifier are trained for each split and we propose the mean of the ten recognition accuracies, which are measured by test samples annotated with emotion labels. Training samples are used to build the map between emotion type and feature histogram, which focus on training a mRVM classifier. The example of the map is shown in Fig. 4.

We use following two indexes, named as Precision and Recall [23] to verify the performance of the algorithm. There are three parameters using for computing those

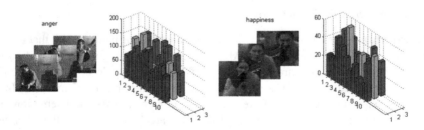

Fig. 4. A map between emotion type and feature histogram. Figure on the left shows relationship between "anger" and its feature histograms, the figure on the right shows feature histograms of "happiness".

two indexes. C is the number of videos which are correctly recognized. F is the number of videos which are labeled wrong emotion type. M is the number of videos which are miss in the classification system.

$$Precision = \frac{C}{C+F} \tag{19}$$

$$Recall = \frac{C}{C+M} \tag{20}$$

The precision and recall of the emotion type classification can be calculated according to the Eqs. (19) and (20). The results are shown as following confusion matrix with Precision and Recall in Fig. 5. "Happiness" has the top recognition rate. The recognizing accuracies of "anxiety", "happiness" and "sadness" are no less than 90% and the average precision is 83.2%, which is satisfactory because we have only 1080 training samples to build our map between emotion types and feature histograms.

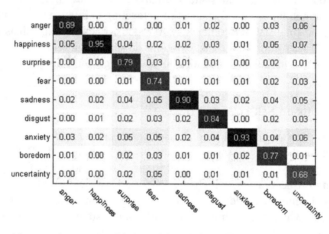

Fig. 5. Recognition results measured by precision and recall. The mean of precision and recall of our method with "anger", "happiness", "surprise", "fear", "sadness", "disgust", "anxiety", "boredom", "uncertainty" listed in sequence, from left to right.

4.2   Experiments on FABO Dataset

In this experiment, we try to compare the performance of our method with that of other novel frameworks. Since videos in FABO Dataset are pre-segmented activity video clips and the gestures are extracted already, this dataset is made up of professionally captured videos and we only use it as an evaluation criterion for our methods. In the observer labeling data of FABO, for each video, number of votes for its emotion label is a important parameter when it was labeled. There are 6 annotators. If the label get 6 votes, it means this annotation gets high credibility. So we only select videos whose label get more than 3 votes as our data for training and testing. The example of the map is shown in Fig. 6.

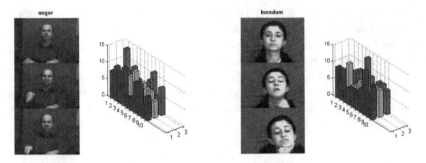

Fig. 6. Feature histograms on some samples in FABO Dataset. Figure on the left shows feature histograms of anger category and right one shows feature histograms of boredom category.

Table 1 shows comparison of recognition accuracy of different methods. To our knowledge, human actions extracted from videos have never been used for emotion recognition. Then we just compare our method with FSPCA [24], tempNorm and BOW we mentioned before and methods just using multimodal data and skeleton data are not included. We can clearly see that our method can perform as well as other methods. In Table 1, highest testing recognition accuracies are highlighted.

Table 1. Performance comparisons on FABO data sets

Emotion Type	Precision (%)			
	FSPCA	tempNorm	BOW	Ours
Anger	85.4	82.1	84.8	**88.3**
Happiness	86.0	81.4	80.9	**91.5**
Surprise	48.2	47.1	44.6	**65.8**
Fear	52.4	53.9	48.8	**59.4**
Sadness	**69.7**	62.2	60.2	55.7
Disgust	54.4	52.0	**56.3**	53.4
Anxiety	79.1	72.6	70.6	**89.6**
Boredom	66.8	64.4	62.5	**70.5**
Uncertainty	55.0	53.2	51.7	**63.0**
Overall	66.33	63.21	62.27	**70.8**

Restricted range of videos, coupled with small-scale training dataset, leading to a result that these four methods in the Table 1 all perform not well. But from this result, we still can draw a conclusion that our method is consistently the top performers. Since movements of people in "happiness" are much more vigorous and dense, we have the best recognition rate on this category. In "sadness" and "disgust", other methods perform better than ours because in this dataset, the differences on body motion frequency and range are hard to find. We review the videos after the experiment and find that it is hard to distinguish these two kind of emotions even for ourselves. The average recognition rate of our method is 70.8% and that of others are 66.33% (FSPCA),

63.21% (tempNorm) and 62.27% (BOW) respectively. To compare the recognition rate of our method on our self-collected dataset and that on FABO, we find that the former is better the latter. The reason is that we always select videos where human express their emotions with exaggerated actions which are easier to be recognized by our method. Besides, the annotators of FABO determine the emotion type depend on not only the body motions but also the face expressions, which can lead to the situation that two similar actions are classified into two emotions and this is also a limitation of our framework.

5 Conclusion

In this paper, we address a novel method for video emotion recognition by exploring action-focused human emotion classification, rather than the affective computing based on facial recognition. The method can identify video emotion patterns through spatio-temporal feature histograms which extends the previous spatio-temporal volumes. Before building 3D volumes, we segment videos into short action units by computing global motion shift and the results agree with physical principles. With mRVM first introduced into emotion analysis in this paper, we relate feature histograms to emotion type. The experiments analysis shows the performances of the algorithm still see room for growth. In the future, audio, interactive objects and other semantic attributes can be modeled through spatial pattern mining to implement scenario understanding for emotion recognition.

References

1. Yan, L., Wen, X., Zhang, L., Son, Y.: The application of unascertained measure to the video emotion type recognition. In: Signal Processing Systems (ICSPS) (2010)
2. Irie, G., Satou, T., Kojima, A., Yamasaki, T.: Affective audio-visual words and latent topic driving model for realizing movie affective scene classification. IEEE Trans. Multimed. **12**(6), 523–535 (2010)
3. Hanjalic, A., Xu, L.Q.: Affective video content representation and modeling. IEEE Trans. Multimed. **7**(1), 143–154 (2005)
4. Kleinsmith, A., Bianchi-Berthouze, N.: Affective body expression perception and recognition: a survey. IEEE Trans. Affect. Comput. **4**(1), 15–33 (2013)
5. Karg, M., Samadani, A., Gorbet, R., Kuhnlenz, K., Hoey, J., Kulić, D.: Body movements for affective expression: a survey of automatic recognition and generation. IEEE Trans. Affect. Comput. **4**(4), 341–359 (2013)
6. Samadani, A.-A., Gorbet, R., Kulić, D.: Affective movement recognition based on generative and discriminative stochastic dynamic models. IEEE Trans. Hum.-Mach. Syst. **44**(4), 454–467 (2014)
7. Barrett, L.F.: Solving the emotion paradox: categorization and the experience of emotion. Pers. Soc. Psychol. Rev. **10**(1), 20–46 (2006)
8. Binali, H., Potdar, V.: Emotion detection state of the art. In: Proceedings of the CUBE International Information Technology Conference, pp. 501–507 (2012)

9. Gunes, H., Pantic, M.: Automatic dimensional and continuous emotion recognition. Int. J. Synth. Emot. **1**(1), 68–99 (2010)
10. Calvo, R.A., D'Mello, S.: Affect detection: an interdisciplinary review of models, methods, and their applications. IEEE Trans. Affect. Comput. **1**(1), 18–37 (2010)
11. Plutchik, R.: The nature of emotions. Am. Sci. **89**(4), 344–350 (2001)
12. Gunes, H., Piccardi, M.: A bimodal face and body gesture database for automatic analysis of human nonverbal affective behavior. IEEE Comput. Soc. **1**(1), 1148–1153 (2006)
13. Binali, H., Perth, W.A., Wu, C., Potdar V.: Computational approaches for emotion detection in text. In: IEEE International Conference on Digital Ecosystems and Technologies (DEST) (2010)
14. Aldelson, E., Bergen, J.R.: Spatiotemporal energy models for the perception of motion. J. Opt. Soc. Am. **2**(2), 284–299 (1985)
15. Korimilli, K., Sarkar, S.: Motion segmentation based on perceptual organization of spatio-temporal volumes. In: Proceedings of the 15th International Conference on Pattern Recognition (ICPR) (2000)
16. Yilmaz, A., Shah, M.: Actions sketch: a novel action representation. In: IEEE Conference on Computer Vision and Pattern Recognition (CVPR) (2005)
17. Ryoo, M.S.: Human activity prediction: early recognition of ongoing activities from streaming videos. In: IEEE International Conference on Computer Vision (ICCV) (2011)
18. Wang, J., Xu, Z., Xu, Q.: Video volume segmentation for event detection. In: International Conference on Computer Graphics, Imaging and Visualization (CGIV) (2009)
19. Loog, M., Lauze, F.: The improbability of Harris interest points. IEEE Trans. Pattern Anal. Mach. Intell. **32**(6), 1141–1147 (2010)
20. Liu, J., Luo, J., Shah, M.: Recognizing realistic actions from videos "in the wild". In: Proceedings of IEEE Conference on Computer Vision and Pattern Recognition (CVPR) (2009)
21. Damoulas, T., et al.: Inferring sparse kernel combinations and relevance vectors: an application to subcellular localization of proteins. In: Proceedings of the 7th International Conference on Machine Learing and Applications, pp. 577–582. IEEE Computer Society (2008)
22. Jiang, Y.G.: Predicting emotions in user-generated videos. In: AAAI Conference on Artificial Intelligence (2014)
23. Lefevre, S., Vincent, N.: Efficient and robust shot change detection. J. Real-Time Image Proc. **2**(1), 23–24 (2007)
24. Samadani, A., Ghodsi, A., Kulić, D.D.: Discriminative functional analysis of human movements. Pattern Recogn. Lett. **34**(15), 1829–1839 (2013)

Rocchio-Based Relevance Feedback in Video Event Retrieval

G.L.J. Pingen[1,2(✉)], M.H.T. de Boer[2,3], and R.B.N. Aly[1]

[1] University of Twente, P.O. Box 217, 7500 AE Enschede, The Netherlands
r.aly@utwente.nl
[2] TNO, P.O. Box 96864, 2509 JG The Hague, The Netherlands
{geert.pingen,maaike.deboer}@tno.nl
[3] University of Nijmegen, P.O. Box 9010, 6500 GL Nijmegen, The Netherlands

Abstract. This paper investigates methods for user and pseudo relevance feedback in video event retrieval. Existing feedback methods achieve strong performance but adjust the ranking based on few individual examples. We propose a relevance feedback algorithm (ARF) derived from the Rocchio method, which is a theoretically founded algorithm in textual retrieval. ARF updates the weights in the ranking function based on the centroids of the relevant and non-relevant examples. Additionally, relevance feedback algorithms are often only evaluated by a single feedback mode (user feedback or pseudo feedback). Hence, a minor contribution of this paper is to evaluate feedback algorithms using a larger number of feedback modes. Our experiments use TRECVID Multimedia Event Detection collections. We show that ARF performs significantly better in terms of Mean Average Precision, robustness, subjective user evaluation, and run time compared to the state-of-the-art.

Keywords: Information retrieval · Relevance feedback · Video search · Rocchio · ARF

1 Introduction

Finding occurrences of events in videos has many applications ranging from entertainment to surveillance. A popular way to retrieve video events from a given collection is to combine detection scores of related concepts in a ranking function. However, selecting related concepts and defining query specific ranking functions without any examples is challenging. Text retrieval, which is the corresponding problem in the text domain, strongly benefits from relevance feedback, which defines ranking functions based on a set of examples with assumed relevance status [15]. While the basic relevance feedback principle recently also gained popularity in video retrieval, the methods of adapting ranking functions are generally developed from scratch. However, from a scientific standpoint it would be more desirable to transfer the knowledge gained for text retrieval into approaches in video retrieval, especially if those methods show stronger performance. Therefore, this paper derives a novel concept-based

© Springer International Publishing AG 2017
L. Amsaleg et al. (Eds.): MMM 2017, Part II, LNCS 10133, pp. 318–330, 2017.
DOI: 10.1007/978-3-319-51814-5_27

relevance feedback algorithm that is derived from a proven relevance feedback algorithm in text retrieval.

The state-of-the-art relevance feedback algorithms in video retrieval update the ranking function based on few examples of the positive and the negative class [5,6]. However, especially in pseudo relevance feedback, which assumes a set of examples to be relevant without actual feedback, we find this hurts performance for some queries because these assumptions are not met. In text retrieval, on the other hand, relevance feedback algorithms update the ranking function based on the centroids in the relevant and non-relevant examples. We propose that this approach is likely to be more robust, because considering centroids evens out outliers in the examples. Furthermore, adapting ranking functions based on the differences in the centroids is also more likely to generalize to the remaining videos in the collection, potentially improving effectiveness overall.

This paper evaluates the proposed relevance feedback algorithm based on several real and pseudo user feedback modes and investigates whether a different trend between modes of relevance feedback exist.

In the next section, we present related work on relevance feedback. The third section explains our Adaptive Video Event Search system (AVES), which uses the proposed relevance feedback algorithm. The fourth section contains the experimental set-up using the TRECVID Multimedia Event Detection benchmark and the fifth section consists of the results of both simulation and user experiments. We end this paper with a discussion and the conclusions and future work.

2   Related Work

Relevance feedback in video event retrieval is an increasingly active field of research. In video retrieval, Dalton et al. [3] showed that pseudo-relevance feedback can increase Mean Average Precision up to 25%, and with human judgment this number can grow up to 55%. The state-of-the-art methods, such as feature-, navigation-pattern, and cluster-based methods, in image retrieval are explained by Zhou et al. [25] and Patil et al. [14]. Oftentimes the system will actively select the documents that achieve the maximal information gain [18]. Other methods use decision trees, SVM's, or multi-instant approaches are explained in Crucianu et al. [2]. Xu et al. [20] present an interactive content-based video system that incrementally refines the user's query through relevance feedback and model visualization. Their system allows a user to select a subset of relevant retrieved videos and use those as input for their SVM-based video-to-video search model, which has been shown to outperform standard text-to-video models. Yang et al. [21] also introduce an SVM based supervised learning system that uses a learning-to-rerank framework in combination with an adapted reranking SVM algorithm. Tao et al. [17] improves on the SVM-based methods using orthogonal complement component analysis (OCCA). According to Wang et al. [19], SVM-based RF approaches have two drawbacks: (1) multiple feedback interactions are necessary because of the poor adaptability, flexibility and robustness of the original visual features; (2) positive and negative samples are treated equally, whereas the positive and negative examples provided by the relevance feedback often have distinctive properties. Within the pseudo relevance feedback, this second point is taken by Jiang et al. [7–9], who use an unsupervised learning approach in which the 'easy' samples are used to learn first and then the 'harder' examples are increasingly added. The authors define the easy samples as the videos that are ranked highest and have smaller loss. The more easy samples are those that are presumably more relevant, and would be ranked higher than others. The system then iterates towards more complex (lower-ranked) videos. An SVM/Logistic-regression model is trained using pseudo labels initiated with logarithmic inverse ordering. This approach reduces the cost of adjusting the model too much when learning from data that is very dissimilar from the learned model. Experiments show that it outperforms plain retrieval without reranking, and that has decent improvements over other reranking systems.

Another field of study in which relevance feedback is often used is in text retrieval. One of the most well-known and applied relevance feedback algorithms that has its origins in text-retrieval is the Rocchio algorithm [15]. This algorithm works on a vector space model in which the query drifts away from the negatively annotated documents and converges to the positively annotated documents. The Rocchio algorithm is effective in relevance feedback, fast to use and easy to implement. The disadvantages of the method are the parameters that have to be tuned and it cannot handle multimodal classes properly.

Another vector space model uses a k-NN method and is used by Gia et al. [6] and Deselaers et al. [5]. k-NN based methods are shown to be effective, are non-parametric, but run time is slower and it can be very inaccurate when the training set is small.

3   Adaptive Video Event Search

In this section, we explain our Adaptive Video Event Search system, named AVES. All relevant feedback algorithms depend on an *initial ranking*. A user can retrieve this initial ranking by entering a textual query into our search engine. The initial score s_v by which a video v is ranked is defined as:

$$s_v = \sum_{d \in D} w_d \cdot (s_{v,d} - b_d), \tag{1}$$

where d is the concept detector, D is set of selected concept detectors, w_d is the weight of concept detector d, $s_{v,d}$ is the concept detector score for video v and b_d is the average score on the background dataset of concept detector d.

Concept detectors are models that are trained to detect concepts in images or videos based on machine learning techniques, such as neural networks or SVMs. For the definition of D and the initial setting of w_d, we adapt a method proposed in Zhang et al. [23], first comparing the query to each of the pre-trained concepts available to our system. The skip-gram negative sampling Word2vec model from Milokov et al. [12] is used in this comparison. The pre-trained GoogleNews model, which is trained on one billion words, embeds the words into 300 dimensions.

The cosine similarity is used to calculate distances between words. The distance between the label of the concept detector and the user query is used as a weight. The thirty concepts with the highest similarities and a value higher than a threshold of 0.35 keep their weights and the other concepts have a weight of zero. In experiments outside of the scope of this paper we found that including a concept detector background score as prior is beneficial to ranking accuracy.

3.1   Adaptive Relevance Feedback (ARF)

In this section, we explain our relevance feedback algorithm, named *ARF*. This method is inspired by the Rocchio algorithm [15]. Different from other algorithms, we use relevance feedback to update the weights for our concept detectors. By updating the weights using relevance feedback our algorithm is more robust to few or wrong annotations. In k-NN methods, wrong annotations can have a high impact on ranking performance. By taking into account the initial concept detector cosine distance to the query, the proposed algorithm is more robust to this type of relevance feedback.

The weights are updated using the following formula:

$$w'_d = w_d + (\alpha \cdot m_R) - (\beta \cdot m_{NR})$$
$$m_R = \frac{\sum_{v \in R} s_{v,d} - b_d}{|R|}$$
$$m_{NR} = \frac{\sum_{v \in NR} s_{v,d} - b_d}{|NR|},$$

(2)

where v is the considered video, d is the concept detector, R is the set of relevant videos, NR is the set of non-relevant videos, $s_{v,d}$ is the score for concept detector d for video v, w_d is Word2vec similarity between the concept detector d and the query, b_d is the average score on the background dataset of concept detector d, and α and β are Rocchio weighting parameters for the relevant and non-relevant examples respectively.

The adjusted concept detector weight, w'_d, is then plugged back into the ranking function (see 1), where we substitute the original Word2vec score for the adjusted weight. This results in new scores, s'_v, for each video v, which is used to create an updated ranked list of videos.

3.2   Experimental Set-up

We use the MEDTRAIN set (5594 videos, 75.45 keyframes on average) and MEDTEST set (27276 videos, on average 57.11 keyframes) from the TRECVID Multimedia Event Detection benchmark [13] as our evaluation datasets. The datasets from this international benchmark is used because of the challenging events, many videos and wide acceptance. The MEDTRAIN contains relevance judgments for forty events, whereas MEDTEST only contains judgments for twenty events but this set is often used in other papers to report performance on. Different from the MEDTRAIN in which we use the top 30 concept detectors, in the MEDTEST we only use the top 5 concept detectors, because of the bigger imbalance of positive and negative videos in the MEDTEST (20:27276) compared to the MEDTRAIN (100:5494). This imbalanced caused more concepts to a very low initial performance, whereas only 5 concepts had less noise and, therefore, slightly higher performance.

In the MEDTRAIN set only thirty-two events are used, because of the correlation of certain concept detectors to a certain event, resulting in a (near-)perfect retrieval result. Since such retrieval results are not interesting for relevance feedback application purposes, we do not consider them for the purpose of this experiment. The omitted events are the following: *Wedding ceremony*; *Birthday party*; *Making a sandwich*; *Hiking*; *Dog show*; *Town hall meeting*; *Beekeeping*; *Tuning a musical instrument*. In additional subjective user evaluation experiments, we verified that the trends obtained with the thirty-two events are also present with the forthy events, only the MAP is slightly higher.

The BACKGROUND set (5000 videos) from the benchmark is used to obtain the background scores b_d in Eq. 1. A total of 2048 concept detectors (D) were used from the ImageNet (1000) [4], Places (205) [24], SIN (346) [13] and TRECVID MED dataset (497) [13]. The concept detectors from the TRECVID MED are manually annotated on the Research set, comparable to Natarajan et al. [22] and Zhang et al. [23]. The output of the eight layer of the DCNN network trained on the ILSVRC-2012 [4] is used as concept detector score per keyframe/image. This DCNN architecture is fine-tuned on the data in the dataset for SIN, Places and TRECVID MED. For each video in our dataset we have extracted 1 keyframe per 2s uniformly from a video. We use max pooling to obtain a concept detector score per video. We purposely did not use higher level concept detectors, such as those available in the FCVID [10] or Sports [11] dataset, to obtain more interesting experiments using relevance feedback. We, therefore, do not aim at highest possible initial ranking, but at a gain with the use of relevance feedback. We believe this is applicable to real world cases, because relevant high level concepts are not always present.

The ARF parameters α and β were taken to be 1.0 and 0.5, in line with text-information retrieval literature [15]. Variations of these parameters were also investigated, as shown in Sect. 3.3.

We compare our relevance feedback algorithm to (1) a baseline without relevance feedback *Initial*; and (2) a k-NN based relevance feedback algorithm named *RS*. The RS algorithm is well-performing in image retrieval [5,6] and the relevance score $relevance(v)$ of a video v calculated as

$$relevance(v) = \left(1 + \frac{dR(v)}{dNR(v)}\right)^{-1}, \tag{3}$$

where dR is the dissimilarity, measured as Euclidean distance, from nearest video in relevant video set R, dNR is the dissimilarity from nearest video in non-relevant video set NR.

The SVM-based methods are not included in this paper, because preliminary experiments showed that on average performance is poor due to limited amount of positive samples.

Modes. We compared the algorithms with respect to the following feedback modes: (1) the mode *Optimal* uses the ground truth to select all relevant videos, (2) the *Pseudo* relevance feedback mode selects the first 10 videos as positive (and the rest as negative), (3) the *Random* mode selects 10 positive videos at random from the ground truth, and finally (4) the *user* mode uses real relevance feedback from users. The selection is done on the first 20 videos, which is around the number of videos a user would initially consider.

For the *User* mode, the task of a group of participants was to select relevant and non-relevant videos. 24 results were shown initially, and more could automatically be loaded by scrolling to the bottom of the page. Ten male participants (age $= 26.3, \sigma = 1.567$) with mainly Dutch origin and at least a Bachelor's degree or higher without dyslexia, colour-blindness, concentration problems, or RSI problems, voluntarily participated in an experiment. The participants had two conditions, which correspond to the re-ranking results by ARF and RS. In each of the conditions, 16 queries, randomly assigned using a Latin rectangle [1], were presented to the user, after which they performed relevance feedback.

Evaluation. To evaluate our algorithm, performance of the algorithms is measured by the following aspects: (1) accuracy; (2) robustness. For accuracy, we use Mean Average Precision (MAP). MAP is the standard evaluation method in TRECVID MED and is based on the rank of the positive videos. With re-ranking, the videos that are indicated as positive are always on the top of the list, increase MAP. It is, however, also interesting to know whether the algorithm is able to retrieve new relevant videos. This is why we introduce a variant, *MAP**. MAP* calculates MAP disregarding the videos that have been viewed by the user already, which we track in our experiment. We evaluate our method on both metrics.

For robustness, we report the robustness index (RI) [16]:

$$RI = \frac{|Z_P| - |Z_N|}{|Z|}, \tag{4}$$

where Z_P and Z_N are the sets of queries where the performance difference between ARF and RS in terms of MAP was positive for ARF or negative, respectively, and $|Z|$ is the total number of queries.

3.3 Results

Accuracy. Table 1 shows the MAP^*, and MAP for the relevance feedback algorithms in the *Optimal* relevance feedback mode. A Shapiro-Wilk test was used to assess normality of our precision scores, The assumption of normality is

Initial (including relevance selection)   Reranking results using RS   Reranking results using ARF

Fig. 1. Relevance feedback and results for *Working on a woodworking project.*

violated in MEDTRAIN for Initial for MAP ($p < 0.0005$), and for all modes for MAP^* ($p < 0.0005; p = 0.01; p = 0.001$, for Initial, ARF and RS respectively) and for all modes in MEDTEST set ($p < 0.0005$ for MAP and MAP^*). On both the MEDTRAIN and MEDTEST set, the non-parametric Friedman test showed a significant difference for MAP^* ($\chi^2 = 13.528, p = 0.001$ in MEDTRAIN; $\chi^2 = 13.241, p = 0.001$ in MEDTEST) and MAP ($\chi^2 = 18.063, p < 0.0005$ in MEDTRAIN; $\chi^2 = 18.000, p < 0.0005$ in MEDTEST). A Wilcoxon Signed-Ranks Test with Bonferroni correction ($0.05/3$) showed a significant difference between **ARF and Initial** in MEDTRAIN ($Z = -4.135, p < 0.0005$ for $MAP^*; Z = -4.252, p < 0.0005$ for MAP), and between **ARF and RS** in both MEDTRAIN and MEDTEST($Z = -2.450, p = 0.014$ for $MAP^*; Z = -3.123, p = 0.002$ for MAP in MEDTRAIN; $Z = -2.427, p = 0.015$ for $MAP^*; Z = -3.509, p < 0.0005$ for MAP in MEDTEST).

Table 1. %MAP using the *Optimal* mode.

Algorithm	MEDTRAIN		MEDTEST	
	MAP^*	MAP	MAP^*	MAP
Initial	15.24	18.06	3.47	6.28
RS	16.74	20.30	2.49	6.80
ARF	**18.92**	**24.22**	**3.78**	**8.80**

Table 2 shows the MAP^* for the different relevance feedback algorithms for the relevance feedback methods. Significance is in line with MAP type results, except we observe that for MEDTEST with Pseudo-relevance selection, both relevance feedback methods do not improve on Initial ranking.

Table 2. %MAP^* scores for Initial, RS and ARF.

Algorithm	MEDTRAIN			MEDTEST		
	Optimal	Pseudo	Random	Optimal	Pseudo	Random
Initial	15.24	15.69	15.69	3.47	**3.42**	3.42
RS	16.74	14.18	14.35	2.49	2.73	3.15
ARF	**18.92**	**18.11**	**18.15**	**3.78**	3.17	**4.36**

Table 3. User experiment $\%MAP^*$ scores and standard deviations.

Algorithm	MAP^*	σ
Initial	13.09	1.02
RS	10.71	1.98
ARF	**15.32**	**1.55**

Table 3 shows the scores found in the user experiments. A Shapiro-Wilk test showed that the precision score distributions do not deviate significantly from a normal distribution at $p > 0.05$ ($p = 0.813; p = 0.947; p = 0.381$, for Initial, RS, and ARF respectively). A statistically significant difference between groups was determined by a one-way ANOVA (F(2,27) = 18.972, p < 0.0005). A post-hoc Tukey's HSD test was performed to verify intergroup differences. The means of all algorithms differed significantly at $p < 0.05$ ($p = 0.006; p = 0.01; p < 0.0005$, for Initial-RS, Initial-ARF, and RS-ARF, respectively).

In these user experiments, on average, 61.65% of marked relevant results were correct, and 92.71% of marked non-relevant results were correct. Further investigation was done to research the effect of the positive and negative annotations on precision scores. Variations of the α and β parameters were analyzed. While performance decreased slightly when disregarding all positive annotations ($\alpha = 0.0$), it dropped drastically when disregarding the negative annotations ($\beta = 0.0$). In line with relevant literature on Rocchio, $\alpha = 1.15$ and $\beta = 0.5$ provides the highest MAP^*. Visualizations of these results can be found in Fig. 2.

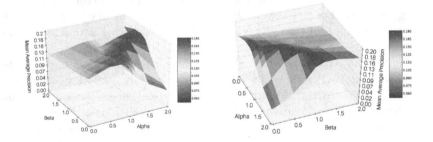

Fig. 2. MAP^* relative to α and β values.

Robustness. To get an overview of the precision per event, we calculated precision averaging over all sessions. A bar plot of RS scores subtracted from ARF scores is shown in Fig. 3. RI was calculated with respect to Initial ranking for ARF and RS, respectively. We see that ARF improves Initial ranking in 71.88% of events, and RS in 37.5%. These scores result in a robustness performance of $RI = 0.4375$ for ARF and $RI = -0.25$ for RS.

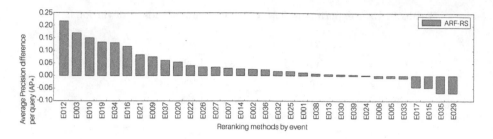

Fig. 3. Average precision difference (AP^*) per event.

Run Time. Since the run-time of the ARF method depends mainly on the size of the video collection $(O(n)$, concept detector weights are updated only based on the selected (non-)relevant results), it is quicker than the RS method whose run-time depends on both total video set size and concept detector set size $(O(n * d))$. Note that the similarity measure applied for RS also factors into this. The average run time for RS and ARF (on an Intel Core i7-4700MQ CPU @ 2.40 GHz x-64 system with 8 GM RAM) is 8003.45 ms and 107.25 ms, respectively.

3.4 Subjective User Evaluation

In additional subjective user evaluation experiments, we also compare ARF results to RS results directly. We asked users (N = 19) to perform relevance selection for as much events as they would like on the initial result set and showed both ARF and RS reranking results. The order in which the events, and reranking results (left or right) were shown was randomized. We then asked users to select the ranking that they thought was the best. Figure 1 shows an example of retrieval results for Initial, RS reranking, and ARF reranking.

We found that ARF is selected 85.96%, of which 83.67% was correct when compared to actual AP scores. RS was preferred in only 14.04% of all queries, of which 25.0% was correct. We also investigated relevance selection per event. An overview of the True Positive (TP), False Positive (FP), True Negative (TN) and False Negative (FN) scores are shown in Fig. 4. Note that this terminology

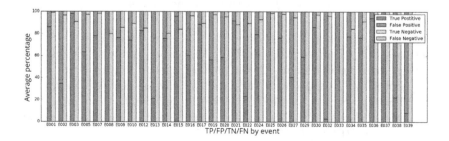

Fig. 4. Percentage relevance selection per event per method.

does not capture the users' beliefs sufficiently, since the user's cannot be wrong in their judgment (if we assume they performed the task honestly). We can state that they are True, or False Positives only relative to the ground truth. A better terminology reflecting the users' honest evaluation might be: *Correct positives*, *Missed negatives*, *Correct negatives* and *Missed positives*.

4 Discussion

Our results show a statistically significant difference ($p < 0.05$) in the means of the MAP^* scores of the ARF algorithm, and that of the Initial and RS algorithm in the user experiments with a relatively small sample size. These results are encouraging, and provide a solid basis for the claim that the ARF algorithm has on average a better performance compared to the RS algorithm. We show that even when discarding events from the MEDTEST set that have a very high accuracy (because we have a high-matching concept detector), we still obtain very reasonable MAP scores. This claim is strengthened by the performance on MAP^*, and MAP in results from experiments with different relevance feedback modes. We see a similar trend between different modes of relevance feedback. Using the *Optimal* relevance feedback mode, ARF performs better than RS and Initial ranking, but not by much. However, when we introduce non-optimal, and perhaps more realistic, relevance feedback modes such as pseudo- or random-relevance feedback, we see ARF performing significantly better. We believe that this effect could be explained by the ARF algorithm being less volatile, and less subject to the effect of misclassification of a single result rippling through to ranking scores. This effect can also be seen in our user experiments, since user relevance feedback is not error-free by a long measure (see Sect. 3.3). In our experiments where we directly let users choose between two reranked sets, they select ARF as the best ranking 85.96% of the time. On the MEDTEST set, we can see comparable results, except that on the *Pseudo* relevance feedback mode, ARF cannot improve on the Initial ranking due to the small number of positive videos at the top of the initially returned results.

Individual events analysis was also performed. There is a clear tendency of the MAP^* to decline when the True Positive (TP) rate drops. While the ARF algorithm obtains the highest MAP^* on average, we see some events on which the RS algorithm, or even Initial ranking performs better. For example, on event *E035, Horse riding competition*, RS outperforms ARF. Manual inspection of the initial video set shows that for these events, relevant and non-relevant videos in the initial set are quite homogeneous regarding concept detector scores, but easily distinguishable by human observers. These findings indicate the absence of good concept detectors that capture this distinction.

Dalton et al. [3] show that pseudo-relevance feedback can increase MAP up to 25%, and up to 55% for real relevance feedback. Though we do not match these numbers on average, for some events we gain a performance boost of up to 54% with pseudo- and up to 25% with real-relevance feedback. One striking observation we make is that although MAP(*) scores show a significant difference between the ARF and RS methods, subjective user evaluation when integrated in a complete video search system does not reflect these differences. However, when asked to compare reranking results directly side-by-side, we do see a preference.

Parameter tuning of the α and β parameters shows the importance of including the non-relevant class in ARF. We see a large decline in performance when disregarding the negative annotations, while this decline was considerably less steep when disregarding the positive annotations (see Sect. 3.3).

5   Conclusion and Future Work

This paper investigated relevance feedback algorithms for video event retrieval. We proposed the adaptive relevance feedback (ARF) algorithm derived from the well-researched Rocchio text retrieval algorithm [15]. While state-of-the-art algorithms in video relevance feedback use few examples from nearest neighbours, ARF updates ranking functions based on the difference between the centroids of the relevant and non-relevant examples.

We investigated several feedback modes, including feedback from real users, on the training set of the Multimedia Event Detection task [13]. We compared ARF against the state-of-the-art algorithms from Gia et al. [5,6], referred to as RS. On the MEDTRAIN and MEDTEST sets, ARF showed stronger average effectiveness compared to RS in terms of MAP^*, and MAP, for a number of different modes of relevance feedback. This effect was also found in subjective user evaluation experiments. Robustness for ARF was also higher compared to RS.

From the above experiments we conclude that there is strong evidence that ARF shows better performance for video event retrieval than the previous state-of-the-art. The performance trends between real and pseudo relevance feedback modes were similar for both tested algorithms, which is a secondary contribution of this paper.

For future work, we propose to evaluate ARF on a broader range of datasets in terms of videos and concept detectors. Furthermore, discovering concept detectors with poor performance through relevance feedback, and adding supplementary concept detectors outside the initial set should be researched. Another interesting avenue for further research is the apparent contrast between objective MAP scores and subjective user evaluation. Do users 'care' about MAP scores?

References

1. Cochran, W.G., Cox, G.M.: Experimental designs (1957)
2. Crucianu, M., Ferecatu, M., Boujemaa, N.: Relevance feedback for image retrieval: a short survey. Report of the DELOS2 European Network of Excellence (FP6) (2004)
3. Dalton, J., Allan, J., Mirajkar, P.: Zero-shot video retrieval using content and concepts. In: Proceedings of the 22nd ACM International Conference on Information and Knowledge Management, pp. 1857–1860. ACM (2013)

4. Deng, J., Dong, W., Socher, R., Li, L.J., Li, K., Fei-Fei, L.: ImageNet: a large-scale hierarchical image database. In: CVPR 2009, pp. 248–255. IEEE (2009)
5. Deselaers, T., Paredes, R., Vidal, E., Ney, H.: Learning weighted distances for relevance feedback in image retrieval. In: 19th International Conference on Pattern Recognition, ICPR 2008, pp. 1–4. IEEE (2008)
6. Gia, G., Roli, F., et al.: Instance-based relevance feedback for image retrieval. In: Advances in Neural Information Processing Systems, pp. 489–496 (2004)
7. Jiang, L., Meng, D., Mitamura, T., Hauptmann, A.G.: Easy samples first: self-paced reranking for zero-example multimedia search. In: Proceedings of the ACM International Conference on Multimedia, pp. 547–556. ACM (2014)
8. Jiang, L., Mitamura, T., Yu, S.I., Hauptmann, A.G.: Zero-example event search using multimodal pseudo relevance feedback. In: Proceedings of the International Conference on Multimedia Retrieval, p. 297. ACM (2014)
9. Jiang, L., Yu, S.I., Meng, D., Mitamura, T., Hauptmann, A.G.: Bridging the ultimate semantic gap: a semantic search engine for internet videos. In: ACM International Conference on Multimedia Retrieval, pp. 27–34 (2015)
10. Jiang, Y.G., Wu, Z., Wang, J., Xue, X., Chang, S.F.: Exploiting feature and class relationships in video categorization with regularized deep neural networks. arXiv preprint arXiv:1502.07209 (2015)
11. Karpathy, A., Toderici, G., Shetty, S., Leung, T., Sukthankar, R., Fei-Fei, L.: Large-scale video classification with convolutional neural networks. In: CVPR (2014)
12. Mikolov, T., Sutskever, I., Chen, K., Corrado, G.S., Dean, J.: Distributed representations of words and phrases and their compositionality. In: Advances in Neural Information Processing Systems, pp. 3111–3119 (2013)
13. Over, P., Awad, G., Michel, M., Fiscus, J., Sanders, G., Kraaij, W., Smeaton, A.F., Quéenot, G., Ordelman, R.: TRECVID 2015 - an overview of the goals, tasks, data, evaluation mechanisms and metrics. In: Proceedings of the TRECVID 2015, p. 52. NIST, USA (2015)
14. Patil, S.: A comprehensive review of recent relevance feedback techniques in CBIR. Int. J. Eng. Res. Technol. (IJERT) 1(6) (2012)
15. Rocchio, J.J.: Relevance feedback in information retrieval (1971)
16. Sakai, T., Manabe, T., Koyama, M.: Flexible pseudo-relevance feedback via selective sampling. ACM Trans. Asian Lang. Inf. Process. (TALIP) 4(2), 111–135 (2005)
17. Tao, D., Tang, X., Li, X.: Which components are important for interactive image searching? IEEE Trans. Circuits Syst. Video Technol. 18(1), 3–11 (2008)
18. Tong, S., Chang, E.: Support vector machine active learning for image retrieval. In: Proceedings of the 9th ACM International Conference on Multimedia, pp. 107–118. ACM (2001)
19. Wang, X.Y., Liang, L.L., Li, W.Y., Li, D.M., Yang, H.Y.: A new SVM-based relevance feedback image retrieval using probabilistic feature and weighted kernel function. J. Vis. Commun. Image Represent. 38, 256–275 (2016)
20. Xu, S., Li, H., Chang, X., Yu, S.I., Du, X., Li, X., Jiang, L., Mao, Z., Lan, Z., Burger, S., et al.: Incremental multimodal query construction for video search. In: Proceedings of the 5th ACM on International Conference on Multimedia Retrieval, pp. 675–678. ACM (2015)
21. Yang, L., Hanjalic, A.: Supervised reranking for web image search. In: Proceedings of the International Conference on Multimedia, pp. 183–192. ACM (2010)
22. Ye, G., Liu, D., Chang, S.F., Saleemi, I., Shah, M., Ng, Y., White, B., Davis, L., Gupta, A., Haritaoglu, I.: BBN VISER TRECVID 2012 multimedia event detection and multimedia event recounting systems

23. Zhang, H., Lu, Y.J., de Boer, M., ter Haar, F., Qiu, Z., Schutte, K., Kraaij, W., Ngo, C.W.: VIREO-TNO@ TRECVID 2015: multimedia event detection
24. Zhou, B., Lapedriza, A., Xiao, J., Torralba, A., Oliva, A.: Learning deep features for scene recognition using places database. In: Advances in Neural Information Processing Systems, pp. 487–495 (2014)
25. Zhou, X.S., Huang, T.S.: Relevance feedback in image retrieval: a comprehensive review. Multimed. Syst. 8(6), 536–544 (2003)

Scale-Relation Feature for Moving Cast Shadow Detection

Chih-Wei Lin[✉]

College of Computer and Information Sciences,
Fujian Agriculture and Forestry University, Fuzhou, China
chihwei1981@ntu.edu.tw

Abstract. Cast shadow is the problem of moving cast detection in visual surveillance applications, which has been studied over years. However, finding an efficient model that can handle the issue of moving cast shadow in various situations is still challenging. Unlike prior methods, we use a data-driven method without the strong parametric assumptions or complex models to address the problem of moving cast shadow. In this paper, we propose a novel feature-extracting framework called Scale-Relation Feature Extracting (SRFE). By leveraging the scale space, SRFE decomposes each image with various properties into various scales and further considers the relationship between adjacent scales of the two shadow properties to extract the scale-relation features. To seek the criteria for discriminating moving cast shadow, we use random forest algorithm as the ensemble decision scheme. Experimental results show that the proposed method can achieve the performances of the popular methods on the widely used dataset.

Keywords: Shadow removal · Scale-relation · Ensemble decision scheme · Random forest

1 Introduction

Visual surveillance related systems (e.g., smart home, security monitoring, and intelligent transportation systems) are important but challenging, and there has been much interest in automatically detecting objects in various environments. Many computer vision researchers propose various background subtraction techniques to address the problem of moving cast detection, where moving cast shadow is the key factor that compromises the accuracy. Due to the properties of moving cast shadow are the same or similar to the moving objects, such as the movement pattern, the shape, the size and the magnitude of intensity change that lead to misclassify of background subtraction.

Many researches had been dedicated to solve the problem of moving cast shadow for the last two decades [1,12]. Among these methods, the algorithm of detecting shadow can be classified as spectral-based, spatial-based and model-based. The spectral-based approaches utilize the attribute of shadow, such as

© Springer International Publishing AG 2017
L. Amsaleg et al. (Eds.): MMM 2017, Part II, LNCS 10133, pp. 331–342, 2017.
DOI: 10.1007/978-3-319-51814-5_28

luminance, chromaticity and physical properties, to detect the shadow directly. The information of HSV, HSI, YUV, C1C2C3, and RGB are considered as the properties of luminance and chromaticity. Two characteristics are utilized: (1) background and shadow are similar in the chromaticity, (2) the shadow has darker illumination compared with background. However, it is parameter tuning. To generate the physical model, two major light sources are considered: ambient light (blue light) and the sun (white light). The physical model performs well in outdoors compared with the other color properties. The attributes of luminance, chromaticity and physical properties are all based on color information, which leads to the misclassification of pixels when the colors of object and shadow are similar.

The attributes of geometry and texture belong to the spatial-based approaches. The geometric attribute considers the orientation, the size, and the shape of moving objects to overcome the drawbacks of the spectral-based methods. However, it has some limitations, for example, the object and shadow must have different orientation and unitary light source. The gradient information is the main property of texture attribute, the performance of texture attribute is better than the spectral-based and geometric attribute when the features of texture are significant and the color information are similar between object and shadow. However, the performance of texture attribute is decreasing when the differences of texture attribute between the object and shadow are not obvious.

Although the spectral-based and spatial-based algorithms provide useful features for shadow detection, they highly depend on parametric assumptions for solving the problem in various environments. The model-based method uses the prior knowledge to generate the model for shadow classification without the parametric assumptions. The prior knowledge includes the geometry of scene, objects class, or light sources. Various properties as mentioned earlier are integrated with mechanism learning algorithm, such as Bayesian network, neural network (NN), and support vector machine (SVM) to generate the model. It is more suitable for various environments compared with spectral-based and spatial-based algorithms.

However, an efficient feature-extracting framework for describing shadow is still needed. Instead of detecting shadow with strong parametric assumptions or complex model, we use a data driven approach and introduce a novel feature-extracting method called Scale-Relation Feature Extracting (SRFE). SRFE leverages the relationship between adjacent scales with various properties of shadow to extract the scale-relation features in scale space. We associate an ensemble decision scheme with scale-relation features to construct the model for moving cast shadow detection.

The rest of this paper is organized as follows. First, we describe the framework of the proposed method, after which we introduce the extraction of scale-relation features of various attributes of shadow in Sect. 3. The scheme of ensemble classifier is presented in Sect. 4. Experimental results are presented in Sect. 5, and Sect. 6 gives the conclusions.

2   Overview of the Proposed Framework

The proposed methodology is designed for handling the problem of moving cast shadow in various environments. To efficiently detect and remove the shadow effect, we propose several tasks, as shown in Fig. 1. The proposed methodology includes four components: (1) Scale space separating, (2) Scale-relation feature extracting, (3) Features assembly, and (4) Ensemble decision.

First, we separate the images with various properties of shadow into several scales by using the cascade filtering approach. The scale space separating can provide abundant information that is useful for handling the problem of shadow removal. In scale space, we generate several octaves for each image with various properties; each octave has several scaled images.

Next, we consider the adjacent scale-relation between two properties of scaled images. We leverage the scale-relation among adjacent scales with various properties of shadow and integrate octaves to extract abundant information. To the best of our knowledge, the attributes of chromaticity, physical properties, geometry, and texture are useful for discriminating the foreground, shadow and background. However, the proposed method is pixel-based. Therefore, the geometric properties are not considered in our work.

To efficiently remove the effect of shadow on various environments, we integrate the scale-relation features with ensemble decision scheme. The ensemble decision scheme uses the concept of majority for discriminating the foreground and shadow.

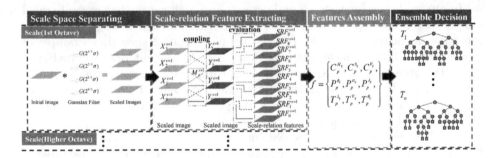

Fig. 1. The framework of the proposed method (Color figure online)

3   Scale-Relation Feature Extracting Scheme

Unlike prior works, which extract a single or assemble multiply features of shadow from a single scale, Scale-Relation Feature Extracting (SRFE) scheme considers not only multi-scale information but also the relationship among adjacent scales. The proposed SRFE mainly aims at providing abundant information with properties of shadow for shadow detection. SRFE involves scale space separating, scale-relation coupling, and scale-relation feature extracting.

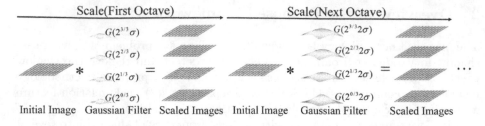

Fig. 2. Example of decomposition of variable-scale on octaves (Color figure online)

3.1 Scale Space Separating

The concept of scale space separating is inspired by [2,9]. According to Koenderink [6] and Lindeberg [8], we use the cascade filtering approach to generate the scaled images. In our work, the Gaussian function is used as the scale-space kernel and to produce the scale space of an image L using the following equation,

$$L(x, y, k\sigma) = G(x, y, k\sigma) * I(x, y) \tag{1}$$

where $I(x, y)$ is an input image, $G(x, y, k\sigma)$ is a variable-scale Gaussian, $*$ is the convolution operation, and

$$G(x, y, k\sigma) = \frac{1}{2\pi(k\sigma)^2} e^{-(x^2+y^2)/2(k\sigma)^2} \tag{2}$$

In our work, we generate several octaves and each octave has several scaled images in the scale space. A constant multiplicate factor k is used to generate the adjacent nearby scales [9] for each octave. Figure 2 shows an example of separating an initial image into scaled images on different octaves. In Fig. 2, an initial image is separated into four scaled image by Gaussian function with various factor k. In our work, we determine the constant multiplicate factor $k = 2^{l/s}$, where $l = 0, 1, 2, 3$, and $s = 3$. We produce $s + 1$ images in the stack of scaled (blurred) images for each octave.

To effectively generate the scaled images, we increase the size of Gaussian filter to substitute the down-sampled image [2] between octaves as illustrated in Fig. 3. In our work, the Gaussian filter, σ, between octaves is increased by factor 2 as shown in Fig. 3. The initial σ of each octave is doubled compared to its previous octave. For example, in the 1st octave, we decompose the initial image

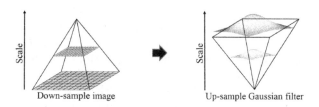

Fig. 3. The selection of scale (Color figure online)

into several blurred images; the scales of the blurred images are between σ and 2σ. In the 2nd octave, the scales of the blurred images are between 2σ and 4σ, as shown in Fig. 2.

3.2 Scale-Relation Coupling

To extract the features with abundant information, we propose the scale-relation coupling scheme for constructing the corresponding relationship between two properties of images. To enhance readability, we denote one property of image as X; the other is denoted as Y.

We firstly apply the rule of separating an image on scale space which is described in Sect. 3.1 to X and Y, respectively. After that, we construct the correspondence relationship between X and Y according to the following equation,

$$M_{ij}^s = \{X_i^s, Y_{j=i+k}^s\}, \quad \forall\, i = 0,1,2,3 \quad k = -1,0,1 \quad s \in S = \{0,1,2,...\} \quad (3)$$

where X_i^s and Y_j^s are the scaled images of X and Y, respectively. S is the level of octave, and i and j are the number of scaled (blurred) images on the octave. M_{ij}^s denotes the coupling of scaled images, X_i^s and Y_j^s. For each octave, we consider the scale-relation of X_i^s to the corresponding adjacent scales Y_j^s with 3 scales. For example, the scale-relation coupling M_{ij}^s of the blurred image X_1^s are $M_{10}^s = \{X_1^s, Y_0^s\}$, $M_{11}^s = \{X_1^s, Y_1^s\}$, and $M_{12}^s = \{X_1^s, Y_2^s\}$, as shown in Fig. 4. The top and bottom scales of X_i^s only have two scale-relation couplings, which are indicated as black and blue lines in Fig. 4.

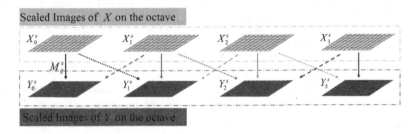

Fig. 4. The scale-relation Coupling (Color figure online)

3.3 Scale-Relation Feature Extracting

We design the Algorithm 1 to evaluate the scale-relation features (SRF) which associate the technique of separating an image with scale-relation coupling. The input data are two images with various properties, X and Y, and one parameter, S, which is described in Sect. 3.2. We firstly separate X and Y into the scaled images, X_i^s and Y_j^s, as described in Sect. 3.1. Then, we consider the corresponding relationship between X_i^s and Y_j^s with 3 scales by using Eq. 3. Next, we calculate the scale-relation features (SRF) for each property of shadow.

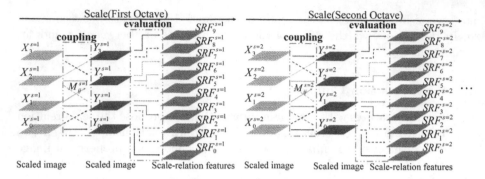

Fig. 5. The proposed framework for generating the scale-relation features (Color figure online)

The framework for generating the scale-relation features is shown in Fig. 5. In our work, we chose three useful attributes of shadow, including chromaticity, physical properties, and texture with various properties to evaluate the scale-relation features (SRF). The details of extracting the scale-relation features of these attributes are described in the following.

To extract the scale-relation features of chromatic properties, we refer to [4] and revise the equations to calculate the properties of chromaticity with scale-relation according to the following equations,

$$C_p^{H_{ij}} = \left| X_p^{H_i^s} - Y_p^{H_j^s} \right|, \tag{4}$$

$$C_p^{S_{ij}} = X_p^{S_i^s} - Y_p^{S_j^s}, \tag{5}$$

$$C_p^{V_{ij}} = X_p^{V_i^s} \Big/ Y_p^{V_j^s}, \tag{6}$$

where X_p and Y_p represent the current and background images at pixel p and the superscript H, S, V describe the properties of hue, saturation and value, respectively. $X_p^{H_i^s}$, $Y_p^{H_j^s}$ are the coupling images M_{ij}^s of current and background images with hue properties in octave s with scale i and j. Similarly, $X_p^{S_i^s}$ and

---

Algorithm 1. The scale-relation feature extraction algorithm

1: **Input:** Two images with various properties X and Y, octaves S
2: **Output:** Scale-relation features, SRF
3: Initializing an empty scale-relation features, SRF
4: Calculating the scaled images of two properties, X_i^s and Y_j^s.
5: Coupling the scaled images of two properties, M_{ij}^s.
6: **for** number of octaves s **do**
7:     **for** each scaled coupling images M_{ij}^s **do**
8:         Calculating the scale-relation features, SRF^s
9:     **end**
10:     $SRF = SRF + SRF^s$
11: **end**

$Y_p^{S_j^s}$ are the coupling images of current and background images with saturation properties, and $X_p^{V_i^s}$ and $Y_p^{V_j^s}$ are the coupling images of current and background images with value properties. $C_p^{H_{ij}}$, $C_p^{S_{ij}}$, and $C_p^{V_{ij}}$ are the scale-relation features of chromaticity. If $C_p^{H_{ij}} > 90$, $C_p^{H_{ij}} = 180 - C_p^{H_{ij}}$. Equations (4) and (5) are the differences between the current and background images by considering the properties of hue and saturation, respectively. Equation (6) is the proportion of intensity of current and background images.

To generate the physical properties, we refer to [5, 12] and revise the equations to calculate the physical properties with scale-relation according to the following equations,

$$P_p^{\phi_{ij}} = \cos^{-1}\left(X_p^{B_i^s} \Big/ \|X_p^{s,j}\| \right), \tag{7}$$

$$P_p^{\theta_{ij}} = \tan^{-1}\left(X_p^{G_i^s} \Big/ \|X_p^{R_j^s}\| \right), \tag{8}$$

$$P_p^{\alpha_{ij}} = \|X_p^{s,i}\| \Big/ \|Y_p^{s,j}\|, \tag{9}$$

where X_p^R, X_p^G and X_p^B describe the components of red, green and blue of the current image at the pixel p. $\|X_p\|$ and $\|Y_p\|$ are the determinants of current and background images at pixel p. P_p^{ϕ} is the angle between blue component and color of current image at pixel p, P_p^{θ} is the angle between green and red components of current image at pixel p, P_p^{α} is the illumination attenuation of the current and background images. $X_p^{B_i^s}$, and $X_p^{s,j}$, are the coupling images M_{ij}^s in octave s with scale i and j. Similarly, $X_p^{G_i^s}$ and $X_p^{R_j^s}$ are the coupling images of green and red channels of current images. $\|X_p^{s,i}\|$ and $\|Y_p^{s,j}\|$ are the coupling images of the determinant of current and background images. $P_p^{\phi_{ij}}$, $P_p^{\theta_{ij}}$, and $P_p^{\alpha_{ij}}$ are the scale-relation features of physical properties.

To calculate the texture features with scale-relation, we revise the widely used method [11, 12] as shown in following,

$$T_p^{\lambda_{ij}} = \left(X_p^{s,i}/Y_p^{s,j} \right) \Big/ \left(X_p^{s,i} - Y_p^{s,j} \right), \tag{10}$$

$$T_p^{\nabla_{ij}} = \sqrt{\left(\nabla X_x^{s,i}\right)^2 + \left(\nabla X_y^{s,i}\right)^2} - \sqrt{\left(\nabla Y_x^{s,j}\right)^2 + \left(\nabla Y_y^{s,j}\right)^2}, \tag{11}$$

$$T_p^{\theta_{ij}} = \tan^{-1}\left(\left(\nabla X_y^{s,i}\right) \Big/ \left(\nabla X_x^{s,i}\right) \right) - \tan^{-1}\left(\left(\nabla Y_y^{s,j}\right) \Big/ \left(\nabla Y_x^{s,j}\right) \right), \tag{12}$$

where X_p and Y_p represent the current and background images at pixel p. ∇X_x and ∇X_y are the horizontal and vertical gradient of the current frame, respectively. ∇Y_x and ∇Y_y are the horizontal and vertical gradient of the background frame, respectively. T_p^{λ} is the difference between the current and background frames. T_p^{∇} and T_p^{θ} are the gradient and orientation of each pixel p, respectively.

$X^{s,i}$ and $Y^{s,j}$ are the coupling images, M_{ij}^s, of the current and background images on octave s with scale i and j. $T_p^{\lambda_{ij}}$, $T_p^{\nabla_{ij}}$, and $T_p^{\theta_{ij}}$ are the scale-relation features of texture properties.

4   Ensemble Decision Scheme

After extracting the scale-relation features, we use the Random Forest technique [3] to be the decision scheme. The random forest algorithm use a set of samples f to construct several decision trees T_i, each decision tree T_i is assembled as a forest F, as shown in Fig. 6. In our work, the set of samples f collects scale-relation features includes $C_p^{H_{ij}}$, $C_p^{S_{ij}}$, $C_p^{V_{ij}}$, $P_p^{\phi_{ij}}$, $P_p^{\theta_{ij}}$, $P_p^{\alpha_{ij}}$, $T_p^{\lambda_{ij}}$, $T_p^{\nabla_{ij}}$, and $T_p^{\theta_{ij}}$.

For each decision tree T_i, it is consist of the non-leaf and leaf nodes in a forest $F = \{T_i\}$, and is constructed by randomly subsampling from the set of scale-relation features (SRF), f. The non-leaf and leaf nodes in each decision tree T_i are indicated as blue circle, and orange rectangle in Fig. 6. To start from the root, each arriving subsample SRF is evaluated and passes to the left or right child node by the splitting criterion of each non-leaf nodes. The training process recursively continues until it achieves the terminal criteria: (1) Reaching the maximum depth; (2) a minimum number of training samples are left.

Each decision tree T_i predicts that an input SRF belongs to foreground or shadow with maximum probability. The forest F integrates each decision tree and predicts with majority voting.

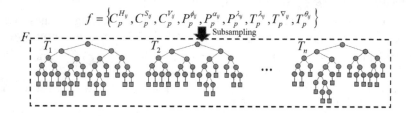

Fig. 6. The framework of an ensemble-decision scheme (Color figure online)

5   Experimental Results

In this section, we first describe the experimental environments and the comparative evaluation. Then, we demonstrate the performance of the proposed method compared with several popular methods.

Table 1. Image sequences used in the comparative evaluation.

Datasets	Sequence type	Image size	Shadow strength	Shadow size	Object size	Noise level	Object speed (pixels)
Campus	Outdoor	352 × 288	Weak	Large	Medium	High	5–10
Hallway	Indoor	320 × 240	Weak	Medium	Variable	Medium	5–15
Highway1	Outdoor	320 × 240	Strong	Large	Large	Medium	30–35
Highway3	Outdoor	320 × 240	Strong	Small	Small	High	20–25
Laboratory	Indoor	320 × 240	Weak	Medium	Medium	Low	10–15
Intelligent room	Indoor	320 × 240	Medium	Medium	Medium	Medium	2–5

5.1   Experimental Environments and Benchmarks

Six videos with benchmark[1] are used to demonstrate the performance of the proposed method as shown in Table 1. Table 1 shows various experimental environments including Campus, Hallway, Highway1, Highway3, Laboratory and Intelligent room. Each video has various conditions: indoor and outdoor environments; various size and strength of shadow; various sizes and speeds of object; different levels of noise. The sizes of shadow and that of object are relative to the frame size. The speeds of object are in pixels per frame. The strength of shadow is relative to the testing sequences.

The Campus sequence is an outdoor environment with particularly noise. The Hallway sequence is an indoor environment with a textured background and it has various sizes of objects and shadows. The Highway1 and Highway3 present two different lighting conditions of traffic environments. The Laboratory and Intelligent room are indoor environments, which show different perspectives and lighting conditions.

5.2   Quantitative Evaluation

We use two metrics, shadow detection rate (η) and shadow discrimination rate (ξ), to evaluate the performance of each method. The shadow detection (η) and shadow discrimination (ξ) rates are proposed by [10], and are calculated according to the following equations,

$$\eta = \frac{TP_s}{TP_s + FN_s} \text{ , and } \quad \xi = \frac{TP_f}{TP_f + FN_f} \tag{13}$$

where s and f refer to shadow and foreground, respectively. TP_s is the number of pixels which are the shadow and be marked as shadow. TP_f represents the number of pixels which are the foreground and be marked as foreground. FN_s represents the number of pixels which are the shadow and be marked as foreground. FN_f represents the number of pixels which are the foreground and be marked as shadow.

---

[1] http://arma.sourceforge.net/shadows/.

5.3   Experiments of the Proposed Methodology

We compare the proposed method with various popular methods, including chromaticity method [4], physical method [5], small region (SR) texture method [7] and large region (LR) texture method [11]. The results of the compared methods are referred to the paper [12].

Figure 7 shows the average shadow detection η and discrimination ξ rates on each testing sequence. The bar with blue, orange, gray, yellow, and purple colors indicate the results of chromaticity, physical, SR texture, LR texture, and the proposed methods. A diamond with blue colors and a triangle with green color indicate the individual detection and discrimination rates, respectively.

From Fig. 7, although the LR texture method has the best performance in Highway3, the proposed method has the best results on the average, and performs well in all datasets. The proposed method has the best performance of shadow detection rate η and has good performance of the shadow discrimination rates ξ in all datasets.

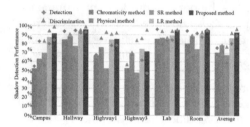

Fig. 7. Experimental results of the proposed method compared with popular methods. (Color figure online)

Fig. 8. Performance of the proposed method with various number of decision trees.

Figure 8 demonstrates the performance of the proposed method with various number of decision tree in ensemble decision scheme. From Fig. 8, although the accuracy of the proposed features has slightly decreased when the number of decision tree is decreased, the entire performance of each testing sequence is stable. In Fig. 8, three testing sequences, Hallway, Laboratory and Intelligent room, perform well and have similar performance. Since the Highway1 and Highway3 sequences have less training data and that make the performance not ideal compared to the other three testing sequences.

Table 2 presents the computational complexity of the proposed method with various number of decision tree. The criterion of average processing time per frame is used to evaluate the computational complexity. The computational time is shown in milliseconds which are obtained on a 64-bit Intel CPU Core i7 running at 2.10 GHz. From Table 2, the larger number of decision trees is, the longer the processing time is. According to Fig. 8 and Table 2, we can chose the most appropriate number of decision trees for shadow detection in the visual surveillance system.

Table 2. Average frame processing time (in milliseconds) per sequence of the proposed method with various number of decision tree.

Datasets	100 DT	50 DT	25 DT	10 DT
Campus	53.95	31.95	21.80	12.48
Hallway	135.90	74.66	44.43	27.86
Highway1	301.66	150.88	90.94	46.15
Highway3	28.92	16.82	11.87	8.79
Laboratory	165.97	96.07	60.61	35.57
Intelligent room	48.51	30.06	18.91	11.99
Average	122.47	66.74	41.42	23.80

Table 3 shows the visualization results of the proposed method. The 1st row is the current frame; the 2nd row is the ground truth where the foreground and shadow pixels are indicated as white and gray, respectively; and the 3rd row is the results of the proposed method. In the 3rd row, the shadow pixels classified as shadow are marked with green; the shadow pixels misclassified as foreground are marked with red; the foreground pixels classified as foreground marked with blue; and the foreground pixel misclassified as shadow are marked with yellow. From Table 3, the false positives, which are foreground pixels and be misclassified as shadow pixels, are occurred around the moving casts. The false negatives, which are shadow pixels and be misclassified as foreground pixels, are occurred inside the moving casts.

Table 3. Visualization experimental results of proposed method

Datasets	Campus	Hallway	Highway1	Highway3	Laboratory	Intelligent room
Current Frames						
Ground Truth						
Proposed method						

6   Conclusion

In this paper, we propose a new feature-extracting framework to associate with ensemble decision classifier to address the problem of moving cast shadow called Scale-Relation Feature Extracting (SRFE). Using the proposed method, we can map various properties into scale-relation space. In SRFE, we extract the features from the attributes of chromacity, physical properties, and texture with scale-relation. Each attribute has several properties. We decompose each image with various properties into several scales using the Gaussian function. The scale-relation features between scaled images with two properties are calculated. We integrate an ensemble decision classifier with scale-relation features to generate the efficient model for shadow detection.

In the experimental results, we demonstrate the performance of the proposed method compared with various popular methods, chromacity, physical, SR texture, and LR texture methods. Comparing the proposed method with these methods, the proposed method performs well in all dataset and has the best performance on the average. The proposed method has the best performance of shadow detection rate η and has good performance of the shadow discrimination rates ξ in all datasets. The results show that the proposed method performs well and it is more suitable for various environments.

Acknowledgements. This work was supported in part by the Natural Science Foundation of Fujian Province of China, under grant 2016J01718.

References

1. Al-Najdawi, N., Bez, H.E., Singhai, J., Edirisinghe, E.A.: A survey of cast shadow detection algorithms. Pattern Recogn. Lett. **33**(6), 752–764 (2012)
2. Bay, H., Ess, A., Tuytelaars, T., Van Gool, L.: Speeded-up robust features (SURF). Comput. Vis. Image Underst. **110**(3), 346–359 (2008)
3. Breiman, L.: Random forests. Mach. Learn. **45**(1), 5–32 (2001)
4. Cucchiara, R., Grana, C., Piccardi, M., Prati, A.: Detecting moving objects, ghosts, and shadows in video streams. IEEE Trans. Pattern Anal. Mach. Intell. **25**(10), 1337–1342 (2003)
5. Huang, J.B., Chen, C.S.: Moving cast shadow detection using physics-based features. In: IEEE Conference on Computer Vision and Pattern Recognition, CVPR 2009, pp. 2310–2317. IEEE (2009)
6. Koenderink, J.J.: The structure of images. Biol. Cybern. **50**(5), 363–370 (1984)
7. Leone, A., Distante, C.: Shadow detection for moving objects based on texture analysis. Pattern Recogn. **40**(4), 1222–1233 (2007)
8. Lindeberg, T.: Scale-space theory: a basic tool for analyzing structures at different scales. J. Appl. Stat. **21**(1–2), 225–270 (1994)
9. Lowe, D.G.: Distinctive image features from scale-invariant keypoints. Int. J. Comput. Vis. **60**(2), 91–110 (2004)
10. Prati, A., Mikic, I., Trivedi, M.M., Cucchiara, R.: Detecting moving shadows: algorithms and evaluation. IEEE Trans. Pattern Anal. Mach. Intell. **25**(7), 918–923 (2003)
11. Sanin, A., Sanderson, C., Lovell, B.C.: Improved shadow removal for robust person tracking in surveillance scenarios. In: 2010 20th International Conference on Pattern Recognition (ICPR), pp. 141–144. IEEE (2010)
12. Sanin, A., Sanderson, C., Lovell, B.C.: Shadow detection: a survey and comparative evaluation of recent methods. Pattern Recogn. **45**(4), 1684–1695 (2012)

Smart Loudspeaker Arrays
for Self-Coordination and User Tracking

Jungju Jee and Jung-Woo Choi[✉]

School of Electrical Engineering, KAIST, Daejeon, Korea
{ji1718,jwoo}@kaist.ac.kr

Abstract. The Internet of Things paradigm aims at developing new services through the interconnection of sensing and actuating devices. In this work, we demonstrate what can be achieved through the interaction between multiple sound devices arbitrarily deployed in space but connected through a unified network. In particular, we introduce techniques to realize a smart sound array through simultaneous synchronization and layout coordination of multiple sound devices. As a promising application of the smart sound array, we show that acoustic tracking of a user-location is possible by analyzing scattering waves induced from the exchange of acoustic signals between multiple sound objects.

Keywords: Self-coordination · Acoustic user detection · Smart sound objects

1 Introduction

Recently many wireless audio players have been introduced in the market. Most of these audio players play a simple role of receiving and reproducing streamed audio signals from host devices. Other applications of such sound devices have been also limited to the multi-room sound production, discrete multichannel surround, and speech recognition. Such wireless audio players, especially wireless loudspeakers, have a great potential to open the new era in home audio services. Suppose that individual loudspeakers include acoustic sensors such as microphones and loudspeakers with intelligence can communicate with each other through a unified sound network. These smart loudspeakers with connectivity can construct a ubiquitous acoustic sensing and transmission architecture. Such an architecture can be realized in form of wireless loudspeakers satisfying a new sound network standard or by devising a new device for wireless communication and acoustic sensing that can be connected to the conventional audio system.

The proliferation of such devices in home environments can realize the Internet of Things (IoT) paradigm [1], in which seamless, mutual interaction between smart objects provides new information and services. When multiple sound devices are incorporated as interconnected smart objects of IoT, they can be a single microphone array collecting information from the environment or can interact with the physical word as a multichannel loudspeaker array controlling a sound field.

For a long time, loudspeaker and microphone arrays have been utilized for various applications. In the context of beamforming, microphone arrays have been used for

© Springer International Publishing AG 2017
L. Amsaleg et al. (Eds.): MMM 2017, Part II, LNCS 10133, pp. 343–355, 2017.
DOI: 10.1007/978-3-319-51814-5_29

sound source localization, noise source identification, and de-reverberation for speech recognition. The use of loudspeaker arrays for discrete multichannel systems (such as 5.1, 7.1 surround) and personal sound systems can be easily found in many commodity devices. Through the combination of microphone and loudspeaker arrays, many acoustic-based indoor localization and surveillance systems have been proposed. All of these are potential applications of an IoT architecture constructed from a smart loud-speaker array and smart sound objects.

Smart connectivity between smart sound objects and environment-aware processing is the key of IoT-based on sound devices. In this work, we attempt to demonstrate promising applications of a smart loudspeaker array through experimental studies in a real environment. In particular, we focus on two functionalities that can be realized through the interaction of multiple sound objects: the self-configuration and user detection abilities.

In contrast to conventional systems, an array consisting of smart sound objects does not have pre-determined configuration or layout. The sound objects can be installed at arbitrary positions in a listening space, so it is natural to assume that objects locations are sparse and can be time varying (Fig. 1). Since smart objects should operate in a seamless way, an adaptive and decentralized control strategy is required to reflect their time-varying locations. In this regard, the first functionality required for smart sound objects is the self-configuration. In-situ coordination of individual sound objects offers a consistent sound rendering and listening experience irrespective of changes in the loudspeaker arrangement.

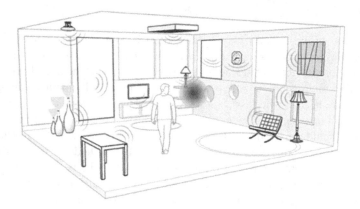

Fig. 1. Conceptual illustration of a smart array consisting of multiple sound objects

For such self-coordination, time-synchronization between multiple objects should be accurate and precise. Inaccurate time synchronization between objects induces audible artifacts, especially on the spatial perception of sound. For example, the human auditory system uses the difference in the time of arrival at both ears to find the location of a sound source. This inter-aural time difference varies in the ± 1 ms range, depending on the sound direction of arrival. Therefore, the clock of each object should be synchronized in the hundreds of microseconds scale at least.

The self-coordination and synchronization of multiple sound objects can be approached in many different ways, e.g., by using radio frequency signals and simplified network time protocol (SNTP). Nevertheless, the SNTP-based solution would be incomplete due to non-identical audio playback latencies of sound systems. In other words, even if tight synchronization is made on the networking layer, different sound systems connected to network devices yield non-identical audio start time because of their different audio playback latencies. In this respect, it is desirable to identify the whole audio latency in real-time by means of acoustic signals. The ubiquitous sensing and processing of IoT paradigm pursuits seamless integration of services, so all these processes have to work in the background, hidden from the user. For this reason, the self-coordination performance of a smart array is investigated using inaudible probing signals in high-frequency range.

Once the stable self-coordination is made, the smart array can operate as a unified sound system. Owing to its sparse arrangement, the smart array may require novel control strategies different from conventional arrays with pre-determined regular geometries. However, on the other hand, the sparse layout can be beneficial for understanding the user context and surrounding environment. As an example, we demonstrate that the user tracking is also possible from further analysis on the IRs measured for the self-coordination.

2 Preliminary Study

Among many studies on self-coordination [2–5], the use of acoustic signals to coordinate between multiple mobile phones can be found in Ref. [6]. The mobile phone is a good candidate as a smart sound object because it has a loudspeaker, embedded microphones, wireless connectivity, and the processing unit required for decentralized signal processing. In this work, a service (Adaptive Mobile Audio Coordination; AMAC) based on new Android API (Sound Pool) was developed (Fig. 2). The API was designed to measure the mutual distance between multiple mobile devices and perform tight synchronization for multichannel sound reproduction. The Sound Pool API determines the playback start time and sound level of each individual mobile phone using measured IRs between every mobile phone pairs. In the preliminary step,

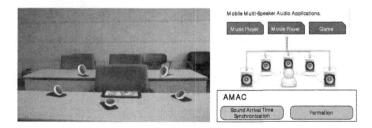

Fig. 2. Overview of the adaptive mobile audio coordination (AMAC); multiple mobile devices arbitrarily positioned in space are controlled to identify their own locations and to produce the 5.1 channel sound

rough synchronization is made through a network-based clock synchronization method (SNTP). Then fine tuning is accomplished through the exchange of acoustic probing signals. The transmission and reception of maximum length sequences between N mobile phones (nodes) provide N^2 responses, whose direct peaks include the information of time differences and mutual distances. The fine synchronization using acoustic signals showed that we can decrease the time mismatch between mobile devices from milliseconds scale (Fig. 3(a)) to hundreds of microseconds scale (Fig. 3(b)), even for the dynamic layout change. The algorithms behind this synchronization are shortly reviewed in the following section, with new algorithms modified to shorten the measurement time and increase the update rate.

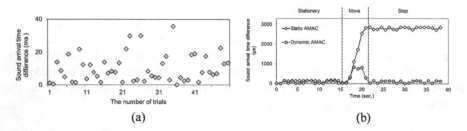

(a)   (b)

Fig. 3. Synchronization between multiple mobile phones. (a) Time difference on the Android mobile phone (b) Time sync. accuracy for dynamically changing loudspeaker layout [6]

3 Self-Coordination of Sound Objects

The self-coordination step encompasses the identification of unknown positions $\mathbf{x}_n = [x_n, y_n, z_n]^T (n = 1, \cdots, N)$ and relative time mismatches between multiple sound objects, from the measurement of mutual distances $r_{mn} = \|\mathbf{x}_m - \mathbf{x}_n\|$ (Fig. 4). The basic coordination principle is similar to that used in the previous study [6], but in here we focus more on the reduction of measurement time.

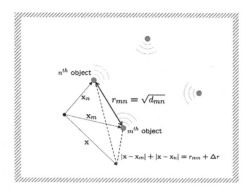

Fig. 4. Position-related variables of multiple sound objects

3.1   Impulse Response Measurement

To identify mutual distances and time mismatches between multiple sound objects, we first need to measure IRs between them. The IRs are measured at each time frame, so the frame's refresh rate can be slow if the measurement time is too long. To shorten the measurement time, devices are simultaneously driven by different chirp signals of non-overlapping frequency bands. Simultaneous acoustic emissions are captured by all microphones and decomposed to extract IRs of multiple devices. In this work, total four loudspeakers are considered (Fig. 5), and logarithmic chirp signals ranging between 18 kHz ∼ 24 kHz were used for making inaudible probing sound signals.

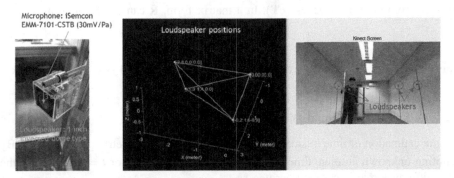

Fig. 5. Loudspeaker and microphone configurations for the impulse response measurement, and self-coordination of four loudspeakers in a room

The simultaneous excitation requires orthogonality between probing signals to avoid mutual interference between different sound emissions from sound objects. In this regard, an OFDM-like architecture is used, for which a probing signal of each sound object has different frequency band from the others. The total frequency band of bandwidth 6 kHz is split into four 1.5 kHz bands, which results in the sampling rate of 3 kHz for each sound object. In the experiment setup shown in Fig. 5, measurements were done in a rectangular room of size $(W \times L \times H = 3.33\,\text{m} \times 7.28\,\text{m} \times 2.58\,\text{m})$ and reverberation time $T_{20} = 0.2$ s. Considering the reverberation time, probing signals of length $T = 50$ ms with a guard time of $T_g = 100$ ms was used. As a result, a frame rate of approximately 6.7 frames/s could be obtained.

For measuring IRs, each microphone was attached close to a loudspeaker with 3 cm distance. The loudspeaker used for the experiment is a 1-in. full-range driver installed in a custom enclosure, and heights of loudspeakers from the floor was 1.6 m (3 loudspeakers) and 1.4 m (1 loudspeaker). Although asynchronous conditions were investigated for the previous study with mobile phones, in this work we will only consider wired and synchronous condition because time mismatches between devices could be compensated with enough accuracy in the previous study [6].

The IRs centered at different frequency bands are demodulated by applying Hilbert transform, to extract envelope profiles $C_{mn}(t)$. Here, n and m are indices of transmitting and receiving objects, respectively. Figure 6(a) depicts 4×4 envelopes of measured IR data, each of which is normalized by the maximum absolute magnitude $\text{Max}(|C_{mn}(t)|)$. Peak locations of direct waves in measured IRs provide the distance and time mismatch information for finding out object locations. In order to increase the detectability of direct peaks, IR spectra are whitened by a linear prediction (LP) filter. Then the peak search is done by an ad-hoc version of Root-MUSIC [8, 9], which directly provides time delay estimates from a set of shifted frequency responses. Among dominant peaks collected from each IR, a leading peak exceeding a pre-defined threshold is chosen as a direct peak that provides delay estimates (Fig. 6(a)).

Estimated delays are then converted into distances by multiplying with the speed of sound (c). The measured distances construct a distance matrix $\mathbf{R}$. Due to time mismatches between devices, the matrix includes true distances ($\mathbf{R}_0$), as well as erroneous distance from time mismatches ($c\mathbf{T}$). In a matrix form, $\mathbf{R}$ can be rewritten as

$$\mathbf{R} = \mathbf{R}_0 + c\mathbf{T} + \mathbf{N}, \tag{1}$$

where the matrix $[\mathbf{T}]_{mn} = \Delta t_{mn}$ denotes the time mismatch, and $\mathbf{N}$ is a noise matrix.

3.2   Estimation of Mutual Distance and Time Mismatch

For the estimation of time mismatch, we utilize the skew-symmetric property of $\mathbf{T}$. By denoting unknown absolute time references of devices as a vector $\mathbf{t} = [t_1, \cdots, t_n]^T$, the time difference matrix can be rewritten as

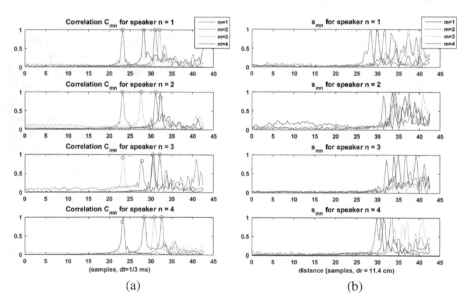

(a)   (b)

Fig. 6. (a) Normalized cross-correlation between four loudspeakers (m: Rx, n: Tx). (b) Successive subtraction $s_{mn}(r)$ obtained from the normalized cross correlation

$$\mathbf{T} = \mathbf{t1}^T - \mathbf{1t}^T, \ (\mathbf{1} = [1, \cdots, 1]^T) \tag{2}$$

which yields the skew-symmetric property. Therefore, the skew-symmetric part $(\mathbf{R}_a)$ of $\mathbf{R}$ gives $c\mathbf{T}$ plus the skew-symmetric part of a noise matrix $(\mathbf{N}_a)$. That is

$$\mathbf{R}_a = \frac{1}{2}(\mathbf{R} - \mathbf{R}^T) = c\mathbf{T} + \mathbf{N}_a. \tag{3}$$

It is well-known that eigenvalues of a skew-symmetric matrix come in complex conjugate pairs [11], and its rank is always even. From the rank theorem $(\mathrm{rank}(\mathbf{A} + \mathbf{B}) \leq \mathrm{rank}(\mathbf{A}) + \mathrm{rank}(\mathbf{B}))$, the rank of the matrix $\mathbf{T}$ is 2 at maximum. To reduce the effect of measurement noises, we can consider the best rank-2 approximation of $\mathbf{R}_a$. For a singular value decomposition $\mathbf{R}_a = \mathbf{U}\mathbf{\Sigma}\mathbf{V}^T$, the best rank-2 approximation is given by collecting two largest singular values and eigenvectors

$$\tilde{\mathbf{T}} = \frac{1}{c}\mathbf{U}_2\mathbf{\Sigma}_2\mathbf{V}_2^T. \tag{4}$$

Likewise, the positions of sound objects are from the symmetric part of $\mathbf{R}$

$$\mathbf{R}_s = \frac{1}{2}(\mathbf{R} + \mathbf{R}^T) = \mathbf{R}_0 + \mathbf{N}_s, \tag{5}$$

where $\mathbf{N}_s = \frac{1}{2}(\mathbf{N} + \mathbf{N}^T)$ is the symmetric part of a noisy distance matrix $\mathbf{N}$. To reduce the influence of noise, Multi-dimensional scaling (MDS) technique [10, 13] is often employed to the Euclidian distance matrix (EDM). An EDM is defined as the squared distance matrix, which can be related to a position matrix

$$\mathbf{X} = [\mathbf{0} \quad \mathbf{x}_2 \quad \cdots \quad \mathbf{x}_N], \tag{6}$$

whose the coordinate origin is set at the location of the first object. The inner product of $\mathbf{X}$ yields a positive semidefinite Gram matrix $\mathbf{G} = \mathbf{X}^T\mathbf{X}$, which can be related to the squared distance d_{mn} from the relation

$$\begin{aligned} d_{mn} &= \|\mathbf{x}_m - \mathbf{x}_n\|^2 = \|\mathbf{x}_m\|^2 + \|\mathbf{x}_n\|^2 - 2\mathbf{x}_m^T\mathbf{x}_n \\ &= d_{m0} + d_{n0} - 2\mathbf{x}_m^T\mathbf{x}_n. \end{aligned} \tag{7}$$

Rewriting Eq. 7 in a matrix form gives

$$\mathbf{G} = -\frac{1}{2}(\mathbf{D} - \mathbf{d1}^T - \mathbf{1d}^T), \tag{8}$$

which implies that the Gram matrix can be derived from the squared distance matrix $\mathbf{D}$ and its first column vector $\mathbf{d}$ multiplied by a vector $\mathbf{1}$. Once the Gram matrix is estimated, it is possible to estimate the position matrix $\mathbf{X}$ as well. The position matrix $\mathbf{X}$ is of rank 3, so SVDs of $\mathbf{X}$ and $\mathbf{G}$ can be expressed by the combination of three largest (nonzero) singular values and singular vectors $(\mathbf{\Sigma}_3, \mathbf{U}_3, \mathbf{V}_3$, respectively).

$$\mathbf{X} = \mathbf{U}_3\mathbf{\Sigma}_3\mathbf{V}_3^H, \quad \mathbf{G} = \mathbf{V}_3\mathbf{\Sigma}_3^2\mathbf{V}_3^H \tag{9}$$

Therefore, for a noise-free EDM, the eigenvalue analysis on Eq. (8) should produce the same eigenvectors $\mathbf{V}_3$ and singular value $\mathbf{\Sigma}_3$. Because the noise matrix that does not follow this rule, it can be filtered out by applying the best rank-3 approximation. After three largest singular values and eigenvectors are identified, the position can be estimated as $\tilde{\mathbf{X}} = \mathbf{\Sigma}_3\mathbf{V}_3^H$ [10].

From this low-rank approximation $\tilde{\mathbf{X}}$, however, one cannot estimate rigid rotation/reflection of the whole geometry expressed by the unitary transform $\mathbf{U}_3$. Therefore, the geometry predicted from each measurement should be rotated/reflected to a known anchor plane such that the estimation is consistent with that from the previous measurement frame. The best rotation/reflection can be found from Kabsch algorithm [14], which searches for the orthogonal transform that minimizes the Frobenius norm of differences between two geometries. Figure 5 shows a single instance of real-time coordination when a user moves a loudspeaker to a different position.

4 User Activity Tracking from Scattering Waves

With stable estimation on object positions, more useful information can be extracted by analyzing IRs between smart objects. Here, we attempt to track a user position in real-time. Since the IRs recorded over a single time frame include scattering information along their propagation paths, we can utilize this information to infer the position or shape of a scattering body. Similar research has been done by Abib *et al.* [7] in terms of RF waves, and here we investigate the detectability using acoustic waves in a reflective room.

4.1 Basic Inference from Measured Impulse Responses

The envelope functions include reflections from the listener's body, but they also include reflections from static scattering objects such as room boundary and furniture. Since reflections from static reflectors remain constant over time, the static profile could be eliminated by subtracting the profile of the previous frame from the current one. Using the envelope profile $C_{mn}^{(i)}(r)$ measured at i^{th} frame, the difference profile can be calculated as

$$s_{mn}^{(i)}(r) = \left| C_{mn}^{(i)}(r) - C_{mn}^{(i-1)}(r) \right|. \tag{10}$$

These difference profiles include scattering information in two different ways. First, the auto-profile $s_{mm}^{(i)}(r)$ recording the sound emission from the mth loudspeaker using the mth microphone measures the scattering profile along the round trip path $2r$ from the mth loudspeaker. On the other hand, the cross-profile $s_{mn}^{(i)}(r) = s_{mn}^{(i)}(r_{mn} + \Delta r)$ between different loudspeakers records the scattering information along the path difference $\Delta r = |\mathbf{x} - \mathbf{x}_m| + |\mathbf{x} - \mathbf{x}_n| - r_{mn}$, where $r_{mn} = |\mathbf{x}_m - \mathbf{x}_n|$ is the direct path between two different loudspeakers. Since the direct path r_{mn} has been identified from the previous step, one can map the profile $s_{mn}^{(i)}(r)$ along an ellipse defined by the path difference Δr. By combining all difference profiles $\mathbf{s}^{(i)}$ $\left([\mathbf{s}^{(i)}] = s_{mn}^{(i)}\right)$, we can define a pseudo-likelihood function that expresses the probability to have a scattering object at a position $\mathbf{x}$. That is,

$$F(\mathbf{x}, \mathbf{s}^{(i)}) = \sum_{m=1}^{M} \sum_{n=1}^{N} 10 \log_{10} s_{mn}^{(i)}(|\mathbf{x} - \mathbf{x}_m| + |\mathbf{x} - \mathbf{x}_n|). \qquad (11)$$

This log-likelihood function expresses the joint probability from different scattering profiles acquired by different microphone-loudspeaker pairs.

The difference profiles $s_{mn}^{(i)}(r)$ extracted from the IR measurements of Fig. 6(a) are shown in Fig. 6(b). For the ease of visualization, all the difference profiles magnitude is normalized to one. These difference profiles show several peaks at different locations from the direct peaks of Fig. 6(a). These peaks are due to the change of a user position. These difference profiles are combined as a form of a likelihood function, to infer the actual user location in 3D space. Figure 7 presents this inference over a horizontal plane in form of a heat map. The rough estimate obtained only with one pair of loudspeakers (Fig. 7(a)) is enhanced as more loudspeaker pairs' data are added (Fig. 7(b, c)). It should be noted here that the user position tracking can be processed simultaneously with the loudspeaker coordination because the coordination and user tracking utilizes different time ranges (direct wave and reflections) among the measured IRs.

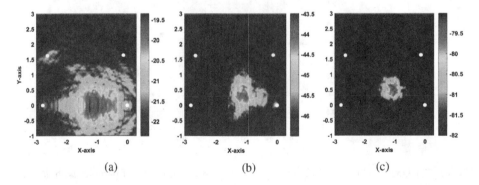

$$\text{(a)} \qquad\qquad \text{(b)} \qquad\qquad \text{(c)}$$

Fig. 7. A log-likelihood function $F(\mathbf{x}_l^{(t)}, \mathbf{s}^{(i)})$ at a single time instance i, constructed from successive subtraction signals $s_{mn}^{(i)}$ between loudspeaker pairs. (a) two Tx-Rx loudspeakers (1, 2) (b) three loudspeakers (1, 2, 3) (c) four loudspeakers (1, 2, 3, 4)

4.2   Tracking with a Particle Filter

Although it is also possible to visualize the likelihood function over a 3D space in form of 3D voxels, the inference over whole 3D space requires a significant amount of computations. Therefore, another means to extract 3D position parameters with reduced computational complexity is required. To this end, we use a particle filter (e.g., [12]) that samples and represents the posterior probability as a finite number of particles positioned at $\mathbf{x}_\ell^{(i)}$ ($\ell = 1, \cdots, L$) for the ith time frame.

For a particle filter, the likelihood function of Eq. (11) is used as a weighting of each particle $\left(w_\ell^{(i)} = F(\mathbf{x}_\ell^{(i)}, \mathbf{s}^{(i)}) \right)$, which is normalized such that total sum of weights is equal to one $\left(\sum_\ell w_\ell^{(i)} = 1 \right)$. In the next step, locations of particles are resampled according to the principle of importance sampling. After the resampling, each particle's dynamics is reflected according to a source dynamics model [12]. From updated particle positions, one can recalculate the particle weight using new measurements and corresponding likelihood function $F(\mathbf{x}_\ell^{(i+1)}, \mathbf{s}^{(i+1)})$. Estimation of a listener's location at each time frame is given by the weighted sum of particle positions, i.e.,

$$\tilde{\mathbf{x}} = \sum_{\ell=1}^{L} w_\ell^{(i)} \mathbf{x}_\ell^{(i)}. \tag{12}$$

Figure 8 shows the particle positions and weights evaluated at the same instance as Fig. 7. Total 300 particles distributed in 3D space is plotted, with the weight of each particle is visualized as the radius of a circle. The green dot indicates the weighted mean position (Eq. (12)), which is taken as the final output from the tracking system.

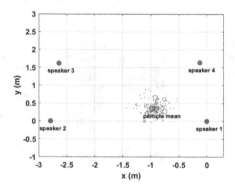

Fig. 8. Tracking of a human location using particle filter. Distribution of particles with different weights $w_\ell^{(i)}$. (green dot) position estimated from the weighted sum of particles (Color figure online)

4.3 Comparison to Vision-Based Tracking

The tracking performance of the proposed system for a single listener was compared with a vision-based tracking method. As mentioned in the previous section, our purpose is not on the tracking of a single position but with visualizing the user's body.

Nevertheless, we can roughly evaluate the accuracy of the proposed system using the mean position obtained from the particle filter.

For comparison, visual tracking data from Microsoft Kinect V2 was recorded at the same time. Then, the head position among the Kinect skeleton tracking data was taken as a position reference (Fig. 9). From the comparison with the Kinect data, it was found that the particle filter could track the single listener position with discrepancy less than 30 cm. However, the particle distribution shows that the scattering is weak along the height direction of the human body. This is mainly due to the mostly horizontal arrangement of loudspeakers, which is not enough to extract scattering information from the vertical direction. Besides of this limitation, the particles could trace the location of the upper body location only with 300 particles.

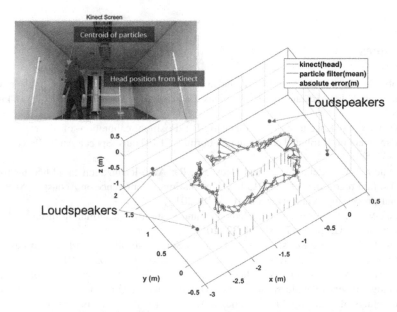

Fig. 9. Comparison of tracked positions and Microsoft Kinect's head position data. (red line) head positions from Kinect (blue line) mean position from a particle filter (green bar) magnitude of position difference (in meter) (Color figure online)

5   Summary and Conclusion

We demonstrated that distributed sound objects can form a unified sound array system through mutual interaction using acoustic signals. Loudspeakers with embedded microphones are taken as an example of smart sound objects, and exchange of inaudible, high-frequency acoustic signals could coordinate objects in 3D space in a seamless, unobtrusive way. As a new functionality of the smart loudspeaker array, a user tracking technique based on acoustic impulse responses was introduced. The user position could be traced by analyzing $N \times N$ scattering signals measured between N

loudspeaker-microphone pairs, and the tracking results were compared to those from a vision-based tracking system.

The current investigation shows acoustic signals measured and identified between multiple sound objects have a potential to extract various information of the sound system itself, the user and environment. This sound-based information tracking may not be superior in terms of accuracy and measurement speed, as compared to radio frequency or vision based systems. However, the major advantage of sound-based systems is that such information tracking is possible only with multiple audio systems, as far as microphones can be embedded to conventional network loudspeakers.

Acknowledgements. Work supported by Samsung Research Funding Centre of Samsung Electronics under Project Number SRFC-IT1301-04 and BK21 project initiated by the Ministry of Education.

References

1. Gubbi, J., Buyya, R., Marusic, S., Palaniswami, M.: Internet of things (IoT): a vision, architectural elements, and future directions. Future Gener. Comput. Sys. **29**(7), 1645–1660 (2013)
2. Raykar, V.C., Kozintsev, I.V., Lienhart, R.: Position calibration of microphones and loudspeakers in distributed computing platforms. IEEE Trans. Speech Audio Process. **13**(1), 70–83 (2005)
3. Gaubitch, N.D., Kleijn, W.B., and Heusdens, R.: Auto-localization in ad-hoc microphone arrays. In: Proceedings of the 2013 IEEE International Conference on Acoustics, Speech and Signal Processing, Vancouver, pp. 106–110 (2013)
4. Leest, A.J.V., Schobben, D.W.: Method of and system for determining distances between loudspeakers, US patent 7,864,631 B2 (2011)
5. Wang, L., Hon, T.K., Reiss, J.D., Cavallaro, A.: Self-localization of ad-hoc arrays using time difference of arrivals. IEEE Trans. Sig. Process. **64**(4), 1018–1033 (2016)
6. Kim, H., Lee, S.J., Choi, J.W., Bae, H., Lee, J.Y., Song, J.H., Shin, I.: Mobile maestro: enabling immersive multi-speaker audio applications on commodity mobile devices. In: Proceedings of Ubicomp 2014, Seattle, WA, USA, 13–17 September 2014
7. Adib, F., Kabelac, Z., Katabi, D.: Multi-person localization via rf body reflections. In: Proceedings of the 12th USENIX Symposium on Networked Systems Design and Implementation, Oakland, USA, 4–6 May (2015)
8. Rao, B.D., Hari, K.V.S.: Performance analysis of root-music. IEEE Trans. Sig. Process. **37**(12), 1939–1949 (1989)
9. Brandstein, M.S., Ward, D.B.: Microphone Arrays: Signal Processing Techniques and Applications. Springer, Berlin (2001)
10. Dokmanic, I.: Listening to distances and hearing shapes: inverse problems in room acoustics and beyond. Ph.D. thesis, Ecole Polytechnique Federale de Lausanne (EPFL) (2015)
11. Ye, G., Liu, D., Jhuo, I.H., Chang, S.F.: Robust late fusion with rank minimization. In: Proceedings of 2012 IEEE Conference on Computer Vision and Pattern Recognition, Providence, RI, pp. 3021–3028 (2012)
12. Ward, D., Lehmann, E., Williamson, R.: Particle filtering algorithms for tracking an acoustic source in a reverberant environment. IEEE Trans. Speech Audio Process. **11**(6), 826–836 (2003)

13. Torgerson, W.S.: Multidimensional scaling: I. theory and method. Psychometrika **17**(4), 401–419 (1952)
14. Kabsch, W.: A discussion of the solution for the best rotation to relate two sets of vectors. Acta Crystallogr. **A34**, 827–828 (1978)

Spatial Verification via Compact Words for Mobile Instance Search

Bo Wang, Jie Shao$^{(\boxtimes)}$, Chengkun He, Gang Hu, and Xing Xu

School of Computer Science and Engineering,
University of Electronic Science and Technology of China,
Chengdu, China
{wangbo,MatthewHe,hugang}@std.uestc.edu.cn,
{shaojie,xing.xu}@uestc.edu.cn

Abstract. Instance search is a retrieval task that searches video segments or images relevant to a certain specific instance (object, person, or location). Selecting more representative visual words is a significant challenge for the problem of instance search, since spatial relations between features are leveraged in many state-of-the-art methods. However, with the popularity of mobile devices it is now feasible to adopt multiple similar photos from mobile devices as a query to extract representative visual words. This paper proposes a novel approach for mobile instance search, by spatial analysis with a few representative visual words extracted from multi-photos. We develop a scheme that applies three criteria, including BM25 with exponential IDF (EBM25), significance in multi-photos and separability to rank visual words. Then, a spatial verification method about position relations is applied to a few visual words to obtain the weight of each photo selected. In consideration of the limited bandwidth and instability of wireless channel, our approach only transmits a few visual words from mobile client to server and the number of visual words varies with bandwidth. We evaluate our approach on Oxford building dataset, and the experimental results demonstrate a notable improvement on average precision over several state-of-the-art methods including spatial coding, query expansion and multiple photos.

Keywords: Mobile instance search · Multiple photos · Spatial verification

1 Introduction

Instance search is a type of retrieval tasks that searches relevant video segments or images of a given instance such as a specific person, object or place. It has drawn many research attention recently. Different from concept-based search, the results of instance search must have the same specific details as query. Meanwhile, different from near-duplicate search, the instance may appear in completely different background and its deformation may be severer in instance search. Due to the peculiarity of instance search, state-of-the-art methods such

© Springer International Publishing AG 2017
L. Amsaleg et al. (Eds.): MMM 2017, Part II, LNCS 10133, pp. 356–367, 2017.
DOI: 10.1007/978-3-319-51814-5_30

as [15–17] focused on spatial relations between local features. The "sketch-and-match" scheme [16] was designed to elastically verify the topological spatial consistency with the triangulated graph. Another method [17] formed a transformation consistency in rotation and scale space by exploring the geometric correlations among local features and incorporating these correlations with each individual match. All these methods proceed with spatial verification and delete wrong matched points after matching SIFT features [7].

Representative visual words can be selected before spatial verification to achieve better performance. Nevertheless, it also brings great challenge. Due to the popularity of mobile devices especially smart phones, in this work we propose a novel approach motivated by a common phenomenon that people normally take a few photographs of a concerned object to choose one with the best quality or save a few photographs of different views for one object. Intuitively, multiple photos can provide more comprehensive and detailed information than a single example to make visual words selected more representative and meaningful.

Figure 1 shows an overview of the proposed approach. It can be divided into two modules including an offline module and an online module. In the offline module, dataset images are described by SIFT features firstly and then hierarchical k-means clustering is applied to generate visual words. In the online module, after an image is submitted as a query, relevant photos are mined from a photo set taken recently and saved in user's mobile device. After described by SIFT features, multi-photos including the query and relevant photos are quantized to groups of visual words. Then, we rank visual words with three criteria: BM25 with exponential IDF (EBM25) [8], significance in multi-photos and separability. In consideration of the limited bandwidth and the instability of wireless network, we only transmit a few high ranking words to server side. Finally, a simple but effective spatial verification method is applied to return images which contain instance in dataset.

The main contributions of our work can be summarized as follows:

- We apply the idea of PageRank algorithm [2] to mine relevant photos which have the same instance with query from user's mobile device.
- We propose a novel solution to answer mobile instance search and adopt three criteria to select representative visual words. Our method requires small bandwidth and is robust with bandwidth change by the compact words selected by our approach.
- We propose an effective spatial verification method to retrieve video segments or images using a few representative visual words.
- We test our approach with a benchmark named Oxford building dataset [10], and show that our approach outperforms three state-of-the-art methods.

The rest of the paper is organized as follows. In Sect. 2, we introduce related work. Section 3 describes our approach in detail. We report our evaluation in Sect. 4, and finally Sect. 5 concludes the paper.

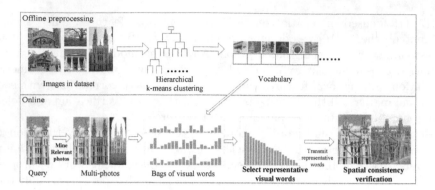

Fig. 1. The flowchart of our proposed approach. In the offline module, SIFT feature and hierarchical k-means clustering are applied to generate vocabulary for the following process. In the online module, relevant photos are firstly mined from user's mobile device after a query is submitted, then compact and representative visual words are selected and transmitted to server side, and finally, the server ranks and returns images by spatial consistency verification with the selected visual words.

2   Related Work

Most researchers have adopted bag-of-words (BOW)-based methods to address instance search. Sivic and Zisserman [12] firstly introduced BOW to object and scene retrieval, which described a set of local features as a histogram to count the occurrence of each visual word. Zhu and Satoh [19] proposed a large vocabulary quantization based on BOW framework to search instance from videos. In [13], an exponential locality similarity function in the feature space and BOW with large vocabularies in Fisher vector [9] and vector of locally aggregated descriptors (VLAD) [6] were applied for general instance search. Murata et al. [8] exploited EBM25 to estimate the importance of visual words in BOW which can suppress the keypoints from frequently occurring background objects. In this work, we also adopt the BOW framework to select compact and representative visual words from vocabulary for its effectiveness and efficiency.

Spatial verification, as an improvement of BOW framework, was widely used to enhance retrieval performance. Weak geometric consistency (WGC) [5] simplified BOW and filtered false matched descriptors using consistency in terms of the angle and the scale. Because of the peculiarity of instance search that retrieved targets have same object with query, more details can be mined by spatial relationships between local features to improve the performance. Zhang and Ngo [15,16] exploited the topological triangulated graph to describe spatial information and verify consistency by estimating the number of common edges between the triangulated graphs. Weak geometric correlation consistency (WGCC) [17] explored the geometric correlations among local features and incorporated these correlations with each individual match to form a transformation consistency in rotation and scale space. The above methods focused on filtering false positives to increase reliability of feature match and decrease computational complexity using spatial consistency. Different from these methods, we propose an effective spatial verification method to retrieve images using a few representative visual words.

Compared with a single query, multiple queries can provide more comprehensive information and suppress noise in single image. Chum et al. [3,4] firstly brought query expansion into visual domain. Given an input query region, query expansion retrieves and forms a richer query by combining the original query with retrieved results. Then, the system retrieves again using the new query and repeats the process as necessary. Arandjelovic and Zisserman [1] exploited the textual query as input of Google image search and regarded the search results as multiple queries to visually retrieve. Different from the above methods, our approach mines relevant images from user's mobile device to form the multiple queries which is called multi-photos in this paper.

3   Our Approach

Given an image containing the user-interested object as a query q, we firstly mine relevant photos for q from user's mobile device. After multi-photos are quantified to bags of visual words, we rank and select visual words based on three criteria including EBM25, significance in multi-photos and separability. Finally, our approach returns images which contain the user-interested object by spatial relations between a few representative visual words. In this section, we will introduce the proposed approach in four parts: preprocessing, mining relevant photos, selecting representative visual words and spatial consistency verification.

3.1   Preprocessing

To achieve better performance, we preprocess images in dataset with the following steps. Firstly, each dataset image is described by SIFT features due to its excellent performance. Secondly, we exploit hierarchical k-means clustering to generate a vocabulary V with regarding each leaf node as a visual word (we set

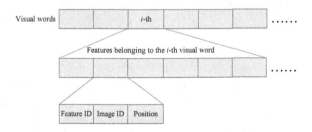

Fig. 2. Inverted file consisting of a set of visual words. Each word contains features which belong to itself. Important attributes of each feature include feature ID, ID of image which the feature belongs to, and the spatial position of the feature.

the size of vocabulary to 65,536 in our implementation). Finally, each image is quantized to a bag of words based on visual words generated previously. Meanwhile, we employ an inverted file as illustrated in Fig. 2 to improve retrieval efficiency. The inverted file consists of 65,536 visual words and each of which contains some features that belong to the corresponding cluster. Each feature links to its necessary information, including the feature ID, the ID of image where it is extracted and its position in the image.

3.2 Mining Relevant Photos

It is a common phenomenon that users often take multiple photos for an object that they are concerned. Therefore, when an image is submitted as a query, there are likely many relevant photos with the query in user's mobile device. In order to select more representative visual words with those relevant photos, we should mine relevant photos first. Figure 3 illustrates the flowchart of this procedure.

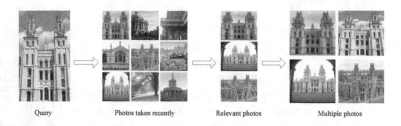

Query   Photos taken recently   Relevant photos   Multiple photos

Fig. 3. Mining relevant photos. After a query is submitted, we get the photos taken recently from the user's mobile device. Then, relevant photos are mined from these photos to generate multiple photos. The multiple photos consist of mined relevant photos and the query.

Given an image as a query, the recent kn_1 photos in user's mobile device will be obtained as candidates. After being described by SIFT features, the query and multiple photos are quantized to groups of visual words. To mine relevant photos from candidates, we propose a method which is similar to the PageRank algorithm [2] to compute relevance R between query and candidates. The main steps are summarized by the following formulas:

$$R^{t+1} = S \times R^t \tag{1}$$

$$R^0(i) = cossim(q, p_i) \tag{2}$$

$$S(i, j) = cossim(p_i, p_j) \tag{3}$$

where $P = [p_1, \cdots, p_i, \cdots, p_{kn_1}]$ denotes a set of candidate photos, t denotes the iterations, and $cossim(p, q)$ denotes the cosine similarity between vectors p and q. As each pair of photos has a cosine similarity, the algorithm converges quickly according to [2]. In our experiment, matrix S is normalized to improve convergence rate and precision before computation. After the algorithm converges, R^t can be regarded as the relevance of candidate photos P. Finally, we sort R^t with descending order and put corresponding top kn_2 photos into multiple photos M.

3.3   Selecting Representative Visual Words

After mining relevant photos, we utilize multiple photos M and three criteria to generate representative visual word set V' from V. In consideration of poor wireless bandwidth and mutability, visual words must be compact and vary with network condition. On the other hand, to achieve better retrieval performance, visual words should be representative and significant. With above requirements, we propose three criteria including EBM25, significance in multiple photos M and separability to select and transmit representative visual words.

BM25 [11] is a state-of-the-art probabilistic model for information retrieval task. Because documents are written with fixed format, non-conceptual keywords tend to occur more frequently. Original IDF is able to suppress the no-conceptual keywords in document processing. However, non-conceptual words in images or videos occur less frequently than in documents, and thus BM25 with original IDF does not get sufficiently suppressed. Murata et al. [8] applied BM25 in instance search domain and modified original BM25 to BM25 with exponential IDF (EBM25) based on properties of instance search. We adopt EBM25 as a criterion to select visual words. For the i-th visual word v_i,

$$\omega_{EBM25_i} = \frac{kf_i}{h1 + kf_i}\omega_i^{exponentialIDF} \tag{4}$$

$$\omega_i^{exponentialIDF} = log(\frac{e^{-n_i/\alpha}}{n_i - e^{-n_i/\alpha}} \times \frac{N - e^{n_i/\alpha} - n_i + e^{-n_i/\alpha}}{e^{n_i/\alpha} - e^{-n_i/\alpha}}) \tag{5}$$

$$kf_i = \frac{k_i}{1 - b + b\frac{vl}{avvl}} \tag{6}$$

where n_i is the number of images containing the i-th visual word v_i, N is the number of images in dataset, k_i is the frequency of occurrence the i-th visual word in query q, vl is the number of features in query q and $avvl$ denotes the average number of features in dataset images. $h1, \alpha$ and b are related parameters which are set to 2, 0.75 and 20 respectively.

To select more representative visual words, significance in multi-photos is considered as an important measurement. In this procedure, we propose an approach called "TF-DF" (term frequency - document frequency), which is opposite to TF-IDF. In the whole dataset, the significance of a word decreases with the number of documents containing the words increasing. It is easy to understand that a word has less specificity if more documents in dataset contain it. However, in our multi-photos M, all images are relevant to query q. Therefore, significance of a visual word is proportional to its occurrence frequency in relevant images. For the i-th visual word v_i, its significance in multi-photos ω_{mp_i} is defined as follows:

$$\omega_{mp_i} = \omega_{tf-df_i} \times \frac{1}{|M|} \sum_{m_j \in M} \frac{k_i^{m_j}}{vl_{m_j}} \tag{7}$$

$$\omega_{tf-df_i} = k_i \times log(\frac{n_i'}{|M|}) \tag{8}$$

where $|M|$ denotes the number of images in multi-photos M, $k_i^{m_j}$ denotes the occurrence frequency of the i-th visual word v_i in m_j, vl_{m_j} denotes the number

---

Algorithm 1. Select representative visual words

Input: q, ω_{EBM25}, ω_{mp}, V, kn_3
Output: V'
initialize $V' = \emptyset$;
for $i = 1 : |V|$ **do**
$\quad | \quad \omega_i^0 = e^{\gamma \times \omega_{EBM25_i}} \times \omega_{mp_i}$;
end
for $j = 0 : kn_3$ **do**
$\quad$ sort ω^j in descending order;
$\quad$ transfer visual words corresponding to ω_1^j from V to V';
$\quad \omega_1^j = 0$;
$\quad$ **for** $i = 1 : |V|$ **do**
$\quad \quad$ compute ω_{sep_i} by Eq. 9;
$\quad \quad \omega_i^{j+1} = e^{\gamma \times \omega_{EBM25_i}} \times \omega_{mp_i} \times \omega_{sep_i}$;
$\quad$ **end**
end
return V';

---

of features in m_j and n_i' denotes the number of images containing the i-th visual word v_i in multi-photos M.

We employ the separability measurement to make visual words compact. To this end, elements in selected visual word set V' should be different from others as much as possible. For the i-th visual word v_i, its separability weight ω_{sep_i} is defined as:

$$\omega_{sep_i} = \prod_{j=1}^{|V'|} e^{\beta \times cossim(v_j', v_i)} \tag{9}$$

where $|V'|$ is the number of visual words selected and v_j' is the j-th visual word in V'. β is the related parameter which is set to -0.04.

As Algorithm 1 demonstrates, our approach selects kn_3 representative visual words with three measurements including ω_{EBM25}, ω_{mp} and ω_{sep}. Firstly, we initialize V' with empty set and ω with ω_{EBM25} being multiplied by ω_{mp}. Then, for each iteration, we compute ω_{sep} based on new V' and calculate ω again, and the visual word corresponding to maximum of ω is added to V'. The iteration stops while there are required kn_3 visual words in V'. Finally, V' is returned as the representative visual word set. In Algorithm 1, the related parameter γ is set to 0.03.

3.4 Spatial Consistency Verification

In this step, we rank and return result images with spatial relations between visual words selected in Sect. 3.3. Different from existing studies that exploit geometric relationships to filter false visual word matches, we use the spatial consistency to compute similarity between query and each image in dataset. In order to simply but effectively achieve retrieval, relative position of the representative visual words is applied to measure the weight of an image. For each image I in dataset D, if relative positions of visual words in I is close to that in query q, we reasonably consider I is similar with q for the robustness of representative visual words and properties of instance search. Figure 4 shows an illustration of spatial consistency verification using four visual words. For any pair of visual words in V', the cosine similarity of their relative positions in q and I is computed and added to the weight of I. The formula of spatial consistency verification is given as follows:

$$Sim(I, q) = \sum_{v'_i, v'_j \in V'} \omega^0_{v'_i} \times \omega^0_{v'_j} \times e^{\delta \times R_{l_I}(i,j)} \tag{10}$$

$$R_{l_I}(i,j) = \max_{f^t_i \in F^i_I, f^c_j \in F^j_I} cossim(L(f^q_i) - L(f^q_j), L(f^t_i) - L(f^c_j)) \tag{11}$$

where v'_i denotes the i-th visual word in V', F^i_I denotes features of the image I which belong to the visual word v'_i and f^t_i is the t-th feature in F^i_I. In our implementation, we set the parameter δ to 0.06. To simplify computation, we choose the first feature in query which belongs to the visual word v'_i as f^q_i. $L(\cdot)$ means the position coordinates of a feature in the corresponding image. We calculate $Sim(I, q)$ as its weight for each image I, and sort them in descending order. Finally, the first kn_4 images are returned as retrieved result.

Fig. 4. Spatial consistency verification of four visual words. For each pair of visual words, we compute the cosine similarity of their relative positions in the query and an image in dataset. The cosine similarities of all pairs of visual words are accumulated to compute the weight of this image.

4   Experimental Evaluation

4.1   Setup

Experimental Dataset. We conduct our experiments on Oxford building dataset [10] which consists of 5,062 images collected from Flickr by searching for particular Oxford landmarks. The collection has been manually annotated to generate a comprehensive ground truth for 11 different landmarks, each represented by 5 possible queries. This gives a set of 55 queries over which an object retrieval system can be evaluated.

Evaluation Metric. In our experiments, average precision at the top N (AP@N) is adopted to evaluate the performance of instance search. The evaluation metric is defined as:

$$AP@N = \frac{1}{K}\sum_{i=1}^{K}(R_i/N) \qquad (12)$$

where K is the number of queries and R_i is the number of relevant images in the top N returned images for the i-th query.

Compared Methods. Three methods are compared with ours:

- Spatial coding (SC) [18]. Spatial coding was proposed to encode the relative positions between each pair of features and filter false matches between images. This method can effectively improve retrieval accuracy and reduce the time cost of partial-duplicate image retrieval.
- Query expansion (QE) [3]. Three improvements of the query expansion were proposed in the BoW-based particular object and image retrieval, which contain preventing TF-IDF failure, the spatial verification and learning relevant spatial context.
- Multiple photos (MP) [14]. This method firstly determines multiple relevant photos and explores the high-level semantic information of an image by finding the contextual saliency from the multiple relevant photos.

4.2   Comparison with Other Methods

In this section, we will compare the performance of our approach with SC, QE and MP. They all use a codebook with 61,724 visual words. In comparison, we set kn_2 to 3 and kn_3 to 30. For each query, we collect five relevant images and five other images from the dataset to compose the candidate photo set P. Table 1 indicates that our approach can mine relevant photos accurately. Because kn_2 is the number of returned images in the process of mining relevant photos, the average precision decreases with kn_2 increasing.

Table 1. AP of mining relevant photos.

kn_2	2	3	4	5
AP	0.9818	0.9818	0.9773	0.9491

Figure 5 shows the average precision of SC, QE, MP and our approach. It is obvious that the proposed approach outperforms other three state-of-the-art methods. Different from the strong spatial constraint of SC, our method is more tolerant of the deformation of instance. Due to multiple photos, we can select more robust visual words. QE cannot achieve good performance when the query

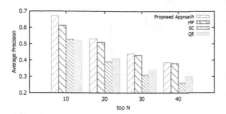

Fig. 5. Comparison in terms of average precision.

has poor quality such as low resolution. Compared with MP, our three criteria can select more compact and representative visual words. Meanwhile, our spatial verification can eliminate noise and rank the relevant image which contains similar spatial relationships with each pair of selected visual words higher. Therefore, our approach is superior to MP in top 1~20. Moreover, our approach achieves 1 for AP@50 in the 5 queries of Radcliffe Camera landmark and 0.9133 of AP@30 in the 5 queries of All Souls landmark. Figure 6 shows two queries and part of retrieval results. It demonstrates that the proposed approach can return the retrieval targets which are rotated, deformed, indistinct and incomplete.

Fig. 6. Part of retrieval results. The top row is a query of Balliol landmark and some returned images, and the bottom row is a query of Hertford landmark and some returned images.

4.3 Effect of Parameters

Effect of kn_2. For the reason that an important criterion of selecting representative visual words in Sect. 3.3 is significance in multiple photos, the number of multiple photos kn_2 has great influence on results. Figure 7(a) shows the effect of kn_2 on average precision with 30 visual words. We can find that the average precision increases with kn_2 increasing. Therefore, the effect of selecting significant words in multiple photos is demonstrated. Meanwhile, our approach still can achieve favorable results when kn_2 is set to 2, which proves the flexibility of our method. When kn_2 reaches 5, the precision after top 20 rises slowly for the reason that the accuracy of mining relevant photos decreases and more noisy images are mined.

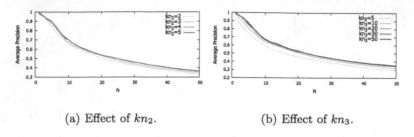

(a) Effect of kn_2.   (b) Effect of kn_3.

Fig. 7. Effect of parameters.

Effect of kn_3. The number of visual words which are used in spatial verification process influences the retrieval precision. AP@10 is 0.5945, 0.6418, 0.6636, 0.6709 and 0.6745, when kn_3 is 5, 10, 20, 30 and 50 respectively. Figure 7(b) demonstrates that using more visual words can achieve better performance. Meanwhile, our approach can achieve comparable performance with only 10 visual words. The visual words are compact enough to adapt to poor wireless network environment.

5 Conclusion

In this paper, we propose a novel approach for mobile instance search. We compute the relevance of recent photos in user's mobile device to generate multi-photos. Then, three criteria including EBM25, significance in multi-photos and separability are applied to select compact and representative visual words. Finally, we adopt a simple but effective spatial verification method to generate the weight of each image and check the representativeness of selected visual words. Our experimental results on Oxford building dataset show the superiority of our approach compared to three state-of-the-art methods.

Acknowledgments. This work is supported by the National Nature Science Foundation of China (grants No. 61672133, No. 61602089, and No. 61632007), and the Fundamental Research Funds for the Central Universities (grants No. ZYGX2015J058 and No. ZYGX2014Z007).

References

1. Arandjelovic, R., Zisserman, A.: Multiple queries for large scale specific object retrieval. In: British Machine Vision Conference, BMVC 2012, Surrey, UK, 3–7 September 2012, pp. 1–11 (2012)
2. Brin, S., Page, L.: The anatomy of a large-scale hypertextual web search engine. Comput. Netw. **30**(1–7), 107–117 (1998)
3. Chum, O., Mikulík, A., Perdoch, M., Matas, J.: Total recall II: query expansion revisited. In: The 24th IEEE Conference on Computer Vision and Pattern Recognition, CVPR 2011, Colorado Springs, CO, USA, 20–25 June 2011, pp. 889–896 (2011)

4. Chum, O., Philbin, J., Sivic, J., Isard, M., Zisserman, A.: Total recall: automatic query expansion with a generative feature model for object retrieval. In: IEEE 11th International Conference on Computer Vision, ICCV 2007, Rio de Janeiro, Brazil, 14–20 October 2007, pp. 1–8. IEEE Computer Society (2007)

5. Jegou, H., Douze, M., Schmid, C.: Hamming embedding and weak geometric consistency for large scale image search. In: Forsyth, D., Torr, P., Zisserman, A. (eds.) ECCV 2008. LNCS, vol. 5302, pp. 304–317. Springer, Heidelberg (2008). doi:10. 1007/978-3-540-88682-2_24

6. Jegou, H., Douze, M., Schmid, C., Pérez, P.: Aggregating local descriptors into a compact image representation. In: The Twenty-Third IEEE Conference on Computer Vision and Pattern Recognition, CVPR 2010, San Francisco, CA, USA, 13–18 June 2010, pp. 3304–3311 (2010)

7. Lowe, D.G.: Distinctive image features from scale-invariant keypoints. Int. J. Comput. Vis. **60**(2), 91–110 (2004)

8. Murata, M., Nagano, H., Mukai, R., Kashino, K., Satoh, S.: BM25 with exponential IDF for instance search. IEEE Trans. Multimed. **16**(6), 1690–1699 (2014)

9. Perronnin, F., Dance, C.R.: Fisher kernels on visual vocabularies for image categorization. In: 2007 IEEE Computer Society Conference on Computer Vision and Pattern Recognition (CVPR 2007), Minneapolis, Minnesota, USA, 18–23 June 2007 (2007)

10. Philbin, J., Chum, O., Isard, M., Sivic, J., Zisserman, A.: Object retrieval with large vocabularies and fast spatial matching. In: 2007 IEEE Computer Society Conference on Computer Vision and Pattern Recognition (CVPR 2007), Minneapolis, Minnesota, USA, 18–23 June 2007 (2007)

11. Robertson, S.E., Zaragoza, H.: The probabilistic relevance framework: BM25 and beyond. Found. Trends Inf. Retr. **3**(4), 333–389 (2009)

12. Sivic, J., Zisserman, A.: Video Google: a text retrieval approach to object matching in videos. In: 9th IEEE International Conference on Computer Vision (ICCV 2003), Nice, France, 14–17 October 2003, pp. 1470–1477 (2003)

13. Tao, R., Gavves, E., Snoek, C.G.M., Smeulders, A.W.M.: Locality in generic instance search from one example. In: 2014 IEEE Conference on Computer Vision and Pattern Recognition, CVPR 2014, Columbus, OH, USA, 23–28 June 2014, pp. 2099–2106 (2014)

14. Yang, X., Qian, X., Xue, Y.: Scalable mobile image retrieval by exploring contextual saliency. IEEE Trans. Image Process. **24**(6), 1709–1721 (2015)

15. Zhang, W., Ngo, C.: Searching visual instances with topology checking and context modeling. In: International Conference on Multimedia Retrieval, ICMR 2013, Dallas, TX, USA, 16–19 April 2013, pp. 57–64 (2013)

16. Zhang, W., Ngo, C.: Topological spatial verification for instance search. IEEE Trans. Multimed. **17**(8), 1236–1247 (2015)

17. Zhang, Z., Albatal, R., Gurrin, C., Smeaton, A.F.: Instance search with weak geometric correlation consistency. In: Tian, Q., Sebe, N., Qi, G.-J., Huet, B., Hong, R., Liu, X. (eds.) MMM 2016. LNCS, vol. 9516, pp. 226–237. Springer, Heidelberg (2016). doi:10.1007/978-3-319-27671-7_19

18. Zhou, W., Lu, Y., Li, H., Song, Y., Tian, Q.: Spatial coding for large scale partial-duplicate web image search. In: Proceedings of the 18th International Conference on Multimedia 2010, Firenze, Italy, 25–29 October 2010, pp. 511–520 (2010)

19. Zhu, C., Satoh, S.: Large vocabulary quantization for searching instances from videos. In: International Conference on Multimedia Retrieval, ICMR 2012, Hong Kong, China, 5–8 June 2012, p. 52 (2012)

Stochastic Decorrelation Constraint Regularized Auto-Encoder for Visual Recognition

Fengling Mao[1], Wei Xiong[1(✉)], Bo Du[1,2], and Lefei Zhang[1]

[1] School of Computer, Wuhan University, Wuhan, China
{mflcs2013,wxiong,zhanglefei}@whu.edu.cn
[2] Collaborative Innovation Center of Geospatial Technology, Wuhan, China
remoteking@whu.edu.cn

Abstract. Deep neural networks have achieved state-of-the-art performance on many applications such as image classification, object detection and semantic segmentation. But the difficulty of optimizing the networks still exists when training networks with a huge number of parameters. In this work, we propose a novel regularizer called stochastic decorrelation constraint (SDC) imposed on the hidden layers of the large networks, which can significantly improve the networks' generalization capacity. SDC reduces the co-adaptions of the hidden neurons in an explicit way, with a clear objective function. In the meanwhile, we show that training the network with our regularizer has the effect of training an ensemble of exponentially many networks. We apply the proposed regularizer to the auto-encoder for visual recognition tasks. Compared to the auto-encoder without any regularizers, the SDC constrained auto-encoder can extract features with less redundancy. Comparative experiments on the MNIST database and the FERET database demonstrate the superiority of our method. When reducing the size of training data, the optimization of the network becomes much more challenging, yet our method shows even larger advantages over the conventional methods.

Keywords: Deep learning · Visual recognition · Auto-encoder

1 Introduction

Big data becomes a hot spot in recent years, owing to the rapid development of the information technologies. Many methods and frameworks have been proposed to deal with big data. Deep learning methods, which can learn high-level representations from the raw data, have proven to be effective to learn representative features on big data [1,2].

Deep learning, including unsupervised methods and supervised methods have achieved excellent performances on various tasks such as image classification, object recognition, etc. However, supervised methods like convolutional neural

This paper is supported in part by the National Natural Science Foundation of China under Grants 61471274, 91338202 and U1536204, and 61401317.

L. Amsaleg et al. (Eds.): MMM 2017, Part II, LNCS 10133, pp. 368–380, 2017.
DOI: 10.1007/978-3-319-51814-5_31

network (CNN) [3] have several shortcomings in dealing with big data. One big disadvantage is the lack of enough labeled data for training the deep supervised models. When the high quality labeled data is inadequate, the models would easily become over-fitted. To relieve overfitting, a possible way is to use unsupervised learning methods. Unsupervised learning methods like k-means clustering [4], sparse auto-encoder and some graphical models [5–8] can extract useful features on massive data automatically, without knowing any supervision information of the training data. In this paper, we focus on the improvement of unsupervised networks for visual recognition tasks.

Even with unsupervised networks, considerable problems still exist. Deep unsupervised networks usually have massive parameters and multiple non-linear hidden layers, which make them very hard to train. Aiming to train the large networks efficiently, several regularizers and optimization methods have been proposed in the literature. Optimization methods like Batch Normalization [9] and Stochastic Gradient Descent (SGD) [10,11], network structures like denoising auto-encoders [12], and regularizers like Dropout [13,14] and DropConnect [15] perform well on benchmark tasks, remarkably improving the networks' performance.

Among these methods, Dropout is the most successful regularizer that have proved to be effective in various types of networks and on various datasets. Dropout is a simple but powerful regularization method proposed by Hinton et al. in [13], which improves the performance of the network by reducing complex co-adaptation of hidden layer activations. In addition, The stochastic process in Dropout can achieve an effect of model averaging, which provides a way of approximately combining an exponential number of different neural network architectures and is easy to approximate the effect of averaging the predictions of all these neural networks, providing a simple and inexpensive way to train large models and reduce overfitting. Dropout fits for various models, such as feedforward neural networks and Restricted Boltzmann Machines (RBM) [16,17], and improves the performance of neural networks in miscellaneous applications, especially for computer vision tasks. However, Dropout optimizes the networks in an implicit way. For instance, Dropout implicitly reduces the complex co-adaption, but cannot quantify the correlation between hidden neurons explicitly.

To solve the problems in the conventional regularizers especially Dropout, in this paper, we propose a novel regularizer called stochastic decorrelation constraint (SDC) to regularize the hidden layers of the network and make attempts to improve the generalization ability of large networks. Specifically, we randomly select a portion of the latent neurons, and then quantify their co-adaption using our proposed regularizer. In our method, we use the correlation between the activations of two individual neurons to express their co-adaption. By constraining the hidden layer, the correlations between those selected neurons are minimized. Compared with Dropout, the proposed SDC quantify the co-adaptations between the hidden neurons explicitly with an elegant objective function. By minimizing the co-adaptions between the selected neurons, the network can learn decorrelated and less redundant features while still preserves the correlations between some of the neurons within the same layer, as these correlations may benefit the optimization of the network. In the meanwhile, SDC has the capacity of model averaging to make the training process much easier. That is to say, by randomly selecting the neurons to be regularized in each training iteration, the network learns a different network each time, as the connections and correlations between the neurons are constrained to be quite different during each iteration. We apply our regularizer to the hidden layer of the auto-encoder to establish a much more efficient unsupervised feature learning system for visual recognition. We call it the stochastic decorrelation constraint regularized auto-encoder.

The primary contributions of this paper are as follows: We proposed a novel regularizer to optimize the unsupervised neural network. We explicitly formulate the co-adaptions between the hidden neurons and reduce them with a clear formula, to force the network to learn less redundant feature representations. The proposed SDC optimizes the training of the network by implicitly averaging over exponentially many networks which are trained with the hidden neurons under different correlation states.

The remainder of this paper is organized as follows: In Sect. 2, we review the related works about stochastic methods and correlation based methods in deep neural networks. Section 3 describes our approach in details. In Sect. 4, we evaluate the performance of the proposed optimization method on several benchmarks, together with experimental analysis and comparison with other typical methods. Section 5 is the conclusion of our work.

2   Related Work

Stochastic Methods. Denoising auto-encoder (DAE) [12] is an unsupervised training network trained with a denoising technique. It can learn robust representations by adding random noise to the input units of an auto-encoder. A common way to perform DAE is to stochastically set partial components of each input units to zero. DropConnect [15] is the an extension of Dropout. Unlike our regularizer on the hidden neurons, DropConnect becomes a sparely connected layer by randomly dropping a portion of connections with a rate p.

Correlation Based Methods. The cross-covariance penalty (XCov) is a decorrelation method, proposed by Cheung et al. [18]. The XCov disentangle the class-relevant variables (such as content of handwritten digits) and the latent variables (such as form or style) in deep auto-encoder by minimizing the correlations between the hidden neurons and the neurons in the classification layer. Differently, our method minimizes the covariance between the hidden neurons to better regularize the network. XCov tries to separate features for classification and reconstruction, while our method aims at regularizing the unsupervised networks to learn decorrelated and non-redundant representations, and optimizing the unsupervised networks.

Another decorrelating regularizer similar to ours is DeCov proposed in [19], which can also reduce overfitting in deep neural networks by learning decorrelated representations. DeCov conducts the decorrelation process on all the neurons of one layer. It forces all the neurons to be independent of each other, but ignores the fact that some neurons may not work effectively without the mutual connection with each other. This is inspired by the biological fact that neurons in the same layer in the brain are partly connected to each other. This drawback may lead to some bad effects on training the network. Compared to our method, another disadvantage of DeCov is the lack of randomness in the training process. Our method regularize the hidden neurons stochastically, which can result in an effect of model averaging.

3 Approach

3.1 Unsupervised Feature Learning Framework

We use unsupervised networks to learn useful features. Specifically, We first extract patches from the images and train a single-layer auto-encoder on the raw patches. Then we reshape the learned filters of the auto-encoder into convolutional kernels and use the kernels to convolve the input images into feature maps, which are further condensed with a pooling operation. We train this network in an unsupervised way to extract meaningful feature representations.

After feature extraction, we use these features and training data labels to train a linear SVM classifier. During the testing procedure, we first extract features of the testing images, and then utilize the linear SVM to classify the learned features.

3.2 The Stochastic Decorrelation Constraint Auto-Encoder

SDC is performed on the hidden activations of the auto-encoder over a minibatch. In our work, we use minibatch SGD [11] method to train the network and the input data are divided into several minibatches. For convenience, in this part we focus on the optimization procedure over one minibatch in a certain iteration.

Suppose we have the input data batch $X = [x^1, x^2, ..., x^M]$, which contains M samples, $x^m \in R^K$ denotes to the m-th sample. Our neural network is a single hidden layer auto-encoder with N units in hidden layer. The batch X is then mapped to the hidden activations $A = [a^1, a^2, ..., a^M]$, where $a^m \in R^N$, a_i^m denotes the activation of unit i and $m \in \{1, 2 \cdots, M\}$ means the m-th sample in a batch. Let J_{SDC} denote the loss of our regularizer (SDC loss), J_{AE} denotes the reconstruction loss of the basic auto-encoder and J denotes the total loss of the network. The process of our method is represented briefly in Fig. 1(a).

Random Neuron Selection. In detail, we set a proper random decorrelated rate r ($0 < r < 1$) and choose $r \cdot N$ hidden units to be decorrelated and temporarily set the other hidden units' activation to zero. Specifically, a random matrix $R \in R^{M \times N}$ is generated, where $R \sim Bernoulli(r)$. Let $\hat{A} = [\hat{a}^1, \hat{a}^2, ..., \hat{a}^M]$

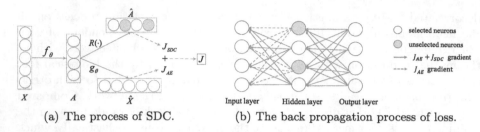

(a) The process of SDC.   (b) The back propagation process of loss.

Fig. 1. (a) The input batch X is mapped to representation A. Then A is mapped back to the input's reconstruction $\hat{X}$. To regularize the hidden layer, We randomly select a portion of the hidden neurons $\hat{A}$ using a random selected matrix R and calculate the covariance (the SDC loss) of the activations of these neurons. The total loss function J is the weighted sum of the SDC loss and the basic auto-encoder loss. (b) The back propagation process of loss: the solid lines mean both the gradient of the J_{AE} loss and the gradient of the J_{SDC} loss, while the dotted lines means only the gradient of the J_{AE} loss.

denotes the matrix of hidden units' activation after the random neuron selection. $\hat{a}_i^m$ correspondingly denotes the activation of the i-th hidden unit for the m-th sample in a particular batch. 'o' means element-wise multiplication. Then we have:

$$\hat{A} = R \circ A. \tag{1}$$

Co-adaption Formulation. We compute the cross-covariance of $\hat{A}$ to get the SDC loss, which can help to disentangle all random selected hidden units and reduce their correlation. Let variable Cov denote the covariance matrix between all activations in hidden layer. Then, we have:

$$Cov = \frac{1}{M} \left(\hat{A} - \mu \right) \cdot \left(\hat{A} - \mu \right)^T$$
$$= \frac{1}{M} \left(R \circ A - \mu \right) \cdot \left(R \circ A - \mu \right)^T \tag{2}$$

where $\mu = \frac{1}{M} \sum_{m=1}^{M} \hat{a}^m$ is the mean of the hidden units's activation over a batch.

Our goal is to minimize the correlation between randomly selected activations. SDC is a sum-squared cross-covariance term between the hidden units' activation over a batch of samples. Since we want to reduce the covariance between different activations, the variance of each hidden unit' activation has no practical use, but increases the value of cross-covariance. So we subtract the diagonal of the Cov matrix. Our final SDC is given in Eq. 3.

$$J_{SDC} = \frac{1}{2} \left(\| Cov \|_F^2 - \| diag(Cov) \|_2^2 \right). \tag{3}$$

When finally adding the SDC loss, the overall objective function used to train the neural network is defined in Eq. 4.

$$J = J_{AE} + J_{SDC}$$
$$= \| \hat{X} - X \|^2 + \frac{\gamma}{2} \left(\| Cov \|_F^2 - \| diag(Cov) \|_2^2 \right). \tag{4}$$

The first loss is the typical squared cost for the auto-encoder and the second loss is our regularization term we add to constrain the hidden neurons. The new parameter γ controls the weight of SDC in the objective function. Figure 1(b) shows how the two losses act on the network through back propagation procedure. As is shown in Fig. 1(b), suppose we select two hidden layer's neurons to decorrelate, the unselected neurons only propagate back the gradients of the J_{AE} loss, without the J_{SDC} loss.

Gradient Propagation. We now describe the back propagation procedure. Since the derivatives of the reconstruction loss term have been provided in previous work, the key step is to compute the partial derivatives of the SDC loss. By referring to Eq. 3, we can compute the derivative of the SDC loss as described in Eq. 5.

$$\frac{\partial J_{SDC}}{\partial \hat{A}} = (Cov - diag(Cov)) \cdot (\hat{A} - \mu)$$
$$= (Cov - diag(Cov)) \cdot (R \circ A - \mu). \tag{5}$$

Theoretical Analysis. In the above deduction, we quantity the correlation of features numerically with cross-covariance. Cross-covariance is a useful measurement for the correlation of two random variables. In neural networks, if the cross-covariance is large, it implies that the two latent features have high linear correlation, which also means there may be large redundancy between features in neural networks. Thus, the SDC loss theoretically reduces the correlation between features. Notice that we randomly select a part of neurons to conduct this decorrelation process. SDC preserves the co-adaption of a few neurons and decorrelates other features. Our method proves to be more effective on improving model performance than full decorrelation in the following contrast experiments. Furthermore, the random selection procedure provides the network with stochasticity, which means we conduct decorrelation on different neurons over different batches. This leads to a result of training an exponential number of networks during the whole training procedure. Hence our regularizer has the ability of model averaging, which makes the model more robust and adaptive to datasets of various distribution.

4   Experiments

In our experiment, we conduct unsupervised feature learning procedure to extract useful features from the raw images. The learning model is a single layer auto-encoder. To evaluate the performance of SDC on network optimization, we compare our method with Dropout, DropConnect and DeCov on two benchmark datasets: the MINST dataset [20] and the FERET dataset [21].

4.1 Datasets Description

The MNIST Dataset. The MNIST handwritten digits database [20] consists a training set of 60000 samples and a test set of 10000 samples. The handwritten digits have been size-normalized and centered in a 28×28 image. These images are labeled with 0–9 (10 classes) different digits. Samples are given in Fig. 2(a).

The FERET Dataset. The facial recognition technology database (FERET) [21] is a standard database of facial images. The whole FERET database consists of 14051 gray scale images of human heads with several different views ranging from left to right profiles. In our work, the FERET dataset we use is a subset of the whole FERET database, which contains 200 identity labels. Each identity has 7 different facial emotion and illumination. The facial images are cropped to a fixed 80×80 size. Examples are shown in Fig. 2(b). We choose four images from each identity class together to form the training dataset, containing totally 800 images. The remaining 600 images are used as the testing dataset.

(a)   (b)

Fig. 2. Sampling images from the databases: (a) MNIST; (b) FERET.

4.2 Model Architecture

The model architecture for the MNIST dataset is a single layer auto-encoder with 500 units in the hidden layer. The kernelsize of convolution is 11×11 and the pooling region size is 6×6. We use the max pooling to subsample the generated feature maps. We set the SDC rate r as 0.4.

The model of the FERET dataset has similar architecture with that of the MNIST dataset. The model architecture for the FERET dataset is a single layer auto-encoder with 1000 units in the hidden layer. The kernel size of convolution is 9×9 and the pooling region dimension is 12×12. We choose the max pooling to subsample the features. We set the SDC rate r as 0.4.

4.3   Experiments and Results

Effect on Decorrelation. In order to verify whether our proposed SDC method can reduce the correlation between hidden neurons, we compute the cross covariance of the activations in the hidden layer using two methods (the proposed SDC, the basic AE) after each iteration. Figure 3 shows the changing curve of the cross covariance.

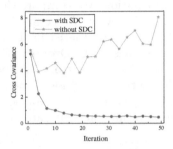

Fig. 3. The cross covariance of hidden neurons changing with iteration when using two methods (SDC and basic AE).

In Fig. 3, the cross covariance with SDC is much lower than that without SDC in the whole process. As the iteration increases, the cross covariance without SDC increases obviously, while it reduces gradually with SDC. This experiment demonstrates that the proposed SDC method can significantly reduce the cross covariance of hidden neurons, which means that SDC has a good effect of decorrelating the hidden neurons.

Effect on Filters. As described in Sect. 3, the proposed SDC have the effect on decorrelating activations in the hidden layer. In this part, we will analyze the effect of SDC on feature representation. To show the impact of SDC on features directly, we visualize the filters learned with or without the proposed SDC method. Figure 4 shows the visualization of filters on the MNIST dataset.

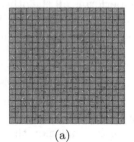
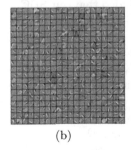

(a)   (b)

Fig. 4. Visualization of filters learned with or without the proposed SDC method on the MNIST database. (a) Filters without SDC; (b) Filters with SDC

The filters in Fig. 4(b) contain much more edge information from the image than that in Fig. 4(a). This contrast experiment shows that decorrelating the hidden neurons have apparent effects on learning good features. Furthermore, it's evident that SDC can detect meaningful features and improve the optimization of the network.

Visual Recognition. In our experiment, we train an auto-encoder with different regularizers for visual recognition. We compare the proposed SDC method with Dropout, DropConnect and DeCov. We conduct visual recognition experiments on both the MNIST dataset and the FERET dataset with the following different methods including the basic AE, Dropout, DropConnect, DeCov and SDC. In order to see the combination effect of different methods, we also combine DeCov and SDC with the typical method Dropout. All hyper-parameters are optimized on the validation set, the validation set is a 10% division of the training set. The classification results on each dataset are given in Table 1.

Table 1. The test accuracy of different methods on MNIST and FERET.

Method	MNIST	FERET
Basic AE	99.02	80.5
Dropout	99.04	80.5
DropConnect	99.21	80.83
DeCov	99.12	80.83
SDC	**99.3**	**81.5**
DeCov + Dropout	99.19	79.5
SDC + Dropout	99.28	81.33

In Table 1, the best performing method for the classification task is the proposed SDC, with the highest testing accuracy of 99.3% for the MNIST dataset and 81.5% for the FERET dataset. Replacing SDC with Dropout reduces the test accuracy to 99.04% for the MNIST dataset and 80.5% for the FERET dataset. It can be concluded that our SDC method outperforms all the other methods on both the MNIST dataset and the FERET dataset, indicating that SDC is effective and workable on different datasets. It also performs well when we combine the SDC and Dropout method.

By carefully comparing the results of different methods in Table 1, we can find out some interesting phenomena. For instance, Dropout method seems not so effective on our single layer auto-encoder. It is clearly shown in Table 1 that the proposed SDC method achieves apparent promotion compared to the basic auto-encoder, while Dropout achieves a little improvement on the two datasets. It indicates that Dropout may not perform well on all the datasets and all the networks. Our regularizer outperforms Dropout in regularizing the auto-encoder.

Effect of Data Size. Since SDC achieves better performance with adequate training data, in this part, we will analyze the performance of SDC with different sizes of training data to see how our method performs with small size's training data. We conduct this experiment with five methods (the basic AE, Dropout, DropConnect, DeCov and SDC) on the subsets of the MNIST dataset and the FERET dataset.

For the MNIST dataset, the number of training images varies in [1000, 3000, 5000, 7000, 9000] and the test dataset remains the same. On each training data size, the classification performance of these methods is evaluated using the test accuracy. Results are shown in Table 2. It's obvious that SDC holds the highest testing accuracy even when training the auto-encoder with insufficient data. For instance, SDC outperforms the basic auto-encoder with the accuracy promotion of 0.54% when the training data size is 1000, while other methods like Dropout have less promotion.

Table 2. The comparison result of four methods on MNIST when training data size varies in [1000, 3000, 5000, 7000, 9000].

Method	1000	3000	5000	7000	9000
Basic AE	95.76	97.58	98.36	98.57	98.78
Dropout	95.92	97.64	98.16	98.7	98.64
DropConnect	96.01	97.66	98.39	98.61	98.76
DeCov	95.79	97.6	98.08	98.61	98.84
SDC	**96.3**	**97.72**	**98.55**	**98.73**	**98.86**

Similarly, for the FERET dataset, the number of training images varies in [200, 400, 600, 800]. For each class of the FERET facial images, we randomly choose 1, 2, 3, 4 images for training, corresponding to the 200, 400, 600, 800 subset size. Table 3 shows the classification result on this dataset. We note that SDC holds the perfect performance as the training data decreases. When we use only 200 training data, replacing the basic auto-encoder with our method achieves the improvement of 3.67%. Adding DropConnect on basic auto-encoder obtains the improvement of 1%. Replacing the basic auto-encoder with DeCov gains a 1.5% promotion. Using Dropout reduces the improvement to 0.5%.

In summary, the proposed SDC method achieves excellent performance with small training data on both the MNIST dataset and the FERET dataset as shown in Tables 2 and 3. These results indicate that when we decrease the training data size, though the training of auto-encoder becomes much harder, SDC method even have larger advantages over the other methods, showing that our method has much better generalization ability and is more robust than the conventional regularizers.

Table 3. The comparison result of four methods on FERET when training data size is 200, 400, 600 and 800.

Method	200	400	600	800
Basic AE	39.33	60.67	71.5	80.5
Dropout	39.83	60.5	71.16	80.5
DropConnect	40.33	58.67	71.67	80.33
DeCov	40.83	61.17	72.5	80.83
SDC	**43**	**62.17**	**73.67**	**81.5**

Hyper-Parameter Analysis. Two new hyper-parameters are introduced to our method, in this part, we give some analysis for the hyper-parameters, as they may have an influence on the network. One parameter is γ (the penalty weight of SDC), which is set to 10 and need be set smaller like 1, 0.1 or 0.01 when decreasing the number of training images in our experiment. The other parameter is the SDC rate r, which controls the number of neurons for decorrelation. We use full dataset of MNIST and FERET to train our models. We let the SDC rate r vary in the range [0.1–0.9] when keeping the other variables constant. The classification results are shown in Fig. 5. From Fig. 5, as the SDC rate changes, the classification performance changes rapidly. The results in Fig. 5 show that the neuron selection rate of SDC has an vital impact on the performance of the network. However, we can still draw an empirical conclusion that SDC gets best performance on both datasets when the SDC rate is 0.4. In real applications, the hyper-parameters can be obtained by grid search on the validation set.

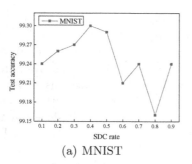

(a) MNIST

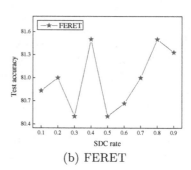

(b) FERET

Fig. 5. The curve of test accuracy when r varies in [0.1–0.9] on different datasets.

5   Conclusion

In this paper, we propose a novel SDC regularizer to optimize the auto-encoder. The SDC method reduces co-adaptions between randomly selected neurons in the hidden layer, which enforces the deep networks to have the stochastic and decorrelated properties. The stochastic property contributes to model averaging, making the network more robust and adaptive to datasets of various distributions. Compared with Dropout, SDC quantifies the correlation of the hidden neurons explicitly. The proposed regularizer enables the network to learn non-redundant representations from raw images. Our experimental results on the benchmarks indicate that our SDC method achieves much better performance than the conventional regularization methods, especially when the training data is inadequate.

References

1. Najafabadi, M.M., Villanustre, F., Khoshgoftaar, T.M., Seliya, N., Wald, R., Muharemagic, E.: Deep learning applications and challenges in big data analytics. J. Big Data **2**, 1–21 (2015)
2. Chen, X.W., Lin, X.: Big data deep learning: challenges and perspectives. IEEE Access **2**, 514–525 (2014)
3. Schmidhuber, J.: Deep learning in neural networks: an overview. Neural Netw. **61**, 85–117 (2015)
4. Hartigan, J.A., Wong, M.A.: Algorithm as 136: a k-means clustering algorithm. J. Roy. Stat. Soc. Ser. C (Appl. Stat.) **28**, 100–108 (1979)
5. Dobra, A., Hans, C., Jones, B., Nevins, J.R., Yao, G., West, M.: Sparse graphical models for exploring gene expression data. J. Multivar. Anal. **90**, 196–212 (2004)
6. Heckerman, D., Geiger, D., Chickering, D.M.: Learning bayesian networks: the combination of knowledge and statistical data. Mach. Learn. **20**, 197–243 (1995)
7. Boykov, Y., Veksler, O., Zabih, R.: Markov random fields with efficient approximations. In: Proceedings of IEEE International Conference on Computer Vision and Pattern Recognition, pp. 648–655 (1998)
8. Hu, X.Y., Eleftheriou, E., Arnold, D.M.: Progressive edge-growth tanner graphs. In: Proceedings of IEEE Global Telecommunications Conference, pp. 995–1001 (2001)
9. Ioffe, S., Szegedy, C.: Batch normalization: accelerating deep network training by reducing internal covariate shift. In: International Conference on Machine Learning (2015)
10. Sohl-Dickstein, J., Poole, B., Ganguli, S.: Fast large-scale optimization by unifying stochastic gradient and quasi-newton methods. In: International Conference on Machine Learning (2014)
11. Zhang, T.: Solving large scale linear prediction problems using stochastic gradient descent algorithms. In: Proceedings of International Conference on Machine Learning, p. 116 (2004)
12. Vincent, P., Larochelle, H., Bengio, Y., Manzagol, P.A.: Extracting and composing robust features with denoising autoencoders. In: Proceedings of International Conference on Machine Learning, pp. 1096–1103 (2008)
13. Srivastava, N., Hinton, G., Krizhevsky, A., Sutskever, I., Salakhutdinov, R.: Dropout: a simple way to prevent neural networks from overfitting. J. Mach. Learn. Res. **15**, 1929–1958 (2014)
14. Hinton, G.E., Srivastava, N., Krizhevsky, A., Sutskever, I., Salakhutdinov, R.R.: Improving neural networks by preventing co-adaptation of feature detectors. arXiv preprint arXiv:1207.0580 (2012)

15. Wan, L., Zeiler, M., Zhang, S., Cun, Y.L., Fergus, R.: Regularization of neural networks using dropconnect. In: Proceedings of International Conference on Machine Learning, pp. 1058–1066 (2013)
16. Hinton, G.E., Osindero, S., Teh, Y.W.: A fast learning algorithm for deep belief nets. Neural Comput. **18**, 1527–1554 (2006)
17. Krizhevsky, A., Hinton, G.: Learning multiple layers of features from tiny images (2009)
18. Cheung, B., Livezey, J.A., Bansal, A.K., Olshausen, B.A.: Discovering hidden factors of variation in deep networks. arXiv preprint arXiv:1412.6583 (2014)
19. Cogswell, M., Ahmed, F., Girshick, R., Zitnick, L., Batra, D.: Reducing overfitting in deep networks by decorrelating representations. arXiv preprint arXiv:1511.06068 (2015)
20. LeCun, Y., Cortes, C., Burges, C.J.: The MNIST database of handwritten digits (1998)
21. Phillips, P.J.: The facial recognition technology (FERET) database (2004)

The Perceptual Lossless Quantization of Spatial Parameter for 3D Audio Signals

Gang Li[1,2,3], Xiaochen Wang[2,3(✉)], Li Gao[2,3], Ruimin Hu[1,2,3],
and Dengshi Li[2,3]

[1] State Key Laboratory of Software Engineering,
Wuhan University, Wuhan, China
ligang10@yeah.net, hurm1964@gmail.com
[2] National Engineering Research Center for Multimedia Software,
School of Computer, Wuhan University, Wuhan, China
clowang@163.com, {gllynnie,reallds}@126.com
[3] Hubei Provincial Key Laboratory of Multimedia and Network Communication
Engineering, Wuhan University, Wuhan, China

Abstract. With the development of multichannel audio systems, the 3D audio systems have already come into our lives. But the increasing number of channels brought challenges to storage and transmission of large amounts of data. Spatial Audio Coding (SAC), the mainstream of 3D audio coding technologies, is key to reproduce 3D multichannel audio signals with efficient compression. Just Noticeable Difference (JND) characteristics of human auditory system can be utilized to reduce spatial perceptual redundancy in the spatial parameters quantization process of SAC. However, the current quantization methods of SAC fully combine the JND characteristics. In this paper, we proposed a Perceptual Lossless Quantization of Spatial Parameter (PLQSP) method, the azimuthal and elevational quantization step sizes of spatial parameters are combined with JNDs. Both objective and subjective experiments have conducted to prove the high efficiency of PLQSP method. Compared with reference method SLQP-L/SLQP-H, the quantization codebook size of PLQSP has decreased by 16.99% and 27.79% respectively, while preserving similar listening quality.

Keywords: 3D audio · Spatial parameters · Quantization · JND

1 Introduction

With the booming of audio playback technologies, audio systems have gone through mono, stereo, to multichannel audio systems. And the typical multichannel audio systems developed from ITU 5.1-channel [1] to NHK 22.2-channel [2], Dolby Atmos 64-channel [3] and even more channels. Owing to more channels set in Three-Dimensional (3D)

This work is supported by National Nature ScienceFoundation of China (No. 61231015, 61671335, 61471271); National High Technology Research and Development Program of China (863 Program) No. 2015AA016306.

L. Amsaleg et al. (Eds.): MMM 2017, Part II, LNCS 10133, pp. 381–392, 2017.
DOI: 10.1007/978-3-319-51814-5_32

space, multichannel audio systems are able to reproduce virtual sound source in 3D space. However the increasing number of channels brought challenges to storage and transmission of large amounts of data at the same time. Therefore, high-efficient compression schemes for multichannel audio coding play an important role on 3D multichannel audio systems.

In recent years, scholars attach much importance to Spatial Audio Coding (SAC) [4] for high-efficient compression. Rather than individually encoding signal in each channel, the multichannel signals are represented as a mono or stereo signals, called downmix signal. To reproduce the multichannel audio signals at the decoder, spatial parameters are extracted in the encoder and transmitted to decoder as side information of downmix signal.

In traditional SAC systems (e.g. MPEG Surround [5], MPEG-H 3D Audio [6]), the Inter-Channel Level Difference (ICLD) is the most important spatial parameter [5], which are extracted between two channel signals. So the spatial sound image must be located between two loudspeakers, no matter how ICLD changes. Although ICLD are certain values without threshold, but the azimuths and elevations of virtual sound source in 3D space has threshold, such as each azimuth can only stay at a 360° circle. Instead of using ICLD between channel pair as spatial parameter, azimuths and elevations of virtual sound source expressing spatial information are more closely related to the spatial location of virtual sound source. Thus, Cheng proposed the S^3AC (Spatially Squeezed Surround Audio Coding) [7, 8] method, which includes Spatial Localization Quantization Point (SLQP) [8] methods for quantizing 3D virtual sound source locations. But SLQP methods have too many codewords that didn't make for high-efficient compression, and it doesn't take spatial psychological acoustics into full account.

Over last decades, many researches in spatial psychological acoustics can be used to encode of SAC. Typically, changing sound source's azimuth/elevation must reach a certain angle to be detected, because of human auditory system's limitation. These required minimum detectable angles is named as Just Noticeable Difference (JND) [9]. If the spatial parameters' quantized distortion can be limited below JND thresholds, the quantization is almost perceptual lossless. In our previous work, Gao proposed a JND Based Spatial Parameter Quantization (JSPQ) method [10, 11], but its codebook only suits to 2D space.

In this paper, we proposed a Perceptual Lossless Quantization of Spatial Parameter (PLQSP) method, the generation of elevations quantization codebook will be described in details, and the generation of azimuths quantization codebook in several planes expand from JSPQ. Both objective and subjective test are included to confirm the efficiency of PLQSP method, and experiment system employed S^3AC coding structure.

2   Overview of Spatial Audio Coding

2.1   The Typical Structures

According to the extraction method of spatial parameter, there are mainly two kinds of coding structures of SAC. One typical method is the tree scheme coding structure used in MPEG surround [5] and MPEG-H 3D [6].

Another typical structure is based on virtual sound source. The multichannel signals can be replaced by a virtual sound source and spatial location information (azimuths, elevations, etc.). The mainly virtual sound estimation techniques are based on Vector Based Amplitude Panning (VBAP) [12], a part techniques of S^3AC [8], etc. After spatial parameters quantization and virtual sound signal quantization, the encoder composes at last. At the decoder, with inverse quantization of virtual sound source and spatial parameters, 3D multichannel signals will be reproduced by general panning techniques such as VBAP or Higher Order Ambisonics (HOA) [13].

2.2   Extracting Virtual Sound Source

In S^3AC structure [8], a loudspeaker is located at azimuth μ_i, elevation η_i, as shown in Fig. 1. The p_i represents the loudspeaker signal, where i is loudspeaker index. And p_i can be decomposed into x-y-z coordinate system as:

$$p_i = g_i \cdot \begin{bmatrix} \cos \mu_i \cdot \cos \eta_i \\ \sin \mu_i \cdot \cos \eta_i \\ \sin \eta_i \end{bmatrix} \tag{1}$$

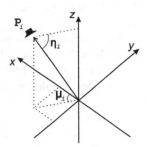

Fig. 1. A Loudspeaker Signal Located at azimuth μ_i, elevation η_i

The g_i is gain of this loudspeaker. All loudspeaker signals N form a virtual sound source, and the resulting source level g will be calculated by orthogonally decomposed components over the three axes as:

$$g^2 = [\sum\nolimits_{i=1}^{N} g_i \cdot |\cos \mu_i \cdot \cos \eta_i|]^2 + [\sum\nolimits_{i=1}^{N} g_i \cdot |\sin \mu_i \cdot \cos \eta_i|]^2 + [\sum\nolimits_{i=1}^{N} g_i \cdot |\sin \eta_i|]^2 \tag{2}$$

The azimuth μ and elevation η of virtual sound source can be estimated by:

$$\tan \mu = \frac{\sum_{i=1}^{N} g_i \cdot \sin \mu_i \cdot \cos \eta_i}{\sum_{i=1}^{N} g_i \cdot \cos \mu_i \cdot \cos \eta_i} \tag{3}$$

$$\tan \eta = \frac{\sum_{i=1}^{N} g_i \cdot \sin \eta_i}{\sqrt{\left[\sum_{i=1}^{N} g_i \cdot \cos \mu_i \cdot \cos \eta_i\right]^2 + \left[\sum_{i=1}^{N} g_i \cdot \sin \mu_i \cdot \cos \eta_i\right]^2}} \tag{4}$$

At the same time, the downmix signal at each frequency bin will be estimated by formula (5), the phase information e^{ϕ_M} is chosen from channels with the highest amplitude.

$$S(k) = \sqrt{g^2(k)} \cdot e^{\phi_M} \tag{5}$$

2.3   Analysis of Spatial Parameters

According to spatial perceptual characteristics of human auditory system, people have different JNDs in different azimuths and elevations. Firstly, the smallest azimuthal JND corresponds to front directions, bigger JND for rear directions and biggest azimuthal JND for side directions [14]. Secondly, the azimuthal JNDs in different elevations have significant differences [14]. With the elevation increasing, the JNDs in the same azimuth but different elevations are also different, and the azimuthal JND is increasing with the elevations' increasing [14]. Thirdly, with the increasing of elevation, elevational JNDs increase too [15].

So the azimuths/elevations with small JNDs allow small quantized errors, the azimuths/elevations with big JNDs allow big quantized errors. In other words, the quantization step sizes of bigger elevations should be bigger; for front/side/rear azimuths, the quantization step sizes of front should be smallest, the quantization step sizes of side should be biggest.

In the structure of S³AC, each azimuth and elevation will be quantized by SLQP methods. The SLQP quantization values distributions are shown in Fig. 2. Each value consists of location information described as azimuth and elevation. The SLQP-L represents quantization values distribution in 3D space with low precision; the SLQP-H

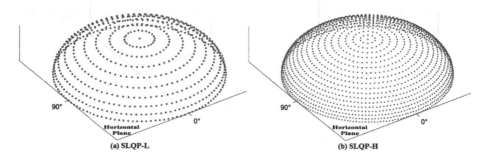

(a) SLQP-L   **(b) SLQP-H**

Fig. 2. Quantization values distributions in SLQP. (a) Quantization of azimuth and elevation with low precision, denoted as SLQP-L; (b) quantization of azimuth and elevation with high precision, denoted as SLQP-H.

represents quantization values distribution in 3D space with high precision. Each point represents a quantization value of azimuth and elevation. Any point located in the interval between two adjacent quantization values is quantified as a value closer to this point. In other words, every quantization interval endpoint between two adjacent points is quantization points' midpoint.

For SLQP-L, a 10° for elevation resolution is utilized, an azimuth precision of 3° is used for horizontal plane, and the quantization precision reduction between layers is 12 values. For example, while the 0° elevation layer has 120 values, the 10° elevation layer has 108 values. For SLQP-H, a 5° for elevation resolution is utilized, an azimuth precision of 2° is used for horizontal plane, and the quantization precision reduction between layers is 10 values.

As show in Fig. 2, the uniform quantization in elevations and azimuths at the same layer didn't follow azimuthal/elevational JND features. Taking horizontal plane's JNDs for example, the minimum JND is about 8.6° [14] at side directions, but the high precision SLQP method (SLQP-H) quantization interval is 1° (Because the distance between two adjacent quantization values are 2°). That caused a lot of perceptual redundant information.

Considering 1024 points 50% overlapping and windowing with Short-Time Fourier Transform (STFT) or Modified Discrete Cosine Transform (MDCT), if spatial parameters are extracted at each frequency bin by SLQP-L/SLQP-H methods in 48 kHz audio, the bit rates are about 450.16 kbps and 517.49 kbps. Obviously, these bit rates are unacceptable.

3 Proposed PLQSP Method of Spatial Parameters

Although spatial parameters quantization is independent of the coding structure, the proposed Perceptual Lossless Quantization of Spatial Parameter (PLQSP) method can be applied to many kinds of coding structures. But in order to effectively evaluate quantization method with SLQP method, S^3AC coding structure is applied. The azimuth and elevation at each frame and each frequency bin can be calculated by formula (3), (4).

3.1 The Design of Elevational Quantization

For any elevation, if elevation is quantized with quantization error below the corresponding elevational JND, the elevation is perceptually lossless quantized. It is principle of proposed spatial parameters quantization method.

Given one elevation C and JND of C, the quantized interval region of C can be calculated. All elevations in the quantized interval region should be quantized as C and all the quantization error will not be bigger than their JNDs. If every elevation is divided into corresponding quantization interval regions, elevations will be perceptually lossless quantized.

The JND data can only be obtained by subjective listening tests. And these data would be different from different listeners. The test environments would also influence the results of one listener. So the final JND data is usually an average from multiple listeners. Besides, it is impractical to get JND data of all azimuths/elevations by listening tests. During the listening tests, a few of typical azimuths/elevations with limited resolution are got. By interpolating to these data, higher resolution JND distributions will be obtained. The elevation JND data are used from [15] with one decimal place.

Given an elevation Λ_i from $0°$ to $90°$ with an increase of λ, it is obvious that:

$$\Lambda_{i+1} = \Lambda_i + \lambda \quad or \quad \Lambda_i = \Lambda_{i+1} - \lambda \tag{6}$$

If λ is just noticeable difference in the listening test, that is to say: $\lambda = JND_{\Lambda_i}$, formula (6) can be express by:

$$\Lambda_{i+1} = \Lambda_i + JND_{\Lambda_i} \quad or \quad \Lambda_i = \Lambda_{i+1} - JND_{\Lambda_i} \tag{7}$$

The following procedures are inspired by formula (7).

First of all, the start value $EV_0 = 0°$ is selected, all quantization values will be obtained by recurrence. For one quantization value $EV_i(i \geq 0)$, the corresponding endpoint EE_i of quantization interval region is expressed as:

$$EE_i = EV_i + JND_{EV_i} \tag{8}$$

After getting EE_i, the EV_{i+1} can be calculated:

$$EV_{i+1} = EE_i + JND_{EE_i} \tag{9}$$

Then, the EE_{i+1} can also be calculated:

$$EE_{i+1} = EV_{i+1} + JND_{EV_{i+1}} \tag{10}$$

As shown in Fig. 3, the interval region of EV_{i+1} is $[EE_i, EE_{i+1})$, and the recurrence ends up when EE_i or EV_i reaches $90°$.

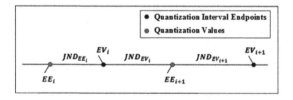

Fig. 3. The demonstration of elevational quantization value generation

Finally, the elevational quantization values distribution is shown in Fig. 4. It has 7 values; in other words, the quantization values in 3D space consist of 7 layers, and azimuth quantization values are in these layers.

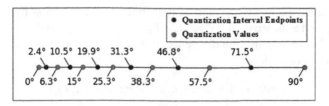

Fig. 4. The distribution of elevational quantization values

3.2 The Design of Azimuthal Quantization

The azimuthal JND data comes from [14]. The quantization of azimuth is similar with the quantization of elevation, but the start value is $AV_0 = 90°$. There are two procedures to obtain other azimuthal quantization values. One procedure is from AV_0 to $0°$, another procedure is from AV_0 to $180°$. Each layer (without elevation = 90°) will have these two procedures, generating azimuthal quantization values in one layer expand from previous research [10].

Considering listening perceptual symmetry, the quantization values distribution from $0°$ to $-180°$ is same with values from $0°$ to $180°$ except the sign. After all the layers (without elevation = 90°) generating azimuthal quantization values, the distribution of quantization values in 3D space are obtained as Fig. 5. There is only a point rather than a circle when elevation = 90°, so the azimuthal quantization value in elevation = 90° is azimuth = 0° in any situation.

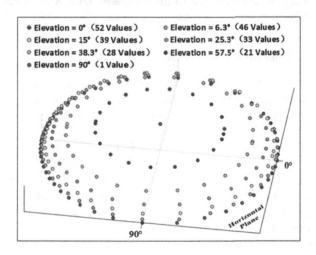

Fig. 5. Distributions of azimuthal quantization values of PLQSP in 3D space

3.3 Compression of Codewords

After azimuths and elevations are respectively quantized as codewords, the differential Huffman coding will be conducted to compress bit rates.

For elevations, the first time frame at each frequency bin coding by binary coding (3 bits), expressed as $index(0, k)$ (or $index(n, k)$, $n = 0$, n is the number of frame). From second to the last time frame, the difference of quantization index in each two adjacent time frames is calculated:

$$diff(n, k) = index(n, k) - index(n+1, k) \qquad (11)$$

Then, according to Huffman codebook of difference of elevational quantization index, each $diff_E(n, k)$ is processed by Huffman coding.

For azimuths, although the different size of codebooks and virtual sound source change among layers, but the Huffman codebook of difference of azimuthal quantization index is universal. In other words, even though two adjacent time frames may come from different layers, but decoder can decode the azimuth information by elevation and difference of azimuthal quantization index. In order to contribute fluctuation range, not every layer's indices starts with $index\_number = 0$. When elevation $= 0°$ layer, the mid codeword is $index\_number = 26$, other layers mid codeword is also defined as $index\_number = 26$. That is to say: except elevation $= 0°$ layer, each layer is not start with $index\_number = 0$. The azimuths difference coding is the same with elevations by using formula (11).

4 Experiments

Both objective and subjective experiments were included to test the performance of PLQSP method. In next paragraphs, the contrast with SLQP in 3D space will be showed. In order to ensure the fairness of the comparison, the S^3AC coding structure is selected as experiment system.

4.1 Analysis of Quantized Distortions

The quantized error of spatial parameters will directly affect the spatial perceptual quality. In this part, the absolute quantized errors of azimuths and elevations were calculated respectively. And absolute quantized error is given in formula (12):

$$E_q = |Ang - C_q| \qquad (12)$$

The Ang represents the azimuths/elevations before quantization, the C_q represents the value of codeword after quantization, and the E_q represents the absolute quantized error in quantization process. In order to facilitate observation, find peaks in E_q data and draw a plot to reflect the trend of distortion.

The absolute quantized error of elevations at $0°$ to $90°$ were shown in the Fig. 6. The JND curve from [15] are utilized to evaluate the perceptual distortion of elevational quantization. When quantized error is bigger than JND, spatially perceptual distortion is inevitable. However, if quantized error is smaller than JND, perceptual redundancy will exist.

Owing to compact quantization step sizes in SLQP-L/SLQP-H, the quantized errors are smaller than PLQSP method except SLQP-L at relatively low elevations. But

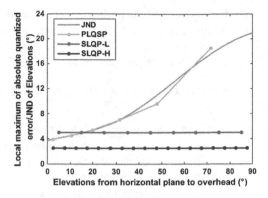

Fig. 6. Quantized errors of elevations in different methods compared with elevational JND data

perceptual redundancies will lead to inefficient compressions of spatial parameters. The error curve of PLQSP is almost in accordance with the JND curve. It will contribute to high efficient compressions of spatial parameters.

Like quantized errors of elevations, SLQP-L/SLQP-H quantized errors of azimuths are smaller than PLQSP that is not good enough to compression. And PLQSP method is still almost in accordance with the JND curve at each layer. More details about analysis azimuthal quantization are shown in [10, 11].

4.2   Analysis of Bit Rates

Since the correspondence between quantization values and codewords, SLQP-L has 658 codewords and SLQP-H has 1729 codewords. As shown in Fig. 5, PLQSP method has 7 layers, the number of quantization codewords is 219 codewords in total.

If using binary coding to encode spatial parameters as one-dimensional table, the comparison of bits rates (with two decimal place) in each spatial parameter is show in Table 1. Compared with SLQP-L, SLQP-H, the bit rate of PLQSP method has decreased by 16.99% and 27.79% respectively.

Table 1. The comparison of binary coding bit rates in each spatial parameter

Methods	Bit Rates (*bits*)
PLQSP	7.77
SLQP-L	9.36
SLQP-H	10.76

Considering 1024 points 50% overlapping and windowing with STFT or MDCT, if spatial parameters are extracted at each frequency bin by PLQSP method in 48 kHz audio, the bit rate is about 373.69 kbps with taking spectral symmetry into account. Obviously, this bit rates is unacceptable in practical application, rather than 450.16 kbps and 517.49 kbps by SLQP-L/SLQP-H methods. So the differential Huffman coding in Sect. 3.3 is used.

Currently, there didn't have published 3D multichannel standard test audio, the test sequences in this paper were produced by VBAP. So bit rates and subjective test will using four 19-channel (22.2-channel without 2 LFEs and 3 bottom loudspeakers) sequences with 48 kHz. The details of sequences are in the Table 2.

Table 2. Details of test audio signal

Names	Duration (s)	Details
Sequence 1	10.73	Moving voice source, birds and running water
Sequence 2	10.97	Noisy crowd street, moving trumpets and other moving musical instruments
Sequence 3	4.79	Fast running racing cars and live broadcast sound
Sequence 4	20.00	Pop song, complex electronic music and moving song source

The differential Huffman coding is only applied in PLQSP method, because SLQP codebooks is too large that differential Huffman encoding is unable to obtain satisfactory results. The bit rates of PLQSP with differential Huffman coding are shown in the Table 3. The average total bit rate is about 254.96 kbps. The comparison of bit rates in different methods (The bit rate of differential Huffman coding is an average.) are shown in Fig. 8. Compared with binary coding PLQSP method, PLQSP with differential Huffman coding has decreased by 31.77%; compared with binary coding SLQP-L/SLQP-H method, it has decreased by 43.36% and 50.73% respectively (Fig. 7).

Table 3. Bit rates of PLQSP with differential Huffman coding

Names	Azimuthal Bit Rates (kbps)	Elevational Bit Rates (kbps)	Total (kbps)
Sequence 1	129.71	81.99	211.70
Sequence 2	177.85	98.09	275.94
Sequence 3	224.44	48.52	272.96
Sequence 4	165.76	93.47	259.23
Average	174.44	80.52	254.96

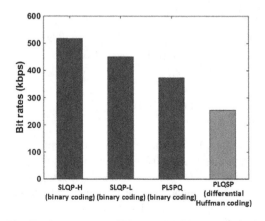

Fig. 7. Comparison of bit rates in different methods

4.3   Subjective Test

Subjective test was intended to evaluate the spatial quality of PLQSP method of spatial parameters. Listeners need to rank the perceived spatial quality of audio sequences with different quantization methods against the original audio sequences. The subjective test scheme MUSHRA of ITU-R BS.1534-1 [16] was employed with a 22.2-channel system (without 2 LFEs and 3 bottom loudspeakers). There were 6 audio sequences used in each MUSHRA test: an original sequence from Table 3 as a reference; an original sequence as a hidden reference; a spatial quality degraded original sequence as a hidden anchor; and test signals coding by PLQSP/SLQP-L/SLQP-H. This test is focused on evaluating virtual sound sources among test sequences.

There were 10 male and female graduated students aged from 20 to 30 taking part in subjective tests, whose research areas are audio signal processing. And were trained before the test. The mean scores with 95% confidence interval of MUSHRA for spatial audio quality of different quantization methods are shown in Fig. 8.

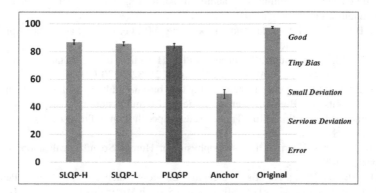

Fig. 8. Subjective test scores for original, anchor sequences and different quantization methods

It is apparently that PLQSP method didn't fade next to SLQP methods in spatial perceptual qualities, and the advantages of SLQP-H/SLQP-L in average scores are only 3.13% and 1.62% respectively.

In total, the proposed PLQSP reduces the bit rates while keeping listening qualities similar with SLQP methods.

5   Conclusion

In this paper, we proposed a Perceptual Lossless Quantization of Spatial Parameter (PLQSP) method for 3D multichannel audio based on human perceptual characteristics just noticeable difference (JND) and differential Huffman coding. Both objective and subjective evaluations experiments are conducted to test the performance of proposed PLQSP method compared with SLQP methods in respect of quantized distortions, bit rates and listening qualities. We found that the proposed PLQSP method had lower bit rates compared with SLQP methods while keeping listening qualities similar with SLQP methods.

For future work, we plan to expand the PLQSP codebook to suit for 3 bottom loudspeakers of 22.2-channle system and apply PLQSP method into practical 3D audio coding systems. On the other hand, we also plan to apply PLQSP into metadata quantization of Spatial Audio Object Coding (SAOC).

References

1. ITU-R BS.775-1. Multichannel Stereophonic Sound System with and with out Accompanying Pictures (1994)
2. Ando, A.: Conversion of multichannel sound signal maintaining physical properties of sound in reproduced sound field. IEEE Trans. Audio Speech Lang. Process. **19**(6), 1467–1475 (2011)
3. Sergi, G.: Knocking at the door of cinematic artifice: Dolby Atmos, challenges and opportunities. New Soundtrack **3**(2), 107–121 (2013)
4. Disch, S., Ertel, C., Faller, C., et al.: Spatial audio coding: next-generation efficient and compatible coding of multi-channel audio. In: Audio Engineering Society Convention 117. Audio Engineering Society (2004)
5. Jeroen, B., Christof, F.: Spatial Audio Processing: MPEG Surround and Other Applications. Wiley, Hoboken (2007)
6. Herre, J., Hilpert, J., Kuntz, A., et al.: MPEG-H audio—the new standard for universal spatial/3D audio coding. J. Audio Eng. Soc. **62**(12), 821–830 (2015)
7. Cheng, B.: Spatial squeezing techniques for low bit-rate multichannel audio coding (2011)
8. Cheng, B., Ritz, C., Burnett, I., et al.: A general compression approach to multi-channel three-dimensional audio. IEEE Trans. Audio Speech Lang. Process. **21**(8), 1676–1688 (2013)
9. Blauert, J.: Spatial Hearing: the Psychophysics of Human Sound Localization. MIT press, Cambridge (1997)
10. Gao, L., Hu, R., Wang, X., et al.: Perceptual Lossless Quantization of Spatial Parameterof multichannel audio signals. EURASIP J. Audio Speech Music Process. **2016**(1), 1–18 (2016)
11. Gao, L., Hu, R., Wang, X., et al.: Effective utilisation of JND for spatial parameters quantisation in 3D multichannel audio. Electron. Lett. **52**(12), 1074–1076 (2016)
12. Pulkki, V.: Virtual sound source positioning using vector base amplitude panning. J. Audio Eng. Soc. **45**(6), 456–466 (1997)
13. Daniel, J., Moreau, S., Nicol, R.: Further investigations of high-order ambisonics and wavefield synthesis for holophonic sound imaging. In: Audio Engineering Society Convention 114. Audio Engineering Society (2003)
14. Heng, W., Cong, Z., Ruimin, H., Weiping, T., Xiaochen, W.: The perceptual characteristics of 3D orientation. In: Gurrin, C., Hopfgartner, F., Hurst, W., Johansen, H., Lee, H., O'Connor, N. (eds.) MMM 2014. LNCS, vol. 8326, pp. 353–360. Springer, Heidelberg (2014). doi:10.1007/978-3-319-04117-9_35
15. Makous, J.C., Middlebrooks, J.C.: Two-dimensional sound localization by human listeners. J. Acoust. Soc. Am. **87**(5), 2188–2200 (1990)
16. Bureau ITU-R. Method for the subjective assessment of intermediate quality level of coding systems. ITU-R Recommendations, Supplement 1 (2014)

Unsupervised Multiple Object Cosegmentation via Ensemble MIML Learning

Weichen Yang, Zhengxing Sun$^{(\boxtimes)}$, Bo Li, Jiagao Hu, and Kewei Yang

State Key Laboratory for Novel Software Technology,
Nanjing University, Nanjing, People's Republic of China
szx@nju.edu.cn

Abstract. Multiple foreground cosegmentation (MFC) has being a new research topic recently in computer vision. This paper proposes a framework of unsupervised multiple object cosegmentation, which is composed of three components: unsupervised label generation, saliency pseudo-annotation and cosegmentation based on MIML learning. Based on object detection, unsupervised label generation is done in terms of the two-stage object clustering method, to obtain accurate consistent label between common objects without any user intervention. Then, the object label is propagated to the object saliency coming from saliency detection method, to finish saliency pseudo-annotation. This makes an unsupervised MFC problem as a supervised multi-instance multi-label (MIML) learning problem. Finally, an ensemble MIML framework is introduced to achieve image cosegmentation based on random feature selection. The experimental results on data sets *ICoseg* and *FlickrMFC* demonstrated the effectiveness of the proposed approach.

Keywords: Multiple foreground cosegmentation · Unsupervised label generation · Saliency pseudo-annotation · Cosegmentation based on MIML learning

1 Introduction

The aim of MFC, proposed by Kim and Xing [1], to extract a finite number of common objects from an image collection, while there only exists an unknown subset of common objects in every single image. Compared with supervised MFC methods, unsupervised MFC method has better flexibility and less restriction, so it has become a new research hotspot. Although several unsupervised MFC approaches have been proposed in the last decade, these approaches still bear obvious defects, which are mainly lied in the following aspects:

Misleading Consistent Information: The key issue of the MFC problem lies in mining consistent information shared by the common objects. The existing methods can not obtain the accurate consistent information, which is mainly due to two defects. Firstly, these methods only utilize the low-level cue to represent proposal objects. For example, Li and Meng in [2] utilized three types of visual

L. Amsaleg et al. (Eds.): MMM 2017, Part II, LNCS 10133, pp. 393–404, 2017.
DOI: 10.1007/978-3-319-51814-5_33

features, i.e., color, texture and shape, to construct the proposals descriptor. Li et al. in [3] proposed a simple but highly effective self-adaptive feature selecting strategy, but only normalized color and bag of feature histograms are used in the strategy. While low-level cue is not suited to represent the semantics of proposal objects directly in fact. Secondly, these methods can not add appropriate constraints in the classification step and reject the noise data in proposal objects. For example, Li and Meng in [2] proposed a robust ensemble clustering scheme. However, there is no clustering constraint in this clustering scheme. It leads to spit the combinational common objects into multiple pieces in the final segmentation. Meng and Li in [4] added several special constraints in their directed graph clustering method to find the noise data. However, there constraints led to loss some common object in the final segmentation.

Ambiguous Segmentation Assist: Existing methods usually adopted an unsupervised approach to rely on manually designed metrics to get the final segmentation result. For example, Li and Fei in [5] constructed a superpixel-level weighted graph for each image, and then adopted multi-label graph cuts with global and local energy to get the final cosegmentation results. Chang and Wang in [6] proposed a novel object detective approach via common object hypotheses, which assigned the label of every segment to the common object that is closest to it. However, unsupervised method is intrinsically ambiguous. Meanwhile, these manually designed metrics are too subjective and can not adapt flexibly to large variation and corruption of common objects, as well as to the inherent ambiguity between complex common objects and background. In summary, ambiguity of the unsupervised approach and manually designed metrics lead to ambiguous segmentation assist, which lead to some segments classify into the wrong class in the final segmentation.

In this paper, a novel multiple foreground cosegmentation framework is proposed. The contributions of the proposed framework compose of following three-folds:

1. **Unsupervised label generation.** Deep semantic feature and a two-stage object clustering method are introduced, which are combined to obtain accurate and rich semantic consistent label between common objects based on object detection. Thus, it solves the misleading consistent information problem unsupervised.
2. **Saliency pseudo-annotation.** A label propagation method is introduced to propagate the object label coming from unsupervised label generation to object saliency coming from saliency detection method. Thus, it makes an unsupervised MFC problem as a supervised MIML learning problem.
3. **Cosegmentation based on MIML learning.** An ensemble MIML framework is introduced to achieve image cosegmentation based on random feature selection, which overcomes the fuzziness and ambiguity of the unsupervised method. Thus, it solves the ambiguous segmentation assist problem.

2   Related Work

Cosegmentation method can be traced back to the early work of Rother and Minka [7] at 2006, which utilized histogram matching to cosegment the common objects from an image pair, and use an approximation approach called submodular-supermodular procedure to minimize the model by max-flow algorithm. After that, scholars along this thought put their effort on finding different kinds of consistency terms in MRF to simplify the optimization [8]. However, these methods can only handle single common object in the image collection, which limits their application. In order to relax this restriction, scholars switch their attention to handle multi-class common object cosegmentation problem. For example, Kim et al. in [9] adopt a method to solve the multi-class cosegmentation problem firstly, by temperature maximization on anisotropic heat diffusion. Yet, these methods restrict that every common object is contained in each image. In order to relax this restriction, Kim and Xing in [1] proposed MFC problem firstly. Then some MFC methods have been proposed successively. These methods are classified into two categories: supervised and unsupervised.

Early MFC methods are mainly supervised methods. These methods utilize user interaction to aid the consistent information obtained. The method in [1] designed an iterative framework that combines foreground modeling and region assignment. The method in [10] established consistent functional maps across the input images and used a formulation that explicitly models partial similarity instead of global consistency. These methods should input the class number of the common objects beforehand. the method in [11] proposed a cosegmentation framework based on standard graph transduction and semi-supervised learning framework, and integrated the global connectivity constraints into the proposed framework. The method in [12] adopted a multiple foreground recognition and cosegmentation problem within a conditional random fields framework. These methods need to annotate some image in input images by users.

Although, Supervised methods can obtain the more accurate number and class of the common object in every single image than unsupervised methods, it is a hard work for user to get the class number of the common objects or annotate labels for the common object in massive images. So recently there are more and more unsupervised methods appeared. the method in [2] proposed a robust ensemble clustering scheme to discover the unknown object-like proposals, and the proposals are then used to derive unary and pairwise energy potentials across all the images. The method in [6] decomposed MFC into three related tasks, i.e., image segmentation, segment matching and figure/ground assignment, and then constructed the foreground object hypotheses, which to determine the foreground objects in each individual image. The method in [4] proposed a new constrained directed graph clustering for the classification step and getting the cluster, then, they extracted the object priors from the clusters, and propagate these priors to all the images to achieve the final multiple objects extraction. The method in [3] proposed a multi-search strategy to obtain the common objects in each image, the strategy extracted each target individually and an adaptive decision criterion is raised to give each candidate a reliable judgment automatically.

3 Overview

There are three parts in our method, i.e., unsupervised label generation, saliency pseudo-annotation and cosegmentation based on MIML learning. The framework of our proposed method is illustrated in Fig. 1.

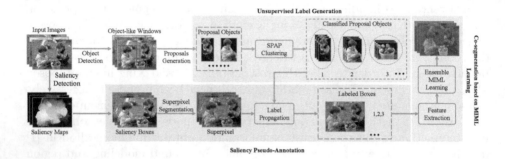

Fig. 1. Framework of our proposed method

In the unsupervised label generation, deep semantic feature is used as a part of feature of the proposal objects. Such feature can better describe the similarity of common objects on semantic level. A two-stage object clustering method is introduced to reject the noise data in proposal objects and mine consistent information shared by the common objects. Based on the above two aspects, accurate consistent label between common objects is obtained without any user intervention.

In the saliency pseudo-annotation, each saliency box coming from saliency detection method is over-segmented to obtain the superpixels. Then, the object label getting from unsupervised label generation is propagated to a saliency box by a label propagation method. Thus, this makes an unsupervised MFC problem as a supervised MIML learning problem.

In the cosegmentation based on MIML learning, an ensemble MIML method is introduced to solve the MIML learning problem coming from saliency pseudo-annotation and obtain the final image cosegmentation based on random feature selection. Using the supervised method, the fuzziness and ambiguity caused by the unsupervised method can effectively be overcame. Accurate image segmentation can be obtained in terms of full usage of consistent information of each common object.

4   Proposed Method

In this section, we will introduce three components of the proposed framework of unsupervised multiple object cosegmentation: unsupervised label generation, saliency pseudo-annotation and cosegmentation based on MIML learning.

4.1   Unsupervised Label Generation

In the unsupervised label generation, firstly, the object detection method in [13] is used to get the proposal objects in each image. Then, in order to extract the feature, each proposal object is divided into local patches on a grid with a sample spacing ten pixels. The local descriptor of each patch is constructed by the low-level features and the high-level features. HOG descriptor is used as the low-level feature in this paper, and it is denoted by f_l. A CNN model [14] is built the high-level features. Firstly, each input image is fed into the per-trained CNN model to get the representations of every pixel from the layer called *crop*, which is a 21-D CNN feature vector. Then the high-level semantic representations of each patch are obtained by a mean pooling operation, and it is denoted by f_h. The final local descriptor f of a patch is generated by concatenating the two-level features, that is

$$f(p) = \frac{1}{2} \left[\mathcal{N}(f_l(p)), \mathcal{N}(f_h(p)) \right] \tag{1}$$

where p denotes a patch, and $\mathcal{N}(\cdot)$ denotes a normalized operator to ensure the descriptor in the range of $[0, 1]$. The final local descriptor f is a 300-D feature vector. All of patches are classified into some clusters using the spectral clustering method. Each cluster center is a code-word. Then the histogram of each proposal object is calculated by counting the number of patches in each cluster. So, the feature of the kth proposal object P_k^m is represented by the clusters of the histogram, and it is denoted by f_k^m.

Lastly, a two-stage object clustering method is introduced to classify proposal objects into several clusters and obtain the consistent label of each proposal object. In first stage, the similarity between proposal objects is calculated in two case. If two proposal objects come from different images, Bhattacharyya distance and Gaussian kernel to calculate their similarity. If their come from the same image, their similarity will be set to 0. Then, the initial clusters $\mathcal{C}' = \left\{ C_1', C_2', \ldots, C_t' \right\}$ are obtained by a new clustering algorithm called SPAP clustering algorithm. In the second stage, the center feature of each initial cluster is calculated, and it denotes by $\mathcal{F}^c = \{f_1^c, f_2^c, \ldots, f_t^c\}$. Then, the similarity of center features is obtained by Bhattacharyya distance and Gaussian kernel, and the final clusters $\mathcal{C} = \{C_1, C_2, \ldots, C_v\}$ are given by the SPAP clustering algorithm. The first times clustering avoids the clustering of the proposal objects from the same image, and the second times clustering can combine the similar clusters that come from the first-time clustering.

Next, our SPAP clustering algorithm is introduced. Spectral clustering makes use of the spectrum of the similarity matrix of the data to perform dimensionality reduction before clustering in fewer dimensions. So, spectral clustering is used as the basis of our SPAP clustering algorithm. Affinity propagation [15] does not require the number of clusters to be determined or estimated before running the algorithm. So, affinity propagation is integrated into our SPAP clustering algorithm to get the number of object classes occur in the image collection. The SPAP clustering algorithm is a novel and easy clustering algorithm. Affinity propagation is used to replace the k-means in the classical spectral clustering. When the Laplacian matrix L is obtained, affinity propagation is used to get the clusters by all eigenvectors. Then E_i and C_i denote the number of eigenvectors that are used in affinity propagation and the number of clusters, respectively, where i denote ith iteration. Let the E_{i+1} equal to C_i until E_i equal to E_{i+1}.

Two comparative experiments are designed to evaluate the performance of unsupervised label generation. The first comparative experiment is designed to evaluate the performance of feature. All proposal objects from three images are classified by our two-stage object clustering method using different features to construct the local descriptor. Figure 2a is the clustering results of using low-level features and the high-level features, and Fig. 2b is the clustering results of only using low-level features. From these results, using the high-level semantic features can get more accurate clustering results.

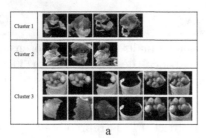 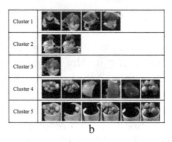

a   b

Fig. 2. Comparison results of different feature

The second comparative experiment is designed to evaluate the performance of SPAP clustering method. All proposal objects from three images are classified by different clustering methods. Figure 3a is the results of the two-stage object clustering method, Fig. 3b is the results of only using the SPAP clustering algorithm, and Fig. 3c is the results of clustering method using affinity propagation. From these results, our clustering scheme succeeds to annotation the label of proposal objects, which belong the part and whole of the same common object. It proves that our SPAP cluster scheme can get the accurate pseudo-label and efficiently solve the problem of misleading consistent information.

4.2   Saliency Pseudo-Annotation

In the saliency pseudo-annotation, the first step is to generate integrated saliency boxes. The saliency map S_m of the image I_m is obtained using the method [16].

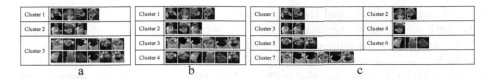

Fig. 3. Comparison results of SPAP clustering scheme

In order to get the foreground of the image through the saliency map, the binaryzation of the saliency map is necessary. The saliency-cut method in [17] performs the binaryzation. In this solution, a loose threshold, which typically results in good recall but relatively poor precision, is used to generate the initial binary mask. Then the method iteratively uses the GrabCut segmentation method to gradually refine the binary mask. The binaryzation of S_m is denoted by S'_m. Then, a box is used to surround every discontinuous saliency region and segment them from the image. Finally, saliency boxes $\mathcal{B}^m = \{B_1^m, B_2^m, \ldots, B_{h_m}^m\}$ are given in the image I_m, where h_m denotes the number of saliency boxes.

The second step is to generate the superpixel of each saliency box. Pixels in the same superpixel have the similar feature and the superpixel can keep the edge of the object, so superpixels satisfy requirements of the instance, which are that features of instance conveniently are extracted and instance only have one label. So method in [18] is utilized to get the superpixel as instance.

The third step is label propagation. Given a saliency box B_k^m, if a proposal object P_t^m fall into this saliency box, the object label propagates from P_t^m to B_k^m. Let $V_{in}(B_k^m, P_t^m)$ and $V_{out}(B_k^m, P_t^m)$ denote proposal object areas, which are inside and outside the saliency box B_k^m, if the condition

$$\frac{V_{in}(B_k^m, P_t^m)}{V_{in}(B_k^m, P_t^m) \cup V_{out}(B_k^m, P_t^m)} > \rho \qquad (2)$$

is satisfied, the proposal object P_t^m is considered falling into the saliency box B_k^m. Thus, an unsupervised MFC problem is made as a MIML learning problem.

4.3 Ensemble MIML Learning

In this section, our designed ensemble MIML learning method based on random feature selection is introduced. The first step extracts the feature of each superpixel, and obtains the feature vector of every superpixel. These feature vectors denote by $\mathcal{F}' = \{f_h^m(s), h = 1, 2, \ldots, h_m, m = 1, 2, \ldots, n, s = 1, 2, \ldots, S\}$, where $f_h^m(s)$ denotes the feature vector of sth superpixel in hth saliency box in mth image. Every feature vector is defined over d-dimensional feature space, $f_h^m(s) \in \mathbf{R}^d$. Using the method in [19], 50 feature sets are constructed by randomly selecting $d' = 0.6d$ features from d-dimensional feature vector. The second step trains a set of weak classifier $\{c_{1b}, c_{2b}, \ldots, c_{fb}\}$ for bth feature set using the method in [20]. The third step trains a weighted SVM to obtain the soft weight of each weak classifier, and then the final classifier $\{c_1, c_2, \ldots, c_f\}$ is obtained. The forth step utilizes the final classifier to classify every superpixel.

A comparative experiment is designed to evaluate the performance of ensemble MIML learning method. The label of every saliency box is given manually. Classical MIML learning method in [20] and our ensemble MIML learning scheme are used to obtain the cosegmentation results, respectively. Figure 4 shows the comparison results. These results prove that our method can get more accurate results than classical method. It proves that our method can efficiently use the consistent information to assist the segmentation.

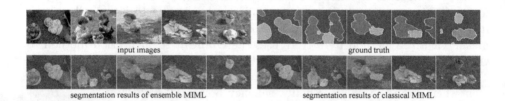

input images   ground truth

segmentation results of ensemble MIML   segmentation results of classical MIML

Fig. 4. Comparison results of ensemble MIML learning scheme

5   Experiments

To evaluate the performance of the proposed method, we conduct comparison experiments on two different available standard datasets *ICoseg* [21] and *FlickrMFC* [1]. The experiments are divided into two parts, which are cosegmentation on image contained single common object and contained multiple common objects. In order to evaluate the accuracy of our proposed method, the F-measure metric is applied. F-measure metric is defined as

$$F - measure = \frac{1}{N} \sum_i^N \frac{\left(1 + \beta^2\right) Pre_i \times Rec_i}{\beta^2 Pre_i + Rec_i} \tag{3}$$

where Pre_i and Rec_i denotes the precision and recall of the ith image. β^2 is a non-negative weight and is set to 0.3 to increase the importance of the precision value. Here N denotes the number of images in the image group. Note that the larger the F-measure is, the more accurate the segmentation result.

5.1   Single Common Object Cosegmentation

Datasets: The *ICoseg* dataset is a fully manually labeled dataset. Only a single common object is presented in every image. The *ICoseg* dataset is used to evaluate the single common object cosegmentation of our proposed method.

Baselines: In this part, we use five recent cosegmentation methods proposed by Joulin et al. [22], Kim et al. [9], Vicente et al. [23], Meng et al. [24],

Rubinstein et al. [25] as baseline. Due to the presence of only a single common object in each image, only proposal objects in the cluster contained most proposal objects are annotated. The others will be abandoned.

Results: Figure 5 shows the evaluation results on some *ICoseg* example taken from four image collections. Eight images per collection are sampled. In each set, the first row is input images, the second row is color-coded cosegmentation results and the third row is ground-truth. These results prove that our method segments out common objects successfully from the images, such as the *bear* in the image collection *Alaskan Brown Bear* is segmented out accurately. Although our method is designed by MFC problem, it also can obtain outstanding results in single common object cosegmentation problem. These results further demonstrate the perfect performance of our proposed method.

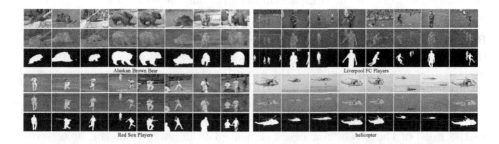

Fig. 5. Evaluation results on *ICoseg* image dataset (Color figure online)

The comparison results on the F-measure metrics are described in Fig. 6. We can see that our method achieves the highest F-measure values in most image collections. For 26 image collections, our method is the best in 9 image collections. The left-most bar set represents the average performance on all the image collections. It is seen that our method outperformed multiple common object cosegmentation method like [9], which results in 0.18 for the F-measure metric on average. At the same time, the comparable segmentation results as the state-of-the-art single common object cosegmentation method of [25] is obtained, which was particularly proposed for cosegmentation of *ICoseg*.

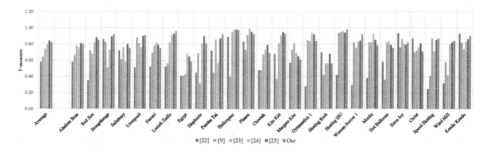

Fig. 6. Comparison results of the F-measure metric on *ICoseg* image dataset

5.2 Multiple Common Object Cosegmentation

Datasets: The *FlickrMFC* dataset is sampled images from Flickr. In each image collection, a fixed number of common objects frequently occur across the collection, but an unknown subset of them appears in every single image. The *FlickrMFC* dataset is utilized to evaluate the multiple common object cosegmentation of our proposed method.

Baselines: In this part, five recent multi-class or multiple foreground cosegmentation methods are used as baseline, which are proposed by Kim and Xing [1], Wang et al. [10], Li and Meng [2], Chang and Wang [6] and Li et al. [3] respectively. The method in [1] has two different versions, supervised setting and unsupervised setting. In order to perform a fair comparison, the method in [1] runs under unsupervised setting. The method in [10] need the total number of common objects. In the comparison, we give out the total number manually for the method in [10].

Results: Figure 7 shows the evaluation results on some *FlickrMFC* examples taken from six image collections. Six images per collection are sampled. In each set, the first row is input images, the second row is color-coded cosegmentation results, where the same class is marked by the same color and the third row is ground-truth. It is seen that only a little of common objects are over segmented, and most of common objects are completely segmented from the image, such as the head of the *girl* is not over segmented from the *girl* in the image collection *apple+picking*, and in the image collection *thinker+Rodin*, the *Rodin* is completely segmented, not segmented into several pieces.

Fig. 7. Evaluation results on *FlickrMFC* image dataset (Color figure online)

Figure 8 describes the comparison results of the F-measure metrics. Our method works best for 7 out of 14 image collections compared with the baselines. On average, our method achieves about 10.93% improvements on the F-measure metrics over the best unsupervised method in the baselines. At the same time, the comparable segmentation results in fully unsupervised are obtained as the supervised multiple common object cosegmentation method of [10], which need user to input the class number of the common objects beforehand.

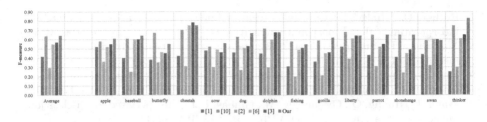

Fig. 8. Comparison results of the F-measure metric on *FlickrMFC* image dataset

6   Conclusion

This paper proposes a framework of unsupervised multiple object cosegmentation, which is composed of three components: unsupervised label generation, saliency pseudo-annotation and cosegmentation based on MIML learning. Based on object detection, unsupervised label generation is done in terms of the two-stage object clustering method, to obtain accurate consistent label between common objects without any user intervention. Then, the object label is propagated to the object saliency come from saliency detection method, to finish saliency pseudo-annotation. This makes an unsupervised MFC problem as a supervised MIML learning problem. Finally, an ensemble MIML framework is introduced to achieve image cosegmentation based on random feature selection. The experimental results on data sets *ICoseg* and *FlickrMFC* demonstrated the effectiveness of the proposed approach.

Acknowledgment. This work is supported by National High Technology Research and Development Program of China (No. 2007AA01Z334), National Natural Science Foundation of China (No. 61321491, 61272219), Innovation Fund of State Key Laboratory for Novel Software Technology (No. ZZKT2013A12, ZZKT2016A11), Program for New Century Excellent Talents in University of China (NCET-04-04605).

References

1. Kim, G., Xing, E.P.: On multiple foreground cosegmentation. In: IEEE CVPR, pp. 837–844 (2012)
2. Li, H., Meng, F.: Unsupervised multiclass region cosegmentation via ensemble clustering and energy minimization. IEEE Trans. Circ. Syst. Video Technol. **24**(5), 789–801 (2014)
3. Li, K., Zhang, J., Tao, W.: Unsupervised co-segmentation for indefinite number of common foreground objects. IEEE Trans. Image Process. **25**(4), 1898–1909 (2016)
4. Meng, F., Li, H.: Constrained directed graph clustering and segmentation propagation for multiple foregrounds cosegmentation. IEEE Trans. Circ. Syst. Video Technol. **25**(11), 1735–1748 (2015)
5. Li, L., Fei, X.: Unsupervised multi-class co-segmentation via joint object detection and segmentation with energy minimization. In: MIPPR, pp. 9812–9814 (2015)
6. Chang, H.S., Wang, Y.C.F.: Optimizing the decomposition for multiple foreground cosegmentation. Comput. Vis. Image Underst. **141**, 18–27 (2015)

7. Rother, C., Minka, T.: Cosegmentation of image pairs by histogram matching-incorporating a global constraint into MRFs. In: IEEE CVPR, pp. 993–1000 (2006)
8. Mukherjee, L., Singh, V., Dyer, C.R.: Half-integrality based algorithms for cosegmentation of images. In: IEEE CVPR, pp. 2028–2035 (2009)
9. Kim, G., Xing, E.P., Fei-Fei, L., Kanade, T.: Distributed cosegmentation via submodular optimization on anisotropic diffusion. In: IEEE ICCV, pp. 169–176 (2011)
10. Wang, F., Huang, Q., Ovsjanikov, M., Guibas, L.J.: Unsupervised multi-class joint image segmentation. In: IEEE CVPR, pp. 3142–3149 (2014)
11. Ma, T., Jan Latecki, L.: Graph transduction learning with connectivity constraints with application to multiple foreground cosegmentation. In: IEEE CVPR, pp. 1955–1962(2013)
12. Zhu, H., Lu, J., Cai, J., Zheng, J., Thalmann, N.M.: Multiple foreground recognition and cosegmentation: an object-oriented CRF model with robust higher-order potentials. In: IEEE WACV, pp. 485–492 (2014)
13. Zitnick, C.L., Dollár, P.: Edge boxes: locating object proposals from edges. In: Fleet, D., Pajdla, T., Schiele, B., Tuytelaars, T. (eds.) ECCV 2014. LNCS, vol. 8693, pp. 391–405. Springer, Heidelberg (2014). doi:10.1007/978-3-319-10602-1_26
14. Long, J., Shelhamer, E., Darrell, T.: Fully convolutional networks for semantic segmentation. In: IEEE CVPR, pp. 3431–3440 (2015)
15. Frey, B.J., Dueck, D.: Clustering by passing messages between data points. Science **315**, 972–976 (2007)
16. Zhu, W., Liang, S., Wei, Y., Sun, J.: Saliency optimization from robust background detection. In: IEEE CVPR, pp. 2814–2821 (2014)
17. Cheng, M.M., Mitra, N.J., Huang, X., Torr, P.H., Hu, S.M.: Global contrast based salient region detection. IEEE TPAMI **37**(3), 569–582 (2015)
18. Achanta, R., Smith, S.A.: SLIC superpixels compared to state-of-the-art superpixel methods. IEEE TPAMI **34**(11), 2274–2282 (2012)
19. Zhou, Z.H., Zhang, M.L.: Multi-instance multi-label learning with application to scene classification. In: NIPS, pp. 1609–1616 (2006)
20. Briggs, F., Fern, X.Z., Raich, R.: Rank-loss support instance machines for MIML instance annotation. In: ACM SIGKDD, pp. 534–542 (2012)
21. Batra, D., Kowdle, A.: iCoseg: interactive co-segmentation with intelligent scribble guidance. In: IEEE CVPR, pp. 3169–3176 (2010)
22. Joulin, A., Bach, F., Ponce, J.: Discriminative clustering for image cosegmentation. In: IEEE CVPR, pp. 1943–1950 (2010)
23. Vicente, S., Rother, C., Kolmogorov, V.: Object cosegmentation. In: IEEE CVPR, pp. 2217–2224 (2011)
24. Rubinstein, M., Joulin, A., Koft, J., Liu, C.: Object co-segmentation based on shortest path algorithm and saliency model. IEEE Trans. Multimedia **14**(5), 1429–1441 (2012)
25. Rubinstein, M., Joulin, A., Kopf, J., Liu, C.: Unsupervised joint object discovery and segmentation in internet images. In: IEEE CVPR, pp. 1939–1946 (2013)

Using Object Detection, NLP, and Knowledge Bases to Understand the Message of Images

Lydia Weiland[1]([✉]), Ioana Hulpus[1], Simone Paolo Ponzetto[1], and Laura Dietz[2]

[1] University of Mannheim, Mannheim, Germany
lydia@informatik.uni-mannheim.de
[2] University of New Hampshire, Durham, NH, USA

Abstract. With the increasing amount of multimodal content from social media posts and news articles, there has been an intensified effort towards conceptual labeling and multimodal (topic) modeling of images and of their affiliated texts. Nonetheless, the problem of identifying and automatically naming the core abstract message (*gist*) behind images has received less attention. This problem is especially relevant for the semantic indexing and subsequent retrieval of images. In this paper, we propose a solution that makes use of external knowledge bases such as Wikipedia and DBpedia. Its aim is to leverage complex semantic associations between the image objects and the textual caption in order to uncover the intended gist. The results of our evaluation prove the ability of our proposed approach to detect gist with a best MAP score of 0.74 when assessed against human annotations. Furthermore, an automatic image tagging and caption generation API is compared to manually set image and caption signals. We show and discuss the difficulty to find the correct gist especially for abstract, non-depictable gists as well as the impact of different types of signals on gist detection quality.

1 Introduction

Recently, much work in image and language understanding has led to interdisciplinary contributions that bring together processing of visual data such as video and images with text mining techniques. Because text and vision provide complementary sources of information, their combination is expected to produce better models of understanding semantics of human interaction [3] therefore improving end-user applications [25].

The joint understanding of vision and language data has the potential to produce better indexing and search methods for multimedia content [11,31]. Research efforts along this line of work include image-to-text [10,14,22,36] and video-to-text [1,5,20] generation, as well as the complementary problem of associating images or videos to arbitrary texts [4,11]. Thus, most previous work concentrates on the recognition of visible objects and the generation of literal, descriptive caption texts. However, many images are used with the purpose of stimulating emotions [26,27], e.g., the image of polar bears on shelf ice. To understand the message behind such images, typically used for writing about complex

© Springer International Publishing AG 2017
L. Amsaleg et al. (Eds.): MMM 2017, Part II, LNCS 10133, pp. 405–418, 2017.
DOI: 10.1007/978-3-319-51814-5_34

406   L. Weiland et al.

A male orangutan waits near a feeding station at Camp Leakey in Tanjung Puting National Park in Central Kalimantan province, Indonesia in this June 15, 2015.	Fight to save Indonesia's wildlife corridors key for endangered orangutan.	Fight to save Indonesia's wildlife corridors key for endangered orangutan.
Gist examples: Mammals of Southeast Asia, Trees, Orangutans, Botany, Plants	*Gist examples:* Habitat, Conservation, Biodiversity, Extinction, EDGE species, Deforestation	*Gist examples:* Habitat Conservation, Biodiversity, Extinction, Politics, Protest
(a) Literal Pairing	(b) Non-literal Pairing	(c) Non-literal Pairing

Fig. 1. Example image-caption pairs sharing either images or captions with their respective gist entities (a, b: http://reut.rs/2cca9s7, REUTERS/Darren Whiteside, c: http://bit.ly/2bGsvii, AP, last accessed: 08/29/2016.

topics like global warming or financial crises, semantic associations must be exploited between the depictable, concrete objects of the image and the potential abstract topics. Current knowledge bases such as Wikipedia, DBpedia, FreeBase can fill this gap and provide these semantic connections. Our previous work [35] introduces such a system for image understanding that leverages such sources of external knowledge. The approach was studied in an idealized setting where humans provided image tags and created the object vocabulary in order to make design choices.

Contribution. Building on top of our previous work, a core contribution of this paper is to study whether the performance of gist detection with external knowledge is impacted when an automatic object detector is used instead of human annotations. We make use of the Computer Vision API[1] from Microsoft Cognitive Services [8] - a web service that provides a list of detected objects and is also capable of generating a descriptive caption of the image. This way, we create a fully automatic end-to-end system for understanding abstract messages conveyed through association such as examples of Fig. 1(b) and (c).

Microsoft Cognitive Service uses a network pre-trained on ImageNet [6]. Additionally it includes a CNN, which is capable of assigning labels to image regions [13], and trains on Microsoft COCO [23] data, thus, resulting in a vocabulary of 2,000 object categories. Together with Microsoft's language generation API, this information is used to generate a caption for an image, that did not have one before.

Our task setup defines the understanding of abstract messages as being able to describe this message with appropriate concepts from the knowledge base - called gist nodes. This way we cast the problem as an entity ranking problem

---

[1] https://www.microsoft.com/cognitive-services/en-us/computer-vision-api.

which is evaluated against a human-generated benchmark of relevant entities and categories.

We study the effects of automatic object detection separately for images with literal descriptive captions and captions that utilize a non-literal (i.e., abstract) meaning. While in both cases reasonable performance is obtained by our approach, experiments point towards room for improvement for object detection algorithms. We identify theoretical limits by analyzing which of the gist nodes represent depictable (e.g. tree) versus non-depictable (e.g., philosophy) concepts. These limits are complemented with experiments considering the signal from either the image or the caption as well as automatically generated captions from the Microsoft API, which ignore the original caption. We demonstrate that understanding the message of non-literal image-caption pairs is a difficult task (unachievable by ignoring the original caption) to which our approach together with MS Cognitive Services provides a large step in the right direction.

2   Problem Statement

We study the problem of identifying the gist expressed in an image-caption pair in the form of an entity ranking task. The idea is that general-purpose knowledge bases such as Wikipedia and DBpedia provide an entry for many concepts, ranging from people and places, to general concepts as well as abstract topics such as "philosophy". Some of these entries represent **depictable** objects, such as "bicycle", "solar panel", or "tree", some could be associated with visual features such as "arctic landscape" or "plants". The task is to identify (and rank) the most relevant concepts (e.g., entities or categories from Wikipedia) that describe the gist of the image-caption pair.

Problem Statement. Given an image with its respective caption as inputs, predict a ranking of concepts from the knowledge base that best represent the core message expressed in the image.

By predicting the most prominent gist of an image-caption pair, these can be indexed by a search engine and provide diverse images in response to concept queries. Our work provides a puzzle-piece in answering image queries also in response to **non-depictable** concepts such as "biodiversity" or "endangered species".

We distinguish two types of image-caption pairs: **Literal pairs**, where the caption describes what is seen on the image. In such cases the gist of the image is often a depictable concept. In contrast, in **non-literal pairs**, image and caption together allude to an abstract theme. These are often non-depictable concepts. Figure 1 displays three examples on the topic of endangered species from both classes with image, caption, and a subset of annotated gist nodes from the knowledge base. The example demonstrates how changing the picture or caption can drastically change the message expressed.

We devise a supervised framework for gist detection that is studied in the context of both styles of image-caption pairs. In particular, we center our study on the following research questions:

RQ0: What is the fraction of depictable concept?
RQ1: Does an automatic image tagging change the prediction quality?
RQ2: Does an automatic caption generation change the prediction quality?
RQ3: Would an automatic approach capture more literal or more non-literal aspects?
RQ4: What is the benefit of joint signals (in contrast to only caption or only image)?

3   Related Work

Especially with the increasing amount of multi-modal datasets, the joint modeling of cross-modal features has gained attention. Different combinations of modalities (audio, image, text, and video) are possible, we focus on those mostly related to our research. Those datasets consist of images with captions and/or textual labeled object regions, e.g., Flickr8k [30] and Flickr30k [37], SBU Captioned Photo Dataset [28], PASCAL 1k dataset [9], ImageNet [21], and Microsoft Common Objects in Context (COCO) [23].

Joint modeling of image and textual components, which utilizes KCCA [15,32] or neural networks [19,33], have shown to outperform single modality approaches. Independent from joint or single modeling, the applications are similar, e.g., multimodal topic generation [34] or retrieval tasks: Generating descriptions for images [7,14,22,29,36] and retrieving images for text [5,11]. The focus in these works lies on the generation of descriptive captions, semantic concept labeling, and depictable concepts [2,15,18,31], which results in literal pairs. In contrast, our approach benefits from external knowledge to retrieve and rank also abstract, non-depictable concepts for understanding both literal and non-literal pairs. Our previous study [35] was conducted on manually given image objects tags and captions. This work studies performance with an automatic object detection system.

4   Approach: Gist Detection

The main idea behind our approach is to use a knowledge base and the graph induced by its link structure to reason about connections between depicted objects in the image and mentioned concepts in the caption. Our basic assumption is that gist nodes may be directly referred in the image or caption or are in close proximity of directly referred concepts. To identify these gist nodes, we propose a graph mining pipeline, which mainly consists of a simple entity linking strategy, a graph traversal and expansion based on a relatedness measure, as shown in Fig. 2. Variations of the pipeline are studied in our prior work [35] but are detrimental to experiments in this work also. We clarify pipeline steps using the running example of a non-literal pair with a typical representative of endangered species in (Fig. 1(b)).

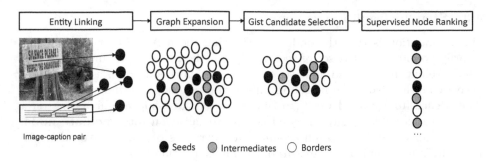

Fig. 2. Gist extraction and ranking pipeline for a given image-caption pair. For simplicity, we omit the edges betwell ween nodes in this figure.

4.1 The Knowledge Graph

Wikipedia provides a large general-purpose knowledge base about objects, concepts, and topics. Furthermore, and even more importantly for our approach, the link structure of Wikipedia can be exploited to identify topically associative nodes. DBpedia is a structured version of Wikipedia. All DBpedia concepts have their source in Wikipedia pages. In this work, our knowledge graph contains as nodes all the Wikipedia articles and categories. As for edges, we consider the following types of relations T, named by their DBpedia link property:

- **dcterms:subject.** The category membership relations that link an article to the categories it belongs to, e.g., Wildlife corridor dcterms:subject Wildlife conservation.
- **skos:broader.** Relationship between a category and its parent category in a hierarchical structure, e.g., Wildlife conservation skos:broader Conservation.
- **skos:narrower.** Relationship between a category and its subcategories, e.g., Conservation skos:narrower Water conservation.

4.2 Step 1: Entity Linking

The first step is to project the image tags from the objects detected in the image as well as the entities mentioned in the caption, e.g., wildlife corridors, onto nodes in the knowledge base, e.g., Wildlife corridor. To obtain the entities mentioned in the caption, we extract all the noun-phrases from the caption text. Each of these noun-phrases and image tags are then linked to entities in the knowledge base as follows: If the noun-phrase/image tag occurs in the knowledge base as an exact name (i.e., title of Wikipedia page, category, or redirect), this entry is selected as unambiguous. However, if it is the title of a disambiguation page, we select the disambiguation alternative with shortest connections to already projected unambiguous concepts (ignoring concepts with more than two hops). In the following, we refer to all the linked knowledge base entities as **seed nodes**.

4.3   Step 2: Graph Extraction

Our assumption is that the nodes representing the message best are not necessarily contained in the set of seed nodes, but lie in close proximity to them.
Thus, we expand the seed node set to their neighborhood graph as follows: We
activate all the seed nodes neighbors on a radius of n-hops, with n = 2 according
to evidence from related work [16,24]. If any of the 2-hop neighbors lies on a
shortest path between any two seed nodes, we call it an **intermediate node**,
and further expand the graph around it on a radius of 2-hops. We name the
resulting graph the **border graph** of the image-caption pair.

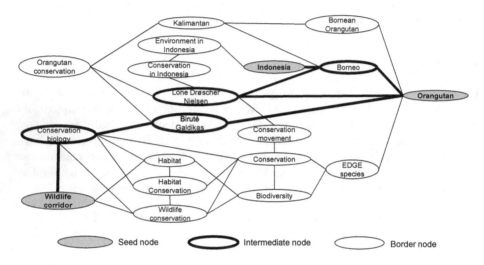

Fig. 3. Example of image-caption graph for the image-caption in Fig. 2.

Example. In Fig. 3, we show a graph excerpt of the image-caption graph
extracted for the image-caption shown in Fig. 1(b). The linked seed nodes are
Indonesia (as extracted from the caption), Orangutan (as extracted from the
image and caption) and Wildlife corridor (as extracted from the caption). In
this example, many suitable gist nodes are among the border nodes, i.e., Conservation and Orangutan Conservation.

4.4   Step 3: Gist Candidate Selection

After obtaining the border graph G, our aim is to select the nodes in this graph
that are best candidates for capturing the message of the image-caption pair.
We estimate for each node x its suitability as gist concept through its average
semantic relatedness $\bar{\sigma}$ to all the seeds (S) and intermediate nodes (I), as shown
in Formula 1.

$$\bar{\sigma}(x, G) = \frac{1}{|S \cup I|} \sum_{y \in S \cup I} \sigma(x, y) \tag{1}$$

In this paper, we use the symbol σ for representing any given relatedness measure. For calculating the relatedness we use the exclusivity-based relatedness measure of Hulpus et al. [17] with their hyperparameter settings ($\alpha = 0.25$, top-3 shortest paths). The measure for an edge is the higher the fewer alternative relations of the same edge type each of its endnode has.

Using the resulting score for each node in the border graph, we select as candidates the top-20 highest scoring nodes. These nodes are eventually ranked in a supervised manner, based on a selection of features that we present in the following step.

4.5   Step 4: Supervised Node Ranking

For each of the candidate nodes, a feature vector is created that is comprised of different measures of proximity between seed nodes and candidate nodes. The effectiveness of these features is studied within a learning-to-rank framework which ranks nodes by their utility of expressing the gist of the image-caption pair. Three boolean features {*Seed, Intermediate, Border*} indicate the expansion stage that included the node. A numerical feature refers to the semantic relatedness of the candidate nodes, to the seeds and intermediates. A retrieval-based feature is computed by considering the content of the Wikipedia articles that corresponds to the graph nodes. The feature aggregates object tags and the caption into a search query to retrieve knowledge based articles from a Wikipedia snapshot. We use a standard retrieval model called query-likelihood with Dirichlet smoothing. Using the entire graph of DBpedia described in Sect. 4.1 global node importance measures can be computed. These features are: (i) the indegree of the node which counts how many DBpedia entities point to the given entity; (ii) the clustering coefficient of the node which measures the ratio of neighbours of the node that are themselves directly connected. This feature was selected under the intuition that nodes denoting abstract concepts tend to have lower clustering coefficient, while nodes denoting specific concepts tend to have higher clustering coefficient; (iii) the node's PageRank as a very popular measure of node importance in graphs, which is also less computationally expensive on large graphs (DBpedia has approx 6 million nodes and 60 million edges) than other node centrality measures such as betweeness.

5   Experimental Evaluation

We begin our evaluation by studying whether the relevant gist nodes (scored 4 or 5 by human annotators) are in general depictable or not (**RQ 0**) by analyzing a subset of the gold standard. Research questions **RQ 1–4** (cf. Sect. 2) evaluate the end-to-end system using Microsoft Cognitive Service in combination with our approach presented in Sect. 4.

Dataset and Gold Standard. To conduct our evaluation we create a dataset[2] containing 328 pairs of images and captions with a balanced amount of literal

---

[2] dataset and gold standard: https://github.com/gistDetection/GistDataset.

and non-literal pairs (164 literal, 164 media-iconic pairs). The non-literal pairs are collected from news portals such as www.theguardian.com. The literal pairs use the same images as the non-literal pairs, but have a descriptive caption that is created by annotators. One of our goals is to be able to evaluate the proposed gist detection approach independently from automated object detection systems, thus annotators also manually assign labels to objects in the images.

The gold standard annotation of gist nodes is conducted by annotators selecting nodes from the knowledge-base representing the gist and assigning ranks to each of the nodes, on a Likert scale ranging from 0 (irrelevant) to 5 (core gist). Nodes with level 4 and 5 are referred to as *relevant gists*. For each pair there is only one gist which is annotated with level 5, following the assumption of having one *core gist* which represents the message best. On a subset, annotators also assessed whether a gist is depictable.

Experimental Setup. The feature set (cf. Sect. 4.5) is used in a supervised learning-to-rank framework (RankLib[3]). As ranking algorithm, we use Coordinate Ascent with a linear kernel. We perform 5-fold cross validation, with training optimized towards the target metric Mean Average Precision (MAP), which is a recall-oriented measure averaging the precision at those ranks where the recall changes. Besides MAP, the evaluations contain normalized discounted cumulative gain (NDCG@10), which is a graded retrieval measure, and Precision (P@10) of the top ten nodes from the ranked lists.

For every image, besides the original and manually set captions and object tags, we also evaluate automatically generated captions and image tags (in the following abbreviated with **MS tags** and **MS captions**, respectively) by using Computer Vision API from Microsoft Cognitive Services.

5.1 RQ 0: Relevant Gists: Depictable or Not?

We study whether gist concepts as selected by annotators tend to be depictable or non-depictable. For a subset of the gold standard pairs, annotators decided for all relevant gist concepts whether this concept is *depictable, not-depictable, or undecided*. On average the fraction of depictable core gists is 88% for literal pairs versus only 39% for the non-literal pairs. On the larger set of all relevant gists, 83% are depictable for literal pairs versus 40% for the non-literal pairs. The decision what an automatic image tagging system might be able to detect is difficult for humans, reflected in an inter-annotator agreement (Fleiss' kappa [12]) of $\kappa = 0.42$ for core gists and $\kappa = 0.73$ for relevant gists.

Discussion RQ 0. These results are in line with our initial assumption that literal pairs tend to have depictable concepts as gist, and the non-literal pairs have a predominant amount of non-depictable concepts as gist (cf. Sect. 2). This finding underlines the fact that the core message of images does not necessarily correspond to objects that are depicted in the image. This reinforces the need for approaches that are able to reason with semantic associations between depictable and abstract concepts.

---

[3] http://lemurproject.org/ranklib.php.

5.2   RQ 1: Manual vs. Automatic Image Tagging

To answer our first research question, we study the performance impact of using an automatic object detector for image tagging, **MS tags**, as opposed to manual tags (**tags**). In both cases we are using the original literal and non-literal captions (**captions**). We refer to the combination of MS tags with original captions as **the realistic, end-to-end approach** considering that images with captions are typically published without image object tags. We compare the performance difference on different stages: First, we note that the manual tags arise from 43 different entities with 640 instances over the complete dataset. The MS tags are from 171 different entities with 957 instances. There are 131 overlapping instances between manual and automatic tags, which amounts to less than one shared tag per image and 20% overlap over the complete dataset.

Second, we compare the performance of the manual and MS tags, both combined with the original captions (cf. Table 1). As expected, a higher performance is achieved with manual tags (Tags&Captions, MAP: 0.74), but the realistic approach achieves a reasonable quality as well (MAP: 0.43).

Discussion RQ 1. The overlap between MS and manual image tags is rather low (20%) and the detected concepts are not always correct (e.g., a polar bear is detected as herd of sheep). However, the MS tags in combination with the original captions achieve a reasonable ranking, which indicates the ability of automatic detectors to find relevant concepts and our method of being capable to handle certain levels of noise.

5.3   RQ 2: Manual vs. Automatic Caption Generation

Spinning the use of automatic detectors further, based on the detected objects, descriptive captions are generated and used as an alternative input. Doing so, we would ignore the original caption and ask, whether this is sufficient. We compare to the same stages as in RQ1. The manual captions use around 300 and 700 different entities (seed nodes) for the literal (l) and non-literal (nl) pairs, respectively. The MS caption results in 130 different entity nodes. 10% (l) and 3% (nl) of the instances overlap between the nodes from the manual and the MS captions across all image-caption pairs. However, the assessment of the non-literal pairs is restricted by the fact that automatic detectors are trained on models with a descriptive purpose.

In the following, we compare the manual captions to the MS captions within our approach. We combine each caption with the manual image tags and provide it as input for the pipeline described in Sect. 4. We study the combinations with respect to the complete dataset (cf. Table 1) and compare again to the pure manual input signals (MAP: 0.74). A better ranking than for the realistic approach can be achieved across all pairs (Tags & MS caption, MAP: 0.48).

In contrast to the strong results of the manual input signals, the pure automatic signals perform worse across all pairs (MAP: 0.14, cf. Table 1).

Discussion RQ 2. The overlap between MS and manual caption is low (3–10%), the MS captions are short, and the focus of the captions does not always match

Table 1. Ranking results (grade 4 or 5) according to different input signals and feature sets. Significance is indicated by * (paired t-test, p-value ≤ 0.05).

	Both				Non-Literal				Literal			
	MAP	Δ%	NDCG@10	P@10	MAP	Δ%	NDCG@10	P@10	MAP	Δ%	NDCG@10	P@10
Tag& Caption	0.74	0.00	0.71	0.71	0.64	0.00	0.59	0.58	0.83	0.00	0.84	0.84
Tag&MS Caption	0.48	−34.75*	0.63	0.56	0.36	−44.44*	0.45	0.38	0.61	−27.33*	0.80	0.73
MS tags& Caption	0.43	−41.67*	0.58	0.53	0.40	−37.55*	0.49	0.44	0.46	−44.95*	0.68	0.61
MS tags&MS caption	0.14	−80.49*	0.28	0.23	0.09	−86.06*	0.17	0.14	0.20	−76.11*	0.39	0.32
Tags only	0.48	−36.79*	0.65	0.57	0.28	−47.25*	0.40	0.33	0.68	−28.29*	0.89	0.82
MS tags only	0.13	−84.02*	0.24	0.20	0.06	−89.50*	0.13	0.11	0.20	−79.83*	0.35	0.29
Caption only	0.38	−49.68*	0.54	0.49	0.31	−51.79*	0.40	0.35	0.45	−47.59*	0.67	0.63
MS caption only	0.07	−91.49*	0.15	0.12	0.05	−93.25*	0.10	0.08	0.09	−89.18*	0.19	0.16

the focus of the manual caption (e.g., example Fig. 1 receives the caption "There is a sign", without considering the orangutan, although it was detected as monkey by the automatic image tagging). However, the ranking of the combined automatic and manual approach with respect to the complete dataset performs reasonably well. This shows promising opportunities for using our approach together with semi-automatic image tagging and/or caption creation in a real-world pipeline. In the following, we study these results in more detail with respect to the distinction between literal and non-literal pairs.

5.4   RQ 3: Literal vs. Non-literal Aspect Coverage by Automatic Detector

Next, we study the input combinations with respect to the non-literal and literal pairs and compare again with the pure manual input (cf. Table 1). Analyzing MS tags&MS captions as input shows a moderate ranking for the literal pairs (MAP: 0.20). However, the performance for the non-literal pairs is bisected (MAP: 0.09). This result is expected because without any context, it is currently impossible for an automatic caption generator to recommend non-literal captions. The realistic approach has a performance decrease of less than 40% (MAP: 0.40 (nl), 0.46 (l)). Substituting the manual captions by the automatic captions results in an even better performance for the literal pairs, but a lower performance than with the realistic approach for the non-literal pairs (Tags&MS caption MAP: 0.36 (nl), 0.61 (l)).

Discussion RQ 3. The evaluation results across all input signal combinations confirm our intuition that gists of non-literal pairs are more difficult to detect. These non-literal pairs, however, are the ones found in news, blogs, and twitter, which are our main interest. Automatic approaches can address descriptive pairs by detecting important objects in the image and describe those in the caption. However, the automatic approaches lack mentioning things that are salient to detect the gist of non-literal pairs. With respect to RQ 3 we have shown that pure automatic detectors achieve fair results for pairs where a descriptive output is wanted. A good performance of automatic object detectors is also achieved within the realistic approach. However, these results indicate that differentiating captions - which is currently done by setting the captions manually - is necessary to detect the gist.

5.5   RQ 4: Comparison of Single Signals vs. Signal Combination

Since experiments from the field of multi-modal modeling have demonstrated improvements by combining textual and visual signals, we study whether this effect also holds for our case—especially with respect to non-literal pairs (cf. Table 1). For a detailed analysis, we also study MS tags Only and MS captions Only.

Given the image tags as input signal only, the literal pairs - apart from ndcg - are nearly as good as using the combined signal as input (MAP: 0.68). In contrast, the non-literal pairs are worse than combining signals ($\Delta\%$: -47.25%). The MS tags have an informative content for the literal, but achieve only a fifth of the performance for the non-literal pairs compared to the manual input (MAP: 0.20 vs. 0.68 (l), 0.06 vs 0.28 (nl)). The same study is conducted on the captions as single input signal (Caption Only and MS caption Only). Interestingly, the caption only performs better than the image tags only for the non-literal pairs. Especially for non-literal pairs the results degrade significantly when the caption is replaced by a MS caption.

Discussion RQ 4. The results of Table 1 show that image-only signals cannot completely convey abstract and/or associative topics and thus, cannot fully address the requirements of non-literal pairs. However, these results prove also another hypothesis, that concrete objects which can be detected in the image, are important pointers towards the relevant gists. We remark that for both types of pairs the performance benefit from the combination of signals. Apart from the manual tags for the literal pairs, we conclude that the gist cannot be detected with only the caption or only the image signal.

6   Conclusion

Our aim is to understand the gist of image-caption pairs. For that we address the problem as a concept ranking, while leveraging features and further gist candidates from an external knowledge base. We compare manually to automatically gathered information created by automatic detectors. The evaluation is conducted on the complete test collection of 328 image-caption pairs, with respect to the different input signals, signal combination, and single signal analysis. Furthermore, we study both, non-literal and literal pairs. Our finding is that combining signals from image and caption improves the performance for all types of pairs. An evaluation of inter-annotator agreement has shown that literal pairs in most of the cases have a depictable gist and non-literal pairs have a non-depictable gist. This analysis result is in line with the finding that non-literal more benefit from the (manual) caption signal, whereas literal more benefit from image signals. Within the realistic scenario, we test the performance of object detectors in the wild, which shows level for improvement of 10%.

Acknowledgements. This work is funded by the RiSC programme of the Ministry of Science, Research and the Arts Baden-Wuerttemberg, and used computational resources offered from the bwUni-Cluster within the framework program bwHPC. Furthermore, this work was in part funded through the Elitepostdoc program of the BW-Stiftung and the University of New Hampshire.

References

1. Barbu, A., Bridge, A., Burchill, Z., Coroian, D., Dickinson, S.J., Fidler, S., Zhang, Z.: Video in sentences out. In: UAI, pp. 102–112 (2012)
2. Bernardi, R., Cakici, R., Elliott, D., Erdem, A., Erdem, E., Ikizler-Cinbis, N., Plank, B.: Automatic description generation from images: a survey of models, datasets, and evaluation measures. arXiv preprint arXiv:1601.03896 (2016)
3. Bruni, E., Uijlings, J., Baroni, M., Sebe, N.: Distributional semantics with eyes: using image analysis to improve computational representations of word meaning. In: MM, pp. 1219–1228 (2012)
4. Das, P., Srihari, R.K., Corso, J.J.: Translating related words to videos and back through latent topics. In: WSDM, pp. 485–494 (2013)
5. Das, P., Xu, C., Doell, R.F., Corso, J.J.: A thousand frames in just a few words: lingual description of videos through latent topics and sparse object stitching. In: CVPR, pp. 2634–2641 (2013)
6. Deng, J., Dong, W., Socher, R., Li, L.J., Li, K., Fei-fei, L.: Imagenet: A large-scale hierarchical image database. In: CVPR (2009)
7. Elliott, D., Keller, F.: Image description using visual dependency representations. In: EMNLP, pp. 1292–1302 (2013)
8. Fang, H., Gupta, S., Iandola, F.N., Srivastava, R., Deng, L., Dollár, P., Zweig, G.: From captions to visual concepts and back. In: CVPR, pp. 1473–1482 (2015)
9. Farhadi, A., Hejrati, M., Sadeghi, M.A., Young, P., Rashtchian, C., Hockenmaier, J., Forsyth, D.: Every picture tells a story: generating sentences from images. In: Daniilidis, K., Maragos, P., Paragios, N. (eds.) ECCV 2010. LNCS, vol. 6314, pp. 15–29. Springer, Heidelberg (2010). doi:10.1007/978-3-642-15561-1_2
10. Feng, Y., Lapata, M.: How many words is a picture worth? Automatic caption generation for news images. In: ACL, pp. 1239–1249 (2010)
11. Feng, Y., Lapata, M.: Topic models for image annotation and text illustration. In: NAACL-HLT, pp. 831–839 (2010)
12. Fleiss, J., et al.: Measuring nominal scale agreement among many raters. Psychol. Bull. **76**(5), 378–382 (1971)
13. Girshick, R.B., Donahue, J., Darrell, T., Malik, J.: Rich feature hierarchies for accurate object detection and semantic segmentation. CoRR (2013)

14. Gupta, A., Verma, Y., Jawahar, C.V.: Choosing linguistics over vision to describe images. In: AAAI, pp. 606–612 (2012)
15. Hodosh, M., Young, P., Hockenmaier, J.: Framing image description as a ranking task: data, models and evaluation metrics. IJCAI **47**, 853–899 (2013)
16. Hulpus, I., Hayes, C., Karnstedt, M., Greene, D.: Unsupervised graph-based topic labelling using DBpedia. In: Proceedings of the WSDM 2013, pp. 465–474 (2013)
17. Hulpuş, I., Prangnawarat, N., Hayes, C.: Path-based semantic relatedness on linked data and its use to word and entity disambiguation. In: Arenas, M., et al. (eds.) ISWC 2015. LNCS, vol. 9366, pp. 442–457. Springer, Heidelberg (2015). doi:10. 1007/978-3-319-25007-6_26
18. Jin, Y., Khan, L., Wang, L., Awad, M.: Image annotations by combining multiple evidence & WordNet. In: MM, pp. 706–715 (2005)
19. Karpathy, A., Li, F.F.: Deep visual-semantic alignments for generating image descriptions. In: CVPR, pp. 3128-3137. IEEE Computer Society (2015)
20. Krishnamoorthy, N., Malkarnenkar, G., Mooney, R., Saenko, K., Guadarrama, S.: Generating natural-language video descriptions using text-mined knowledge. In: AAAI (2013)
21. Krizhevsky, A., Sutskever, I., Hinton, G.E.: Imagenet classification with deep convolutional neural networks. In: NIPS, pp. 1097–1105 (2012)
22. Kulkarni, G., Premraj, V., Dhar, S., Li, S., Choi, Y., Berg, A.C., Berg, T.L.: Baby talk: understanding and generating image descriptions. In: CVPR, pp. 1601–1608 (2011)
23. Lin, T.-Y., Maire, M., Belongie, S., Hays, J., Perona, P., Ramanan, D., Dollár, P., Zitnick, C.L.: Microsoft COCO: common objects in context. In: Fleet, D., Pajdla, T., Schiele, B., Tuytelaars, T. (eds.) ECCV 2014. LNCS, vol. 8693, pp. 740–755. Springer, Heidelberg (2014). doi:10.1007/978-3-319-10602-1_48
24. Navigli, R., Ponzetto, S.P.: Babelnet: the automatic construction, evaluation and application of a wide-coverage multilingual semantic network. Artif. Intell. **193**, 217–250 (2012)
25. Nikolaos Aletras, M.S.: Computing similarity between cultural heritage items using multimodal features. In: LaTeCH at EACL, pp. 85–92 (2012)
26. O'Neill, S., Nicholson-Cole, S.: Fear won't do it: promoting positive engagement with climate change through imagery and icons. Sci. Commun. **30**(3), 355–379 (2009)
27. O'Neill, S., Smith, N.: Climate change and visual imagery. Wiley Interdisc. Rev.: Clim. Change **5**(1), 73–87 (2014)
28. Ordonez, V., Kulkarni, G., Berg, T.L.: Im2text: describing images using 1 million captioned photographs. In: NIPS (2011)
29. Ortiz, L.G.M., Wolff, C., Lapata, M.: Learning to interpret and describe abstract scenes. In: NAACL HLT 2015, pp. 1505–1515 (2015)
30. Rashtchian, C., Young, P., Hodosh, M., Hockenmaier, J.: Collecting image annotations using Amazon's mechanical turk. In: CSLDAMT at NAACL HLT (2010)
31. Rasiwasia, N., Costa Pereira, J., Coviello, E., Doyle, G., Lanckriet, G.R., Levy, R., Vasconcelos, N.: A new approach to cross-modal multimedia retrieval. In: MM, pp. 251–260 (2010)
32. Socher, R., Fei-Fei, L.: Connecting modalities: semi-supervised segmentation and annotation of images using unaligned text corpora. In: CVPR (2010)
33. Socher, R., Karpathy, A., Le, Q.V., Manning, C.D., Ng, A.Y.: Grounded compositional semantics for finding and describing images with sentences. ACL **2**, 207–218 (2014)

34. Wang, C., Yang, H., Che, X., Meinel, C.: Concept-based multimodal learning for topic generation. In: He, X., Luo, S., Tao, D., Xu, C., Yang, J., Hasan, M.A. (eds.) MMM 2015. LNCS, vol. 8935, pp. 385–395. Springer, Heidelberg (2015). doi:10. 1007/978-3-319-14445-0_33
35. Weiland, L., Hulpus, I., Ponzetto, S.P., Dietz, L.: Understanding the message of images with knowledge base traversals. In: Proceedings of the 2016 ACM on International Conference on the Theory of Information Retrieval, ICTIR 2016, Newark, DE, USA, 12–16 September 2016, pp. 199–208 (2016)
36. Yang, Y., Teo, C.L., Daumé III, H., Aloimonos, Y.: Corpus-guided sentence generation of natural images. In: EMNLP, pp. 444–454 (2011)
37. Young, P., Lai, A., Hodosh, M., Hockenmaier, J.: From image descriptions to visual denotations: new similarity metrics for semantic inference over event descriptions. In: ACL, pp. 67–78 (2014)

Video Search via Ranking Network with Very Few Query Exemplars

De Cheng[1,2]($\boxtimes$), Lu Jiang[2], Yihong Gong[1], Nanning Zheng[1],
and Alexander G. Hauptmann[2]

[1] Xi'an Jiaotong University, Xi'an, China
dechengxjtu@gmail.com
[2] Carnegie Mellon University, Pittsburgh, USA

Abstract. This paper addresses the challenge of video search with only a handful query exemplars by proposing a triplet ranking network-based method. Based on the typical scenario for video search system, a user begins the query process by first utilizing the metadata-based text-to-video search module to find an initial set of videos of interest in the video repository. As bridging the semantic gap between text and video is very challenging, usually only a handful relevant videos appear in the initial retrieved results. The user now can use the video-to-video search module to train a new classifier to search more relevant videos. However, since we found that statistically only fewer than 5 videos are initially relevant, training a complex event classifier with a handful of examples is extremely challenging. Therefore, it is necessary to improve video retrieval method that works for a handful of positive training example videos. The proposed triplet ranking network is mainly designed for this situation and has the following properties: (1) This ranking network can learn an off-line similarity matching projection, which is event independent, from other previous video search tasks or datasets. Such that even with only one query video, we can search its relative videos. Then this method can transfer previous knowledge to the specific video retrieval tasks as more and more relative videos being retrieved, to further improve the retrieval performance; (2) It casts the video search task as a ranking problem, and can exploit partial ordering information in the dataset; (3) Based on the above two merits, this method is suitable for the case where only a handful of positive examples exploit. Experimental results show the effectiveness of our proposed method on video retrieval with only a handful of positive exemplars.

Keywords: Video search · Few positives · Partially ordered · Ranking network · Knowledge adaptation

1 Introduction

Large-scale analysis of video data becomes ever more important due to the unprecedented growth of user-generated videos on the internet [10,17]. Video

© Springer International Publishing AG 2017
L. Amsaleg et al. (Eds.): MMM 2017, Part II, LNCS 10133, pp. 419–430, 2017.
DOI: 10.1007/978-3-319-51814-5_35

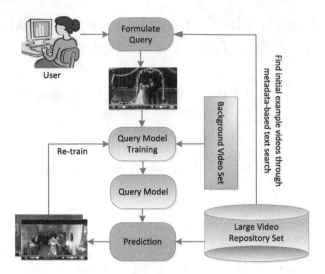

Fig. 1. Illustration of the designed video search system.

search entails retrieving those videos in a collection that meet a user's needs, either based on the provided text metadata or an video itself as the query. It has great potential for a range of applications such as video search, browsing and hyperlink, etc [3].

In the multimedia community, the video search task has been studied in the form of Multimedia Event Detection (MED). The task can be carried out in two main steps: feature extraction and search [17]. The main goal of feature extraction is to extract discriminative feature representations from the video's raw visual and audio channels, while the search phase usually contains a separate "text-to-video" and "video-to-video" phase. The user begins the query process by first utilizing the text-to-video search module to find an initial set of relevant videos. Then based on the user-selected relevant videos from the initial query result set, the user now can train another specific classifier to do video-to-video search to obtain more relevant videos. Almost all classifiers are assuming sufficient positive examples (i.e. SVM). However, the positive videos are very hard to obtain from the initial query result set by the initial classifier in reality. Based on a real-world prototype system, when searching for our favorite event videos, we first use the original text metadata for retrieval, and then collect positive exemplars from the initial retrieved results, aiming to train a specific event detection classifier for obtaining more retrieved results. Based on the statistics from one year of search log files containing more than 40000 queries, we have found that, in most cases, less than 3 positive or relevant exemplars can be collected from the initial result set, to train the video-to-video classifier. Thus, it is necessary for us to develop a video search method which is appropriate for only handful positive exemplars. Although the NIST has defined challenge which obtained respectable detection accuracy when there are only 10 positive examples for training, the

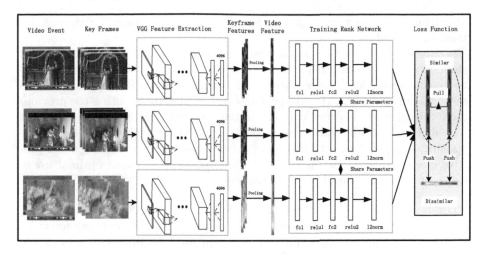

Fig. 2. Illustration of the video-to-video retrieval training framework.

number 10 is somewhat arbitrarily selected. Most importantly, the obtained positive examples are always far fewer than 10 in practice. Therefore, in this paper we focus on video retrieval in a more realistic scenario, that is to say, we want to train an efficient video retrieval classifier with very few positive examples.

To train the classifier for query event videos, we first collect the positive examples from the initial text-to-video retrieved results. Though SVM is effective in most current systems, our experimental results substantiate the observation that its performance would likely be less robust when there are only a few positive training examples. Thus in this paper, we propose a triplet ranking network and cast the video search task as a ranking problem. Based on the triplet ranking network, we can first pre-train a task independent similarity projection, such that even with only one query video example, we can use this similarity function to find its relative examples. Then an event specific similarity function can the trained as more and more relative examples being retrieved, by fine-tuning the model parameters with the retrieved event videos. Obviously, the proposed ranking network can be fine-tuned on top of pre-trained network parameters on other datasets or tasks. Therefore, it would be beneficial to leverage and adapt knowledge from previous, related tasks or other datasets to further improve the search accuracy, in order to overcome the insufficient number of labeled examples [11]. What's more, sometimes the related examples can also be obtained to expand the training data and form some partially ordered examples, which also can be utilized by the proposed ranking network.

We summarize the contributions of the paper for video retrieval as follows:

– First, the approach lets researchers pay much more attention to the video-to-video search method with only a few positive video exemplars, for a practical video search system. Meanwhile, we proposed a triplet ranking network algorithm which is much suitable for video search by using a few positive training exemplars.

- The proposed triplet ranking network can first learn an event independent similarity model from other related tasks or datasets, to do the initial video-to-video search even by one query video example. Then it gradually obtains more and more accurate event-specific search model, by adapting the previous model to the specific event with the initial retrieved relative examples.
- The method cast the video search task as a ranking problem instead of the classification problem, which is more suitable when there are only a few positive examples. Yet, the proposed ranking network can make optimal use of partial ordered information in the training datasets.

The rest of the paper is organized as follows. In Sect. 2, we briefly review the related works. Section 3 introduces our proposed ranking network for video search, and its training algorithm. The experimental results, comparisons and analysis are presented in Sect. 4. Conclusion comes in Sect. 5.

2   Related Work

Video search, which is studied in two forms: one is the text-to-video search, and another is content based video-to-video search. For the text-to-video search, users typically start their search using query words rather than providing video examples. The task is called Zero-Example search [9]. Similarly to text-to-text search, a video can also be treated as a document, based on the results of ASR (Audio Speech Recognition), OCR (Optical Character Recognition), and intermediate concept detection [8]. Obviously, most of these text-to-text video search can directly search for semantic concepts in videos.

For the video-to-video search, it is studied in the form of multimedia event detection [11,13], which is an interesting problem. A number of studies have been proposed to tackle this problem by using several positive training examples (typically 10 or 100) [3,6]. Just as a machine learning algorithm like the support vector machine (SVM [4]), we can use these video query examples to train a video detector. These initial videos can be either uploaded by the user or the results of a text-to-video search. Then, the search results are collected in a list of potential positive examples. The user can skim though these videos and add them into the positive or negative training sets for the model training. Generally, based on one state-of-the-art video search system, the event detection classifiers are trained using both high and low level features, and the final decision is derived from the classification results. Recently, in order to resemble a real-world video search scenario, some studies have proposed to use zeros training examples for event detection [2,7,12,15], which is called "0Ex" or semantic query. This 0Ex makes users usually start the search without any example videos. However, our work is different from all of the above, we focus on video search with only a few query videos (usually less than 5), where the number 5 is based on the statistics from a real-world search system.

Detecting complex video event with handful positive exemplars is more challenge and useful than the existing works which use more than 100 or 10 positive exemplars for training. Particularly, it is very hard for the user to obtain sufficient query examples for a particular event, and even sometimes there isn't any annotated training examples for one specific event. What's more, when there are only a few positive training exemplars, it leads the classifier hard to train. Based on this fact, we fucus on addressing the video event search task with only handful positive examples. As a ranking problem, we aim to make the distance between the same event video closer than that of the different ones, and vice versus. This step is category independent, which means that we only need to catch the similarity properties among all the same event videos. However, in some specifical video event detection, we can use the fine-tuning method to obtain the event-specific search model on top of the pre-trained similarity model.

In the following sections, we will present our proposed method and the experiments in detail.

3   The Proposed Algorithm

3.1   Motivations

Motivated by the facts mentioned in Sect. 1, the triplet neural network could be a natural choice [5] for the following reasons: (1) Although there are only a handful of positive exemplars, the number of possible triplet inputs would be very large, which is very appropriate for the case with only a few positive training examples; (2) The proposed triplet neural network can utilize the partially ordered training examples obtained by the user. For instance, denoting the query video as x, the obtained positive video as x^+, the relevant video as x^r, and the negative video as x^-, then we have the distance rank orders illustrated as Eq. 1, where $d(x, y)$ means the distance between x and y.

$$d(x, x^+) < d(\phi(x, x^r) < d(x, x^-).$$ (1)

The partially ordered training examples illustrated in Eq. 1 can be captured by the following three triplets input,

$$\begin{aligned} d(x, x^+) &< d(x, x^r) \\ d(x, x^r) &< d(x, x^-) \\ d(x, x^+) &< d(x, x^-). \end{aligned}$$ (2)

(3) Since neural networks are non-convex and get easily stuck in a local minimum, then by using the fine-tuning way, we can transfer the similarity properties from other training datasets of different event videos to the current event-specific video retrieval system.

For example, in the MED13 dataset, we first train the triplet ranking network on other dataset, such as Yahoo Flickr Creative Common (YFCC100M) dataset [14], which aims to learn the overall similarity projection of the same event videos and we call it category independent. Since the neural network are non-convex and easy to fall into the local minimum, then we fine tune the network using one specific event videos in MED13 dataset to get this category dependent similarity model. By using the above two step ranking network, we can adapt the similarity properties from other training data of the different event videos to one specific event video retrieval system. What's more, since one can generate vary large scale triplet input even with only a few positive exemplars, it is very suitable for the case with only a handful positive training examples.

3.2 Video Retrieval Pipeline

Based on the merits of the triplet ranking network, a video search system was designed in our research group as illustrated in Fig. 1. Given the query example video, we just need to compute the cosine distance between every video in the video repository set with the query example videos, and then based on these distance ranking orders, we can retrieve more relevant videos of interest according to the top ranked videos.

More specifically, the user first use the off-line trained video search model to get an initial ranked list, and assume that the top-ranked videos in the retrieved ranked list are highly likely to be correct, then the fine-tuning framework can obtain more training data for the query model by selecting the positive examples from the top-k-ranked videos. These positive examples are added into the example video set, such that a better and specific query-based retrieval model could be trained with these query and newly added positive examples. In our frame work, we are aiming to use these newly generated triplet examples to get more accurate and specific retrieval model on top of the pre-trained model, thus the fine-tuning process converges very quickly, which only needs several hundred iterations. Finally, we can obtain a query-based specific video retrieval model. This has become common practice that the query model training step isn't trained directly from the image pixels, but from VGG features extracted (or a comparable deep network) previously by our system, which makes the query model training process only take several minutes.

3.3 The Proposed Triplet Ranking Network

As illustrated in Fig. 2, the proposed video retrieval training framework includes two steps: First, we build a video representation using VGG features. In order to represent a event video, we extract the keyframes of the video sequence, then extract the VGG features for each frame which is 4096 dimension, and finally use the average pooling to get the event video representation from all the keyframe features. Second, we train the triplet ranking network based on the generated VGG features for each event video. In this step, the input triplet examples should be generated well in advance. The $i - th$ input triplet example is created by $x_i = <x_i, x_i^+, x_i^->$, which is formed by three input VGG features of the event videos, where (x_i, x_i^+) are videos features from the same event, and (x_i, x_i^-) belong to different event videos. Each element in the triplet example is put into each channel of the triplet network. Through the three channels that share the same parameter w and b, i.e., weights and biases, we map triplets x_i from the VGG feature space into another learned feature space, where x_i is represented as $\phi_w(x_i) = <\phi_w(x_i^o), \phi_w(x_i^+), \phi_w(x_i^-)>$. When the triplet ranking network is trained, the learned feature space will have the property that the distance between the matched event video feature pair $(\phi_w(x_i^o), \phi_w(x_i^+))$ is much closer than that of the mismatched event video feature pair $(\phi_w(x_i^o), \phi_w(x_i^-))$.

The network architecture in each channel includes two fully connected layers, two relu layers, and one $L2$ norm layer. And each of the fully connected layers has 4096 neurons. The following section describes the triplet loss layer, and we define the loss function on this projected feature space.

Ranking Loss Function. As stated in Sect. 3.3, given one triplet input $x_i = <x_i^o, x_i^+, x_i^->$, the ranking network model maps the triplet input x_i into its corresponding feature space $\phi_w(x_i) = <\phi_w(x_i^o), \phi_w(x_i^+), \phi_w(x_i^-)>$. The distance between two event videos can be donated as $d_w(x_i^o, x_i^+) = d(\phi_w(x_i^o), \phi_w(x_i^+))$ and $d_w(x_i^o, x_i^-) = d(\phi_w(x_i^o), \phi_w(x_i^-))$, where they should satisfy the following constraint:

$$d(\phi_w(x_i^o), \phi_w(x_i^+)) < d(\phi_w(x_i^o), \phi_w(x_i^-)).$$

Therefore, the loss of our ranking model can be defined by the hinge loss as follows,

$$L(x_i^o, x_i^+, x_i^-, w) = max\{0, d_w(x_i^o, x_i^+) - d_w(x_i^o, x_i^-) + \tau\} \tag{3}$$

where τ represents the gap parameter between two distance, which is set to $\tau = 1$ in the experiment. Then our objective function for the training can be represented as Eq. 4,

$$min_w \frac{\lambda}{2}|w||_2^2 + \frac{1}{N}\sum_{i=1}^{N} max\{0, d_w(x_i^o, x_i^+) - d_w(x_i^o, x_i^-) + \tau\} \tag{4}$$

where w is the parameter weight of the network, N is the number of the triplets of the samples. λ is a constant representing weight decay, which is set to $\lambda = 0.0005$.

In the loss function, the distance between two video features is measured by the cosine distance as illustrated in Eq. 5 in our experiments.

$$d_w(x_i^o, x_i^+) = 1 - \frac{\phi_w(x_i^o) \cdot \phi_w(x_i^+)}{\sqrt{\phi_w(x_i^o) \cdot \phi_w(x_i^o)^T} \cdot \sqrt{\phi_w(x_i^+) \cdot \phi_w(x_i^+)^T}} \tag{5}$$

In order to simplify the cosine function, we have split the cosine distance in Eq. 5 into the $l2$ normalization layer and the dot product layer. Then after the $l2$ normalization layer, the final loss layer can be written as Eq. 6.

$$L(x_i^o, x_i^+, x_i^-, w) = max\{0, \phi_w(x_i^o) \cdot (\phi_w(x_i^-) - \phi_w(x_i^+)) + \tau\}. \tag{6}$$

We use the stochastic gradient decent algorithm to train the proposed neural network architecture with the triplet loss function as defined in Eq. 6. From the above loss function, it is clear that the gradient on each input triplet can be easily computed given the values of $\phi_w(I_i^o)$, $\phi_w(I_i^+)$, $\phi_w(I_i^-)$ and $\frac{\partial L}{\partial w}, \frac{\partial L}{\partial w}, \frac{\partial L}{\partial w}$, which can be obtained by separately running the standard forward and backward propagations for each video feature in the triplet examples. Our implementation is based on the "caffe" platform, we only add the l2 normalization layer and the cosine distance based triplet loss layer in the platform.

Triplet Example Generation. Supposed that we have a labeled multimedia event detection dataset with C classes of event, and that each class of event have N labeled videos. The number of all possible matched pairs is $CN(N-1)$. Since in the MED13 dataset, C equals 20, and we have used a few positive exemplars in our framework ($N \leq 10$). Then we have generated all the possible matched pairs, and the mismatched one is randomly selected from the background videos. In the triplet examples, we select M(=2000) background videos for each matched pair to constitute the M triplet training examples. In the training process, for each batch of 200 instances, the selected triplet examples should contain different classes of event videos in each iteration. This policy ensures that large amount of distance constraints are posed on a small number of event videos. When in the fine-tuning process, since there is only one class of event videos, the number of background videos should be larger ($M = 5000$).

4   Experiments and Results

Datasets and Evaluation: We tested the performance of the proposed method on the TRECVID MED13 datasets and evaluation protocol, and we also use the YFCC100M dataset to pre-train our similarity function to test the adaptation properties of the proposed method. The MED13 datasets have been introduced by NIST for all participants in the TRECVID competition and research community to conduct experiments on. For this dataset, there are 20 complex events. Event names include Birthday party, Bike trick, etc. Refer to [1] for the complete list of event names. In the training section, there are approximately 100 positive exemplars per event, and all events share negative exemplars with about 5,000 videos. The testing section has approximately 23,000 search videos. The total duration of videos in each collection is about 1,240 hours. We used the 10 exemplar(10Ex) query settings in the system, while the experiments with less query examples are done by randomly selecting the corresponding number of query examples in the 10Ex setting. The evaluation metric used is the Mean Average Precision(MAP).

4.1   Compare with the Baseline Methods

As illustrated in Fig. 3, we did experiments on MED13 dataset using various numbers of positive exemplars from 1 to 10 without using the pre-trained model,

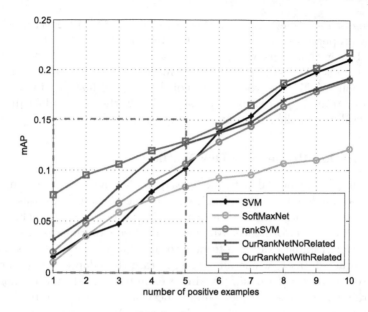

Fig. 3. Comparison on MED13Test with various number of positive examples.

and we have compared our proposed method with other methods, i.e.,SVM, rankSVM, and two-layer soft-max network(SoftMaxNet). It can be clearly seen that our proposed ranking network is much better than other methods with the number of training examples bellow 6, while slightly better than other methods with increasing numbers of positive training exemplars. As the proposed ranking network can use the partially ordered information as illustrated in Eq. 1, we have used some related examples in the training process. Then the performance can get much better as shown in Fig. 3. This clearly illustrates that the proposed triplet ranking network is very suitable for video event retrieval task with only a few positive training exemplars.

Table 1. Comparison with state-of-the-art methods with 10Ex on MED13Test.

Method	LCD_{VLAD} [16]	rankSVM	SVM	Ours	OurFusion	YFCC-Adapt
mAP	.250	.173	.210	.217	.248	**.253**

4.2   Compare with State-of-the-Art on 10Ex

We also compared our method with state-of-the-art method [16] on MED13 10Ex. When we consider the output of the second fully connected layer of our proposed ranking network as the projected VGG (PVGG) feature, and then combine this with the original input VGG features, we can obtain comparable results to the state-of-the-art method [16] as shown in Table 1. This reveals that the complementary properties of the projected VGG features. Although the method [16] is slightly better than ours without adaptation on YFCC100M dataset, their learned feature is 25,088-D, which is much higher than our used(4,096D). Since the input of our ranking network consists of the features instead of image pixels, such that our training process is very fast, and only the dot product distance is needed to evaluate in the search process.

4.3   Adaptation Property of Our Method

Since optimizing the neural network is easy to fall into the local minimum, giving an reasonable parameter initialization trained from other dataset or task can transfer other information to the specific video event retrieval task, and thus further improve the performance. We have conducted experiments based on various number of training examples from 1 to 5 videos on top of the parameters trained on other dataset, which aims to illustrate the adaptation properties of the proposed method with handful training examples. As shown in Table 2, we first pre-train our proposed triplet ranking network on the YFCC100M dataset to learn the similarity projection for video search. With the obtained parameters as our initialization, we then fine-tune on the MED13 dataset (denoted as YFCC-Adapt) on each specific video event. It can be clearly seen that much improvement can be obtained for the final search performance. This shows that we can adapt other previous knowledge to each specific query model to further improve the search accuracy. Obviously, we can clearly see that, when only with one query video for one event and some background negative videos to train the ranking model, its performance is much higher than any other methods. Therefore, giving reasonable parameter initialization trained from other dataset or task can transfer expanded information to the specific event video retrieval task.

Table 2. Experiment results on MED13Test with adaptation training from YFCC dataset.

Method	1Ex	2Ex	3Ex	4Ex	5Ex	10Ex
SVM	.0149	.0352	.0474	.0795	.1020	.210
OurRankNet	.0758	.0951	.1061	.1199	.1289	.217
YFCC-Adapt	**.0861**	**.1011**	**.1132**	**.1254**	**.1362**	**.253**

5   Conclusion

This paper has addressed the challenge of Multimedia Event Detection with only a few query examples, and proposed a ranking network method by casting the event video search task as the ranking problem. The proposed method can achieve better search performance than other methods with a handful of query examples. This method can be very suitable to be used in a practical video search system when only very few positive examples can be obtained (often less than 5). What's more, the proposed ranking network can be used to transfer other retrieval information to the specific event video, to further improve the search performance. In the future, we will focus on transferring knowledge from other larger datasets or tasks to the specific video event detection task.

Acknowledgement. This work was supported by the National Basic Research Program of China (Grant No.2015CB351705), the State Key Program of National Natural Science Foundation of China (Grant No.61332018).

References

1. Trecvid med 13. http://www.nist.gov/itl/iad/mig/med13.cfm
2. Apostolidis, E., Mezaris, V., Sahuguet, M., Huet, B., Červenková, B., Stein, D., Eickeler, S., Redondo Garcia, J.L., Troncy, R., Pikora, L.: Automatic fine-grained hyperlinking of videos within a closed collection using scene segmentation. In: Proceedings of the 22nd ACM International Conference on Multimedia, pp. 1033–1036. ACM (2014)
3. Bhattacharya, S., Yu, F.X., Chang, S.-F.: Minimally needed evidence for complex event recognition in unconstrained videos. In: ICMR, p. 105. ACM (2014)
4. Chang, C.-C., Lin, C.-J.: LIBSVM: a library for support vector machines. ACM Trans. Intell. Syst. Technol. (TIST) **2**(3), 27 (2011)
5. Cheng, D., Gong, Y., Zhou, S., Wang, J., Nanning, Z.: Person re-identification by multi-channel parts-based CNN with improved triplet loss function. In: CVPR (2016)
6. Gkalelis, N., Mezaris, V.: Video event detection using generalized subclass discriminant analysis and linear support vector machines. In: ICMR, p. 25. ACM (2014)
7. Habibian, A., Mensink, T., Snoek, C.G.: Composite concept discovery for zero-shot video event detection. In: ICMR, p. 17. ACM (2014)
8. Hauptmann, A.G., Christel, M.G., Yan, R.: Video retrieval based on semantic concepts. Proc. IEEE **96**(4), 602–622 (2008)
9. Jiang, L., Meng, D., Mitamura, T., Hauptmann, A.G.: Easy samples first: self-paced reranking for zero-example multimedia search. In: Proceedings of the 22nd ACM International Conference on Multimedia, pp. 547–556. ACM (2014)
10. Jiang, L., Yu, S.-I., Meng, D., Mitamura, T., Hauptmann, A.G.: Bridging the ultimate semantic gap: a semantic search engine for internet videos. In: ICMR (2015)
11. Ma, Z., Yang, Y., Sebe, N., Hauptmann, A.G.: Knowledge adaptation with partiallyshared features for event detectionusing few exemplars. PAMI **36**, 1789–1802 (2014)
12. Mazloom, M., Li, X., Snoek, C.G.: Few-example video event retrieval using tag propagation. In: Proceedings of International Conference on Multimedia Retrieval, p. 459. ACM (2014)
13. Tamrakar, A., Ali, S., Yu, Q., Liu, J., Javed, O., Divakaran, A., Cheng, H., Sawhney, H.: Evaluation of low-level features and their combinations for complex event detection in open source videos. In: 2012 IEEE Conference on Computer Vision and Pattern Recognition (CVPR), pp. 3681–3688. IEEE (2012)

14. Thomee, B., Shamma, D.A., Friedland, G., Elizalde, B., Ni, K., Poland, D., Borth, D., Li, L.-J.: YFCC100M: the new data in multimedia research. Commun. ACM **59**(2), 64–73 (2016)
15. Wu, S., Bondugula, S., Luisier, F., Zhuang, X., Natarajan, P.: Zero-shot event detection using multi-modal fusion of weakly supervised concepts. In: CVPR, pp. 2665–2672 (2014)
16. Xu, Z., Yang, Y., Hauptmann, A.G.: A discriminative CNN video representation for event detection. In: CVPR (2015)
17. Yu, S.-I., Jiang, L., Xu, Z., Yang, Y., Hauptmann, A.G.: Content-based video search over 1 million videos with 1 core in 1 second. In: ICMR (2015)

Demonstrations

A Demo for Image-Based Personality Test

Huaiwen Zhang[1,2](✉), Jiaming Zhang[3], Jitao Sang[1], and Changsheng Xu[1,2]

[1] Institute of Automation, Chinese Academy of Sciences, Beijing, China
{huaiwen.zhang,jtsang,csxu}@nlpr.ia.ac.cn
[2] University of Chinese Academy of Sciences, Beijing, China
[3] Shandong University of Technology, Zibo, China

Abstract. In this demo, we showcase an image-based personality test. Compared with the traditional text-based personality test, the proposed new test is more natural, objective, and language-insensitive. With each question consisting of images describing the same concept, the subjects are requested to choose their favorite image. Based on the choices to typically 15–25 questions, we can accurately estimate the subjects' personality traits. The whole process costs less than 5 min. The online demo adapts well to PCs and smart phones, which is available at http://www.visualbfi.org/.

1 Introduction

Personality affects man's attitude and social behavior such as "patterns of thought, occupational choice, criminal activity, and political ideology". One of the most popular personality models is Big Five (BF) or Five-Factor Model (FFM) [3], which defines personality along five dimensions, i.e., Openness (O), Conscientiousness (C), Extraversion (E), Agreeableness (A) and Neuroticism (N).

The most common personality questionnaire is Big Five Inventory (BFI), which is text-based and designed by psychological experts. The limitations of text-based personality test are three-fold: (1) Complicated and burdensome. The subjects need to read and understand each question thoroughly before making responses. (2) Prone to bias. The response is essentially based on the subjects' subjective perception of their own psychological property, which is easily debiased by self-enhancement. (3) Language-sensitive. Language-specific models cannot be directly translated into a destination language, but need to be carefully developed by experts.

Multimedia, especially the images and videos which are shared by users on the social network, conveys rich emotions and personality preferences. Cristani et al. uses the images posted as favorite to predict the personality traits of 300 Flickr users [1]. Sharath Chandra et al. extend the features in [1] and provide the personality modeling based image recommendation [2]. Inspired by this, we propose to research towards an image-based personality test.

© Springer International Publishing AG 2017
L. Amsaleg et al. (Eds.): MMM 2017, Part II, LNCS 10133, pp. 433–437, 2017.
DOI: 10.1007/978-3-319-51814-5_36

2   System Overview

Language psychology shows that the choice of words is driven not only by the meaning, but also by psychological characteristics of speakers/writers such as emotions and personality traits [4]. In other words, different psychological characteristics could lead to different word choices for the same meaning. Therefore, we are motivated to make an analogy in the scenario of image choice, and predict the subjects' personality traits by investigating their preferences for images belonging to the same concept.

2.1   Architecture

Different from text-based personality test, image-based personality test is data-driven rather than designed by psychological experts. This demo is based on the PsychoFlickr dataset provided in [1]. This dataset comprises of 300 Flickr users, with each user having 200 favorite images and self-assessed personality traits. 82-d content and aesthetic features have been extracted for each image.

The first step is concept extraction and expansion. For each image, GoogleNet is employed to obtain the confidence score over 1,000 ImageNet categories and the top-5 categories with the score larger than 0.1 are remained. Hypernyms of the 1,000 concepts are traced with WordNet and added into the included concepts for corresponding images. Totally 1,789 concepts at four levels are obtained to construct the candidate concept set.

The next step is exploiting users' interactions with the 1,789 concepts by observing their favorite images, to predict their personality traits. We introduce a view-based GBDT (vGBDT). View here corresponds to concept. For each round of base regressor, vGBDT first select the most discriminative view from the candidate concept sets, and then tunes the optimal partitions and the output leaf value based on the users' image features under this view. Assuming M base regressors (concepts) are considered for each personality trait, a personality questionnaire consisting of $M * 5$ questions will be designed.

The base regressor of vGBDT naturally divides the images from same concept into J parts. Affinity Propagation is used to obtain several image clusters for a concept. Within the largest image cluster, the one which is nearest to the cluster center in its own part, is selected as the option image. Finally $M * 5$ questions with each question consisting of J options are determined.

2.2   User Interface

A set of "choose-your-favorite-image" questions is designed, with each question corresponding to one concept and options for each question corresponding to different patterns of images under this concept. Subjects are requested to select one favorite image from each question. Typically M is set as 3 or 5, with each questionnaire totally consisting of 15 or 25 questions, the entire test can be completed in 3–5 min. As showed in Fig. 1, the whole system UI consists of four parts: (1) Welcome page, in which features of image-based personality are

Fig. 1. User interface of image-based personality test

shown. (2) Guide page, briefly introducing how to conduct the test. (3) Main questionnaire page, containing J image-options for each question, a progress bar and a button to next question where subjects could tap to select the image they favor. (4) Result page, presenting the estimated personality results for the subjects with detailed explanation for each trait.

This image-based personality test can effectively resolve the problems in text-based test: (1) Natural. Image is recognized as more natural interaction means. (2) Objective. The intent behind choosing images is not clear, so subjects can make objective responses according to their realistic perceptions. (3) Language-insensitive. Human perception of the visual information is universal regardless of their mother tongue.

3   Real-World Experiment

We conducted a real-world evaluation by recruiting 67 master workers from Amazon mechanical Turk (MTurk). This test is available at http://test.visualbfi. org. As illustrated in Fig. 2, each subject was asked to answer five questionnaires in order: BFI-10, vBFI_1, vBFI evaluation, BFI-44, vBFI_2. For each subject, we examined his/her credibility by calculating the difference between the calculated personality scores of BFI-10 and BFI-44. 40 subjects with RMSE lower than 1.2 are remained.

The experimental results are shown in Table 1. Comparison between BFI-10 and vBFI_1 leads to the accuracy evaluation of the image-based personality test. We can see that Visual BFI (vBFI_1) and BFI-10 has a RMSE between 1.5 to 2.0. This result is comparable to that obtained by CG+LASSO [1]. Note that in CG+LASSO, for each subject all his/her 200 favorite images are examined. In our proposed test, only $M * 5$ responses of each subject are needed to derive the personality trait. This demonstrates that by examining users favorite images belonging to few selective concepts, we can achieve comparable, if not better prediction accuracy than that based on much more unorganized favorite images.

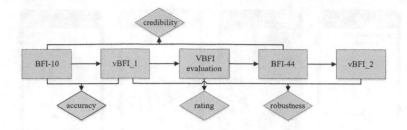

Fig. 2. Real-world experiment flow

The goal of image-based personality test is not to fit with the text-based test results, but to match with the subjects' own perception. After finishing vBFI_1, we show the estimated personality results to the subjects with a detailed explanation of each trait. Each subject rates how accurately the derived traits are on a seven-likert scale (1 being worst while 7 being best). The mean of the resultant ratings is 5.150 (std = 1.494), suggesting that the derived traits from vBFI generally match well with their own perceptions.

Table 1. Real-world evaluation results from MTurk.

Trait	vBFI_1 v.s. BFI-10 (RMSE)	Rate (mean/std.)	vBFI_1 v.s. vBFI_2 (RMSE)
O	1.647	5.150/1.494	0.872
C	1.859		1.004
E	2.059		1.101
A	1.506		0.866
N	2.075		1.300

We further compare the derived traits from two versions of visual BFI (vBFI_1 and vBFI_2) to examine the robustness of image-based personality test. Different from vBFI_1, vBFI_2 is constructed with image-options selected from the second-largest cluster. Results from Table 1 show a relatively low RMSE around 1.0, which demonstrate the feasibility and robustness of visual BFI in practical personality test.

4   Conclusion

In this paper, we present a demo for the image-based personality test, which is a data-driven questionnaire on personality. By exploiting users' favorite interactions with images, the image-based personality test is demonstrated more natural, objective, and language-insensitive compared to the traditional text-based personality test.

Acknowledge. This work is supported by National Natural Science Foundation of China (No. 61432019, 61225009, 61303176, 61272256, 61373122, 61332016).

References

1. Cristani, M., Vinciarelli, A., Segalin, C., Perina, A.: Unveiling the multimedia unconscious: implicit cognitive processes and multimedia content analysis. In: Proceedings of the 21st ACM International Conference on Multimedia, pp. 213–222. ACM (2013)
2. Guntuku, S.C., Roy, S., Weisi, L.: Personality modeling based image recommendation. In: He, X., Luo, S., Tao, D., Xu, C., Yang, J., Hasan, M.A. (eds.) MMM 2015. LNCS, vol. 8936, pp. 171–182. Springer, Heidelberg (2015). doi:10.1007/978-3-319-14442-9_15
3. McCrae, R.R., John, O.P.: An introduction to the five-factor model and its applications. J. Pers. **60**(2), 175–215 (1992)
4. Tausczik, Y.R., Pennebaker, J.W.: The psychological meaning of words: LIWC and computerized text analysis methods. J. Lang. Soc. Psychol. **29**(1), 24–54 (2010)

A Web-Based Service
for Disturbing Image Detection

Markos Zampoglou[1]([⊠]), Symeon Papadopoulos[1], Yiannis Kompatsiaris[1],
and Jochen Spangenberg[2]

[1] CERTH-ITI, Thessaloniki, Greece
{markzampoglou,papadop,ikom}@iti.gr
[2] Deutsche Welle, Berlin, Germany
jochen.spangenberg@dw.com

Abstract. As User Generated Content takes up an increasing share of
the total Internet multimedia traffic, it becomes increasingly important
to protect users (be they consumers or professionals, such as journalists)
from potentially traumatizing content that is accessible on the web. In
this demonstration, we present a web service that can identify disturbing
or graphic content in images. The service can be used by platforms for filtering or to warn users prior to exposing them to such content. We evaluate the performance of the service and propose solutions towards extending the training dataset and thus further improving the performance of
the service, while minimizing emotional distress to human annotators.

1 Introduction

With the proliferation of social media and capturing devices, Internet users are
increasingly exposed to User Generated Content (UGC) uploaded by other users.
In such environments, it is possible that a user may join a public platform and
upload content that can be traumatizing to others, such as pornographic or
violent imagery. While most platforms provide reporting mechanisms that allow
users to request the removal of such content, users can only report content after
they have been exposed to it. Furthermore, professionals such as journalists or
police officers spend a lot of time browsing and collecting UGC in search of
information. Due to the nature of their work, such content may often be violent
or disturbing, and prolonged exposure to such content can be psychologically
traumatizing. Automatically detecting such content and forewarning the user
before displaying it could help protect both professional and casual users, not as
a form of censorship but as a personal tool for preventing emotional trauma.

2 Background

We use the term *disturbing images* to refer to *depictions of humans or animals
subjected to violence, harm, and suffering, in a manner that can cause trauma to
the viewer.* Research suggests that users who are systematically exposed to such
material are in danger of serious psychological harm [2][1]. In research literature,

---

[1] Also see http://dartcenter.org/.

© Springer International Publishing AG 2017
L. Amsaleg et al. (Eds.): MMM 2017, Part II, LNCS 10133, pp. 438–441, 2017.
DOI: 10.1007/978-3-319-51814-5_37

Fig. 1. Sample *violent* images that do not qualify as *disturbing* for our task.

the term *violence detection* refers to a very active field, e.g. [1, 7]. Yet such works usually concern video (exploiting audio and motion cues), and the content they aim to detect is quite different from the one discussed in this work, as their definitions of violence often include, e.g. individuals fighting or weapons being fired. A recent attempt at detecting violent images [8] is based on a dataset collected using Google image search with violence-related keywords, many of which *violent* images would not be considered *disturbing* given our definition (Fig. 1). Finally, another recent work aims at detecting *horror* images [5], which however often lack the realism of real-world *disturbing* UGC.

3   Disturbing Image Detection Service

3.1   Back-End

Dataset Collection and Organization. As, to our knowledge, this is the first effort in tackling this task, we had to build an annotated dataset in order to train and evaluate our service. The dataset consists primarily of images collected from the web. As the manual dataset generation process is psychologically demanding for annotators we used a two-step approach: An initial set was collected manually from the web, depicting war zones, accidents and other similar situations, with and without disturbing content. Adding around 100 non-disturbing images from the UCID dataset [6] and following manual annotation, we ended up with 990 images. We then used this initial set to build the rest of the dataset semi-automatically: we trained a classifier using the approach described below and employed automated scripts to download images from a number of websites specializing in disturbing and/or graphic pictures. Images were automatically classified as `disturbing` or `non-disturbing`, and the annotators only had to visually correct the results. This process proved a lot less stressful than manually annotating all images. The final dataset contains 5401 images, 2043 of which are labeled as `disturbing`. Due to their graphic nature we decided against demonstrating image samples, but will share the dataset upon request.

Classification Approach. With respect to classification, we decided to use Convolutional Neural Networks (CNN), as they have exhibited exceptional performance in many image classification challenges in recent years. However, given the relatively small size of our dataset, we cannot hope to train such a classifier from scratch. In such cases we can take a model that is pre-trained on

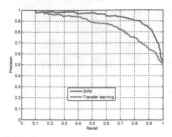 

Fig. 2. Left: Results of classification using the two approaches. Right: A screenshot from the developed web User Interface.

a general set of concepts (e.g. publicly available networks trained on the Ima-geNet dataset) and either use the output of the second-to-top layer as an image descriptor for a standard classifier (e.g. SVM), or re-train only the upper layers of the network using our data, in an approach known as *transfer learning*. We attempted both approaches using the *BVLC Reference CaffeNet* network[2] [4] from the Caffe framework [3]. Figure 2 (left) shows that, given our current dataset, SVM classification with CNN features worked best, reaching 0.864 Precision at 0.868 Recall, versus 0.769 Precision and 0.763 Recall by the transfer learning-based CNN model. Thus we use the SVM approach for our demo.

Web Service. We used Caffe to make the classification model accessible via a REST service. The back-end is built in Python (PyCaffe) and Java. It accepts an image URL, downloads and runs the image through the classifier, and returns the prediction as JSON. If the incoming load overcomes the system limits, a queuing mechanism allows the service to respond asynchronously to meet the demands of batch classifying images in large collections. The API also offers a feedback mechanism, which accepts an image URL and the correct annotation for an image. This feature can be used both to improve classifier performance, but also as a first step towards a personalized classifier that can take subjective preferences into account. It should be noted that in its current form the system accuracy drops significantly when classifying thumbnails (<200 pixels in either dimension), thus it is currently aimed at images above that size.

3.2 Front-End

The web front-end provides a simple interface for the classification and feedback services[3]. A user submits an image URL and the service visualizes the classifier prediction as a percentage on a bar, indicating the probability that the image has disturbing content. The bar allows the user to submit their own perception on what the prediction should have been, in case it differs from the one provided.

---

[2] Downloaded from http://caffe.berkeleyvision.org/model_zoo.html.
[3] The demo is available at http://reveal-mklab.iti.gr/reveal/disturbing/.

In that case the image is stored alongside the value to be used in refining the classifier. Figure 2 (right) shows the classification result for a web image.

4   Conclusions and Future Steps

We presented a public service for the analysis and potential filtering of disturbing images using state-of-the-art technologies on a novel dataset, with feedback mechanisms to further improve future results by incorporating user annotations. In order to further refine our dataset, given the problems of manually annotating large numbers of disturbing images, we are trying to incorporate semi-automatic and crowd-sourcing methods to disperse the emotional burden of the task. Given the -partly- subjective nature of the task, we are also considering the potential for personalized classification. We also believe there is significant application potential if the interface could take the form of a browser plug-in that would be able to analyze all images to be displayed on the browser. Finally, we are considering the possibility of a localization system that can isolate only the offending part of the image -however, manually annotating such a dataset for training and evaluation would incur considerable emotional toll on annotators, and how to reduce this toll through semi-automatic means is an open issue.

Acknowledgements. This work is supported by the REVEAL and InVID projects, partially funded by the European Commission under contract numbers 610928 and 687786. In addition, we would like to acknowledge the support that NVIDIA provided us through the GPU Grant Program.

References

1. Déniz, O., Serrano, I., et al.: Fast violence detection in video. In: Battiato, S., Braz, J. (eds.) VISAPP 2014, vol. 2, pp. 478–485. SciTePress, Setúbal (2014)
2. Dubberley, S., Griffin, E., Bal, H.M.: Making secondary trauma a primary issue: a study of eyewitness media and vicarious trauma on the digital frontline (2015). http://eyewitnessmediahub.com/uploads/browser/files/Trauma%20Report.pdf
3. Jia, Y., Shelhamer, E., et al.: Caffe: convolutional architecture for fast feature embedding. arXiv preprint arXiv:1408.5093 (2014)
4. Krizhevsky, A., Sutskever, I., Hinton, G.E.: Imagenet classification with deep convolutional neural networks. In: NIPS 2012, pp. 1097–1105 (2012)
5. Li, B., Xiong, W., et al.: Horror image recognition based on context-aware multi-instance learning. IEEE Trans. Image Process. **24**, 5193–5205 (2015)
6. Schaefer, G., Stich, M.: UCID: an uncompressed color image database. In: Storage and Retrieval Methods and Applications for Multimedia, vol. 5307, pp. 472–480. SPIE (2004)
7. Sjberg, M., Ionescu, B., et al.: The MediaEval 2014 affect task: violent scenes detection. In: MediaEval (2014)
8. Wang, D., Zhang, Z., Wang, W., Wang, L., Tan, T.: Baseline results for violence detection in still images. In: AVSS, pp. 54–57 (2012)

An Annotation System
for Egocentric Image Media

Aaron Duane, Jiang Zhou, Suzanne Little[(⊠)],
Cathal Gurrin, and Alan F. Smeaton

Insight Centre for Data Analytics, Dublin City University, Dublin, Ireland
aaron.duane4@mail.dcu.ie,
{suzanne.little,cathal.gurrin,alan.smeaton}@dcu.ie

Abstract. Manual annotation of ego-centric visual media for lifelogging, activity monitoring, object counting, etc. is challenging due to the repetitive nature of the images especially for events such as driving, eating, meeting, watching television, etc. where there is no change in scenery. This makes the annotation task boring and there is danger of missing things through loss of concentration. This is particularly problematic when labelling infrequently or irregularly occurring objects or short activities. To date annotation approaches have structured visual lifelogs into events and then annotated at the event or sub-event levels but this can be limited when the annotation task is labelling a wider variety of topics-events, activities, interactions and/or objects. Here we build on our prior experiences of annotating at event level and present a new annotation interface. This demonstration will show a software platform for annotating different levels of labels by different projects, with different aims, for ego-centric visual media.

Keywords: Lifelog · Annotation

1 Introduction

One of the most intriguing forms of lifelog data which people can accumulate is visual data from wearable cameras, recording either continuous video or frequent images taken at regular intervals [2]. This is now referred to as ego-centric image media, reflecting the fact that it is usually taken from the first person viewpoint and can suffer from image quality issues due to camera shake because of movement of the wearer. For the most part, such visual lifelogs are indexed and become searchable based on their metadata, the date, time and perhaps location at which the images were taken. Increasingly we are realising that in addition to this we need to analyse and index visual lifelogs based on content using either manually annotated tags, or automatic detection of semantic concepts. For either of these approaches there is a need for a software tool to allow end users to manually annotate lifelog images, either to train a machine learning classifier to recognise concepts, or to be used directly as lifelog descriptors.

© Springer International Publishing AG 2017
L. Amsaleg et al. (Eds.): MMM 2017, Part II, LNCS 10133, pp. 442–445, 2017.
DOI: 10.1007/978-3-319-51814-5_38

In this paper we introduce a software platform for manual annotation of visual lifelogs. A short demonstration video can be seen at https://goo.gl/ru2ZxZ. We describe the system in terms of the different kinds of user roles involved in the annotation, and then we describe the annotation process itself. This is followed by a short user feedback and evaluation of the interface looking at how it supports different annotation strategies and the requirements of the system.

2   Annotation System

2.1   Users

The annotation system consists of three different user types, the uploader, the annotator and the administrator (or assigner). The *uploader*s role is to be the camera wearer, the source of the lifelog data that is to be annotated within the system. Uploaders load their data to the database and the system in turn automatically segments the data into 'photo packages'. Upon logging in, the *administrator* has two primary roles, creating new *annotators* and assigning annotators to photo packages which have been uploaded. From the annotator's control panel the administrator can see a list of all annotators created within the system. The administrator has the option to edit or delete an annotator via this interface and to view the total number of images an annotator has been assigned to annotate. On selecting an annotator's assign button, the administrator is brought to the assignment interface where they see a list of all photo packages uploaded to the system.

It is important to note that multiple annotators can be assigned to the same photo package simultaneously. This is useful if annotators specialise or more than one annotator is needed to divide the workload of a large photo package. During the development of the system, the requirement for separate annotations from different projects or perspectives, yet on the same photo packages, became apparent. For example, lifelogs may be annotated by a group interested in food intake while another set of annotators may be interested in the wearers' exercise habits. To address these requirements, the concept of a *project* was introduced. When assigning an annotator to a package, it is necessary to choose from which project they are being assigned to. This means that different annotators can work on the same photo package under different project guidelines and there will be no risk of interfering with another project's annotations.

When the annotator is ready, they simply press the annotate button next to the package assigned to them and are immediately taken to the primary annotation screen (see Fig. 1). This annotation interface contains three sections, the calendar, the ontology and the annotatable images. The calendar is used to navigate between different days within a photo package if a package spans multiple days. These days are in turn segmented into hours for the purpose of navigability and convenience. When the annotator has targeted the day and hour they want to work on, they can begin selecting the photos they wish to annotate. In Fig. 1 we can see, from the blue border surrounding the images, that currently 10 images have been selected for annotation.

The annotation system imposes a 3 level hierarchical ontology of terms with terms being selectable via a one-click concertina navigation on the left side of the screen. Level 1 of the ontology appears in blue, level 2 in orange and level 3 in white. At this point the annotator will navigate through the ontology and choose which leaf node annotation terms they wish to attach to the selected images. Upon choosing an annotation term, the selected images are highlighted with a green overlay to indicate that they have been annotated. In the top left corner of each image, the annotator can see how many terms are currently attached.

After adding a term to a single or to a group of images, the annotator can remove the annotation by clicking on the term a second time in the ontology. For this option to be available, all selected images must contain the annotation already, otherwise clicking on the term in the ontology will simply add the annotation to any images that do not already have it.

Fig. 1. Annotating images using the ontology. (Color figure online)

If the annotator needs to examine an image closely or to see a list of all annotations attached to an image, they have the option of clicking the magnifying glass beneath each image. Upon doing so, an overlay will appear over the screen containing a larger version of the image and beneath that, all the annotation terms attached to this particular image (Fig. 2). It is important to note that the annotations from all projects and all annotations appear in this overlay. The annotator cannot interact with another projects annotation terms and they do not count toward the total terms attached to the image within their project.

The visibility of other projects in the annotation space was introduced because it is often beneficial for an annotator to see what other annotation terms different projects have attached to an image. If this is not the case, the annotator can elect to ignore other terms and focus on their own annotations which are highlighted in green and which they have the option to remove. If the annotator needs to see a full resolution version of the selected image, they can

Fig. 2. Exploring details of a single image.

click the image from within this overlay and a new tab will open containing the image at its native resolution.

3   Demonstration Proposal

The proposed demonstration will use data gathered live at the conference and engage with the users to rapidly annotate sample data. Users will gain an understanding of the unique requirements for annotating lifelog data and how this may be used in a variety of applications. In addition the demonstration will provide the opportunity for further observations as to the annotation strategy and possible future improvements for the interface.

The annotation tool introduced in this paper offers an alternative to the lifelog browser tool described in [1] and over which it has several advantages including a more flexible framework to ingest and annotate with structured levels of labels and enabling multiple annotations of images based on individual projects. Labelling ego-centric visual media for both information retrieval or higher-level data analytics at the event, activity and object level will influence the annotation strategy and hence the future development of specialised annotation interfaces.

Acknowledgements. This paper is based on research supported by Science Foundation Ireland under grant number SFI/12/RC/2289 and Health Research Council of New Zealand.

References

1. Doherty, A.R., Smeaton, A.F.: Automatically segmenting lifelog data into events. In: Ninth International Workshop on Image Analysis for Multimedia Interactive Services, WIAMIS 2008, pp. 20–23 (2008)
2. Gurrin, C., Smeaton, A.F., Doherty, A.R.: Lifelogging: personal big data. Found. Trends Inf. Retr. **8**(1), 1–125 (2014)

DeepStyleCam:
A Real-Time Style Transfer App on iOS

Ryosuke Tanno, Shin Matsuo, Wataru Shimoda, and Keiji Yanai(✉)

Department of Informatics, The University of Electro-Communications,
1-5-1 Chofugaoka, Chofu-shi, Tokyo 182-8585, Japan
{tanno-r,yanai}@mm.inf.uec.ac.jp

Abstract. In this demo, we present a very fast CNN-based style transfer system running on normal iPhones. The proposed app can transfer multiple pre-trained styles to the video stream captured from the built-in camera of an iPhone around 140ms (7fps). We extended the network proposed as a real-time neural style transfer network by Johnson et al. [1] so that the network can learn multiple styles at the same time. In addition, we modified the CNN network so that the amount of computation is reduced one tenth compared to the original network. The very fast mobile implementation of the app are based on our paper [2] which describes several new ideas to implement CNN on mobile devices efficiently. Figure 1 shows an example usage of DeepStyleCam which is running on an iPhone SE.

1 Introduction

In 2015, Gatys et al. proposed an algorithm on neural artistic style transfer [3,4] which synthesizes an image which has the style of a given style image and the contents of a given content image using Convolutional Neural Network (CNN) as shows in Fig. 2. This method enables us to modify the style of an image keeping the content of the image easily. It replaces the information which are degraded while the signal of the content image goes forward through the CNN layers with style information extracted from the style image, and reconstructs a new image which has the same content as a given content images and the same style as a given style image. In this method, they introduced "style matrix" which was presented by Gram matrix of the feature maps, that is, correlation matrix between feature maps in CNN.

However, since the method proposed by Gatys et al. required forward and backward computation many times (in general several hundreds times), the processing time tends to become longer (several seconds) even using GPU.

Then, several methods using only feed-forward computation of CNN to realize style transfer have been proposed so far. One of them is the method proposed by Johnson et al. [1]. They proposed perceptual loss functions to train the ConvDeconvNetwork as a feed-forward style transfer network. The ConvDeconvNetwork consists of down-sampling layers, convolutional layers and up-sampling layers,

© Springer International Publishing AG 2017
L. Amsaleg et al. (Eds.): MMM 2017, Part II, LNCS 10133, pp. 446–449, 2017.
DOI: 10.1007/978-3-319-51814-5_39

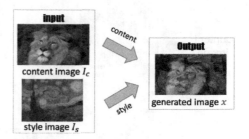

Fig. 1. "DeepStyleCam" running on an iPhone SE.

Fig. 2. "Neural style transfer" which creates a novel image by mixing the content and the style of two given images.

which accepts an image and outputs an modified image. This networks is commonly used for super-resolution [5] and coloring of gray-scale images [6]. In their method, they train a ConvDeconvNetwork so that the style matrix of its output image becomes closer to the style matrix of the given fixed style image and the CNN features of the input image leaves unchanged by using the proposed perceptual losses.

However, the ConvDeconvNetwork trained by their method can treat only one fixed style. If transferring ten kinds of styles, we have to train ten different ConvDeconvNetwork independently. This is not good for mobile implementation in terms of required memory size. Then, we modified Johnson et al.'s method so that one ConvDeconvNetwork can train multiple styles at the same time.

2   Proposed System

We modified the ConvDeconvNetwork used in [1] by adding a fusion layer and a style input stream as shown in Fig. 3. This is inspired by Iizuka et al's CNN-based coloring work [6]. They proposed adding a contextual stream to the ConvDeconvNetwork. With this improvement, they achieved coloring depending on the content of a target image.

We propose a style transfer network with style input as shown in Fig. 3. When training, we provide sample images to the content stream and style images to the style stream. The training method is the same as [1]. Please refer the detail on training to [1].

When transferring images, we input a target image to the content stream and one of the style images used in the training phrase to the style stream. Then, we can obtained the results corresponding on the selected style images by using only one trained network as shown in the top row of Fig. 4.

In addition, we shrunk the ConvDeconvNetwork compared to [1] to save computation costs. We added one down-sampling layer and up-sampling layer, replaced 9×9 kernels with smaller 5×5 kernels in the first and last convolutional

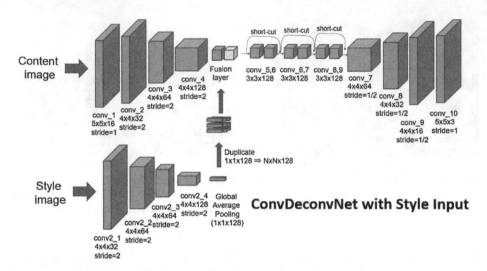

Fig. 3. Style transfer network with style input.

Fig. 4. Examples of the results. (Top) normal mode, (bottom) color preserving mode. (Color figure online)

layers, and reduced five Residual Elements into three. We confirmed that these network shrinking did not harm the quality of outputs significantly.

Regarding mobile implementation, we followed our work on Efficient Mobile CNN Implementation [2]. In the method, CNN networks are directly converted to a C source code which utilized multi-threading and iOS Accelerate Framework for CNN computation.

In addition, we implemented color preserving mode [7] which transfers a selected style only into gray-scale elements of an input image. The results in the color preserving mode as shown in the bottom row of Fig. 4. It can keep color of the content image while changing only the intensity of pixels, which is especially suitable for food images.

3   Demo Video and App on the iOS App Store

We prepared the videos recorded that the DeepStryleCam app was running in the practical settings. We released the app on the iOS app store. You can try "DeepStyleCam" on your iPhone or iPad. Note that we strongly recommend to use iPhone 6s/7/SE or iPad Pro for this app, because the app requires much computational power.

- Demo video of the DeepStyleCam.
  http://foodcam.mobi/deepstylecam/
- iOS app on the App Store, "RealTimeMultiStyleTransfer"
  https://itunes.apple.com/jp/app/realtimemultistyletransfer/id1161707531

References

1. Johnson, J., Alahi, A., Fei, L.F.: Perceptual losses for real-time style transfer and super-resolution. In: Proceedings of European Conference on Computer Vision (2016)
2. Yanai, K., Tanno, R., Okamoto, K.: Efficient mobile implementation of a CNN-based object recognition system. In: Proceedings of ACM Multimedia (2016)
3. Gatys, L.A., Ecker, A.S., Bethge, M.: A neural algorithm of artistic style. arXiv:1508.06576 (2015)
4. Gatys, L.A., Ecker, A.S., Bethge, M.: Image style transfer using convolutional neural networks. In: Proceedings of IEEE Computer Vision and Pattern Recognition (2016)
5. Dong, C., Loy, C.C., He, K., Tang, X.: Learning a deep convolutional network for image super-resolution. In: Fleet, D., Pajdla, T., Schiele, B., Tuytelaars, T. (eds.) ECCV 2014. LNCS, vol. 8692, pp. 184–199. Springer, Heidelberg (2014). doi:10.1007/978-3-319-10593-2_13
6. Iizuka, S., Simo-Serra, E., Ishikawa, H.: Let there be color!: joint end-to-end learning of global and local image priors for automatic image colorization with simultaneous classification. ACM Trans. Graph. **35**(4), 110 (2016). (Proceedings of SIGGRAPH 2016)
7. Gatys, L.A., Bethge, M., Hertzmann, A., Shechtman, E.: Preserving color in neural artistic style transfer. arXiv:1606.05897 (2016)

V-Head: Face Detection and Alignment for Facial Augmented Reality Applications

Zhiwei Wang$^{(\boxtimes)}$ and Xin Yang

Shool of Electronic Information and Communications,
Huazhong University of Science and Technology, Wuhan, China
{zhiweiwang,xinyang2014}@hust.edu.cn

Abstract. Efficient and accurate face detection and alignment are key techniques for facial augmented reality (AR) applications. In this paper, we introduce *V-Head*, a facial AR system which consists of three major components: (1) joint face detection and shape initialization which can efficiently localize facial regions based on the proposed face probability map and a multipose classifier and meanwhile explicitly produces a roughly aligned initial shape, (2) cascade face alignment to locate 2D facial landmarks on the detected face, and (3) 3D head pose estimation based on the perspective-n-point (PnP) algorithm so as to overlay 3D virtual objects on the detected faces. The demonstration can be accessed from https://drive.google.com/open?id=0B-H2fYiPunUtRHBFTDRzRkZvVEE.

Keywords: Face detection · Face initialization · Face alignment · AR

1 Introduction

Face alignment aims at locating points with facial semantic meanings in the face image, e.g. pupil, mouth corners and nose bridges. An efficient and accurate face alignment system is highly demanded by many facial AR applications, such as eyewear virtual try-on and virtual makeover.

Most existing face alignment systems rely on a face detection step which provides a bounding box around a face region and then place an initial shape, which could be obtained by averaging all shapes in a training dataset or randomly selected from a training set, in the bounding box. The initial shape is progressively updated so as to match the actual shape in the image. However, the performance of those systems is usually limited by an alignment-unfriendly face detector, i.e. shifting and scaling of a face bounding box could lead to non-trivial discrepancy between the final alignment result and the actual shape. Unsupervised initialization schemes, e.g. a mean shape or a random shape, can hardly guarantee a good performance of face alignment especially for faces with large pose.

To address the above problems, the authors in [2] propose an alignment-friendly face detector by utilizing the shape-indexed features [1] and combining face detection and face alignment. However, their method still utilizes an

© Springer International Publishing AG 2017
L. Amsaleg et al. (Eds.): MMM 2017, Part II, LNCS 10133, pp. 450–454, 2017.
DOI: 10.1007/978-3-319-51814-5_40

unsupervised initialization scheme, which is sensitive to large pose variations. Recently, a supervised initialization method [5] utilizes a ConvNet framework to estimate head pose, which is then used to generate a reasonable initial shape according to detected face bounding box's size and location. However, a small variation of the location and size of the detected bounding boxes greatly affects the performance of the supervised initialization, degrading the robustness of the initialization scheme.

In this paper, we introduce *V-Head*, a facial AR system which can simultaneously detect face and initialize facial shape at an ultrafast speed. Meanwhile, the initial shape can always guarantee a fast and accurate convergence to the actual shape. In particular, our system differs from the existing solutions in two aspects:

– **Ultrafast Face Proposal:** Most if not all facial AR applications capture face information in the form of color images by a frontal camera. Based on this fact, we propose an ultrafast face proposal method based on a two-level classifier: a face probability map (FPM) is first generated from a color image by applying a Naive Bayesian Model (NBM) to enhance face information, followed by a boosted classifier based on FPM to eliminate most non-face regions. The proposed face proposal method facilitates the real-time performance of the overall system (i.e. 32 fps on a single thread CPU of an x86 computer).
– **Joint Face Detection and Initialization:** Based on the shape-indexed features and cascade face detection framework, we propose a new joint face detection and initialization method, which concurrently obtain a facial bounding box and initial facial landmarks (i.e. shape) in one step. Our joint method guarantees an alignment-friendly face bounding box and a more reliable initial shape, enhancing the robustness of the subsequent face alignment.

2   System Implementation

2.1   Core Components

Figure 1 illustrates our face alignment system which consists of three key modules: joint face detection and initialization, cascade face alignment and head pose estimation.

(1) **Face Detection and Initialization:** For each image, we first convert it to a FPM via a NBM. FPM is a mixture of skin probability map (SPM), eyes probability map (EPM) and mouth probability map (MPM), which are three most distinctive components in face region. Specifically, we build SPM in the $YCbCr$ color space by applying Bayes rule based on skin pixel' and non-skin pixels' (Cb,Cr) value. Each pixel of a SPM represents its likelihood to being belong to skin. Similarly, we construct an EPM and an MPM, and each probability map is one of FPM's three channels. FPM construction can be implemented very efficiently via lookup tables. Then we implement a

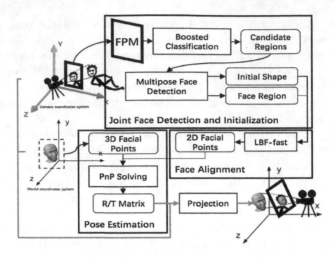

Fig. 1. System framework for *V-Head* demo

boosted classifier based on LBP features extracted from FPM to eliminate most non-face regions. As FPM can greatly enhance face regions and suppress background clutters thus most non-facial regions can be excluded at very early stage of the boosted classifier, significantly increasing the detection speed.

Given the remaining face-like regions obtained by the face proposal, we train a multipose face detector to further detect and initialize faces. We utilize the shape-indexed features instead of Haar-like features and for each candidate face region we overlay several different poses on the region and extract features relative to each head pose. Facial regions, which contain a face and are represented by shape-indexed features extracted from shapes roughly aligned with the face, will be labeled positive. Otherwise, they will be labeled as negative samples. A random forest is learned utilizing the RealBoost algorithm as our classifier. During the testing phase, we try each shape, which is used for extracting features when training, on a face-like region, if the extracted features get the highest score from our classifier, the corresponding shape can be considered as an initialization for the following face alignment.

(2) **Cascade Face Alignment:** In our demo, we implement local binary features based face alignment method (LBF) [3] with the same settings as LBF-fast and use initial shapes generated by our joint face detection and initialization method. LBF shares the same shape-indexed features with our joint method and achieves ultrafast speed (3000 fps on PC) benefiting from linearly regressing binary features extracted by random forests. We collect 3837 annotated face images from 300W [4], and use original images and their mirror version to train LBF model.

(3) **Head Pose Estimation:** We pre-build a 3D head model and extract a set of 3D points from it, including outer corners of eyes, tip of nose, corners of mouth, point of jaw and points of ears, which present sufficient 3D structure information of human head. By correlating extracted 3D facial points and aligned 2D facial points in the image and calibrating intrinsic matrix of the camera, we solve PnP problem to estimate the rotation (roll, pitch, and yaw) and 3D translation of the camera with respect to the world coordinate system. Our pre-build 3D virtual head can be projected to the 2D image by estimated camera pose.

2.2   User Interface

We use Cocos2d-x game engine to build our demo. Figure 2 illustrates the main UI of our system. In the right area, the user can choose 3D head, glasses and mask to virtually try on. In the center area, we display video frames captured by a frontal camera and project user's selected 3D object on user's face in the frame. 3D object's rotating and translating synchronize with user's so that the object looks like being *virtually weared* by user.

Fig. 2. The UI of *V-Head*

References

1. Cao, X., Wei, Y., Wen, F., Sun, J.: Face alignment by explicit shape regression. Int. J. Comput. Vis. **107**(2), 177–190 (2014)
2. Chen, D., Ren, S., Wei, Y., Cao, X., Sun, J.: Joint cascade face detection and alignment. In: Fleet, D., Pajdla, T., Schiele, B., Tuytelaars, T. (eds.) ECCV 2014. LNCS, vol. 8694, pp. 109–122. Springer, Heidelberg (2014). doi:10.1007/978-3-319-10599-4_8
3. Ren, S., Cao, X., Wei, Y., Sun, J.: Face alignment at 3000 FPS via regressing local binary features. In: The IEEE Conference on Computer Vision and Pattern Recognition (CVPR), June 2014

4. Sagonas, C., Tzimiropoulos, G., Zafeiriou, S., Pantic, M.: 300 faces in-the-wild challenge: the first facial landmark localization challenge. In: Proceedings of the IEEE International Conference on Computer Vision Workshops, pp. 397–403 (2013)
5. Yang, H., Mou, W., Zhang, Y., Patras, I., Gunes, H., Robinson, P.: Face alignment assisted by head pose estimation. In: Proceedings of the British Machine Vision Conference (BMVC), pp. 130.1–130.13. BMVA Press, September 2015

Video Browser Showdown

Collaborative Feature Maps for Interactive Video Search

Klaus Schoeffmann[1(✉)], Manfred Jürgen Primus[1], Bernd Muenzer[1],
Stefan Petscharnig[1], Christof Karisch[1], Qing Xu[2], and Wolfgang Huerst[3]

[1] Institute of Information Technology, Klagenfurt University, Klagenfurt, Austria
{ks,mprimus,bernd,spetsch}@itec.aau.at, ckarisch@outlook.com
[2] School of Computer Science and Technology, Tianjin University, Tianjin, China
qingxu@tju.edu.cn
[3] Information and Computer Sciences, Utrecht University, Utrecht, The Netherlands
huerst@uu.nl

Abstract. This extended demo paper summarizes our interface used for the Video Browser Showdown (VBS) 2017 competition, where visual and textual known-item search (KIS) tasks, as well as ad-hoc video search (AVS) tasks in a 600-h video archive need to be solved interactively. To this end, we propose a very flexible distributed video search system that combines many ideas of related work in a novel and collaborative way, such that several users can work together and explore the video archive in a complementary manner. The main interface is a perspective Feature Map, which shows keyframes of shots arranged according to a selected content similarity feature (e.g., color, motion, semantic concepts, etc.). This Feature Map is accompanied by additional views, which allow users to search and filter according to a particular content feature. For collaboration of several users we provide a cooperative heatmap that shows a synchronized view of inspection actions of all users. Moreover, we use collaborative re-ranking of shots (in specific views) based on retrieved results of other users.

Keywords: Video retrieval · Interactive search · Collaboration

1 Introduction and Related Work

The Video Browser Showdown (VBS) is an annual live evaluation competition of interactive video search tools. It started in 2012 with visual known-item search (KIS) tasks in single videos, randomly selected from a set of 30 videos that were about one-hour long in duration, and has became increasingly challenging over the years. In 2014 the tasks were selected from 76 videos and extended by textual KIS tasks, where a textual description about the target scene is presented as a query instead of a visual excerpt [7]. The collection to search further increased in 2015 and 2016 to about 100 h and 250 h, respectively [2]. VBS 2017 (this year) is particularly challenging, because the data set increased even further, to 4593 video files with about 600 h of content. In addition to the increase in size, there is

© Springer International Publishing AG 2017
L. Amsaleg et al. (Eds.): MMM 2017, Part II, LNCS 10133, pp. 457–462, 2017.
DOI: 10.1007/978-3-319-51814-5_41

also a second type of querying, namely ad-hoc video search (AVS), which is the interactive version of AVS from TRECVID [6]. In contrast to KIS tasks – which requests participants to find one particular 20 seconds long segment – AVS tasks may have many results across the whole data set.

We approach the VBS 2017 challenge with a highly flexible interactive video search system that combines several ideas from previous years and integrates collaborative search features. The main interface is a similarity-based map of keyframes (called *Feature Map*), which uses hierarchical refinement to provide an overview of keyframes with different levels of granularity, similar to the one used by the winner of VBS 2016 [1]. This Feature Map is presented in 3D perspective to better exploit the screen real estate and show more images at once. The last iterations of the VBS showed that none of the many different interfaces worked well for all of the tasks and session. Particularly text-based KIS tasks are very hard to solve with color-based search only. Similarly, it is often hard to derive the best matching semantic concept from a given visual example. Also, sometimes users need the temporal context of keyframes within the corresponding video sequence/file, or would rather like to temporally browse through the data instead of searching. Therefore, in addition to the Feature Map, our search system provides several other views, which can be used for search and filtering according to some specific content feature, for web-based search by example, or for simple temporal browsing of keyframes. All of these views are designed in a way that allows several users to search simultaneously (and cooperatively, if desired).

This paper describes the general architecture of our interface, but omits many details due to the page limit. An elaborate technical paper detailing all the different parts of the interface and the underlying content analyses is under preparation for submission.

2 Proposed Approach

As already mentioned in the introduction, we propose a collaborative video search system that uses several different interfaces (called *views*), where the user can select the most-appropriate one for the current search task and intent.

2.1 Feature Map

The main user interface of our tool is the *Feature Map* (see Fig. 1). It is basically a two-dimensional grid of keyframes which are arranged based on similarity. The underlying user-selectable similarity metric can be any combination of the following four criteria (i.e., in total 15 different map arrangements can be selected):

- Concept similarity (CNN-Features)
- Color similarity (Feature Signatures)
- Texture similarity (Histogram of Oriented Gradients)
- Motion similarity (Motion Histogram)

Fig. 1. Feature Map: the main view of our interface arranges keyframes according to a similarity criteria, which can be selected on the left. The current hierarchical layer is visualized by a pyramidal indicator at the bottom left. The minimap below indicates the currently visible section and includes a heatmap that summarizes the current search activity of all users. Filter Views can be selected via a navigation menu at the top. (Color figure online)

The Feature Map is shown in a configurable 3D perspective view, which allows a better overview over a larger area than a flat 2D view. The Feature Map is built up hierarchically. The top layer shows approximately 40×40 frames. Each of the four subjacent layer shows four times the amount of keyframes than its overlying layer. The lowest layer shows all the keyframes of the entire video collection, which are 300000 in total.

2.2 Browse and Filter Views

The Feature Map is intended to be used as main interface, where users browse keyframes according so some similarity and refine their search over time. However, for some search tasks it might be inappropriate to start with this view. Therefore, we provide several complementary views that display a list of keyframes according to a filtering criterion. Each of these keyframes can then be used as starting point in the Feature Map. Due to space limitations we, however, show only a screenshot of the first view (the Storyboard) and omit others.

– **Storyboard**: In this view all videos are shown in a sequential list (Fig. 2). Each video is represented by uniformly sampled frames that are coherently visualized for fast human inspection, as described in [4]. The list is re-ranked according to the search activities of the collaborators. For example, if other

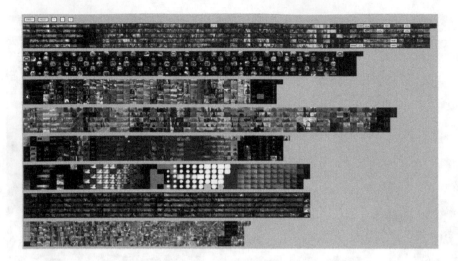

Fig. 2. The storyboard shows a coherent visualization of keyframes temporally sampled from the videos at equidistant positions (other browse & views are not shown due to space limitations).

users inspect many shots of a video in the Feature Map, it is up-ranked in the storyboard. Similarly, if other searchers are also browsing in the storyboard, already inspected videos are down-ranked (we call this *context-sensitive collaborative re-ranking*). This is an advancement of our previous approach [3].

- **Color Filter**: Here the user can choose different hue, saturation, and value areas from the HSV color space, to filter for matching keyframes.
- **Concept Filter**: This is a text-based search for semantic content classes, detected by convolutional neural networks (CNNs). We use two different CNNs for that purpose: (i) the well-known AlexNet [5], and (ii) a self-trained version of AlexNet using a manually selected large set of images from ImageNet (419630 images of 77 classes).
- **Web Example**: Here, a search engine is provided to gather appropriate images from the web (e.g., Bing or Google). The user can select an image from the result set and directly analyze it on-the-fly through a web service running on our content analysis server. The result of this analysis is used for similarity search in the Feature Map.

2.3   Architecture

Our collaborative video search system uses two different servers: a video server and an interaction server. The video server performs several types of video content analysis (see top left in Fig. 3) and stores all the results as well as the videos, and makes them accessible via a web server and web services. It uses a content-sensitive shot detection algorithm that builds on motion flow analysis and comparison of edge histograms and color histograms for selecting the

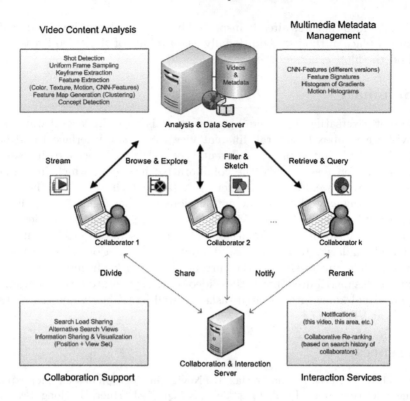

Fig. 3. Collaborative video search system

most representative keyframe. The video server provides a web interface with several different views, which can be used simultaneously by several users that are additionally connected to a collaboration and interaction server.

The interaction server uses a WebSocket connection among all clients and can actively contact clients for forwarding interaction data of a specific user. Such interaction data could be a notification (e.g., a hint to another user to inspect a specific area or video) or inspection information that provides the basis for collaboration features such as collaborative re-ranking or information sharing.

2.4 Collaboration

Special attention was paid to the support of collaborative features to allow for cooperative work. One element that supports collaboration is a heatmap that is shown in the lower left corner of Fig. 1. Red parts are locations of potentially correct keyframes, marked by other users. The green color highlights areas where users have looked into without finding the correct shot. The heatmap is influenced by the filtering and search actions of all participants. A further feature that is influenced by the work of the users is the ranking of the video list shown in Fig. 2. A video that was already inspected by a different user is shifted down to a rearward place in the list. Additionally, the ranking of the video list is rebuilt based on the filter events of the users.

3 Summary

We present a versatile video search system that is novel in several ways. First, it provides a toolbox of several different views (i.e., sub-interfaces) and hence is a flexible tool that supports different types of search. Second, it uses several collaboration features like the collaborative heatmap, which immediately shows which areas were inspected intensively (and which were not), the collaborative re-ranking as well as specific notifications to other users. Finally, it uses a hierarchically refineable Feature Map with a changeable underlying feature for similarity arrangement of keyframes. This enables users to quickly switch from color-based similarity to texture-based similarity, or motion- or concept-based similarity, while always keeping the currently selected keyframe in center view. We expect this distributed interactive video search system to be efficient for textual and visual known-item search tasks, as well as ad-hoc video search tasks, as issued at VBS 2017.

References

1. Barthel, K.U., Hezel, N., Mackowiak, R.: Navigating a graph of scenes for exploring large video collections. In: Tian, Q., Sebe, N., Qi, G.-J., Huet, B., Hong, R., Liu, X. (eds.) MMM 2016. LNCS, vol. 9517, pp. 418–423. Springer, Heidelberg (2016). doi:10.1007/978-3-319-27674-8_43
2. Cobârzan, C., Fabro, M., Schoeffmann, K.: Collaborative browsing and search in video archives with mobile clients. In: He, X., Luo, S., Tao, D., Xu, C., Yang, J., Hasan, M.A. (eds.) MMM 2015. LNCS, vol. 8936, pp. 266–271. Springer, Heidelberg (2015). doi:10.1007/978-3-319-14442-9_26
3. Hudelist, M.A., Cobârzan, C., Beecks, C., Werken, R., Kletz, S., Hürst, W., Schoeffmann, K.: Collaborative video search combining video retrieval with human-based visual inspection. In: Tian, Q., Sebe, N., Qi, G.-J., Huet, B., Hong, R., Liu, X. (eds.) MMM 2016. LNCS, vol. 9517, pp. 400–405. Springer, Heidelberg (2016). doi:10.1007/978-3-319-27674-8_40
4. Hürst, W., Werken, R., Hoet, M.: A storyboard-based interface for mobile video browsing. In: He, X., Luo, S., Tao, D., Xu, C., Yang, J., Hasan, M.A. (eds.) MMM 2015. LNCS, vol. 8936, pp. 261–265. Springer, Heidelberg (2015). doi:10.1007/978-3-319-14442-9_25
5. Krizhevsky, A., Sutskever, I., Hinton, G.E.: Imagenet classification with deep convolutional neural networks. In: Pereira, F., Burges, C., Bottou, L., Weinberger, K. (eds.) Advances in Neural Information Processing Systems 25, pp. 1097–1105. Curran Associates Inc. (2012)
6. Over, P., Awad, G., Michel, M., Fiscus, J., Sanders, G., Shaw, B., Kraaij, W., Smeaton, A.F., Quénot, G.: TRECVID 2012 – an overview of the goals, tasks, data, evaluation mechanisms and metrics. In: Proceedings of TRECVID 2012 (2012)
7. Schoeffmann, K.: A user-centric media retrieval competition: the video browser showdown 2012–2014. IEEE MultiMed. **21**(4), 8–13 (2014)

Concept-Based Interactive Search System

Yi-Jie Lu$^{(\boxtimes)}$, Phuong Anh Nguyen, Hao Zhang, and Chong-Wah Ngo

Department of Computer Science,
City University of Hong Kong, Kowloon, Hong Kong
{yijie.lu,panguyen2-c,hzhang57-c}@my.cityu.edu.hk, cscwngo@cityu.edu.hk

Abstract. Our successful multimedia event detection system at TREC-VID 2015 showed its strength on handling complex concepts in a query. The system was based on a large number of pre-trained concept detectors for textual-to-visual relation. In this paper, we enhance the system by enabling human-in-the-loop. In order to facilitate a user to quickly find an information need, we incorporate concept screening, video reranking by highlighted concepts, relevance feedback and color sketch to refine a coarse retrieval result. The aim is to eventually come up with a system suitable for both Ad-hoc Video Search and Known-Item Search. In addition, as the increasing awareness of difficulty in distinguishing shots of very similar scenes, we also explore the automatic story annotation along the timeline of a video, so that a user can quickly grasp the story happened in the context of a target shot and reject shots with incorrect context. With the story annotation, a user can refine the search result as well by simply adding a few keywords in a special "context field" of a query.

Keywords: Video search · Known-Item Search · Concept bank · Semantic query · Video reranking · Story annotation

1 Introduction

In TRECVID 2015, we developed a multimedia event detection system for zero-example event detection that achieved the best performance [4]. The core of the system is a large concept bank that contains about 2,800 pre-trained concept detectors covering common objects, actions, scenes and everyday activities. To perform a text query search in an unannotated video corpus, the crux of the system is to solve the textual-to-visual relation using the concept bank as a knowledge base.

We have studied several facts which significantly impact retrieval performance. Such facts include the number of concepts, concept specificity, and concept discriminativeness regarding the query. However, the performance of an automatic video retrieval system is still far from perfection, especially when no precisely matched concepts can be found in the concept bank. In this case, the system would propose concepts with the smallest word distance towards the query. This metric, however, often suggests off-topic concepts due to a lack of

© Springer International Publishing AG 2017
L. Amsaleg et al. (Eds.): MMM 2017, Part II, LNCS 10133, pp. 463–468, 2017.
DOI: 10.1007/978-3-319-51814-5_42

common sense that can distinguish a concept from the context of the query topic. A feasible solution is to employ a human evaluator to quickly adjust the result by screening the machine-proposed concepts. On the other hand, although our existing system can be adapted to Ad-hoc Video Search, it is inefficient for Known-Item Search. This is because a text query is insufficient to describe the fine details which are required to mine the exact query clip from a number of clips sharing the same semantic content. Hence, a human needs to painstakingly dig into hundreds of results to find the correct match even if the top results are all relevant. We, therefore, seek help from an interactive search where a user can refine a first-time search result with different methods so that the correct match has a higher chance to show up.

Video Browser Showcases [8] in previous years suggest using high-level visual concepts [5–7] and low-level visual descriptors [1,2] as two lines of approach. For Known-Item Search, the systems using low-level features generally have an advantage over those using high-level concepts. It is worth to mention that a color sketch search method was shown to be very effective in 2014 and 2015 [1,3]. But as low-level features do not contain semantic information, the systems with high-level concepts have their inherent benefit on Ad-hoc Video Search where queries are only formed by text. In this paper, we tend to integrate both methods into an interactive search system. The concept-based search system is mainly used for generating the first-time search result. Then, we implement different reranking techniques to incorporate the strength of both high-level concepts and low-level features. Specifically, highlighted concept reranking is a simple and quick method for a user to emphasize a particular characteristic in the query. When a user finds one or more visually relevant clips in the search result, either relevance feedback or color sketch can be further exploited to refine the result so that the user has a better chance to hit the correct answer. Furthermore, there is an increasing awareness of difficulty in distinguishing shots sharing very similar scenes in the search result [1]. We recount the dominant concepts along the timeline of a video to facilitate video browsing so that a user can quickly grasp the context of a target shot even though the shot itself is not distinctive. We also implement a *context field* in the query to quickly refine the result in this scenario. The following sections detail each component of our system.

2   Concept-Based Video Search System

We adapt our zero-example event search system to general-purpose video search. The search system is backed by a large concept bank which contains thousands of concept detectors for textual-to-visual relation. The most important module in our system is called *semantic query generation* which generates the internal query representation by calculating the distance from a query to each concept. The internal query, a.k.a. the *semantic query* is formed by a number of selected concepts with their weights. The weight calculation and concept selection are discussed in our paper [4].

Fig. 1. (a) An example of semantic query editing. (b) Training examples for the concept "swimming" vs. the meaning of the term *swimming* under the context of the query.

As the queries in Ad-hoc Video Search are generally more specific and shorter, the term weights based only on the query are unreliable compared to the prolonged event query used in multimedia event detection. We, therefore, shed more light on semantic query editing by involving human-in-the-loop. As illustrated in Fig. 1a, given a text query *"a diver wearing a diving suit and swimming under water"* without any further editing, the system first generates a list of candidate concepts[1] loosely relevant to the query. The weight of each concept is indicated by a weight bar. A user then can quickly refine the concept list by removing wrong and non-discriminative [4] concepts, and watch the search result change at the same time. Figure 1b shows a typical wrong concept "swimming" which is easily identified by a human but difficult by a machine algorithm. As in a human's sense, the term *swimming* under the context of the query means *underwater diving* which is visually different from the sport *swimming* the concept automatically proposed. Furthermore, we also allow a user to adjust a concept's weight in order to strengthen or weaken the concept. For example, the concept "person" is not important in most query examples because the term is too common. While in some rare cases, such as *"a person sitting beside a laptop,"* the concept "person" should not be depreciated. A user thus can manually increase the weight for "person."

3   Video Reranking

In order to facilitate Known-Item Search, we implement three methods for video reranking. A user can adjust the scope of a reranking method. By default, a reranking is only performed within the top videos recommended by the concept-based video search. For example, *highlighted concept reranking* is most effective

---

[1] The tag in the brackets of Fig. 1a denotes the dataset from which the concept comes.

Fig. 2. (a) Concept reranking by emphasizing concepts about "outdoor". (b) The top results change accordingly.

in the scope of the top 300–500 videos. This limitation ensures that the algorithm is not applied to the semantically irrelevant videos at the bottom of the rank list.

Highlighted concept reranking is a simple and quick reranking approach used to highlight particular characteristics in a semantic query. Figure 2a emphasizes the concept "outdoor" in the query *"a person playing guitar outdoors."* As shown in Fig. 2b, the top retrieval result of the original semantic query mixes guitar playing both indoors and outdoors. It is reasonable to highlight the concept "outdoor." But, if we simply increase the weight of "outdoor" in the semantic query, it would pull up noisy outdoor activities which do not contain guitar playing at all. A feasible way is thus to rerank only within the clips about guitar playing. Figure 2b shows the reranking result in the scope of the top-500 clips.

Relevance feedback is used when a user identifies one or more visually relevant clips. Even with highlighted concept reranking, the retrieval result is still diverse if the search system is only based on the high-level semantic concepts. Once a user has identified some relevant clips, these clips can be served as training examples having fine-grained visual details. We intuitively want to refine the result using these visual details. We train SVM classifiers for the user picked clips and rerank the result according to this feedback. The new result is expected to be much more specific and focused on the visually similar clips according to the user's choice.

Color sketch was a very successful approach in Video Browser Showcase 2014 and 2015 [1,3]. Basically, color sketch uses position-color features. These low-level features characterize the colors with their positions on a keyframe. A user can perform the search by simply drawing a few color circles on the empty canvas. We incorporate this search approach to be a reranking alternative mainly for its accuracy on Known-Item Search. In our system, the user can not only draw a new sketch but also use the color sketch automatically extracted from several marked clips in the search result for reranking.

4   Context Annotation

Video Browser Showcase 2014 raised a critical problem in Known-Item Search that it was difficult to distinguish the shots with very similar scenes in a search

Fig. 3. The context view for a search result. (Color figure online)

result. The problem was noticeable when a large portion of the query clip was about a TV studio scene [1]. To tackle this problem, we automatically annotate the master shots along the whole timeline of a video by its dominant concepts, then fold the adjacent shots sharing the same dominant concept. This process is called *story annotation*. With this annotation, we can enhance the result presentation by showing a *context view* of a target shot. Figure 3 is an example. When the query is a common concept/scene, such as an anchor man, we expect multiple relevant shots of the similar scene to appear in the search result (red centered in Fig. 3). Although hardly any decision can be made by the shots themselves, by expanding the result to a *context view* (images with black text underneath), we can easily grasp the story around each shot and thus distinguish these shots. The benefit of story annotation is not limited to the result presentation. In addition, we implement a special *context field* in the query for quickly screening the search result. A user may simply add a few keywords in the context field of the query to refine the search results, eventually coming up with the shots having matched context only. For instance, when querying a report of a flooded village but the query clip is mostly an anchor person in a news studio, other than describing the exact query clip in the system query, we tend to add keywords like *flood*, *rooftop*, *rescue man*, and even *river* (which is visually similar to flood) in the context field.

Acknowledgments. The work described in this paper was supported by two grants from the Research Grants Council of the Hong Kong Special Administrative Region, China (CityU 11210514 and CityU 11250716).

References

1. Blažek, A., Lokoč, J., Matzner, F., Skopal, T.: Enhanced signature-based video browser. In: He, X., Luo, S., Tao, D., Xu, C., Yang, J., Hasan, M.A. (eds.) MMM 2015. LNCS, vol. 8936, pp. 243–248. Springer, Heidelberg (2015). doi:10.1007/978-3-319-14442-9_22
2. Cobârzan, C., Hudelist, M.A., Fabro, M.: Content-based video browsing with collaborating mobile clients. In: Gurrin, C., Hopfgartner, F., Hurst, W., Johansen, H., Lee, H., O'Connor, N. (eds.) MMM 2014. LNCS, vol. 8326, pp. 402–406. Springer, Heidelberg (2014). doi:10.1007/978-3-319-04117-9_46
3. Lokoč, J., Blažek, A., Skopal, T.: Signature-based video browser. In: Gurrin, C., Hopfgartner, F., Hurst, W., Johansen, H., Lee, H., O'Connor, N. (eds.) MMM 2014. LNCS, vol. 8326, pp. 415–418. Springer, Heidelberg (2014). doi:10.1007/978-3-319-04117-9_49
4. Lu, Y.J., Zhang, H., de Boer, M., Ngo, C.W.: Event detection with zero example: select the right and suppress the wrong concepts. In: ACM ICMR (2016)
5. Moumtzidou, A., Mironidis, T., Apostolidis, E., Markatopoulou, F., Ioannidou, A., Gialampoukidis, I., Avgerinakis, K., Vrochidis, S., Mezaris, V., Kompatsiaris, I., Patras, I.: VERGE: a multimodal interactive search engine for video browsing and retrieval. In: Tian, Q., Sebe, N., Qi, G.-J., Huet, B., Hong, R., Liu, X. (eds.) MMM 2016. LNCS, vol. 9517, pp. 394–399. Springer, Heidelberg (2016). doi:10.1007/978-3-319-27674-8_39
6. Ngo, T.D., Nguyen, V.H., Lam, V., Phan, S., Le, D.-D., Duong, D.A., Satoh, S.: NII-UIT: a tool for known item search by sequential pattern filtering. In: Gurrin, C., Hopfgartner, F., Hurst, W., Johansen, H., Lee, H., O'Connor, N. (eds.) MMM 2014. LNCS, vol. 8326, pp. 419–422. Springer, Heidelberg (2014). doi:10.1007/978-3-319-04117-9_50
7. Rossetto, L., Giangreco, I., Heller, S., Tănase, C., Schuldt, H., Dupont, S., Seddati, O., Sezgin, M., Altıok, O.C., Sahillioğlu, Y.: IMOTION – searching for video sequences using multi-shot sketch queries. In: Tian, Q., Sebe, N., Qi, G.-J., Huet, B., Hong, R., Liu, X. (eds.) MMM 2016. LNCS, vol. 9517, pp. 377–382. Springer, Heidelberg (2016). doi:10.1007/978-3-319-27674-8_36
8. Schoeffmann, K.: A user-centric media retrieval competition: the video browser showdown 2012–2014. IEEE MultiMed. **21**(4), 8–13 (2014)

Enhanced Retrieval and Browsing
in the IMOTION System

Luca Rossetto[1(✉)], Ivan Giangreco[1], Claudiu Tănase[1], Heiko Schuldt[1],
Stéphane Dupont[2], and Omar Seddati[2]

[1] Databases and Information Systems Research Group,
Department of Mathematics and Computer Science, University of Basel,
Basel, Switzerland
{luca.rossetto,ivan.giangreco,c.tanase,heiko.schuldt}@unibas.ch
[2] Research Center in Information Technologies, Université de Mons, Mons, Belgium
{stephane.dupont,omar.seddati}@umons.ac.be

Abstract. This paper presents the IMOTION system in its third version. While still focusing on sketch-based retrieval, we improved upon the semantic retrieval capabilities introduced in the previous version by adding more detectors and improving the interface for semantic query specification. In addition to previous year's system, we increase the role of features obtained from Deep Neural Networks in three areas: semantic class labels for more entry-level concepts, hidden layer activation vectors for query-by-example and 2D semantic similarity results display. The new graph-based result navigation interface further enriches the system's browsing capabilities. The updated database storage system $ADAM_{pro}$ designed from the ground up for large scale multimedia applications ensures the scalability to steadily growing collections.

1 Introduction

In this paper we introduce the 2017 version of the IMOTION system which is the third iteration (after [11,13]) of the system participating in the Video Browser Showdown [2].

We provide a brief overview of the overall architecture of the system in Sect. 2, and elaborate in greater detail on the improvements made since the previous version in Sect. 3. Section 4 concludes.

2 The IMOTION System

2.1 Overview

The IMOTION system is a sketch-based video retrieval system which supports a large variety of query paradigms, including query-by-sketch, query-by-example, query-by-motion and querying using semantic concepts. It allows to search using multiple query containers, e.g., a still image, a user-provided sketch, the specification of motion via flow fields or by denoting a semantic concept. The IMOTION system is built in a flexible and modular way and can easily be extended to support further query modes or feature extractors.

© Springer International Publishing AG 2017
L. Amsaleg et al. (Eds.): MMM 2017, Part II, LNCS 10133, pp. 469–474, 2017.
DOI: 10.1007/978-3-319-51814-5_43

2.2 Architecture

The 2017 IMOTION system is based on the ADAM$_{pro}$ database [3] and the Cineast retrieval engine [12] which are both part of the vitrivr[1] open-source content-based multimedia retrieval stack [14]. The IMOTION system has a custom browser-based front end which communicates with the storage and retrieval back-end via a web server which also serves the static content such as videos and preview images. Figure 1 shows an overview of the architecture of the IMOTION system.

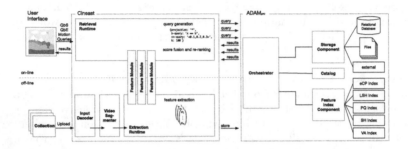

Fig. 1. Architectural overview of the IMOTION system.

3 New Functionality

3.1 Concept Detection

Since the last edition, we have expanded the set of semantic features supported by IMOTION. All these features are based on Deep Neural Network classifiers:

- We have extracted semantic categories representing entry-level labels of environments from the Places2 dataset. Classification was performed using the pre-trained VGG16-places365 network [18].
- We have trained image-level classifiers for the 80 classes of the MS COCO Detection challenge [9]. The feature data is obtained from the last fully connected layer ("fc7") of a VGG convolutional network. The model is trained on the MS COCO train2014 data and it learns the 80 labels independently using multinominal logistic regression.
- We kept the 325 semantic entry-level categories obtained from n-grams from last year [11].

Given the participation in this year's TRECVID Ad Hoc Search task[2], which also operated on the IACC.3 data, we integrated the result scores for our estimated best run into the search engine. We have extended the list of 30 AVS

---

[1] https://www.vitrivr.org/.
[2] http://www-nlpir.nist.gov/projects/tv2016/tv2016.html#avs.

textual queries with several queries we consider useful for browsing e.g., "shots with two people", "shots showing cartoons", etc.

As in our previous system, we use multiple ConvNets for feature extraction and object/action recognition. We replaced the temporal ConvNet trained on dense optical flow maps with ConvNets that are able to recognize visual actions that may be detected from single images. In order to train these ConvNets, we used the two databases Stanford 40 [17] with 40 categories of actions and COCO-a with 140 categories [10].

We also use a modified version of the DenseCap [7] language model (LM). We use a beam search approach in order to keep multiple results at each generated word. We hence end up with a number of alternatives sentences for each region of interest. From these sentences, we recover a set of words corresponding to objects and attributes. We also use downsampled (bilinear sampling) features extracted with DenseCap ConvNet. This ConvNet was trained on the Visual Genome [8] dataset.

3.2 Semantic Class Selection

As with the previous version of the system, one supported query mode is to search for instances of detected semantic concepts. In the 2016 IMOTION system [11] we implemented the interface for the selection of these concepts as a list of icons which could be added to a canvas via drag and drop. Figure 2 shows an example of this UI element. The new selection interface for VBS 2017, depicted in Fig. 3, uses a text box with an auto-complete feature to select semantic classes. Every class adds a weight slider by which the importance of this class with respect to the query can be specified.

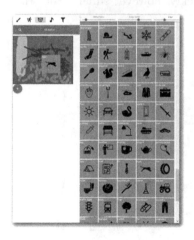

Fig. 2. Semantic class selection in the 2016 IMOTION system.

Fig. 3. New semantic class selection in the 2017 IMOTION system.

3.3 Result Presentation and Browsing

In addition to the existing querying capabilities, for the 2017 version of the system we put additional emphasis on exploratory search and browsing capabilities. In a manner similar to several of the 2016 VBS systems (e.g., [1]), we have implemented a similarity-based navigation interface. The new interface allows to navigate through the resulting grid by panning and zooming as it places visually and semantically similar results close to each other.

3.4 Text-Based Retrieval

At VBS 2017, we use traditional text retrieval based on Lucene to search in the text extracted from the ASR (as provided with the video data), and captions extracted from the keyframes using DenseCap [7].

3.5 ADAM$_{pro}$

In the most current version, the IMOTION system uses the new ADAM$_{pro}$ database. The ADAM$_{pro}$ database [3] is geared towards offering storage and retrieval capabilities for multimedia objects and the corresponding metadata. To this end, it supports both Boolean retrieval and k nearest neighbour similarity searches in the vector space retrieval model and is particularly tailored to support large multimedia collections. ADAM$_{pro}$ comes with various index structures that are very different in their nature: Locality-Sensitive Hashing [5] and Spectral Hashing [16] are hash-based methods and form together with Product Quantization [6] and extended Cluster Pruning (eCP) [4] a group of indexes which support a rather coarse retrieval which can be executed very quickly, however suffers from the

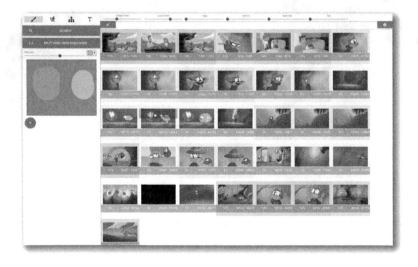

Fig. 4. Screenshot of the 2017 IMOTION system UI.

fact that it may miss result candidates as they are pruned by mistake from the candidate list. The Vector Approximation-File (VA-File) index [15], on the other hand, may degenerate to a sequential scan in worst case; however, it will not prune by mistake a true result candidate. Finally, $ADAM_{pro}$ supports sharding a collection to multiple nodes to increase the retrieval efficiency.

4   Conclusions

The 2017 version of the IMOTION system has received significant upgrades over previous versions in both indexing and browsing (Fig. 4). Compared to last year's version, we have tripled the number of semantic classes and improved the class selection mechanism. In agreement with video browsing state of the art, the results browsing interface features semantic-based arrangement, which is supposed to significantly reduce the interaction overhead for browsing and near-hit search. Finally, the new version of IMOTION is backed up by the new $ADAM_{pro}$ storage system, which comes with a large variety of indexing structures to decrease query latency.

Acknowledgements. This work was partly supported by the Chist-Era project IMO-TION with contributions from the Belgian Fonds de la Recherche Scientifique (FNRS, contract no. R.50.02.14.F) and the Swiss National Science Foundation (SNSF, contract no. 20CH21_151571).

References

1. Barthel, K.U., Hezel, N., Mackowiak, R.: Graph-based browsing for large video collections. In: He, X., Luo, S., Tao, D., Xu, C., Yang, J., Hasan, M.A. (eds.) MMM 2015. LNCS, vol. 8936, pp. 237–242. Springer, Heidelberg (2015). doi:10.1007/978-3-319-14442-9_21
2. Cobârzan, C., Schoeffmann, K., Bailer, W., Hürst, W., Blažek, A., Lokoč, J., Vrochidis, S., Barthel, K.U., Rossetto, L.: Interactive video search tools: a detailed analysis of the video browser showdown 2015. Multimedia Tools Appl., 1–33 (2016). doi:10.1007/s11042-016-3661-2
3. Giangreco, I., Schuldt, H.: ADAMpro: database support for big multimedia retrieval. Datenbank-Spektrum **16**(1), 17–26 (2016)
4. Gudmundsson, G., Jónsson, B., Amsaleg, L.: A large-scale performance study of cluster-based high-dimensional indexing. In: Proceedings of the International Workshop on Very-Large-Scale Multimedia Corpus, Mining and Retrieval (VLS-MCMR 2010), Firenze, Italy, pp. 31–36. ACM (2010)
5. Indyk, P., Motwani, R.: Approximate nearest neighbors: towards removing the curse of dimensionality. In: Proceedings of the Symposium on the Theory of Computing, Dallas, Texas, USA, pp. 604–613. ACM (1998)
6. Jegou, H., Douze, M., Schmid, C.: Product quantization for nearest neighbor search. IEEE Trans. Pattern Anal. Mach. Intell. **33**(1), 117–128 (2011)
7. Johnson, J., Karpathy, A., Fei-Fei, L.: Densecap: fully convolutional localization networks for dense captioning. In: Proceedings of the IEEE Conference on Computer Vision and Pattern Recognition (2016)

8. Krishna, R., Zhu, Y., Groth, O., Johnson, J., Hata, K., Kravitz, J., Chen, S., Kalantidis, Y., Li, L.-J., Shamma, D.A., et al.: Visual genome: connecting language and vision using crowdsourced dense image annotations. arXiv preprint arXiv:1602.07332 (2016)

9. Lin, T.-Y., Maire, M., Belongie, S., Bourdev, L., Girshick, R., Hays, J., Perona, P., Ramanan, D., Zitnick, C.L., Dollár, P.: Microsoft COCO: common objects in context. ArXiv e-prints, May 2014

10. Ronchi, M.R., Perona, P.: Describing common human visual actions in images. In: Jones, M.W., Xie, X., Tam, G.K.L. (eds.) Proceedings of the British Machine Vision Conference (BMVC 2015), pp. 1–12. BMVA Press, Norwich (2015)

11. Rossetto, L., et al.: IMOTION – searching for video sequences using multi-shot sketch queries. In: Tian, Q., Sebe, N., Qi, G.-J., Huet, B., Hong, R., Liu, X. (eds.) MMM 2016. LNCS, vol. 9517, pp. 377–382. Springer, Heidelberg (2016). doi:10.1007/978-3-319-27674-8_36

12. Rossetto, L., Giangreco, I., Schuldt, H.: Cineast: a multi-feature sketch-based video retrieval engine. In: 2014 IEEE International Symposium on Multimedia (ISM), pp. 18–23. IEEE (2014)

13. Rossetto, L., Giangreco, I., Schuldt, H., Dupont, S., Seddati, O., Sezgin, M., Sahillioğlu, Y.: IMOTION — a content-based video retrieval engine. In: He, X., Luo, S., Tao, D., Xu, C., Yang, J., Hasan, M.A. (eds.) MMM 2015. LNCS, vol. 8936, pp. 255–260. Springer, Heidelberg (2015). doi:10.1007/978-3-319-14442-9_24

14. Rossetto, L., Giangreco, I., Tanase, C., Schuldt, H.: vitrivr: a flexible retrieval stack supporting multiple query modes for searching in multimedia collections. In: Proceedings of the 2016 ACM on Multimedia Conference, pp. 1183–1186. ACM (2016)

15. Weber, R., Schek, H.-J., Blott, S.: A quantitative analysis and performance study for similarity-search methods in high-dimensional spaces. In: Proceedings of the International Conference on Very Large Data Bases (VLDB 1998), New York, USA, pp. 194–205 (1998)

16. Weiss, Y., Torralba, A., Fergus, R.: Spectral hashing. In: Proceedings of the Annual Conference on Neural Information Processing Systems (NIPS 2008), Vancouver, Canada, pp. 1753–1760 (2008)

17. Yao, B., Jiang, X., Khosla, A., Lin, A.L., Guibas, L., Fei-Fei, L.: Human action recognition by learning bases of action attributes and parts. In: 2011 International Conference on Computer Vision, pp. 1331–1338. IEEE (2011)

18. Zhou, B., Lapedriza, A., Xiao, J., Torralba, A., Oliva, A.: Learning deep features for scene recognition using places database. In: Advances in Neural Information Processing Systems, pp. 487–495 (2014)

Semantic Extraction and Object Proposal for Video Search

Vinh-Tiep Nguyen[1](✉), Thanh Duc Ngo[2], Duy-Dinh Le[2], Minh-Triet Tran[1],
Duc Anh Duong[2], and Shin'ichi Satoh[3]

[1] University of Science, Vietnam National University-HCMC,
Ho Chi Minh City, Vietnam
{nvtiep,tmtriet}@fit.hcmus.edu.vn
[2] University of Information Technology, Vietnam National University-HCMC,
Ho Chi Minh City, Vietnam
{thanhnd,ldduy,ducda}@uit.edu.vn
[3] National Institute of Informatics, Tokyo, Japan
satoh@nii.ac.jp

Abstract. In this paper, we propose two approaches to deal with the
problems of video searching: ad-hoc video search and known item search.
First, we propose to combine multiple semantic concepts extracted from
multiple networks trained on many data domains. Second, to help user
find exactly video shot that has been shown before, we propose a sketch
based search system which detects and indexes many objects proposed
by an object proposal algorithm. By this way, we not only leverage the
concepts but also the spatial relations between them.

Keywords: Semantic extraction · Object proposal · Sketch based
search

1 Introduction

With the rapid growth of video data from many sources such as social sites,
broadcast TVs, films, one of the most fundamental needs is to help users find
exactly what they are looking for in video databases. In this scenario, people
did not see any target video shots before. The input query could be a text with
ad-hoc description about the content they want to search. Here is an example
of this query type: finding shots of a man lying on a tree near a beach. In the
second scenario, people already saw the target video shot and the task of the
system is to find exactly that one. In the Video Browser Showdown 2017, the
dataset contains 4593 videos collected from the Internet with 144 GB in storage
and 600 h in duration. The participants need to solve two tasks: Ah-hoc Video
Search (AVS) and Known-Item Search (KIS) corresponding to two types of query
as mentioned above.

To deal with AVS query type, where users try to describe what they are
looking for using verbal description, high level based features are extracted to

© Springer International Publishing AG 2017
L. Amsaleg et al. (Eds.): MMM 2017, Part II, LNCS 10133, pp. 475–479, 2017.
DOI: 10.1007/978-3-319-51814-5_44

match with human language. Moreover, the result of last year Video Browser Showdown has shown that, leveraging high level feature using deep convolutional neural network (CNN) is one of the state of the art methods [1]. Although the performance of these networks are increasing every year, the number of concepts used for training is limited. On the other hand, query topics given by users are unpredictable. In this paper, we combine multiple concepts from multiple datasets including ImageNet [7], Visual Genome [4], MIT Places [9] and SUN attribute [5] to hopefully cover most popular topics that users may be interested in.

For KIS task, in case of finding a video shot that has been shown to a user, the system must point out exactly. Using previous approach is not suitable for this case due to scenes of nearly similar concepts. To overcome this problem, we propose to leverage spatial relation between objects in a video frame. To enhance the performance of searching we detect candidate objects using a object proposal algorithm and index them for searching in later. To search a frame with objects, users use a canvas to describe their query. To further improve the search result, users may use text based query to further filter out irrelevant objects.

2   Semantic Extraction

In this section, we propose to extract semantic features to match with ad-hoc query given by users. Because the users may pay attention to any aspects of a video frame, the set of semantic concepts is unknown. Figure 1 shows many aspects of a frame that people may be interested in from simple objects: the man, the beach, the coconut tree to their complex relations: the man lying on the tree, the tree next to the beach.

Fig. 1. Many aspects of a picture that people may pay attention to.

Since the number of concepts is unlimited and the query of the user is unpredictable, to increase the recall of the system, we propose to extract as much background and foreground information as possible. The proposed system includes

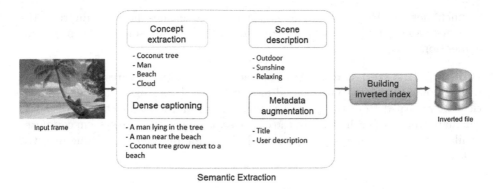

Fig. 2. Proposed system for searching based on semantic description.

two main parts: semantic extraction using deep models trained on large scale datasets and semantic features indexing using inverted file. Figure 2 illustrates our proposed framework with two main stages.

Semantic Extraction. This is the most important part of our proposed system to detect main concepts in a video frame. Inspired by recent successes of deep learning techniques, in this paper, we attempt to leverage the powerful of deep feature in semantic search task. Compared to low level feature based approach, deep feature (also known as high level feature) is closer to semantic based query representation and has lower storage cost. In this system, semantic aspects that we are interested in includes:

- Main objects: ones that appear in a large enough region of the video frame with assumption that the higher salient object gives the higher score from the output activation of the pre-trained deep convolutional neural network. In this paper, we use VGG-16 network proposed by Simonyan and Zisserman [8] to extract main objects. This is one of the state of the art models for object classification task on ImageNet dataset. We sample original video frame to overlapping 224×224 patches then transfer to the pre-trained feed forward network. Feature maps from the output activation are aggregated together using average pooling approach. Five objects which give highest score will be used to represent a video frame.
- Scene attributes: includes indoor/outdoor labels, main topic of the image such as building, park, kitchen. In our system, the attributes is extracted from the state of the art model trained on MIT scene and SUN attribute dataset [9].
- Object relations: to describe densely relations between objects, we propose to use dense captioning approach which is based on a Convolutional Neural Network-Recurrent Neural Network (CNN-RNN) to generate many sentences from the detected objects [3].
- Metadata: provided by video sharing user includes title, summary content, tags. These data often mention about the main topic of video but not too

much details. However, these information sometimes help to improve the performance of the system by combining with other semantic concepts as mentioned above.

Building Inverted Index. After extracting semantic features, the searching task is now equivalent to text based retrieval task. This stage is to index semantic text returned from the previous stage. A standard *tf-idf* scheme is used to calculate weight of each word. In the online searching stage, the system computes similarity scores between query text and video semantic features using inverted index structure.

3   Searching with Objects

Finding an exact scene requires specific and discriminative cues. One of the most discriminative cues in a scene is related to object instances. We therefore propose to employ such cue in our system. Object in scenes (i.e. video frames) are detected and indexed for searching. Particularly, we apply an object proposal approach to locate top 50 object proposals in the scene. With each object, we extract its features for indexing. The features include size of the bounding box, position of the center, its edge-based presentation (e.g. HOG or shape context [2]), and the dominant color. All objects in all frames will be indexed for searching.

To search a frame with objects, users use a canvas to describe their query. First, they draw a sketch of the object. The dominant color of the object can be selected from a color panel. They then move the sketch to an appropriate position in the frame. Based on its position, color, and edge-based feature, a set of relevant frames are returned and displayed in the main panel. The results is also changed as users re-select the color, re-draw the sketch, or change the position of the object in the frame. Object position can be changed by dragging mouse.

Using object proposal is mainly to deal with generic object classes. To handle some specific classes with high precision, we employ YOLO for object detection [6]. All instances of 20 object classes in the frames will be detected and indexed. By doing this, we enable users to search for a frame with an object (belongs to one of the 20 classes) and its exact position. The users prepare a query by indicating the object class and point its position in the canvas.

4   Dealing with Duplicate Scenes

There are duplicate scenes in a video clips e.g. a TV show may have lots of similar scenes. This causes difficulty in selecting the correct one. To deal with such problem, we rely on scene transitions. Particularly, if a scene can not be differentiated from its duplicates, we further check its neighborhood. By showing adjacent scenes, users are given more information to select. We enable this function in our system by using thumbnails and video player. When a user select a frame representing a scene in the main panel, a thumbnail is displayed on the top right of the frame. The thumbnail includes 4 images corresponding to 4 keyframes of 4 adjacent scenes (2 scenes before and 2 scenes after the current scene). The video player is used to play the video. This is necessary if users need to more information, for example, 4 frames of 4 adjacent scenes is not enough to recognize.

Acknowledgement. This research is funded by Vietnam National University HoChiMinh City (VNU-HCM) under grant number B2013-26-01. We are thankful to our colleagues Sang Phan, Yusuke Matsui, Benjamin Renoust who provided their source code to make our system more efficient.

References

1. Barthel, K.U., Hezel, N., Mackowiak, R.: Navigating a graph of scenes for exploring large video collections. In: Tian, Q., Sebe, N., Qi, G.-J., Huet, B., Hong, R., Liu, X. (eds.) MMM 2016. LNCS, vol. 9517, pp. 418–423. Springer, Heidelberg (2016). doi:10.1007/978-3-319-27674-8_43
2. Dalal, N., Triggs, B.: Histograms of oriented gradients for human detection. In: Proceedings of the 2005 IEEE Computer Society Conference on Computer Vision and Pattern Recognition (CVPR 2005), vol. 1, pp. 886–893. IEEE Computer Society, Washington, DC (2005). http://dx.doi.org/10.1109/CVPR.2005.177
3. Johnson, J., Karpathy, A., Fei-Fei, L.: Densecap: fully convolutional localization networks for dense captioning. In: Proceedings of the IEEE Conference on Computer Vision and Pattern Recognition (2016)
4. Krishna, R., Zhu, Y., Groth, O., Johnson, J., Hata, K., Kravitz, J., Chen, S., Kalanditis, Y., Li, L.J., Shamma, D.A., Bernstein, M., Fei-Fei, L.: Visual genome: connecting language and vision using crowdsourced dense image annotations (2016)
5. Patterson, G., Hays, J.: SUN attribute database: discovering, annotating, and recognizing scene attributes. In: Proceeding of the 25th Conference on Computer Vision and Pattern Recognition (CVPR) (2012)
6. Redmon, J., Divvala, S.K., Girshick, R.B., Farhadi, A.: You only look once: unified, real-time object detection. CoRR abs/1506.02640 (2015). http://arxiv.org/abs/1506.02640
7. Russakovsky, O., Deng, J., Su, H., Krause, J., Satheesh, S., Ma, S., Huang, Z., Karpathy, A., Khosla, A., Bernstein, M., Berg, A.C., Fei-Fei, L.: ImageNet large scale visual recognition challenge. Int. J. Comput. Vis. (IJCV) **115**(3), 211–252 (2015)
8. Simonyan, K., Zisserman, A.: Very deep convolutional networks for large-scale image recognition. CoRR abs/1409.1556 (2014)
9. Zhou, B., Lapedriza, A., Xiao, J., Torralba, A., Oliva, A.: Learning deep features for scene recognition using places database. In: Ghahramani, Z., Welling, M., Cortes, C., Lawrence, N.D., Weinberger, K.Q. (eds.) Advances in Neural Information Processing Systems, vol. 27, pp. 487–495. Curran Associates, Inc., Red Hook (2014)

Storyboard-Based Video Browsing Using Color and Concept Indices

Wolfgang Hürst[1](✉), Algernon Ip Vai Ching[1], Klaus Schoeffmann[2], and Manfred J. Primus[2]

[1] Utrecht University, Utrecht, Netherlands
huerst@uu.nl, A.A.K.IpVaiChing@students.uu.nl
[2] Klagenfurt University, Klagenfurt, Austria
{ks,mprimus}@itec.aau.at

Abstract. We present an interface for interactive video browsing where users visually skim storyboard representations of the files in search for known items (known-item search tasks) and textually described subjects, objects, or events (ad-hoc search tasks). Individual segments of the video are represented as a color-sorted storyboard that can be addressed via a color-index. Our storyboard representation is optimized for quick visual inspections considering results from our ongoing research. In addition, a concept based-search is used to filter out parts of the storyboard containing the related concept(s), thus complementing the human-based visual inspection with a semantic, content-based annotation.

Keywords: Video browsing · Visual inspection · Concept-based indexing

1 Motivation

Previous years of the Video Browser Showdown (VBS) competition have shown that optimized storyboard representations are an effective way for quickly skimming the content of reasonably sized video archives. For example, in 2015, we participated in the VBS with a storyboard-based visualization of a 100-hour video database [2]. Despite no content analysis, the optimized storyboard layout enabled users to quickly and efficiently search for known items, resulting in the 3rd place of the overall competition. Likewise, concept-based search, where a set of predefined and trained semantic concepts is used to search in the videos, has shown to work particularly well with larger databases. For example, in 2016, when the size of the database was increased to 200 h, the concept-based search in our system performed particularly well in the part of the VBS competition where targets have been presented by textual descriptions [1], resulting in the 2nd place of the overall competition. Our contribution for this year, where the data size has been significantly increased to 600 h, therefore aims at combining the advantages of both approaches into one system. In the following, we summarize the preprocessing of the data and indexing – including storyboard layout, color index, and concept detection. Then, we describe a common search process, also illustrating the basic idea behind our approach, and conclude with a short discussion.

© Springer International Publishing AG 2017
L. Amsaleg et al. (Eds.): MMM 2017, Part II, LNCS 10133, pp. 480–485, 2017.
DOI: 10.1007/978-3-319-51814-5_45

2   Indexing

To visually inspect and find a target scene, we represent video content via a storyboard, i.e., a temporarily sorted set of thumbnails of frames extracted from the videos. The layout of this temporal arrangement is optimized based on our research about visual perception [3, 4] and experience from previous years' VBS participations [1, 2]. Figure 1 illustrates the basic design.

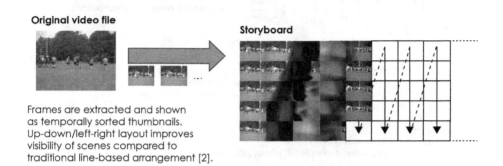

Original video file

Frames are extracted and shown as temporally sorted thumbnails. Up-down/left-right layout improves visibility of scenes compared to traditional line-based arrangement [2].

Storyboard

Fig. 1. Storyboard layout. (Color figure online)

In the original storyboard version from the 2015 competition [2], we used a large storyboard containing all video files sorted by their duration. Given the significantly increased dataset of 600 h for this year's event, single video files are now no longer represented as a whole but instead separated into segments created during indexing. Figure 2 summarizes this indexing process. First, a storyboard for each video is created by extracting frames and aligning them in the layout introduced in Fig. 1. These individual storyboards for each video in the database are later also used in the search process (cf. Sect. 3).

Then, each video is split into segments of 25 frames. This can be done in a brute force manner by just combining 25 consecutive frames. Alternatively, content analysis, such as automatic shot detection can be used. In the latter case, segments are formed by taking 25 equally distributed frames from a shot to form the scene. Experiments comparing the impact of these two and other content-related segmentations on search performance are part of our ongoing research.

In the next step, an index is created for each segment. First, an identifying color is assigned by analyzing the histograms of the 25 frames. A related score expresses how much this color represents the whole scene depicted by these frames. Then, semantic concepts are assigned to the center frame using comparable techniques like in our system from the 2016 VBS competition [1]. Similar to the dedicated color, each assigned concept has an associated confidence score indicating the likelihood of that concept appearing in that particular video frame.

Based on this index, an initial storyboard containing the whole database is created. To do this, segments are sorted by color, and then by their related color confidence score. This color-sorted storyboard is then used as starting point for the search, as we will describe in the next section.

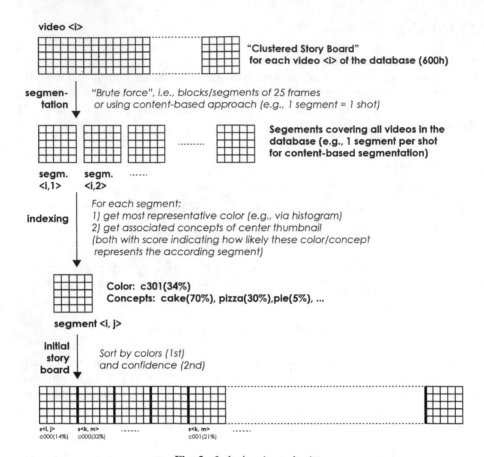

Fig. 2. Indexing (overview).

3   Searching

The initial storyboard of color-sorted segments created in the indexing process is shown on the screen at the beginning of the search. Users can skim it by scrolling up and down. Direct access to particular parts is possible via a color index on the left side. Concept-based browsing is done via text entry or selection of concepts from a list via a popup menu. In the following we describe a typical search process as it appears in our system for known item search (KIS) and ad-hoc video search (AVS) – the two tasks evaluated at the VBS. Figure 3 gives an overview of the procedure for color-based search.

For KIS tasks, a 20 s clip is played, which represents the target, i.e., the "known item" that the user has to find. In our approach, the user identifies a scene or shot that has a dominating color – this might be the most frequent one (e.g., green from the grass of a football field in a soccer video) or a "color tone" reflected in the whole image (e.g., a landscape colored in orange due to the setting sun). Clicking on the corresponding

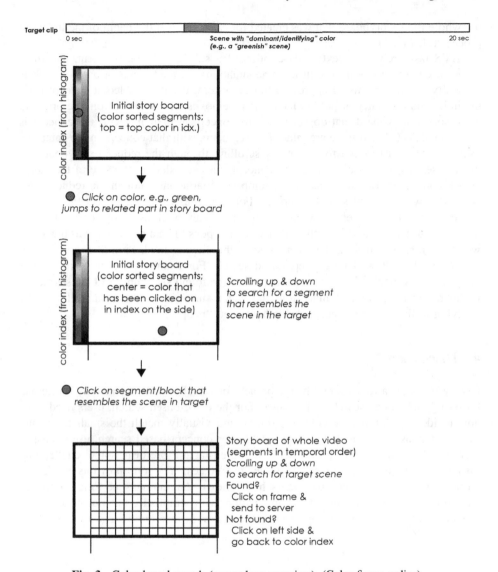

Fig. 3. Color-based search (exemplary overview). (Color figure online)

color in the color bar on the left repositions the storyboard around the segments that have this color assigned in their index. Users can then visually inspect the data by scrolling the storyboard up and down. Of course, we cannot assume that the segment is directly visible. First, identifying the dominating color is not an easy task for humans, and second, there might be many segments with the same or very similar colors. Yet, the optimized visualization and scrolling mechanisms also used in our system from 2015 [2] should enable users to quickly find segments similar to the searched target. Once such a target candidate has been identified, users can click on it to get a view of the whole file's storyboard that was created for each video at the begin of the indexing process (cf. Fig. 1).

AVS tasks rely on a textual description of a subject, object, or event, and thus, color-based search is only useful in some situations. Instead, concept-based search is generally a more promising approach. In our system, users can select a concept from the index via text entry or pop-up menu. This removes all segments from the original, color-sorted storyboard that do not have this concept in their index. If the concept is very sophisticated or if there are rather few segments with that concept in the database, users can then just search for targets by scrolling through the reduced storyboard. In other cases, users can select a second concept, further reducing the size of the story-board, and so on. Likewise, users can jump to a particular color in the reduced sto-ryboard – which is still sorted by colors. For example, a user can select the concept "soccer", which will likely show many segments with green as dominating color, but then jump to the color white if the target scene happens to be a game that was played in winter with snow covering the green grass of the playing field. Similarly, color-based search can be followed by concept-based search. For example, a scene with people partying on the deck of a boat on the ocean can be found searching for blue segments, if the dominating color is blue from the water surrounding the boat, followed by applying a filter for the concepts "ship/boat" and "people".

4   Discussion

Our system uses a rather simplified approach for search, and relies heavily on human browsing and human search performance. For the color-based search, users need to be able to identify dominating colors in scenes and visually match the small thumbnail representations of segments to target scenes. For concept-based search, users have to identify the concepts in the system and then again preform some challenging visual inspections of the presented target candidates. Yet, our previous work and participations from the last two years have shown that humans perform extremely well for such tasks, and thus the combination with simplistic, but powerful search functionality is a promising approach that we now further optimize with this year's system design.

References

1. Hudelist, M.A., Cobârzan, C., Beecks, C., Werken, R., Kletz, S., Hürst, W., Schoeffmann, K.: Collaborative video search combining video retrieval with human-based visual inspection. In: Tian, Q., Sebe, N., Qi, G.-J., Huet, B., Hong, R., Liu, X. (eds.) MMM 2016. LNCS, vol. 9517, pp. 400–405. Springer, Heidelberg (2016). doi:10.1007/978-3-319-27674-8_40
2. Hürst, W., van de Werken, R.: Human-based video browsing-investigating interface design for fast video browsing. In: 2015 IEEE International Symposium on Multimedia (ISM), pp. 363–368. IEEE, December 2015

3. Hürst, W., Darzentas, D.: Quantity versus quality: the role of layout and interaction complexity in thumbnail-based video retrieval interfaces. In: Proceedings of the 2nd ACM International Conference on Multimedia Retrieval (ICMR 2012), Article 45, p. 8. ACM, New York (2012)
4. Hürst, W., Snoek, C.G.M., Spoel, W.-J., Tomin, M.: Keep moving! Revisiting thumbnails for mobile video retrieval. In: Proceedings of the International Conference on Multimedia (MM 2010), pp. 963–966. ACM, New York (2010)

VERGE in VBS 2017

Anastasia Moumtzidou[1(✉)], Theodoros Mironidis[1],
Fotini Markatopoulou[1,2], Stelios Andreadis[1], Ilias Gialampoukidis[1],
Damianos Galanopoulos[1], Anastasia Ioannidou[1], Stefanos Vrochidis[1],
Vasileios Mezaris[1], Ioannis Kompatsiaris[1], and Ioannis Patras[2]

[1] Information Technologies Institute/Centre for Research and Technology
Hellas, 6th Km. Charilaou - Thermi Road, 57001 Thermi-Thessaloniki, Greece
{moumtzid, mironidis, markatopoulou, andreadisst,
heliasgj, dgalanop, ioananas, stefanos,
bmezaris, ikom}@iti.gr
[2] School of Electronic Engineering and Computer Science, QMUL, London, UK
i.patras@qmul.ac.uk

Abstract. This paper presents VERGE interactive video retrieval engine, which is capable of browsing and searching into video content. The system integrates several content-based analysis and retrieval modules including concept detection, clustering, visual similarity search, object-based search, query analysis and multimodal and temporal fusion.

1 Introduction

VERGE interactive video search engine is capable of retrieving and browsing video collections by integrating multimodal indexing and retrieval modules. VERGE has evolved to support Known Item Search (KIS), Instance Search (INS) and Ad-Hoc Video Search tasks (AVS). The aforementioned tasks require the incorporation of browsing, exploration, or navigation capabilities of the video or image collection.

The VERGE search engine was evaluated by participating in several video retrieval related conferences and showcases such as TRECVID, VideOlympics and Video Browser Showdown (VBS). Specifically, ITI-CERTH participated with consistency in several TRECVID Search tasks including the KIS task and the INS task for consecutive years starting from 2007. Moreover, it has participated in the VideOlympics event, and in VBS competition starting from 2014. The proposed version of VERGE aims at participating to the KIS and AVS tasks of VBS [1].

2 Video Retrieval System

VERGE combines advanced browsing and retrieval functionalities with a user-friendly interface, and supports the submission of queries and the accumulation of relevant results. The following indexing and retrieval modules are integrated in the developed search application: (a) Visual Similarity Search; (b) Object-based Visual Search; (c) High Level Concepts Retrieval; (d) Automatic Query Formulation and Expansion;

© Springer International Publishing AG 2017
L. Amsaleg et al. (Eds.): MMM 2017, Part II, LNCS 10133, pp. 486–492, 2017.
DOI: 10.1007/978-3-319-51814-5_46

(e) ColorMap Clustering; (f) CNN-based visualization; and (g) Multimodal and Temporal Fusion and Search. The above modules allow the user to search through a collection of images and/or video keyframes. Figure 1 depicts the general framework.

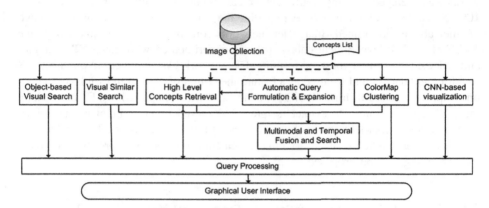

Fig. 1. Framework of VERGE.

2.1 Visual Similarity Search

This module performs content-based retrieval, also known as query by image content, by using deep convolutional neural networks (DCNNs). Specifically, we have trained GoogleNet [2] on 5055 ImageNet concepts, and we used the output of the last pooling layer, with dimension 1024, as a global keyframe representation. In order to achieve fast retrieval of similar images, we constructed an IVFADC index for database vectors and then computed K-Nearest Neighbours from the query file [3].

2.2 Object-Based Visual Search

This module performs instance-based object retrieval using two different methods. The first one is based on the Bag-Of-Word model [4]. An inverted index is built for searching the image database BoW vectors, while tf-idf weights and the position of each frame in the retrieved list are used for ranking. The second one relies on Convolutional Neural Networks (CNNs). Several pre-trained CNNs are explored in order to represent each frame with features extracted either from a fully-connected or a convolutional layer. Similarity between the query and the database images is measured based on an appropriate distance. In both methods, the query can be either the keyframe or any cropped part of it.

2.3 High Level Concepts Retrieval

This module indexes the video shots based on 1000 ImageNet and 346 TRECVID SIN high level concepts (e.g. water, aircraft). The rationale for selecting a different network compared to Sect. 2.1 is that the retrieval accuracy of the Sect. 2.4 module that uses these concepts is higher in terms of MAP compared to the complete set of 5055 concepts. To obtain scores regarding the 1000 ImageNet concepts, we applied five pre-trained ImageNet DCNNs on the AVS test keyframes. The output of these networks was averaged in terms of arithmetic mean to obtain a single score for each of the 1000 concepts. To obtain the scores regarding the 346 concepts we fine-tuned (FT) two of the above pre-trained ImageNet networks on the 346 concepts using the TRECVID AVS development dataset [4]; we experimented with many FT strategies and selected the single best performing FT network. We applied it on the AVS development dataset and we used as a feature the output of the last hidden layer to train one Support Vector Machine (SVM) per concept. Then, we applied this FT network on the AVS test keyframes to extract features, and served them as input to the trained SVM classifiers in order to gather scores for each of the 346 concepts. The final step of high-level concepts retrieval was to refine the calculated detection scores by employing the re-ranking method proposed in [5].

2.4   Automatic Query Formulation and Expansion Using High Level Concepts

This module formulates and expands an input query in order to translate it into a set of high level concepts. First, we check if the entire query is included in the available pool of high-level concepts. If the query is found, then no further action is necessary. Otherwise, we transform the original query to a set of elementary "subqueries", using Part-of-Speech tagging and a task-specific set of NLP rules. For example, if the original query contains a sequence in the form "Noun – Verb – Noun", this triad is considered to be a "subquery"; the motivation is that such a sequence is much more characteristic of the original query than any of these three words alone would be, and at the same time it is easier to find correspondences between this and the concepts in our pool, compared to doing so for a very long and complex query. Subsequently, we check if any of the "subqueries" are included in our concept pool. Otherwise, the original query and the subqueries are used as input to the zero-example event detection pipeline [6] and the most relevant concepts are identified. In contrast, if at least one of the subqueries is included in the pool, then we select the corresponding concepts and, we use the semantic relatedness measure [7] to select the single most semantically-relevant concept for each of the remaining subqueries. Either way, the results is a set of high-level concepts that are much related and describe well the input query given the relatively limited number of concepts in our pool.

2.5   ColorMap Clustering

Motivated by [8, 9], video keyframes are clustered by color using Self Organizing Maps (SOM) into color classes. Each color class is represented in the GUI by the most representative image within the color class. All representative images are determined by their distances to the SOM's best matching unit per color class. Each particular image I in the collection is represented in RGB form, and indexed as a vector $(s_R, s_G, s_B)_I$, where s_R, s_G and s_B is the similarity score between the image I and the pure red (s_R), pure green (s_G) and pure blue (s_B) image, respectively. The similarity score is based on pixel-by-pixel comparisons and averaged over all pixels of the image I. In this way, VERGE clusters all images into color classes and offers fast browsing in the collection of video keyframes.

2.6 CNN-Based Visualization

CNNs are proposed for the effective visualization of datasets, given that they can be interpreted as gradually transforming the images into a representation in which the classes are separable by a linear classifier. The method tested is the t-SNE method [10] which has shown very satisfactory results. The procedure followed involves taking a set of images and extracting CNN codes. These codes are plugged into t-SNE and a 2-dimensional vector is produced for each image. Finally, the corresponding images are visualized in a grid.

2.7 Multimodal and Temporal Fusion and Search

This module fuses the visual descriptors of Sect. 2.1, the concepts of Sect. 2.3 and the color features of Sect. 2.5. Given a query shot and its central keyframe in the time domain, this module retrieves similar shots by performing center-to-center comparisons among video shots. On the top-k retrieved shots, re-ranking is performed, taking into account the adjacent keyframes of the top-retrieved shots.

More specifically, as depicted in Fig. 2, given a query shot, its color features, its concepts and its DCNN descriptors are extracted from the central frame, which is

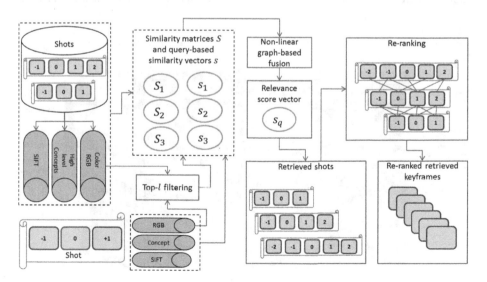

Fig. 2. The VERGE multimodal fusion and search module.

marked by zero in the figure. An initial filtering stage keeps only the top-l relevant-to-the-color shots and then computes an $l \times l$ similarity matrix and an $l \times l$ similarity vector per modality. The similarity matrices and vectors are then fused in a non-linear and graph-based way, following [11], providing a fused relevance score vector s_q for the retrieved shots. Pairwise comparisons on the keyframes of the top-k retrieved shots result to the final list of re-ranked retrieved keyframes. In the initial top-l filtering stage we used the color features as the dominant modality, but this could be set on demand. The final re-ranking stage involves the computation of mutual similarities among not necessarily central keyframes between any two shots.

3   VERGE Interface and Interaction Modes

This year the retrieval utilities incorporated into the system are more than ever, offering a multitude of search options. Thus, the user interface has to give the end user an intuitive and effective way to run queries fast and obtain the best possible results.

A novel component of the interface is the General purpose search input field, common to the user's experience of other search engines. The user can initiate the search procedure by simply describing the shot he/she is looking for, in natural language. The system analyses the text in an intelligent manner, using the *Automatic Query Formulation and Expansion using High-Level Concepts* module, and returns a single or more combined concepts. The user can edit this proposed list by adding or removing concepts and perform a concept-based search. Another novelty is the *Multimodal and Temporal Fusion* that the user can easily invoke by clicking on one or more shots to serve as query, and selecting the dominant modality for each shot.

Describing the user interface (Fig. 3), there is a toolbar with many useful options on the top. In detail, from left to right, a burger icon opens a toggle menu that contains the different search capabilities, namely the *Concept-* and *Topic-based* search, and the

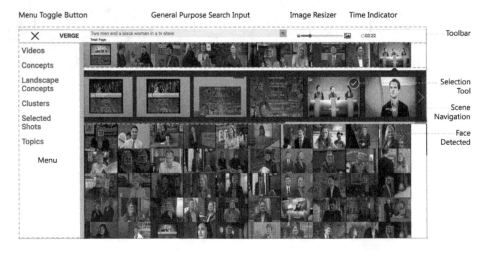

Fig. 3.   Screenshot of VERGE video retrieval engine.

ColorMapClustering. The menu also includes the user's selected shots and the total set of video shots. Next to the application's logo, the General purpose search input field can be seen, followed by the Image Resizer that modifies the amount of results in the viewport, by changing the size of the shots. The last toolbar component applies only to the contest and shows the remaining time for the submission, accompanied by an animated red line on the top of the screen. The central component of the interface includes a shot-based representation of the video results in a grid-like view. Clicking on a shot allows the user to navigate through the whole scene where this frame belongs, displaying the related shots in a chronological order. Moreover, each shot supports tools to run the *Visual Similarity* and the *Object-based Visual* modules. Finally, all selected images are saved in a deposit that can be quickly accessed for further searching or just for the submission.

To illustrate the functionality of the VERGE interface[1], we describe a simple usage scenario. Supposing that the user is interested in finding a clip of two men and a black woman in a TV show (Fig. 3), he/she can begin with the General purpose search. An appropriate selection of proposed concepts is received (e.g. *Female_Person*, *Two_People*), that the user is able to edit before performing the concept-based search. If a relative image is found during this step, the user can continue with all the above-mentioned retrieval modules to collect more similar images, as well as browse the complete scene to find previous or next shots.

4 Future Work

Future work includes applying multimodal temporal fusion and search on multiple queries. The retrieved results can then be fused and the user will be presented with a single list. Another feature would be to allow the user to create a more complicated query using as base a shot, and describe it by considering multiple modalities.

Acknowledgements. This work was supported by the EU's Horizon 2020 research and innovation programme under grant agreements H2020-687786 InVID, H2020-693092 MOVING, H2020-645012 KRISTINA and H2020-700024 TENSOR.

References

1. Cobârzan, C., et al.: Interactive video search tools: a detailed analysis of the video browser showdown 2015. In: Multimedia Tools and Applications, pp. 1–33 (2015)
2. Szegedy, C., et al.: Going deeper with convolutions. In: Proceedings of IEEE Conference on Computer Vision and Pattern Recognition, pp. 1–9 (2015)
3. Jegou, H., Douze, M., Schmid, C.: Product quantization for nearest neighbor search. IEEE Trans. Pattern Anal. Mach. Intell. **33**, 117–128 (2011)

---

[1] http://mklab-services.iti.gr/vss2016.

4. Markatopoulou, F., et al.: ITI-CERTH participation to TRECVID 2015. In: TRECVID 2015 Workshop, Gaithersburg, MD, USA (2015)
5. Safadi B., Quénot, G.: Re-ranking by local re-scoring for video indexing and retrieval. In: 20th ACM International Conference on Information and Knowledge Management, pp. 2081–2084 (2011)
6. Tzelepis, C., Galanopoulos, D., Mezaris, V., and Patras, I.: Learning to detect video events from zero or very few video examples. In: Image and Vision Computing (2015)
7. Gabrilovich, E., Markovitch, S.: Computing semantic relatedness using wikipedia-based explicit semantic analysis. IJCAI 7, 1606–1611 (2007)
8. Barthel, K. U., Hezel, N., and Mackowiak, R.: ImageMap - Visually Browsing Millions of Images. In: MultiMedia Modeling, pp. 287–290 (2015)
9. Moumtzidou, A., et al.: A multimedia interactive search engine based on graph-based and non-linear multimodal fusion. In: CBMI 2016 International Workshop, pp. 1–4 (2016)
10. t-SNE visualization of CNN codes. http://cs.stanford.edu/people/karpathy/cnnembed/
11. Gialampoukidis, I., et al.: A hybrid graph-based and non-linear late fusion approach for multimedia retrieval. In: CBMI 2016 International Workshop, pp. 1–6. IEEE (2016)

Video Hunter at VBS 2017

Adam Blažek$^{(\boxtimes)}$, Jakub Lokoč, and David Kuboň

SIRET Research Group, Department of Software Engineering,
Faculty of Mathematics and Physics, Charles University, Prague, Czech Republic
{blazek,lokoc}@ksi.mff.cuni.cz, kubondavid@seznam.cz

Abstract. After almost three years of development, the Video Hunter tool (formerly the Signature-Based Video Browser) has become a complex tool combining different query modalities, multi-sketches, visualizations and browsing techniques. In this paper, we present additional improvements of the tool focusing on keyword search. More specifically, we present a method relying on an external image search engine and a method relying on ImageNet labels. We also present a keyframe caching method employed by our tool.

1 Introduction

Known-item search tasks represent a challenging retrieval scenario, where users search for a memorized scene. As the users cannot directly formulate their search intents, the known-item search systems provide interactive interfaces [11] enabling iterative querying, browsing and refinement. The users and their ability to identify correct results are indispensable for the whole retrieval process, unlike classical query by example model where the retrieval process is fully automatic for a given query. In order to find the most promising approaches for known-item search tasks, events like Video Browser Showdown [3,10] (VBS) are organized. Such events help to recognize successful approaches implemented by various video retrieval tools. As demonstrated at previous Video Browser Showdowns, a clear winner outperforming all the other approaches in all the retrieval tasks was not identified yet. However, the most successful tools show some promising directions.

The top three teams [1,5,6] competing at VBS 2016 shared several strategies. All the teams used a color sketching canvas enabling the definition of a memorized color layout [7]. Such query initialization helps to localize the scenes of interest and then to continue with other search options. Another shared feature is the temporal context of the matched keyframes that differentiate similar scenes from different parts of the video. All the three tools employed also features from deep convolutional neural networks, however, each system in a different way. The winning tool [1] used the features to construct a navigation graph that connects visually and semantically similar keyframes. The graph browsing then helps in the initial phase of the retrieval process. The tool at the second position [5] used deep learning to identify concepts in the video collection and enabled keyword search. Another unique feature of the tool is the collaboration model, where an

L. Amsaleg et al. (Eds.): MMM 2017, Part II, LNCS 10133, pp. 493–498, 2017.
DOI: 10.1007/978-3-319-51814-5_47

operator of the main tool sends ranked video lists to a tablet application. The tablet application is based on sequential scanning relying just on human-based visual inspection. The third (our) tool [6] used DeCAF features [4] for similarity search in the collection of extracted keyframes. In order to initialize the query, the tool relies on multi-sketch approach enabling the combination of edges and a color layout.

In this paper, we present the extended version of our tool integrating also keyword search options. Since the video collection consistently doubles its size every year, we also discuss a keyframe caching method used by our tool.

2   Video Hunter in a Nutshell

The Video Hunter is a content-based video retrieval and exploration tool. Before the tool can be used for a new video, a video preprocessing phase is necessary. In the preprocessing, keyframes from the video are extracted (uniformly sampled), and feature descriptors from the keyframes are extracted. So far, the tool considers position-color feature signatures, edge histograms and DeCAF features [4]. For the details of the retrieval models based on the extracted features and efficient indexing, we refer the reader to our previous works [2,6,7]. In addition, the new version of our tool extracts also concepts as described in Sect. 3.

The tool interface (see Fig. 1) contains a query initialization/formulation panel on the right and a larger visualization/browsing panel on the left. Users can initialize the query by drawing a colored sketch approximating a memorized color layout. In the same sketch, users can draw also edges to further restrict

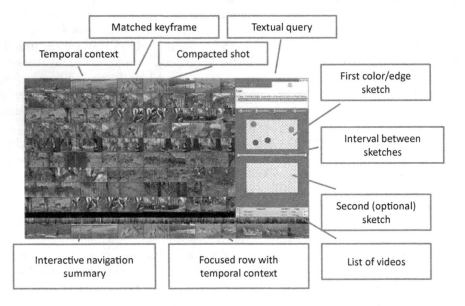

Fig. 1. The Video Hunter tool interface.

the query sketch. Since the searched scenes often contain two distinct shots, users can specify also a second sketch for refining the search. The importance of particular features can be fine tuned by weights. A novel query option is the keyword search, where users can specify concepts identified in the indexed videos or search for relevant images in third party servers. The images then serve as a candidate query example for similarity search using DeCAF features.

The visualization part consists of the panel presenting results of the querying, the interactive navigation summary and one additional wide row of focused results. The results are presented in a compact form with highlighted matched keyframes. Each matched keyframe is surrounded by its temporal context from the video. The panel supports coarse to fine visualizations, depicting either one result or more results per row. With more results on one row, the temporal context shrinks. Users can browse the results in several ways. Both vertical (more pages) and horizontal (more temporal context) scrolling is enabled. Users can pick selected colors from the results or start similarity search browsing by selecting a keyframe of interest.

3   Video Hunter Keyword Search

For smaller video collections, known-item search tasks were manageable by purely content-based retrieval and browsing. Note that in the early years of VBS, keyword search was even prohibited. However, with the growing size of the collections (doubles every year), another intuitive query initialization/restriction interface became necessary and so the keyword search was permitted. The videos without annotation represent still an issue for effective keyword search, partially facilitated by recent developments in deep learning. In the following, we describe two approaches how to incorporate keyword search to the Video Hunter tool.

3.1   External Image Search Engine

Image keyword search is a well established method adopted by many successful Internet services (e.g., Google images). After years of development, the services provide impressive results when matching query keywords to indexed images, employing annotations extracted from surrounding texts and also statistics from billions of searching users. Hence, without an external knowledge about the inner workings of the services, novice users of a video retrieval system can still easily search for a candidate query object at a third party server. From the selected object, DeCAF features can be extracted and used for similarity search in the collection of extracted keyframes. The Video Hunter provides an interface for querying a third party server, selecting a suitable query object and the consecutive similarity search processing.

Although such a method is straightforward, it brings also several shortcomings. The method relies on the Internet connection, service availability and also assumes that the user can select a sufficiently good example query object.

3.2   ImageNet Labels

The second approach is based on labels automatically assigned to keyframes using an arbitrary ImageNet [9] classification model (we use the Deep Convolutional Neural Network by K. Simonyan and A. Zisserman [12]). Since the classification models are usually restricted to a limited set of labels (e.g., 1000), the Video Hunter provides two additional processing pipelines to connect the labels to user queries.

First, top five labels are extracted with their probabilities for each key-frame and a tree of their hypernyms is constructed. The nodes of the tree represent labels (or synsets) and edges represent WordNet hypernym–hyponym relations between them. The probability of an inner node of the tree is a sum of probabilities of its children. Hence, also labels not present in the employed classification model can be considered. All the synsets are stored in an inverted file index for efficient retrieval.

Second, a user query is preprocessed in the following way. The query is filtered just to nouns that are transformed into their basic form (ImageNet comprises just nouns). The meaning of the nouns is explored using WordNet [8] and a set of synsets is assigned to the nouns. Then, each query synset is iteratively generalized to its hypernym until it is present in the database. Hence, the system searches for the most specific synsets present in the inverted file index. Furthermore, rankings for synset s are weighed by $1 - avgp(s)$, where $avgp(s)$ is the average probability of synset s over all the dataset key-frames, i.e., rarely occurring concepts are preferred.

4   Key-Frames Visualization

In user-aided search in video the way of displaying results is just as crucial as the utilized search techniques. Up to date, Video Hunter offers total of three ways of displaying results each serving a different scenario.

We believe that in the early search stages it is desirable to offer coarse overview of the indexed data, i.e., explore the dataset. For this purpose, Video Hunter displays up to hundreds of matched key-frames at once in a 2D grid grouping visually similar key-frames together (similarly to Barthel et al. [1]). Users are able to zoom in to a particular area and display additional key-frames visually similar to the selected ones. Alternatively, the matched key-frames might be organized in the grid according to their ranking rather than visual similarity.

Later on, the search is typically narrowed down and discriminating similar key-frames becomes the challenge. Often, there is no other option than examining the neighborhood of the matched key-frames, i.e., the following and preceding frames. We allow this in the last of our key-frames visualization wherein results are displayed in rows, each row contains one of the matched key-frames as well as its temporal context. All the visualization options are depicted in Fig. 2.

Fig. 2. Video Hunter visualization options. From left to right: matched key-frames with high context, matched key-frames with no context sorted by their ranking and 2D image map.

4.1 Key-Frames Caching

With datasets of hundreds of hours of video content both filtering and displaying results become a challenge. We are utilizing a variety of indexes, approximations and heuristics to keep the ranking fast enough to retain real-time responses. All for nothing, however, if the displaying of the key-frame thumbnails takes a long time.

For these reasons, we incorporated three-level key-frames cache wherein blocks of consecutive images are stored in jpegs on hard-drive (1st level), loaded on-demand to main memory (2nd level) and finally decoded to bitmaps (3rd level). Blocks are freed and reused with least-recently-used strategy ensuring low memory footprint. Furthermore, we employ several heuristics such as "load neighborhood blocks as well" in order to cover common user browsing patterns and possibly load needed key-frames in advance.

5 Conclusion

In this paper, we present the Video Hunter tool, a follow up of the previously introduced Signature-based Video Browser. We present two novel extensions, keyword search and additional results visualization options. The keyword search techniques are divided into two approaches – first relying on an external image search engine and second relying on ImageNet labels. In the future, we would like to further extend database inspection options of our tool and employ more sophisticated classification models.

Acknowledgments. This research was supported by grant SVV-2016-260331 and GAUK project no. 1134316. We would also like to thank Jan Pavlovský for helping us with 2D image maps.

References

1. Barthel, K.U., Hezel, N., Mackowiak, R.: Navigating a graph of scenes for exploring large video collections. In: Tian, Q., Sebe, N., Qi, G.-J., Huet, B., Hong, R., Liu, X. (eds.) MMM 2016. LNCS, vol. 9517, pp. 418–423. Springer, Heidelberg (2016). doi:10.1007/978-3-319-27674-8_43

2. Blažek, A., Lokoč, J., Skopal, T.: Video retrieval with feature signature sketches. In: Traina, A.J.M., Traina, C., Cordeiro, R.L.F. (eds.) SISAP 2014. LNCS, vol. 8821, pp. 25–36. Springer, Heidelberg (2014). doi:10.1007/978-3-319-11988-5_3

3. Cobârzan, C., Schoeffmann, K., Bailer, W., Hürst, W., Blažek, A., Lokoč, J., Vrochidis, S., Barthel, K.U., Rossetto, L.: Interactive video search tools: a detailed analysis of the video browser showdown 2015. Multimedia Tools Appl., 1–33 (2016)

4. Donahue, J., Jia, Y., Vinyals, O., Hoffman, J., Zhang, N., Tzeng, E., Darrell, T., Decaf: a deep convolutional activation feature for generic visual recognition. CoRR, abs/1310.1531 (2013)

5. Hudelist, M.A., Cobârzan, C., Beecks, C., Werken, R., Kletz, S., Hürst, W., Schoeffmann, K.: Collaborative video search combining video retrieval with human-based visual inspection. In: Tian, Q., Sebe, N., Qi, G.-J., Huet, B., Hong, R., Liu, X. (eds.) MMM 2016. LNCS, vol. 9517, pp. 400–405. Springer, Heidelberg (2016). doi:10.1007/978-3-319-27674-8_40

6. Kuboň, D., Blažek, A., Lokoč, J., Skopal, T.: Multi-sketch semantic video browser. In: Tian, Q., Sebe, N., Qi, G.-J., Huet, B., Hong, R., Liu, X. (eds.) MMM 2016. LNCS, vol. 9517, pp. 406–411. Springer, Heidelberg (2016). doi:10.1007/978-3-319-27674-8_41

7. Lokoč, J., Blažek, A., Skopal, T.: Signature-based video browser. In: Gurrin, C., Hopfgartner, F., Hurst, W., Johansen, H., Lee, H., O'Connor, N. (eds.) MMM 2014. LNCS, vol. 8326, pp. 415–418. Springer, Heidelberg (2014). doi:10.1007/978-3-319-04117-9_49

8. Miller, G.A.: Wordnet: a lexical database for english. Commun. ACM 38(11), 39–41 (1995)

9. Russakovsky, O., Deng, J., Hao, S., Krause, J., Satheesh, S., Ma, S., Huang, Z., Karpathy, A., Khosla, A., Bernstein, M., Berg, A.C., Fei-Fei, L.: Imagenet large scale visual recognition challenge. Int. J. Comput. Vis. 115(3), 211–252 (2015)

10. Schoeffmann, K.: A user-centric media retrieval competition: the video browser showdown 2012–2014. IEEE MultiMedia 21(4), 8–13 (2014)

11. Schoeffmann, K., Hudelist, M.A., Huber, J.: Video interaction tools: a survey of recent work. ACM Comput. Surv. 48(1), 14 (2015)

12. Simonyan, K., Zisserman, A.: Very deep convolutional networks for large-scale image recognition. CoRR, abs/1409.1556 (2014)

Author Index

Printed in the United States
By Bookmasters